MATERIALS, CIRCUITS & DEVICES SERIES 29

Nano-CMOS and Post-CMOS Electronics

Other volumes in this series:

Volume 2	**Analogue IC Design: The Current-mode Approach** C. Toumazou, F.J. Lidgey and D.G. Haigh (Editors)
Volume 3	**Analogue-Digital ASICs: Circuit Techniques, Design Tools and Applications** R.S. Soin, F. Maloberti and J. France (Editors)
Volume 4	**Algorithmic and Knowledge-based CAD for VLSI** G.E. Taylor and G. Russell (Editors)
Volume 5	**Switched Currents: An Analogue Technique for Digital Technology** C. Toumazou, J.B.C. Hughes and N.C. Battersby (Editors)
Volume 6	**High-frequency Circuit Engineering** F. Nibler *et al.*
Volume 8	**Low-power High-frequency Microelectronics: A Unified Approach** G. Machado (Editor)
Volume 9	**VLSI Testing: Digital and Mixed Analogue/Digital Techniques** S.L. Hurst
Volume 10	**Distributed Feedback Semiconductor Lasers** J.E. Carroll, J.E.A. Whiteaway and R.G.S. Plumb
Volume 11	**Selected Topics in Advanced Solid State and Fibre Optic Sensors** S.M. Vaezi-Nejad (Editor)
Volume 12	**Strained Silicon Heterostructures: Materials and Devices** C.K. Maiti, N.B. Chakrabarti and S.K. Ray
Volume 13	**RFIC and MMIC Design and Technology** I.D. Robertson and S. Lucyzyn (Editors)
Volume 14	**Design of High Frequency Integrated Analogue Filters** Y. Sun (Editor)
Volume 15	**Foundations of Digital Signal Processing: Theory, Algorithms and Hardware Design** P. Gaydecki
Volume 16	**Wireless Communications Circuits and Systems** Y. Sun (Editor)
Volume 17	**The Switching Function: Analysis of Power Electronic Circuits** C. Marouchos
Volume 18	**System on Chip: Next Generation Electronics** B. Al-Hashimi (Editor)
Volume 19	**Test and Diagnosis of Analogue, Mixed-signal and RF Integrated Circuits: The System on Chip Approach** Y. Sun (Editor)
Volume 20	**Low Power and Low Voltage Circuit Design with the FGMOS Transistor** E. Rodriguez-Villegas
Volume 21	**Technology Computer Aided Design for Si, SiGe and GaAs Integrated Circuits** C.K. Maiti and G.A. Armstrong
Volume 22	**Nanotechnologies** M. Wautelet *et al.*
Volume 23	**Understandable Electric Circuits** M. Wang
Volume 24	**Fundamentals of Electromagnetic Levitation: Engineering Sustainability through Efficiency** A.J. Sangster

Nano-CMOS and Post-CMOS Electronics: Devices and Modelling

Edited by
Saraju P. Mohanty
and Ashok Srivastava

The Institution of Engineering and Technology

Published by The Institution of Engineering and Technology, London, United Kingdom

The Institution of Engineering and Technology is registered as a Charity in England & Wales (no. 211014) and Scotland (no. SC038698).

First published 2016

The Institution of Engineering and Technology
Michael Faraday House
Six Hills Way, Stevenage
Herts, SG1 2AY, United Kingdom

www.theiet.org

British Library Cataloguing in Publication Data
A catalogue record for this product is available from the British Library

ISBN 978-1-84919-997-1 (hardback)
ISBN 978-1-84919-998-8 (PDF)

Typeset in India by MPS Limited
Printed in the UK by CPI Group (UK) Ltd, Croydon

Contents

Preface **xi**

1 High-κ dielectrics and device reliability **1**
1.1 Introduction 1
1.2 Alloying HfO_2 and ZrO_2 2
1.3 Advanced ALD process: intermediate treatment 4
1.4 SPA plasma 5
1.5 Cyclic deposition and SPA plasma treatment to ALD $Hf_{1-x}Zr_xO_2$ 5
1.5.1 Impact of Zr addition and SPA plasma on electrical properties 8
1.5.2 Reliability study by constant voltage stress 9
1.6 Al incorporation into HfO_2 13
1.6.1 $HfAlO_x$ alloy structures 14
1.6.2 Al_2O_3/HfO_2 bilayer structures 15
1.6.3 Problems with excess Al incorporation 16
1.6.4 Extremely low Al incorporation in HfO_2 17
1.7 Conclusion 27
Acknowledgment 28
References 28

2 High mobility n and p channels on gallium arsenide and silicon substrates using interfacial misfit dislocation arrays **35**
2.1 Introduction 35
2.2 IMF versus pseudomorphic growth 38
2.3 III-Sb on GaAs substrates 39
2.4 III-Sb on silicon substrates 42
2.4.1 Lattice mismatch solution: IMF layer 43
2.4.2 Antiphase domains (APDs) 47
2.4.3 Thermal expansion coefficient 50
2.5 GaSb membranes 51
2.5.1 Substrate removal technique 51
2.5.2 ELO technique 53
2.6 InAs and InGaSb channels on GaAs 54
2.7 Conclusions 57
References 57

3 Anodic metal-insulator-metal (MIM) capacitors **61**
3.1 Introduction 61

3.2 MIM capacitor 63
3.3 Anodization for nanoelectronics 64
3.4 Anodic alumina MIM capacitors 66
3.4.1 Fabrication process flow and crystalline properties 67
3.4.2 Capacitance and voltage linearity 67
3.4.3 Leakage characteristics and conduction mechanisms 70
3.5 Anodic titania MIM capacitors 73
3.5.1 Fabrication process, oxide formation, and crystallization 74
3.5.2 Capacitance, voltage linearity, and leakage characteristics 76
3.6 Anodic bilayer MIM capacitors 79
3.6.1 Fabrication process flow 80
3.6.2 Formation of bilayer and crystallization 81
3.6.3 Capacitance, voltage linearity, and leakage characteristics 82
3.7 Modeling of high-*k* MIM capacitors 84
3.7.1 Modeling the voltage linearity 85
3.7.2 Macroscopic model 86
3.7.3 Microscopic model 88
3.7.4 Model verification 89
3.8 Conclusion 90
References 92

4 Graphene transistors—present and beyond 99
4.1 Introduction 99
4.2 Fabrication of graphene 100
4.3 Properties of graphene 101
4.3.1 Band structure 101
4.3.2 Carrier density 103
4.3.3 Ambipolar field effect 104
4.3.4 Conductivity 104
4.3.5 Scattering mechanism 106
4.3.6 High-field transport 108
4.3.7 Low-field mobility 109
4.3.8 Substrate and gate dielectrics 110
4.3.9 Joule heating 112
4.3.10 Contact resistance 112
4.3.11 Quantum capacitance 113
4.4 Modeling and simulation 114
4.4.1 Classical transport 114
4.4.2 Semiclassical transport 115
4.4.3 Quantum transport 117
4.5 GNR FET 119
4.5.1 Graphene nanoribbon 119
4.5.2 Device structure 123
4.5.3 Device performance metrics 123
4.6 Conclusion 128
References 128

5 Junction and doping-free transistors for future computing **139**
5.1 Introduction 139
5.2 JLFET limitations 143
5.3 Dopingless FET 146
5.4 Junction and doping-free FET 148
5.4.1 Junction and doping-free DG FET 149
5.4.2 Dopingless BJT 159
5.5 Conclusion 166
References 166

6 Nanoscale high-κ/metal-gate CMOS and FinFET based logic libraries **169**
6.1 Introduction 169
6.2 Summary of this chapter 172
6.3 HKMG bulk MOSFET 174
6.3.1 HKMG device structure 175
6.3.2 HKMG device modeling 176
6.4 DG-FinFET device 178
6.4.1 DG-FinFET device structure 178
6.4.2 DG-FinFET device modeling 180
6.5 The proposed methodology for logic library creation 183
6.5.1 Sources of variation and nature of variability 183
6.5.2 Statistical logic library characterization flow 184
6.6 Power, leakage, and delay models for HKMG and DG-FinFET technology 187
6.6.1 For HKMG-based technology 187
6.6.2 For DG-FinFET-based technology 190
6.7 Device level characterization of high-κ and FinFET 191
6.7.1 For HKMG CMOS 191
6.7.2 For DG-FinFET 199
6.8 PVT-aware logic level characterization 204
6.9 Conclusion and directions for future research 205
Acknowledgment 206
References 206

7 FinFET and reliability considerations of next-generation processors **213**
7.1 Introduction 213
7.2 Background 215
7.2.1 NBTI degradation mechanism 215
7.2.2 Target GPU architecture 216
7.3 Hybrid-device warp scheduler 217
7.3.1 Opportunity for improvement 217
7.3.2 Two-stage scheduling 219
7.4 Hybrid-device sequential-access L2 cache 221
7.5 Experimental setup 222

7.6 Result analysis 224
7.6.1 Warp scheduler 224
7.6.2 L2 cache 228
7.7 Related work 232
7.7.1 NBTI mitigation 232
7.7.2 Characterization of FinFET reliability 233
7.7.3 Hybrid-device design 233
7.8 Conclusion 233
References 234

8 Multiple-independent-gate nanowire transistors: from technology to advanced SoC design 237
8.1 Introduction 237
8.2 Multiple-independent-gate field-effect transistors 238
8.2.1 TIG device overview and operation 238
8.2.2 Device fabrication and electrical characterization 240
8.2.3 Physical understanding 243
8.2.4 Performance predictions 245
8.3 Circuit design opportunities 247
8.3.1 Generalities 247
8.3.2 Compact data path design 249
8.3.3 Advanced low-power techniques 251
8.3.4 Memory opportunities 254
8.3.5 Case study: implementation of a Polar code decoder with MIGFETs 257
8.4 Summary and conclusions 259
Acknowledgment 260
References 260

9 Exploration of carbon nanotubes for efficient power delivery 265
9.1 Introduction 265
9.2 Modeling of CNTs 266
9.3 CNTs for 2D power delivery network 269
9.3.1 Branch analysis with CNTs 270
9.4 CNTs for 3D power delivery network 274
9.4.1 CNT TSV analysis 278
9.4.2 Voltage drop analysis on a 3D PDN 280
9.5 Conclusion 284
Acknowledgment 284
References 284

10 Timing driven buffer insertion for carbon nanotube interconnects 287
10.1 Introduction 287
10.2 Problem formulation 290
10.3 CNT interconnects 290
10.3.1 Resistance for CNT 291

10.3.2 Capacitance for CNT 292
10.3.3 Inductive impact is not important 293
10.3.4 Elmore delay model for bundled SWCNTs interconnects 294
10.4 Timing buffering for CNT interconnects 294
10.4.1 Add wire 294
10.4.2 Add buffer 296
10.4.3 Branch merge 296
10.4.4 Add driver 297
10.4.5 Pruning 298
10.5 An example 299
10.6 Experimental results 302
10.6.1 Experimental setup 302
10.6.2 Experimental results 304
10.7 Conclusions 308
Acknowledgment 309
References 309

11 Memristor modeling – static, statistical, and stochastic methodologies 313
11.1 Introduction 313
11.2 Static modeling 315
11.2.1 TiO_2 thin-film memristor 315
11.2.2 Memristor static (bulk) model 316
11.3 Statistical modeling 316
11.3.1 Theoretical analysis 316
11.3.2 3D device sample generation flow 319
11.3.3 The impact of process variations 322
11.4 Stochastic modeling 324
11.4.1 ON and OFF static states 324
11.4.2 Dynamic switching process 324
11.4.3 Stochastic model verification 327
11.5 Robustness of a neuromorphic system 329
11.6 Conclusion 332
Acknowledgment and disclaimer 332
References 332

12 Neuromorphic devices and circuits 337
12.1 Introduction 337
12.2 Emerging memory technologies 338
12.3 Memristor and resistive memory 340
12.3.1 Memristor 341
12.3.2 Switching mechanisms 342
12.3.3 Plasticity 345
12.3.4 Memristor integration 346
12.4 Memristive synapse circuits: current-mode design 347
12.4.1 Overview 347

12.4.2 Area and power consumption 349
12.5 Application: image clustering 351
12.5.1 Algorithm overview 351
12.5.2 Hardware design 352
12.5.3 Clustering MNIST images 354
12.6 Summary 355
Acknowledgment 355
References 355

Index **357**

Preface

We live in the era of smart cities, social computing, and cloud computing in which hardware and software work together to provide the users meaningful systems that allow them to do their jobs more efficiently, with minimal human intervention. A very good example is the manufacturing of semiconductor chips in ultra clean room environments with the slightest human–machine interaction for efficient production and reliability. Smart cities are needed to meet the demands of excessive growth of population in which the limited earthly resources are over utilized. For example, a smart health care system such as the body-area network is able to provide better health care to patients even when doctors cannot be present but are available remotely. A smart transport system is able to provide real-time locations of the transportation system. A smart city is able to handle its street lights more efficiently. A waste management system is able to handle large amounts of waste in a city with minimal effort. In general, the information and communication technology (ICT) is the core of such smart systems from a small to a large scale implementation.

The recent evolution of a technology called Internet of Things (IoT) which can often process big data, though it still remains vaguely defined, is being explored as a next generation solution for building and interconnecting such smart systems. No doubt, the end users of these smart systems deal with a system using some form of software as the front end. However, in general a combination of hardware and software can implement such systems through the IoT. But software does need some hardware as a base to be executed on. Hardware in general can be sensors of every type, analog integrated circuits, digital integrated circuits, or even mixed-signal integrated circuits. These are designed by design engineers at various levels of abstraction depending on their nature, analog or digital. The overall design process is, however, based on a specific process technology to manufacture integrated circuits and systems. Current chip manufacturing technology uses nanometer scale CMOS (nano-CMOS) which is generally known as nanoelectronic technology. Nanotechnology, of which nanoelectronics is a subset, has the potential to revolutionize numerous industry sectors including information technology, energy, medicine, homeland security, and transportation. An estimation shows that the total market for nanotechnology related products would be close to few trillion dollar and the industry would need several million personnel.

The demand for ever smaller and portable electronic devices such as smart mobile phones and tablets has ultimately driven the scaling of CMOS to nano-CMOS to reach its physical limits with the smallest possible feature sizes. The feature sizes of CMOS transistors have dramatically shrunk with technology scaling and the gate oxide thickness (T_{ox}) has reached the range of 12–16 Å which is just a few monolayers of SiO_2.

The scaling of CMOS transistors presents various short-channel issues including process variations, high-leakage power, low reliability, and thermal effects. The use of alternatives for SiO_2 as the gate dielectric as well as multiple gates has been used as solutions. Moore's law predicts its end within a decade and the International Technology Road Map for Semiconductors (ITRS) in its past several reports suggested several nonclassical devices, based on novel emerging materials, as a possible replacement of silicon. This has led to the evolution of several nonclassical devices.

The use of high-κ dielectrics serves the dual purpose of scaling the device as well as reducing gate leakage. As an alternative to silicon various nonclassical devices, including spintronics, carbon nanotube transistors, graphene transistors, tunnel transistors and memristors, have been introduced that could replace the traditional and ubiquitous silicon transistor. The generic term post-CMOS is used to represent these technologies. Graphene transistors (GFETs) can operate at high frequencies (e.g., 100 GHz and above) and have potential for high-speed nanoelectronics. The memristor combines the behavior of memory and a resistor and remembers its most recent resistance when the voltage is turned off, until the next time the voltage is turned on. The memristor is receiving significant attention due to it promising properties. With the development of emerging nanoscale devices, their design and manufacturing processes need to develop and mature so that nanoscale device-based systems can be efficiently built. Detailed discussion of these issues and their corresponding solutions is lacking in existing texts and existing curricula in academia. Most importantly the design engineers' task has been severely complicated due to the emergence of these issues which has led to longer design cycle time. As a consequence, yield loss and high chip cost are common and the overall impact is the increased cost of consumer electronics. There is a large gap between the skill of design engineers and understanding of these devices and their integration in design methodologies. However, existing books are typically based on traditional CMOS devices and do not cover the detailed modeling and design aspects of post-CMOS devices. As a consequence, existing books do not train engineers in emerging nanoscale device-based electronics and do not catalyze the growth of nanoelectronics. The traditional literature does not cater to the expectations of the emerging nanotechnology industry; however, this work will meet the demand of training engineers in emerging nanoelectronics.

In order to train students, practicing engineers, and researchers, this book presents 12 carefully chosen chapters to cover the complete spectrum of nanoscale devices of CMOS, post-CMOS, and nanoelectronics. Chapter 1 titled "High-κ Dielectrics and Device Reliability" by Bhuyian and Misra suggests that with the sub-45nm technology node, high-κ gate dielectrics such as HfO_2 have emerged. This chapter focuses on HfO_2 dielectrics with particular emphasis on the most important characteristics and especially reliability. Chapter 2 which is titled as "High Mobility n- and p-Channels on Gallium Arsenide and Silicon Substrates using Interfacial Misfit Dislocation Arrays" authored by Renteria, Addamane, Shima, and Balakrishnan also discusses a material level solution of technology scaling. This chapter describes the use of compound semiconductors with Si to solve the challenges of technology scaling. Chapter 3 titled "Anodic Metal-Insulator-Metal (MIM) Capacitors" by Kannadassan, Mallick, and Baghini discusses the efficient realization of on-chip capacitors. It is a fact that the capacitor is an important passive element in various integrated circuits and system applications. This chapter introduces efficient MIM capacitor realizations using nanostructured anodic high-κ. Chapter 4 authored by Srivastava and Banadaki, titled "Graphene Transistors—Present and Beyond", discusses the state-of-art of graphene-based transistors with a prediction for its future directions. Chapter 5, titled "Junction and Doping-Free Transistors for Future Computing", by Singh and Sahu presents new types of transistors for efficient implementation of future generation integrated circuits. In Chapter 6, Yanambaka, Mohanty, Kougianos, and Ghai present logic libraries which and can be used in automatic synthesis flows of digital integrated circuits under the title "Nanoscale High-κ/Metal-Gate CMOS and FinFET based Logic Libraries". Chapter 7 titled "FinFET and Reliability Considerations of Next-Generation Processors" by Zhang, Chen, Peng, and Chen, presents reliability solutions for FinFET-based processors. Chapter 8 titled "Multiple-Independent-Gate Nanowire Transistors: From Technology to Advanced SoC Design" by Gaillardon, Zhang, Amaru, and De Micheli discusses a novel class of device, the multiple-independent-gate field-effect transistor (MIGFET) for better functionality and flexibility as compared to the classic MOSFETs. Chapter 9 titled "Exploration of Carbon Nanotubes for Efficient Power Delivery" by Todri-Sanial, presents the idea of using carbon nanotubes (CNTs) for reliable power delivery networks in 2D and 3D integrated circuits. Chapter 10 titled "Timing Driven Buffer Insertion for Carbon Nanotube Interconnects" by Liu, Zhou, and Hu discusses the use of CNT as a potential high-speed high-performance interconnect as compared to metal interconnects. Chapter 11 titled "Memristor Modeling—Static, Statistical, and Stochastic Methodologies" by Li, Hu, and Liu present the electrical properties of the memristors using the most popular TiO_2 thin film device as a case study. In Chapter 12, Kudithipudi, Merkel, and Kurinec propose methods for memristor-based efficient neuromorphic realizations under the title "Neuromorphic Devices and Circuits."

The editors hope that graduate students (Ph.D./M.S.), researchers, and practicing engineers are the primary readership of this book and would be greatly benefitted by its content. In addition, senior undergraduate students will also benefit from the content of this book. The book addresses many of the nanoelectronics device level issues and solutions to aid in the design of integrated circuits. Special features of this book include the following:

- Coverage of various nano-CMOS based devices, issues, and modeling.
- Coverage of post-CMOS nanoelectronic technologies instead of just nano-CMOS only focus.
- Coverage of new materials and devices like graphene and memristor and their modeling.
- Coverage of carbon nanotube (CNTs) and nanowire transistors as well as interconnects.
- Coverage of key issues, challenges, and solutions of nanoelectronic challenges that the industry is striving to address.
- Coverage of design methods accounting for nanoscale issues and challenges.

- Practicing engineers will learn various nanoscale device structures and modeling.
- The book can serve as reference for senior undergraduate as well as graduate (Ph.D./M.S.) students.

The book will be appealing to many institutions, research organizations, and semiconductor companies in the world where the following disciplines are taught and/or are being researched: electrical engineering, computer engineering, electronics engineering, electronic communication engineering, electrical and electronics engineering. The editors hope that faculty from academic institutions in various countries will use the book in their teaching, research, and teaching.

The editors would like to thank the authors of the individual chapters without whom the book would not have been happened. The editors would like to thank the IET staff who tremendously in bringing this book to excellent shape. The help of anonymous reviewers who reviewed the book proposal and provided us with many constructive feedbacks is also greatly appreciated.

Saraju P. Mohanty
Professor, University of North Texas, USA.
saraju.mohanty@unt.edu

Ashok Srivastava
Professor, Louisiana State University, USA.
eesriv@lsu.edu

Chapter 1

High-κ dielectrics and device reliability

M.N. Bhuyian[1] and D. Misra[1,2]

Technology scaling continues to be driven by the cost per function due to proliferation of mobile computing. With sub-45-nm technology node, high-κ gate dielectrics such as HfO_2 have emerged. This chapter is dedicated to high-κ dielectrics with particular emphasis to most important characteristics the reliability.

1.1 Introduction

Microelectronics has been the most important driving force for almost all kind of technology evolutions in the past five decades [1, 2]. Continuous device scaling leads to a decrease in cost per function of technology and improves the economic productivity and the quality of life through proliferation of computers, communication, and other industrial and consumer electronics. Modern microprocessors used in today's world consist of hundreds of millions of metal-oxide semiconductor field effect transistors (MOSFETs) [3]. With the scaling of devices below the 45-nm technology node, high-κ gate dielectric materials emerged as a replacement of SiO_2, and metal gate has replaced polysilicon gate in the high-performance logic family and low standby power logic family [3, 4]. HfO_2-based dielectric materials have been considered as the most promising alternative of SiO_2 in the complementary metal oxide semiconductor (CMOS) technology because of their quality superior to other high-κ dielectrics considering CMOS compatibility, higher dielectric constant, suitable band offset with Si, and good thermal stability with Si [4, 5].

According to the International Technology Roadmap for Semiconductor (ITRS) 2013 updates [6], many physical dimensions of transistors are expected to be crossing the 10-nm threshold in the years 2020–2025. Research on high-κ/metal gate (HK/MG) stack is continuing to scale the equivalent oxide thickness (EOT) to sub-0.7 nm as well to have better quality dielectrics [4–6]. Various interleaved treatments in the atomic layer deposition process of Hf-based high-κ dielectrics have attracted tremendous attention in order to enhance the quality of dielectrics for CMOS technology [7–13]. Recently, it has been reported that the poor dielectric characteristics

[1]Department of Electrical and Computer Engineering, New Jersey Institute of Technology, Newark, NJ 07102, USA
[2]Center for Nano Science and Engineering (CeNSE), Indian Institute of Science, Bangalore, 560012, India

of chemical vapor deposited (CVD) and atomic layer deposited (ALD) grown silicon oxide films can be improved by exposing them to a slot-plane-antenna (SPA) plasma with various gases such as O_2/Ar, Ar, and O_2/He [14–16]. The SPA plasma provides a high-density plasma with low electron temperature, where the radicals diffuse from the plasma generation region to the wafer surface [17]. Further investigations were carried out with incorporating other materials like Zr or Al into HfO_2 in order to foster device scaling and their performance enhancement [18–35]. The addition of Zr into HfO_2 was shown to be beneficial for better EOT downscaling by several reports [18–22]. The incorporation of aluminum into HfO_2 by forming $(HfO_2)_{1-x}(Al_2O_3)_x$ films [23–32] or HfO_2/Al_2O_3 bilayers [24,33,34] was reported to be promising for high-κ on silicon and high-mobility substrates. Recently, Tapily *et al.* [35] reported a mixed structure of tetragonal and monoclinic phase formation for ALD $Hf_{1-x}Al_xO_y$ ($x = 0$–0.25) with 2 Å lower EOT and one order of magnitude reduced gate leakage current.

Reliability is a critical concern for high-κ dielectrics in order to integrate them into mainstream commercial integrated circuits. Gate stack reliability can be evaluated by understanding the charge trapping behavior of the dielectric and its response to the electrical stress [36–38]. HfO_2 has been widely studied for its reliability under different stress conditions [36–40]. Since Zr or Al incorporation in HfO_2 and interleaved treatment processes is promising, the reliability of these dielectrics needs to be investigated. The knowledge of stress-induced defects, defect activation energy, and charge to breakdown can improve the understanding of their effects on device reliability. Also, it is known that even though the EOT is successfully scaled in some processes, the performance of the metal-oxide semiconductor (MOS) device strongly depends on the quality of the interface between the silicon substrate and the interfacial layer (IL) [41–43]. In addition, the process-induced interface traps also significantly influence the long-term reliability of the devices. Interface traps, the result of a structural imperfection, act as generation/recombination centers with an energy distribution throughout the silicon band gap. When the device is in operation, electrons or holes occupy interface traps and contribute to the threshold voltage shift. They also contribute to leakage current, low-frequency noise, reduced mobility, drain current, and transconductance [43]. Density of interface states, D_{it}, versus energy, E, at the Si/IL interface provides a comprehensive understanding of the impact of various process conditions on interface defects. Also, an understanding of the impact of electrical stress on interface state generation for these dielectrics will help their integration in future CMOS technology.

1.2 Alloying HfO_2 and ZrO_2

HfO_2 and ZrO_2 have been investigated extensively as possible alternatives to SiO_2-based options due to their relatively higher dielectric constants and larger band gap [4, 5]. Both Zr and Hf are in group IV in the periodic table and are considered to be the two most similar elements in the periodic table. HfO_2 and ZrO_2 have similar properties and are completely miscible in solid state [18]. The different crystalline phases of HfO_2 and ZrO_2 are cubic, tetragonal, orthorhombic, and monoclinic. The

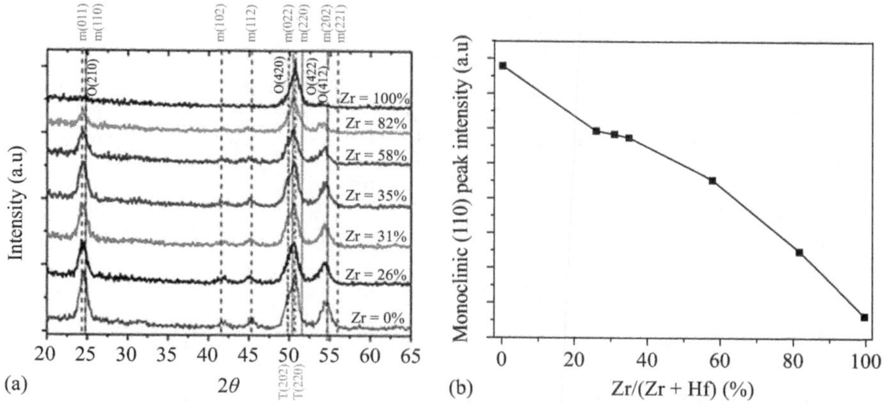

Figure 1.1 *(a) GIIXRD spectra of $Hf_{1-x}Zr_xO_2$ as a function of Zr% in the dielectrics and (b) intensity of monoclinic (1 1 0) diffraction peak as a function of Zr/(Zr + Hf)% in the dielectrics*

Source: © 2012 The Electrochemical Society. Reprinted, with permission, from K. Tapily, S. Consiglio, R.D. Clark, R. Vasić, E. Bersch, J.J. Sweet, I. Wells, G.J. Leusink, and A.C. Diebold, "Texturing and tetragonal phase stabilization of ALD $Hf_xZr_{1-x}O_2$ using a cyclical deposition and annealing scheme," ECS Trans. 2012, vol. 45(3), pp. 411–420

monoclinic phase is thermodynamically the most stable phase for both ZrO_2 and HfO_2 in bulk form, but possesses the lowest κ value [5]. The amorphous phase also exhibits a similar κ value as the monoclinic phase [44]. For both ZrO_2 and HfO_2 the cubic phase has a higher κ value than monoclinic while the tetragonal phase has the highest κ value due to the lower phonon frequencies and higher effective charges [5, 44]. Tetragonal stabilization of HfO_2 by the addition of ZrO_2 was reported by several groups [18–22, 45].

Figure 1.1(a) shows the grazing incident in plane X-ray diffraction (GIIXRD) spectra for the $Hf_{1-x}Zr_xO_2$ alloy with different Zr% in the dielectrics. Figure 1.1(b) shows the intensity of monoclinic (1 1 0) diffraction peak as a function of Zr/(Zr + Hf)% in the dielectrics [45]. The observed results in Figure 1.1 clearly show the stabilization of higher-κ tetragonal phase with addition of ZrO_2 into HfO_2.

Figure 1.2(a) shows typical capacitance voltage characteristics for ALD $Hf_{1-x}Zr_xO_2$ deposited by using $HfCl_4$, $ZrCl_4$, and H_2O precursors on in-situ steam grown SiO_2/Si interface [22] and annealed in nitrogen ambient at 1050°C. Inset in Figure 1.2(a) shows an increase in C_{ox} or decrease in EOT with increase in Zr percentage. Figure 1.2(b) shows the comparison of gate leakage current density for $Hf_{1-x}Zr_xO_2$ with different Zr/(Hf + Zr) contents from Reference 22. A slight increase in J_g with increasing Zr content reveals that Zr incorporation into HfO_2 modifies the band gap and the band offset, as ZrO_2 has comparatively a lower value of the band gap and the conduction band offset with Si [4, 5].

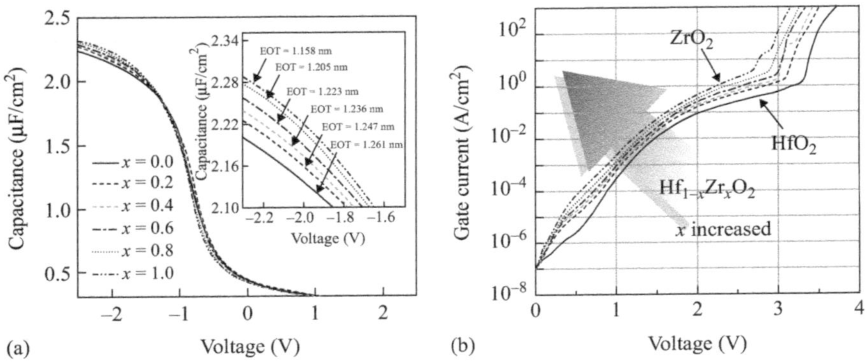

Figure 1.2 *(a) Typical C–V plot for $Hf_{1-x}Zr_xO_2$ for different Zr/(Hf + Zr) contents. Inset shows the magnified C–V plot to demonstrate the relationship between EOT and Zr content. (b) Comparison of gate leakage current for $Hf_{1-x}Zr_xO_2$ with different Zr/(Hf + Zr) contents*

Source: © 2012 IEEE. Reprinted, with permission, from C.K. Chiang, J.C. Chang, W.H. Liu, C.C. Liu, J.F. Lin, C.L. Yang, J.Y. Wu, C.K. Chiang, and S.J. Wang, "A comparative study of gate stack material properties and reliability characterization in MOS transistors with optimal ALD zirconia addition for hafina gate dielectric," in Proc. IEEE IRPS 2012, pp. GD. 3.1–3.4

1.3 Advanced ALD process: intermediate treatment

High-κ dielectrics deposited with various intermediate treatments were shown to be beneficial in several reports [7–13]. Multiple deposition and annealing of HfO_2 film deposited with metal-oxide chemical vapor deposition was reported by Yeo *et al.* [46] (700°C anneal) and Ishikawa *et al.* [47] (750–950°C anneal).

For ALD dielectrics, Nabatame *et al.* [48] have demonstrated a device performance benefit from performing an in-situ annealing (650°C) after each ALD cycle during the growth of $HfAlO_x$ films, deposited using an Hf-alkylamide precursor. Delabie *et al.* [12] also reported that intermediate thermal treatments (420–500°C), applied to the $HfCl_4/H_2O$ process, led to a significant reduction in in-film Cl content, whereas a post-deposition annealing (PDA) treatment led to no Cl reduction. Clark *et al.* [13] observed almost tenfold reduction in gate leakage current using HfO_2 gate oxide with multiple intermediate thermal treatments as compared to a single PDA. Apart from thermal treatment, an interleaved treatment in the atomic layer deposition process by using room temperature ultraviolet ozone [7–9], D_2O radical [10, 11], and remote microwave N_2O plasma [12] was reported to enhance the device performance. There are several reports showing performance improvement achieved by using SPA plasma in dielectric processing. Nagata *et al.* [14] demonstrated that SPA plasma (Ar/O_2) treatment results in better densification of CVD SiO_2. Kobayashi *et al.* [49] used SPA radical oxidation to produce improved GeO_2 interfacial layer growth with no substrate orientation dependence. Decrease in gate leakage current and trap density due to SPA plasma exposure was also reported by Kawase *et al.* [15]. Tanimura *et al.* [16] reported a reduction in impurity concentration of ALD SiO_2 by exposing it to SPA plasma. Unlike N_2O or ultraviolet ozone, SPA Ar plasma does not induce interfacial oxide growth that limits scaling potentials of such processes [7, 12, 17].

1.4 SPA plasma

The SPA plasma system can provide a large diameter plasma as required by 300-mm wafer fabrication process. Although conventional plasma sources, such as electron cyclotron resonance plasma, helicon plasma, and inductively coupled plasma can provide plasma with sufficiently low electron temperature in the wafer region, the damages caused by these conventional plasma sources are significant considering the strict requirement of integrated circuit processing. SPA plasma on the other hand causes very little damage to the wafer, can handle high power, and operates in the over dense regime. Radicals diffuse from the plasma generation region to the wafer surface in SPA plasma process [17].

Figure 1.3 shows an SPA plasma system. Some important features of SPA plasma are:

- high density ($\sim 10^{12}/cm^3$);
- low electron temperature (0.7–1.5 eV);
- wide process window (7–1000 Pa);
- optional bias to accelerate ions.

The planar antenna structure of SPA plasma system is advantageous in realizing a compact apparatus for semiconductor processing.

1.5 Cyclic deposition and SPA plasma treatment to ALD $Hf_{1-x}Zr_xO_2$

In this research, ALD $Hf_{1-x}Zr_xO_2$ with $x = 0$, 0.31, and 0.8 have been deposited in the MOS capacitor structure with a TiN metal gate. Samples were subjected to SPA Ar plasma in a cyclic deposition and plasma treatment (DSDS) process as shown in Figure 1.4(a).

Figure 1.4(a) schematically shows the DSDS process. Figure 1.4(b) compares the dielectric thickness and IL thickness for dielectric with DSDS process and standard as-deposited process (As-Dep) without any plasma treatment.

ALD $Hf_{1-x}Zr_xO_2$ is deposited in a 300-mm TEL Trias™ cleanroom tool by using tetrakis (ethylmethylamido) hafnium (TEMAH) as the Hf precursor, tetrakis (ethylmethylamido) zirconium (TEMAZ) as the Zr precursor, and H_2O as the oxidant at a deposition temperature of 250°C. Details of the device fabrication process will be found in Reference 50. After initial cleaning, a sacrificial oxide layer was grown and then removed during gate pre-clean step. The pre-clean included a rinse with ozone/deionized water that resulted in a SiO_2 IL on the order of 0.6–0.8 nm

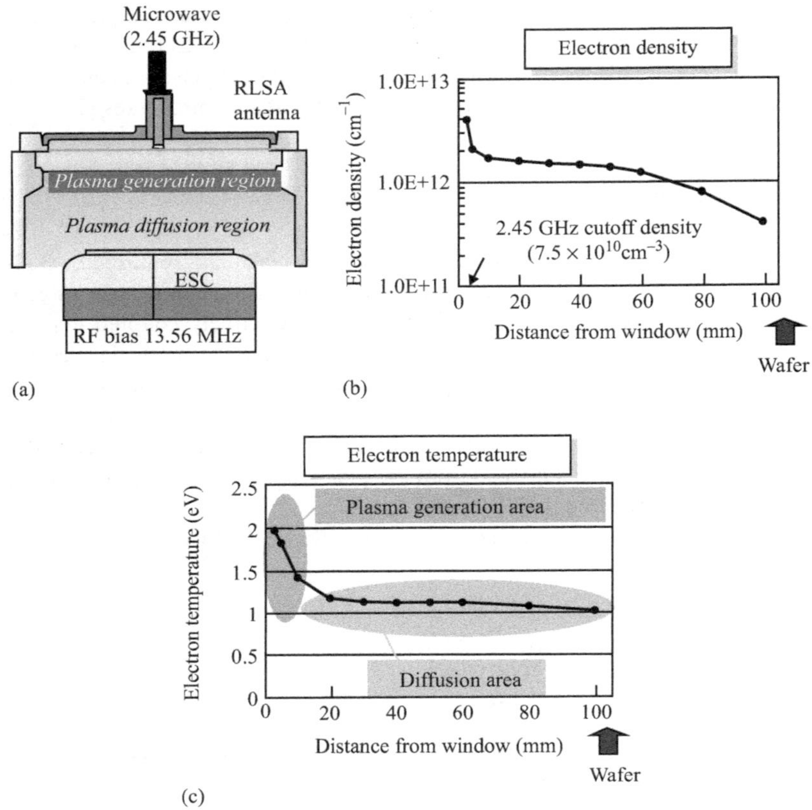

Figure 1.3 (a) A typical SPA plasma system and (b and c) electron density and electron temperature in the plasma generation area and plasma diffusion area in an SPA plasma system. Abbreviations: ESC, electrostatic chuck; RF, radio frequency; RLSA, radial line slot antenna

in thickness. The SiO_2 layer was then subjected to a radical flow nitridation (RFN) process that slightly increased the IL thickness and results in an approximate 0.7–0.9 nm of SiON IL. The $Hf_{1-x}Zr_xO_2$ films were then deposited by precisely controlling the individual HfO_2 and ZrO_2 ALD cycles contained within each super cycle of the ALD process. Two different Hf precursor to Zr precursor pulse ratio of 3:1 and 1:3 was used for $Hf_{1-x}Zr_xO_2$ whereas samples with only HfO_2 were deposited without any Zr precursor. The samples were subjected to Ar plasma in the SPA system after every 22 ALD cycles, and the deposition and plasma exposure process were repeated

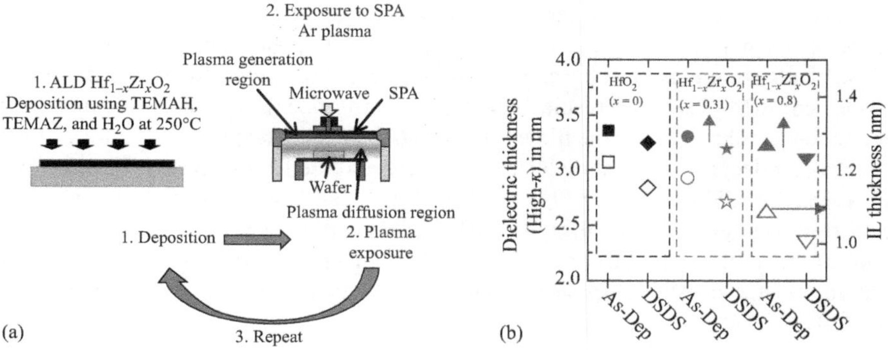

Figure 1.4 *(a) Schematic representation of DSDS process using an SPA plasma in a cluster tool system. (b) Dielectric thickness (filled symbols on left scale) and IL thickness (open symbols on right scale) for MOSCAPs with DSDS and As-Dep HfO_2 ($x = 0$), $Hf_{1-x}Zr_xO_2$ ($x = 0.31$), and $Hf_{1-x}Zr_xO_2$ ($x = 0.8$). Abbreviations: MOSCAP, metal oxide semiconductor capacitor; DT, direct tunneling; RF, radio frequency; RLSA, radial line slot antenna; TEMAZ, tetrakis (ethylmethylamido) zirconium; ESC, electrostatic chuck; F–N, Fowler–Nordheim; P–F, Poole–Frenkel*

cyclically to get a desired thickness. The DSDS process is shown in Figure 1.4. For both DSDS and As-Dep $Hf_{1-x}Zr_xO_2$, a total of 44 ALD cycles were used to deposit the dielectrics, and none of them were subjected to any PDA. Following the high-κ gate oxide deposition, the metal gate was formed by 50 nm of CVD TiN. The metal oxide semiconductor capacitors (MOSCAPs) were then formed by initially depositing a hard mask followed by patterning, etching, cleaning and stripping the hard mask subsequently.

Hf and Zr composition for the different ALD films was measured by X-ray photoelectron spectroscopy (XPS) by using Thermo Fisher Theta Probe™ XPS system. High-κ film thickness values for DSDS and As-Dep processed films were measured by X-ray reflectivity (XRR) on an in-line 300-mm fab Rigaku MFM65 system. IL (SiON) thickness values were estimated by a combination of spectroscopic ellipsometry (SE) with XRR. HfO_2 and ZrO_2 showed identical growth rate (~0.7 Å/cycle) at 250°C. Also, all films remained amorphous, as the samples were not subjected to any PDA treatment. Synchrotron grazing incidence X-ray diffraction (XRD) measurements confirmed no crystalline phase formation (not shown). It was determined from the XPS measurements that ALD $Hf_{1-x}Zr_xO_2$ deposited with Hf precursor to Zr precursor ratio of 3:1 and 1:3 resulted in $x = 0.31$ and $x = 0.8$, respectively. Figure 1.4(b) shows the comparison of dielectric thickness and IL thickness for DSDS and As-Dep processed high-κ dielectrics with different Zr percentages. It was observed that samples with higher Zr percentages have lower IL thickness (Figure 1.4(b)). It is known that chemically grown oxide has much higher percentage of oxygen deficiency centers [51] and also HfO_2 can supply more oxygen to the IL as compared to $Hf_{1-x}Zr_xO_2$

because of higher amount of incorporated oxygen in HfO_2 [52]. Exposure to intermediate plasma increases the film density by reducing the impurity concentration [16]. In addition, plasma suppresses thermal-induced oxygen diffusion to the IL for oxide regrowth [53]. As a result, both the dielectric thickness and the IL thickness reduction are observed for the films subjected to intermediate plasma (Figure 1.4(b)). The largest decrease in total dielectric thickness and the IL thickness was, therefore, observed for DSDS $Hf_{1-x}Zr_xO_2$ ($x = 0.8$). PDA can lead to possible transformation from one phase to another [43–46]. Since no PDA was used for our samples and no crystalline structure was observed by synchrotron grazing incidence XRD measurements, possible decrease in thickness due to volume variation can be ruled out [54].

1.5.1 Impact of Zr addition and SPA plasma on electrical properties

For electrical characterization, we used a 4284A LCR meter for capacitance voltage (C–V) measurements and a 4156B Semiconductor Parameter Analyzer for current voltage (I–V) measurements in a Cascade Micro-chamber with a precision probe station. The C–V measurement frequency used in this work was 250 kHz, and the device size was 40 μm × 40 μm. EOT and flat-band voltage of the devices were estimated from the C–V curves using the NC State CVC program [55]. Gate leakage current per unit area (J_g) was measured at $-1V + V_{FB}$ in the negative gate bias region. Interface state density (D_{it}) in Si band gap was estimated by using the conductance method [56]. The DC gate voltage was varied from 1 V to −2 V with −20 mV voltage step, and the measurement frequencies used for D_{it} estimation were from 100 Hz to 1 MHz. Automatic measurement programs were used to limit the de-trapping behavior and improve efficiency in data collection and formulation.

From Figure 1.5(a–c), the impact of Zr addition and cyclic SPA plasma exposure on EOT (Figure 1.5(a,b)), flat-band voltage (Figure 1.5(a)), gate leakage current density, J_g (Figure 1.5(b)), and interface state density, D_{it} (Figure 1.5(c)) is observed. With increasing percentage of Zr in the gate oxide, EOT downscaling is possible and is directly related to the physical thickness variations as shown in Figure 1.4(b). Intermediate SPA plasma exposure further lowers the EOT (Figure 1.5(a,b)) with more influence on devices with lower Zr percentages, as addition of Zr reduces available oxygen in the film which is responsible for IL regrowth. It was further observed in Figure 1.5(a) that devices with higher Zr percentage had comparatively higher initial flat-band voltage values possibly due to the detrimental effect of Zr on the SiON interface [18]. It is known that ZrO_2 has comparatively lower electron affinity than HfO_2 [57], which results more positive charge formation in the dielectrics for devices with $x = 0.8$. Also, SPA plasma contributes to slight increase in initial flat-band voltage values (Figure 1.5(a)) due to increase in interface state density (Figure 1.5(c)) because of limited regrowth of IL thickness (Figure 1.4(b)). It is known that the interface state density increases for thinner ILs due to an increased stress in Si/IL interface [58]. Figure 1.5(b) reveals that the addition of Zr into HfO_2 increases the gate leakage current density by increasing

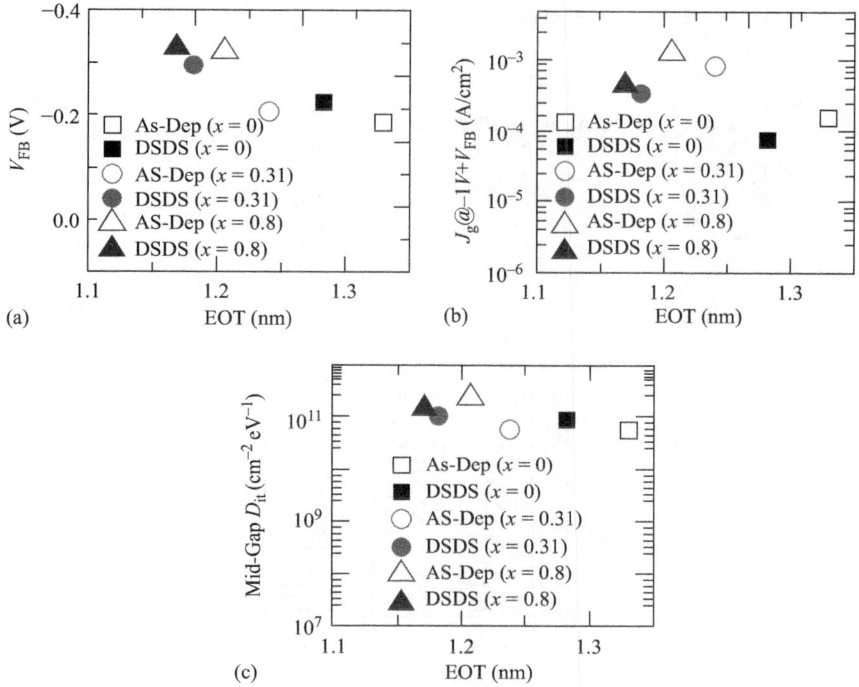

Figure 1.5 (a) Flat-band voltage as a function of EOT for DSDS and As-Dep $Hf_{1-x}Zr_xO_{2-}$ with $x = 0, 0.31$, and 0.8, (b) comparison of gate leakage current density sensed at $-1V + V_{FB}$, and (c) comparison of the mid-gap level D_{it} as a function of EOT for different dielectrics

the conduction band offset with Si [57]. However, cyclic SPA plasma exposure significantly contributes to gate leakage current reduction for these dielectrics (Figure 1.5(b)).

1.5.2 Reliability study by constant voltage stress

When devices are subjected to a constant voltage stress, defects such as interface states, electron traps, and positively charged donor-like states are generated in the bulk and at interface [36–38]. Stress-induced flat-band voltage shifts [40], stress-induced leakage currents (SILCs) [59], and stress-induced interface state generation [42] are the typically observed behaviors for reliability study. In addition, during the constant voltage stress, a critical density of generated traps leads to the dielectric breakdown [60]. In this research, the constant voltage stress was implemented by applying a negative bias to the gate while keeping the substrate grounded. The applied stress voltage was varied from −3 V to −3.4 V for different dielectrics according to their EOT and V_{FB} variation to have an equal stress field across all dielectrics.

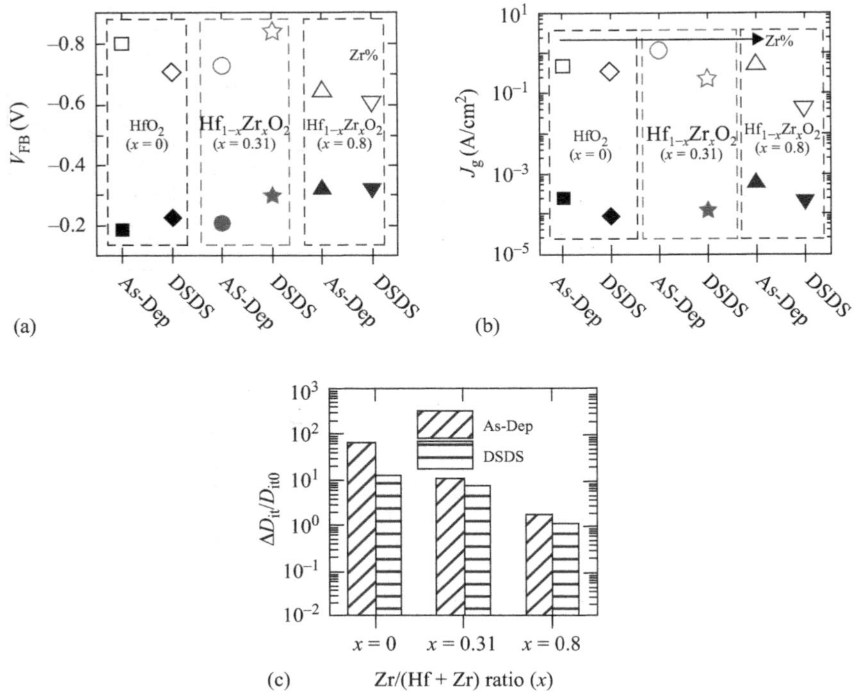

Figure 1.6 *(a,b) Flat-band voltage shift and gate leakage current density for unstressed devices (filled symbols) and for devices after the application of a constant voltage stress in the gate injection mode for 1000 s (open symbols) and (c) change in the mid-gap D_{it} for As-Dep and DSDS $Hf_{1-x}Zr_xO_2$ with $x = 0, 0.31$, and 0.8. The applied stress voltage was varied from -3 V to -3.4 V for different dielectrics according to their EOT and V_{FB} variation to have an equal stress field across all dielectrics*

1.5.2.1 Impact of stress on flat-band voltage, leakage current, and interface state density

Figure 1.6 shows the impact of constant voltage stress on the flat-band voltage shift, V_{FB} (Figure 1.6(a)), gate leakage current density, J_g (Figure 1.6(b)), and interface state density, D_{it} (Figure 1.6(c)) for DSDS and As-Dep $Hf_{1-x}Zr_xO_2$ for $x = 0$, 0.31, and 0.8. When devices were stressed in the gate injection mode, electrons are injected into the high-κ layer first and then subsequently tunnel through the IL to reach the conduction band of silicon [50]. Under electrical stress, the creation of fixed positive charge centers in the dielectrics and stress-induced interface state generation contributed to the flat-band voltage shift [39, 61]. It was observed that, Zr addition in HfO_2 improves stress-induced flat-band voltage shift for both As-Dep and DSDS processing while additional benefits came from intermediate plasma (Figure 1.6(a)). When the flat-band voltage shifts of HfO_2 and $Hf_{1-x}Zr_xO_2$ with $x = 0.8$ were compared with unstressed devices after 1000-s stress (Figure 1.6(a)), the impact of SPA plasma exposure was clear. For $Hf_{1-x}Zr_xO_2$ with $x = 0.31$, the shifts in flat-band voltage values due to stress are identical for both As-Dep and DSDS processing (Figure 1.6(a)). Despite higher initial flat-band voltage value in the negative direction, DSDS HfO_2 showed 21% improvement in stress-induced flat-band voltage shift as compared to As-Dep HfO_2 when devices were compared after 1000-s stress. On the other hand, an improvement of 12% in stress-induced flat-band voltage shift was observed for DSDS $Hf_{1-x}Zr_xO_2$ with $x = 0.8$ content as compared to As-Dep $Hf_{1-x}Zr_xO_2$ with similar Zr percentage. Based on this observation, it is inferred that SPA plasma exposure reduces the impurity content and provides significant immunity to stress-induced trap center formation in the dielectric thereby improves the reliability of the gate dielectric [16].

It is observed from Figure 1.6(b) that DSDS $Hf_{1-x}Zr_xO_2$ with $x = 0.8$ has one order of magnitude lower value for J_g after 1000-s stress as compared to As-Dep $Hf_{1-x}Zr_xO_2$ with $x = 0.8$, while little improvement is observed due to SPA plasma exposure to devices with HfO_2 ($x = 0$). As explained earlier, the SPA plasma reduces the number of impurities in the oxide film, which act as trap centers [16]. In addition, the SPA plasma is capable of growing an atomically flat surfaces and interfaces [62], which helps in gate leakage current reduction. Additionally, the conduction process is entirely through the high-κ layer. When the device is under negative bias condition, electrons first injected to the high-κ layer, hopping through the high-κ layer by trap-assisted tunneling (TAT) subsequently, reach the conduction band of Si by direct tunneling through the IL [40, 50]. This suggests that SPA plasma exposure enhances the quality of high-κ film by reducing the number of traps in the film and Zr addition enhances this improvement further. In this work, measurements were taken at room temperature, and the sense voltage did not exceed 1.5 V for any devices. Therefore, we believe Fowler–Nordheim tunneling, and Schottky emission may not contribute significantly for these devices. Also, gate leakage current density measured at different temperatures did not show the signature of Poole–Frenkel mechanism at this sense field (not shown). We believe the TAT is the main mechanism that contributes to the conduction mechanism in the high-κ layer.

From Figure 1.6(c), a gradual reduction in D_{it} generation in the mid-gap has been found when the Zr percentage increases which is further reduced due to the SPA plasma exposure. DSDS $Hf_{1-x}Zr_xO_2$ with 80% Zr/(Hf + Zr) showed around two orders of magnitude reduction in the $\Delta D_{it}/D_{it0}$ as compared to As-Dep HfO_2 deposited without an SPA plasma exposure. As described earlier, the SPA plasma has a low electron energy. The use of Ar plasma reduces the free radical concentration in the plasma leaving only ground-state low-energy radicals which primarily interact with the exposed dielectric during plasma treatment [16, 17]. Since the first plasma exposure was after 22 ALD cycles, the total dielectric thickness from the SiON/Si interface would be approximately 2.5 nm including ~1 nm of IL. This thickness might allow low-energy radicals to reach the interface. It was previously reported that an SPA plasma exposure can reduce impurity concentration in the dielectrics and thereby increases film density [14, 16]. As explained earlier, the SPA plasma exposure helps to grow atomically flat surfaces and interfaces [62] and improves bonds in

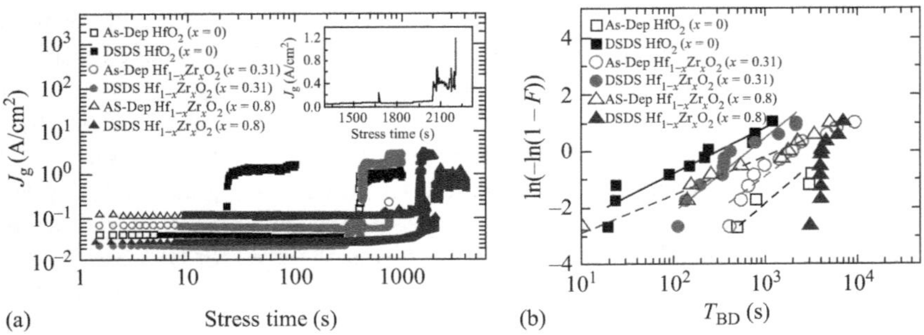

Figure 1.7 (a) Breakdown characteristics during gate injection stress showing electron trapping, SBD, and HBD. Inset shows PBD prior to HBD for DSDS $Hf_{1-x}Zr_xO_2$ ($x = 0.8$) and (b) Weibull plot of time to breakdown (T_{BD}) for DSDS and As-Dep $Hf_{1-x}Zr_xO_2$ with different Zr percentages

the interface by providing energy from low-energy radicals [15]. Therefore, in addition to suppressed trap formation in the bulk high-κ dielectrics, DSDS $Hf_{1-x}Zr_xO_2$ with 80% Zr/(Hf + Zr) demonstrated a suppressed interface state generation after the application of a constant voltage stress [50].

1.5.2.2 Effect of Zr addition and SPA plasma on TDDB

To further evaluate the reliability, the time-dependent dielectric breakdown (TDDB) study was conducted by subjecting the devices to a constant voltage stress in the gate injection mode. Figure 1.7 shows the change in the current density as a function of time till breakdown for DSDS and As-Dep $Hf_{1-x}Zr_xO_2$ with different Zr percentages. All devices demonstrated a slim decrease in the gate leakage current density due to electron trapping followed by the soft breakdown (SBD), the progressive breakdown (PBD) (inset of Figure 1.7), and subsequently, the hard breakdown (HBD) as stress continued. This behavior is in accordance with previous reports of the gate dielectric degradation mechanism [59, 60]. A critical number of traps generated in different locations between the anode and the cathode result in an SBD, and as the stress continued, an increased energy dissipation of these localized areas drives the device into the thermal runway or the HBD [60, 63]. However, devices with As-Dep $Hf_{1-x}Zr_xO_2$ with $x = 0.31$ and $x = 0$ have the SBD and the PBD regions for very short duration as compared to other devices which is due to their comparatively higher IL thickness [60]. Before the first SBD, a decreased gate leakage during the stress is observed for DSDS processed devices as compared to As-Dep devices for all Zr percentages (Figure 1.7) which further supports the SILC characteristics (Figure 1.6(b)).

Figure 1.7(b) shows the Weibull plot of time to breakdown, T_{BD} for both DSDS and As-Dep processed $Hf_{1-x}Zr_xO_2$ with different Zr percentages for a constant field stress in the gate injection mode. A decrease in the time to breakdown was observed for As-Dep $Hf_{1-x}Zr_xO_2$ with increasing Zr percentage, whereas opposite behavior was observed for DSDS $Hf_{1-x}Zr_xO_2$. This followed the same trend as observed earlier for the gate leakage current J_g value after 1000-s stress for both DSDS and

Table 1.1 Weibull slope for As-Dep and DSDS $Hf_{1-x}Zr_xO_2$ with different Zr percentages

	As-Dep HfO_2 ($x = 0$)	**DSDS HfO_2 ($x = 0$)**	**As-Dep $Hf_{1-x}Zr_xO_2$ ($x = 0.31$)**	**DSDS $Hf_{1-x}Zr_xO_2$ ($x = 0.31$)**	**As-Dep $Hf_{1-x}Zr_xO_2$ ($x = 0.8$)**	**DSDS $Hf_{1-x}Zr_xO_2$ ($x = 0.8$)**
Weibull slope β	1.4	0.7	1.0	1.1	0.6	3.9

As-Dep processed devices (Figure 1.6(b)). DSDS $Hf_{1-x}Zr_xO_2$ with $x = 0.8$ shows a minimum time to breakdown T_{BDmin} of 3000 s, while DSDS HfO_2 shows a T_{BDmin} of 20 s (Figure 1.7(b)).

Table 1.1 shows the comparison of the Weibull slope, β for different dielectrics. A higher rate of early breakdown is observed for As-Dep $Hf_{1-x}Zr_xO_2$ with $x = 0.8$ and DSDS HfO_2 with $x = 0$, as they have $\beta < 1$ [64]. Since thinner oxides require few traps to form a conductive breakdown path and consequently, they have a lower value of β due to a larger statistical spread on the average density to form such a conductive path as compared to thicker oxides [64]. As Zr addition results in a lower IL thickness and a lower dielectric thickness (Figure 1.4(b)), a reduction in the Weibull slope was observed with an increase in Zr percentage for As-Dep devices (Table 1.1). In contrast, an opposite trend in β is observed for DSDS processed devices with the highest $\beta = 3.9$ for DSDS $Hf_{1-x}Zr_xO_2$ with $x = 0.8$ and the lowest $\beta = 0.7$ for DSDS HfO_2 with $x = 0$ in this work. This can be explained from the difference in the electronic structure of HfO_2 and ZrO_2. Zheng *et al.* [57] reported that neutral HfO_2 and ZrO_2 are highly polar and HfO_2 has a higher electron affinity as compared to ZrO_2. It is possible that the SPA plasma introduces excess electrons to HfO_2 and ZrO_2 during the processing of DSDS $Hf_{1-x}Zr_xO_2$. A higher percentage of excess electrons in DSDS HfO_2 contributes to form an early percolation path during the stress showing higher degradation due to stress-induced trap generation as compared to devices with higher Zr percentage. This is because the concentration of excess electron reduces with increase in Zr content. Therefore, the breakdown characteristics in Figure 1.7 further confirm the SILC characteristics observed earlier (Figure 1.6(b)). In addition, the reduction of intrinsic traps for DSDS $Hf_{1-x}Zr_xO_2$ ($x = 0.8$) as compared to As-Dep $Hf_{1-x}Zr_xO_2$ with $x = 0.8$ is depicted as a higher Weibull slope for DSDS $Hf_{1-x}Zr_xO_2$ ($x = 0.8$) due to an exposure to the cyclic SPA plasma [59, 60].

1.6 Al incorporation into HfO_2

Scaling below 22-nm technology node requires gate dielectric materials with properties superior to those of conventional high-κ materials. Al_2O_3 has been used to improve the thermal stability of high-κ HfO_2 films [27, 28]. It was found that Al incorporation into HfO_2 results increase in transition temperature from amorphous to polycrystalline state [23, 27–30, 35]. HfO_2 has comparatively lower crystallization

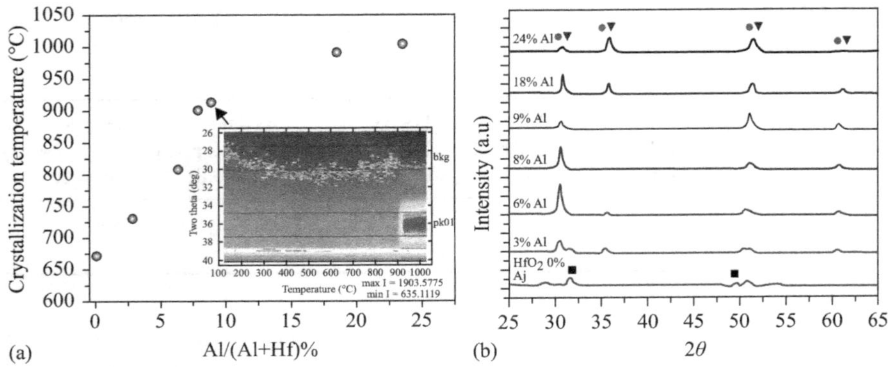

Figure 1.8 *(a) Crystallization temperature of $Hf_{1-x}Al_xO_y$ as a function of Al/(Al + Hf)% in the dielectrics. Inset shows XRD plot at the onset of crystallization for $Hf_{1-x}Al_xO_y$ with $x = 0.09$. (b) Synchronous grazing in plane XRD pattern for ALD $Hf_{1-x}Al_xO_y$ as a function of Al/(Al + Hf)%*

Source: © 2014 The Electrochemical Society. Reprinted, with permission, from K. Tapily, S. Consiglio, R.D. Clark, R. Vasić, C.S. Wajda, J.J. Sweet, G.J. Leusink, and A.C. Diebold, "Higher-K formation in atomic layer deposited $Hf_{1-x}Al_xO_y$," ECS Trans. 2014, vol. 64(9), pp. 123–131

temperature than Al_2O_3 [23, 28]. Therefore, HfO_2 allows lower thermal budget after its deposition, as polycrystalline grain boundary–induced leakage current and lateral nonuniformity increases after PDA at high temperature [23, 28].

1.6.1 $HfAlO_x$ alloy structures

Figure 1.8 shows the effect of Al incorporation in HfO_2 on the crystallization temperature [35]. It was found that HfO_2 starts to crystallize at around 680°C, while $Hf_{1-x}Al_xO_y$ with 25% Al/(Al + Hf) starts to crystallize at around 1000°C. It was found that Al acts as a network modifier and stabilizes the amorphous phase of the metal oxides [23]. In addition, enhancement in the dielectric constant of ALD HfO_2 due to Al incorporation by inserting few Al–O ALD cycles in the ALD process was also reported [26, 35].

Figure 1.8(b) shows that ALD $Hf_{1-x}Al_xO_y$ after annealing has peak position shift toward a larger 2θ value in the XRD pattern with the addition of Al. The shift in the peak is revealed to be due to the tetragonal crystalline phase formation with the addition of Al into HfO_2 [35]. HfO_2 without any Al content showed peaks due to the monoclinic phase in the XRD pattern in Figure 1.8(b). In other words, the addition of Al into HfO_2 increases the dielectric constant after annealing by stabilizing the tetragonal phase [26, 35].

In addition, the incorporation of Al into HfO_2 was found to limit the interfacial SiO_x regrowth by suppressing the oxygen diffusion down to the interface [65]. Therefore, the addition of Al into HfO_2 further helps the EOT downscaling by limiting the low-κ SiO_x layer growth [23, 24]. In addition, a reduced gate leakage current, a lower

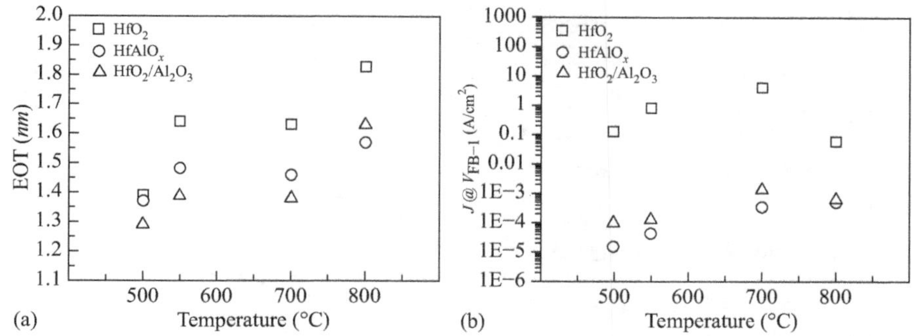

Figure 1.9 *(a) The variation of EOT value of pure HfO_2, $HfAlO_x$ alloy, and Al_2O_3/HfO_2 stack as a function of PDA temperatures and (b) the variation of current density at $V_{FB} - 1\,V$ as a function of PDA temperatures for pure HfO_2, $HfAlO_x$ alloy, and Al_2O_3/HfO_2 stack on Si*
Source: © 2007 The Electrochemical Society. Reprinted, with permission, from Y.-K. Chiou, C.-H. Chang, C.-C. Wang, K.-Y. Lee, T.-B. Wu, R. Kwo, and M. Honga, "Effect of Al incorporation in the thermal stability of atomic-layer-deposited HfO_2 for gate dielectric applications," J. Electrochem. Soc. 2007, vol. 154(4), pp. G99–G102

hysteresis value, and an improvement in interface property were also observed for Al-doped HfO_2 [23, 24, 30, 35].

1.6.2 Al_2O_3/HfO_2 bilayer structures

In addition of getting better thermal stability by Al incorporation in the form of $HfAlO_x$ alloy, benefits from Al incorporation in the form of Al_2O_3/HfO_2 stack structure were also reported [24, 33, 34]. Chiou *et al.* [24] compared the thermal stability of gate dielectrics for ALD HfO_2, $HfAlO_x$ alloy, and Al_2O_3/HfO_2 bilayer stack on a p-type Si (1 0 0) substrate in relation to their structural and electrical properties. In comparison to ALD HfO_2 and Al_2O_3/HfO_2 bilayer stack, $HfAlO_x$ alloy showed the superior characteristics in terms of the gate leakage current reduction and the EOT downscaling ability as well as a reduced interface state density [24]. It was found that bond structure variation in the stack form and in the alloy form is responsible for better performance in case of ALD $HfAlO_x$ alloy as compared to others. Also, it was observed that the HfO_2 film began to crystallize around 600°C, but the HfO_2 sublayer in the Al_2O_3/HfO_2 stack became crystallized around 700°C. The $HfAlO_x$ alloy on the other hand remained amorphous even after a rapid thermal annealing in N_2 atmosphere at 1000°C for 30 s [24].

Figure 1.9(a) shows the comparison of the EOT for pure HfO_2, $HfAlO_x$ alloy, and Al_2O_3/HfO_2 stack for different PDA temperatures as observed by Chiou *et al.* [24]. It was found that $HfAlO_x$ alloy and Al_2O_3/HfO_2 stack were able to control the EOT, when high-temperature annealing was done in contrast to pure HfO_2, which showed significant increase in the EOT level as the PDA temperature was increased. The increment of the EOT value against the PDA treatment was the lowest for the $HfAlO_x$

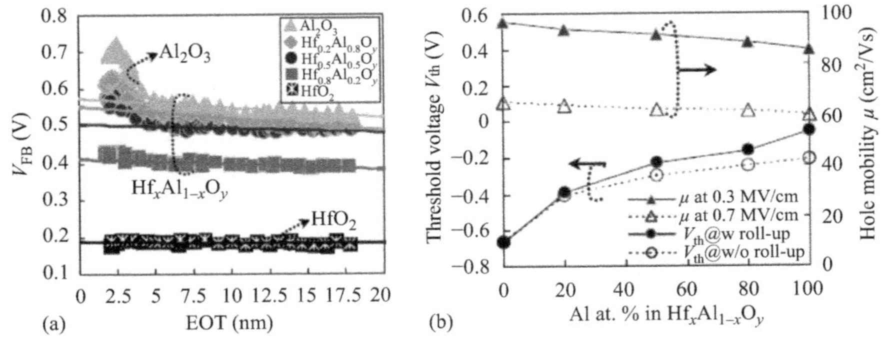

Figure 1.10 *(a) V_{FB}–EOT plots for NiSi/$Hf_xAl_{1-x}O_y$/terraced SiO_2/n-Si p-MOSFETs. All samples were subjected to PDA at 800°C and post-Si deposition annealing at 1000°C. (b) Dependencies of V_{th} and hole mobility at both 0.3 and 0.7 MV/cm on Al content in $Hf_xAl_{1-x}O_y$ for NiSi/$Hf_xAl_{1-x}O_y$/SiO_2 (1 nm)/Si p-MOSFETs. The closed circles stand for experimental V_{th} that takes V_{FB} roll-up effects into consideration, and the open circles stand for estimated V_{th} that excludes V_{FB} roll-up effects, which have been calculated qualitatively by comparing measured data with fitted linear V_{FB}–EOT relationship in (a)*
Source: © 2009 AIP Publishing LLC. Reprinted, with permission, from W. Wang, K. Akiyama, W. Mizubayashi, T. Nabatame, H. Ota, and A. Toriumi, "Effect of Al-diffusion-induced positive flatband voltage shift on the electrical characteristics of Al-incorporated high-k metal-oxide-semiconductor field-effective transistor," J. Appl. Phys. 2009, vol. 105(6), pp. 064108-1–6

alloy, followed by Al_2O_3/HfO_2 stack, and the highest was for HfO_2. In addition to improved EOT value for $HfAlO_x$ alloy, reduction in the interface state density and the gate leakage current density was also reported in the above study due to the Al incorporation into HfO_2. It is known that Al_2O_3 has comparatively a higher band gap and band offset with Si as compared to HfO_2 and thus Al incorporation into HfO_2 can reduce the tunneling leakage current [24]. Figure 1.9(b) shows the comparison of the leakage current density as a function of the PDA annealing temperature for pure HfO_2, $HfAlO_x$ alloy, and Al_2O_3/HfO_2 stack on Si. It was found that the leakage current density increased with the annealing temperature in 500–700°C range, while 800°C PDA resulted decrease in the leakage current density especially for PDA HfO_2. The decrease in J_g for high-temperature PDA was directly related to the thickening of IL at 800°C, while increase of the leakage current in 500–700°C range was attributed to the local enhancement of current emission due to increase in the interface roughness with an increased PDA temperatures [24].

1.6.3 Problems with excess Al incorporation

It was observed that Al incorporation into HfO_2 shifts the flat-band voltage toward positive direction due to Al diffusion to the interfacial SiO_x, which affects fixed charges near the Si/SiO_x interface [66]. Figure 1.10(a) shows the V_{FB}–EOT plots for the NiSi/$Hf_xAl_{1-x}O_y$/terraced SiO_2/n-Si p-MOSFETs for different Al contents in the $Hf_xAl_{1-x}O_y$ dielectrics [66]. It was found that the scaling of EOT to a significantly lower value by thinning the IL results in a flat-band voltage roll-up, which was mainly attributed to the atom diffusion–induced charge formation. Reduction of IL thickness helps to change in the co-ordination number of Al^{3+} from six to four. This increases negative fixed charge, and consequently a positive flat-band voltage shift is observed for an increased Al incorporation in HfO_2. Also, both the annealing temperature and the annealing time were found to have a significant effect on the charge formation in the dielectric because of the Al diffusion [66]. In addition, more dipole formation in the high-κ/SiO_x interface is also reported for the Al incorporation into HfO_2 [67, 68], which contributes to a positive flat-band voltage shift. However, the effect of Al incorporation into HfO_2 on hole mobility was found to be insignificant at both high and low effective field regions [66]. Figure 1.10(b) shows the change in the threshold voltage and the hole mobility for a p-MOSFET with $Hf_xAl_{1-x}O_y$ dielectrics. Ota *et al.* [69] also reported that Al incorporation has little effect on the electron mobility, when n-MOSFET with different Al percentages in the dielectrics was investigated. It was found that in case of the Al profiled $HfAlO_x$ gate stacks, the electron mobility at the higher field was as high as the universal curve and the influence of Al profiles on the electron mobility was restricted to the low effective field region [69].

1.6.4 Extremely low Al incorporation in HfO_2

Although Al has been added in different concentrations in different stack structures, they could not meet the reliability challenges, because of higher Al to Hf ratio (>6%) obtained in the dielectrics [32, 67, 68, 70–73]. These studies did not explain the impact of Al incorporation on either the dielectric/metal gate interface or the Si/SiO_x interface. A detailed study of reliability for HK/MG stack with variation in Al concentration near the HK/MG interface and Si/SiO_x interface is, therefore, required. Furthermore, the understanding of interface state degradation under electrical stress can benefit the integration of Al-doped HfO_2 into future CMOS technology. Additionally, since the standard thermal process required for source/drain activation in CMOS devices can be as high as 1000°C [58], PDA temperature variation can, therefore, impact the dielectric.

In this work, for the first time, we have developed a process technology to incorporate extremely low Al in HfO_2 (Figure 1.11(a,b)). By depositing ALD $HfAlO_x$ along with ALD HfO_2 in a layered structure (see Figure 1.11(a,b)), we were able to achieve Al/(Hf + Al)% in the range <1% to ~7% in the dielectrics where lower leakage current and EOT reduction were observed for low Al percentage. Aluminum concentration variation near the HK/MG interface and near the Si/SiO_2 interface was modulated by varying the thickness and location of the $HfAlO_x$ layer in a multi-layered gate stack. While all the dielectrics structures used in this work were subjected to a PDA at 800°C one set of the multi-layered stacks was also annealed at 680°C and 700°C. We have chosen these three annealing temperatures because for 0 to 7% Al/(Al + Hf) in the dielectrics crystallization temperature increases from 680°C to 800°C [35].

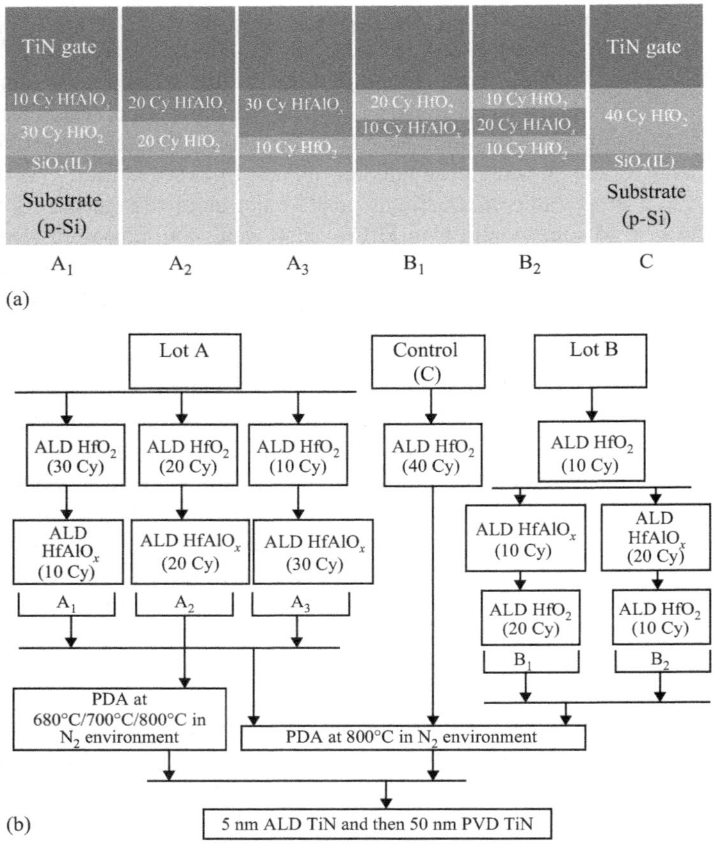

Figure 1.11 *(a) Device structures for Al incorporation in HfO_2. (b) Device fabrication process flow*

1.6.4.1 Device fabrication: ALD multi-layered structure

Two different lots, Lot A (A_1, A_2, and A_3) having $HfAlO_x$ as the top layer in a bilayer structure and Lot B (B_1 and B_2) having $HfAlO_x$ in a sandwiched structure, were deposited by ALD process.

The atomic layer depositions were conducted in a 300-mm TEL Trias™ clean-room tool. The starting substrate surface used in this study is a 300-mm Si(0 0 1) wafer. The native oxide on the 300-mm wafers was removed using a TEL Certas™ chemical oxide removal process [35]. A thin SiO_2 IL was then formed using vapor ozone. ALD HfO_2 and $HfAlO_x$ were deposited by using TEMAH and trimethylaluminum as Hf and Al precursors, respectively, with H_2O as the co-reactant at 250°C deposition temperature.

Figure 1.11(a) shows device structures for Al incorporation into HfO_2 in different locations in the dielectrics, whereas Figure 1.11(b) shows device fabrication process

Table 1.2 Composition of Hf and Al from XPS and dielectric thickness measured by SE

Split	Annealing T (°C)	Al/(Hf + Al)%	High-κ + IL thickness (Å)
C – HfO_2	800	0	35.54
A_1 – 10 Cy $HfAlO_x$	800	2.38	35.80
A_2 – 20 Cy $HfAlO_x$	680	4.97	38.61
	700	4.49	38.18
	800	4.19	35.58
A_3 – 30 Cy $HfAlO_x$	800	6.66	37.23
B_1 – 10 Cy $HfAlO_x$	800	0.57	34.34
B_2 – 20 Cy $HfAlO_x$	800	2.56	35.38

flow for Lot A (A_1, A_2, and A_3), Lot B (B_1 and B_2), and the control sample, C. For Lot A aluminum concentration was varied by depositing (i) 30 ALD cycles of HfO_2 and 10 ALD cycles of $HfAlO_x$ (A_1), (ii) 20 ALD cycles of HfO_2 and 20 ALD cycles of $HfAlO_x$ (A_2), and (iii) 10 ALD cycles of HfO_2 and 30 ALD cycles of $HfAlO_x$ (A_3). For Lot B, the dielectrics were formed by depositing (i) 10 ALD cycles of HfO_2, 10 ALD cycles of $HfAlO_x$, and 20 ALD cycles of HfO_2 (B_1) and (ii) 10 ALD cycles of HfO_2, 20 ALD cycles of $HfAlO_x$, and 10 ALD cycles of HfO_2 (B_2). A total of 40 cycles were used for the entire deposition process for all the samples. The control sample in this study was deposited with 40 ALD cycles of HfO_2 (C). All samples except A_2 were subjected to PDA in N_2 environment at 800°C in a clustered rapid thermal chamber without breaking vacuum. For A_2 with 20 ALD cycles (Cy) $HfAlO_x$ dielectrics were annealed at 680°C, 700°C, and 800°C in N_2 environment. The metal gate for these devices was formed by growing 5-nm ALD TiN followed by a 50-nm PVD TiN.

1.6.4.2 Physical properties of $HfAlO_x$ dielectrics

Hf and Al compositions for the different ALD films were measured by XPS by using Thermo Fisher Theta Probe™ XPS system. Dielectric thicknesses for different films were measured by SE. Table 1.2 summarizes the compositions of Hf and Al measured by XPS and dielectric thicknesses measured by SE for different dielectrics.

Dielectrics from Lot B (B_1 and B_2) showed comparatively lower aluminum concentration as compared to the dielectrics from Lot A (A_1 and A_2) where an identical number of ALD $HfAlO_x$ cycles were used (Table 1.2). Since the $HfAlO_x$ layer in Lot A is the top layer and in Lot B is in the middle (Figure 1.11(a)), the observed lower Al percentage for Lot B samples is possibly due to weaker XPS signal intensity from Al source, further away from the top surface of the devices. Note that the increase in $HfAlO_x$ layer thickness for Lot A devices brings $HfAlO_x$ layer closer to the IL (Figure 1.11(a)). During PDA process done prior to metal gate deposition, aluminum can diffuse toward the IL through the HfO_2 layer in case of Lot A. On the other hand, Al can diffuse in both directions in case of Lot B (B_1 and B_2), and the distance between $HfAlO_x$ and the IL remains constant. Also, it is possible that in case of Lot A the presence of more Hf facilitated an increased incorporation of Al per cycle as compared to Lot B [27]. Therefore, sample B_1 demonstrates the minimum Al/(Hf + Al)% (~0.6%), while sample A_3 demonstrates the maximum Al/(Hf + Al)% (~7%) in the dielectrics.

Comparison of dielectrics (A_2) annealed at 680°C, 700°C, and 800°C showed that with increase in annealing temperature, Al concentration slightly reduced which could be due to out-diffusion of Al at high temperature [34]. It was found that dielectrics with ~7% Al/(Al + Hf) start to crystallize at 800°C annealing temperature while crystallization temperature decreases for dielectrics with lower Al percentage [35]. Therefore, A_3 with 30 Cy $HfAlO_x$ showed ~2 Å higher film thickness because of a mixed structure of amorphous and crystalline phase formation as compared to other dielectrics (Table 1.2). When dielectrics have 4–5% Al, they remain amorphous even after annealing at 700°C [35]. As a result, A_2 with 20 Cy $HfAlO_x$ annealed at 680°C, and 700°C showed ~3 Å higher thickness as compared to the dielectric annealed at 800°C (Table 1.2). B_1 having the lowest Al percentage showed the lowest dielectric thickness (Table 1.2).

1.6.4.3 Comparison of EOT, V_{FB}, and J_g

Figure 1.12(a) plots the flat-band voltage (V_{FB}) as a function of EOT for all devices subjected to an annealing at 800°C. Figure 1.12(b) shows gate leakage current density sensed at $-1V + V_{FB}$ for these dielectrics. V_{FB} and leakage current for dielectrics (A_2) annealed at 680°C, 700°C, and 800°C are shown in Figure 1.12(c,d). For possible variation analysis, three devices from each device type are presented in Figure 1.12(a–d). From Figure 1.12(a), an EOT reduction due to Al incorporation in HfO_2 is observed for both lots as compared to the control device, C. For Lot A with $HfAlO_x$ as the top layer in a bilayer structure, dielectrics with 2–4% Al/(Al + Hf) showed the EOT downscaling potential with an increased flat-band voltage shift (Figure 1.12(a)). On the other hand, B_1 and B_2 from Lot B with $HfAlO_x$ in a sandwiched structure showed a comparable flat-band voltage with the control device C, while they showed reduction in the EOT due to Al incorporation (Figure 1.12(a)). Dielectrics having 10 cycles of $HfAlO_x$ from both lots showed significant reduction in the average EOT (18% for A_1 and 14% for B_1) as compared to the control device C, with HfO_2 only (Figure 1.12(a)).

On the other hand, A_3 with 30 Cy $HfAlO_x$ showed only 6% reduction in the average EOT. A_3 also showed more EOT variation, while other device types showed minimal variation for three identical devices (Figure 1.12(a)). It was found that after 800°C annealing, dielectrics with <2% Al/(Al + Hf) have higher crystallization with a mixed structure of monoclinic and tetragonal phase formation, while increased Al incorporation inhibits the crystallization process [35]. On the other hand, the control device C with HfO_2 crystallizes into monoclinic phase which is thermodynamically stable phase for HfO_2 [35]. It is known that tetragonal stabilization of HfO_2 results in a higher dielectric constant [26, 35]. Therefore, dielectrics with 10 Cy $HfAlO_x$ from both lots showed a higher EOT downscaling potential because of higher crystallization and tetragonal stabilization (Figure 1.12(a)). The observed higher flat-band voltage for A_1 and A_2 can be attributed to grain boundary–related fixed charges, due to higher crystallization as compared to A_3 [74]. In addition, with increase in Al concentration,

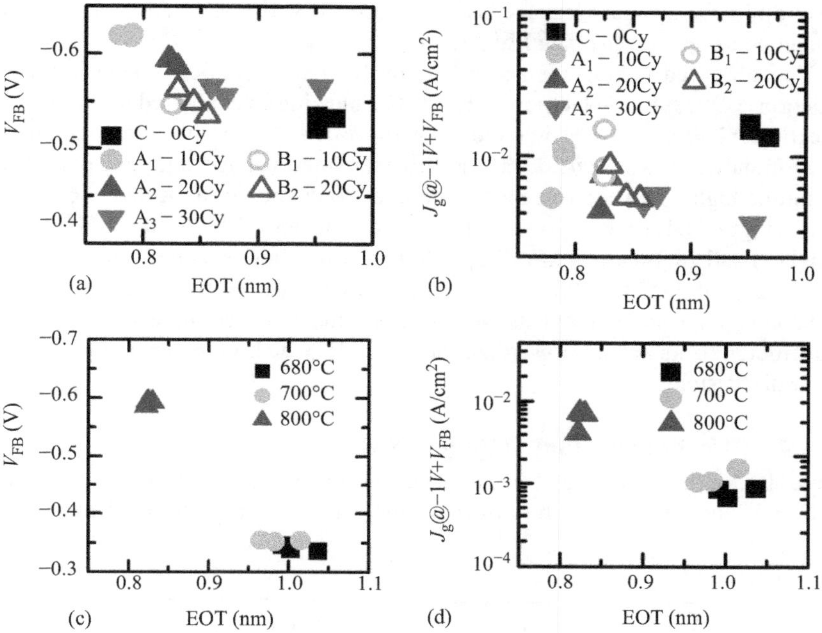

Figure 1.12 *(a,b) Flat-band voltage, V_{FB} and gate leakage current, J_g sensed at $-1V + V_{FB}$ as a function of EOT for dielectrics annealed at 800°C. Comparison for A_2 with 20 Cy $HfAlO_x$ annealed at 680°C, 700°C, and 800°C in (c,d). For possible variation analysis, three devices from each device types are presented (a–d)*

more dipole formation at high-κ/IL interface can result positive flat-band voltage shift for A_2 and A_3 as compared to A_1 [67, 71]. On the other hand, in case of B_1 and B_2 with intermediate $HfAlO_x$ layer, dipoles formed in the opposite interfaces can cancel each other, and therefore they showed comparable flat-band voltage with the control device [68, 71]. Less crystallization for A_3 with ~7% Al resulted higher EOT with more variation due to enhanced spatial nonhomogeneity [35].

It is clear from Figure 1.12(b) that the presence of Al in the dielectric reduces the gate leakage current, which is consistent with the previous reports [23, 24, 26]. Al incorporation in HfO_2 was found to reduce gate leakage current due to an increase in band gap and band offset with Si [23, 24, 26]. Furthermore, Al addition into HfO_2 leads to a smoother dielectric film surface, which in turn contributes to smaller leakage current [75]. Therefore, the lowest value of gate leakage current is observed for the devices with the highest concentration of aluminum (sample A_3) which is in average 70% lower than the control sample. This can be attributed to the inhibited crystallization with an increase in Al content. Since the dielectric remained more amorphous, leakage current reduced. In addition of having reduced EOT, A_1 showed 41% reduction in the average gate leakage current density as compared to the control device C (Figure 1.12(b)). B_1 having the lowest Al percentage showed 32%

reduction in average gate leakage current, while B_2 with 2.56% Al in HfO_2 showed 59% reduction in average gate leakage current.

The observed characteristics in Figure 1.12(c,d) revealed a significant EOT reduction (~20%) for dielectrics with 800°C annealing as compared to 680°C, while annealing at 700°C showed slight reduction (~5%) in EOT. Also, dielectrics annealed at 800°C showed more than 200 mV negative flat-band voltage shift, and one order of magnitude higher leakage current as compared to dielectrics annealed at 680°C and 700°C (Figure 1.12(c,d)). As discussed earlier, dielectrics with 20 Cy $HfAlO_x$ had partial crystallization after annealing at 800°C, while they remained amorphous after annealing at 680°C and 700°C. Therefore, the observed EOT reduction, increased flat-band voltage, and higher leakage current for the dielectric annealed at 800°C can be attributed to its partial crystallization [35, 74], which is in accordance with our earlier discussions.

1.6.4.4 Effect of constant voltage stress

Figure 1.13(a,b) shows the stress-induced flat-band voltage shift (ΔV_{FB}) and the SILC as a function of the stress time for dielectrics with different Al percentages.

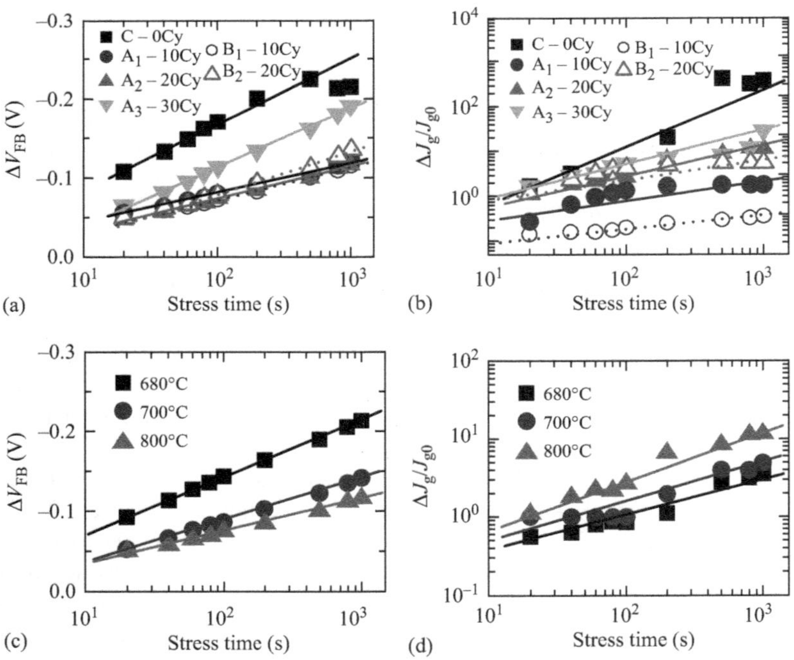

Figure 1.13 (a,b) Stress-induced flat-band voltage shift and SILC as a function of stress time for dielectrics annealed at 800°C. Evolution of ΔV_{FB} and SILC as a function of time for A_2 annealed at 680°C, 700°C, and 800°C in (c,d)

Comparison for dielectrics (A_2) annealed at different temperatures are shown in Figure 1.13(c,d). It is observed that the presence of Al in HfO_2 reduces the stress-induced trap generation for both Lot A and Lot B compared to the control sample C (Figure 1.13(a)). The observed characteristics further reveal that by adding a small percentage of Al (<4% in this study), stress-induced trap formation can be minimized significantly. It was previously observed that Al interacts with the native defect states in HfO_2 which leads to the passivation of the charged oxygen vacancy–induced defect bands [76, 77]. Therefore, dielectrics with 10 Cy $HfAlO_x$ (sample A_1 and B_1) and 20 Cy $HfAlO_x$ (A_2 and B_2) showed an improvement in the quality of the dielectrics due to the addition of an optimal percentage of Al in HfO_2 lower than earlier reported. On the other hand, higher percentage of incorporated Al might have originated more charged dipoles in dielectrics with 30 Cy $HfAlO_x$ (A_3) due to a compositional phase separation and hence, sample A_3 showed degradation in flat-band voltage shift [67, 68, 71, 78]. Figure 1.13(b) shows the normalized SILC for both Lot A and Lot B for Al incorporation in HfO_2. The SILC was measured at low sense voltage region ($V_{sense} < 0.5\,V$) in the negative bias condition. In this region, trap generation at and near the interface has the major contribution in the observed SILC [79–81]. All devices with an $HfAlO_x$ layer showed improvement as compared to the control device C. However, with increase in Al concentration in the dielectrics, an increase in SILC was observed (Figure 1.13(b)). This can be attributed to the degradation of SiO_2 IL and Si/SiO_2 interface due to Al diffusion from $HfAlO_x$ after annealing [66].

Even though annealing at 800°C was found to degrade the SILC characteristics (Figure 1.13(d)) as compared to 680°C and 700°C annealing, in contrast, the stress-induced flat-band voltage shift showed improvement for these dielectrics (Figure 1.13(c)). After 1000-s stress duration, dielectrics annealed at 800°C showed around 100 mV lower flat-band voltage shift as compared to the dielectrics annealed at 680°C. As explained earlier, both bulk and interface charge generation due to stress contributes to flat-band voltage shift, while SILC in the low sense voltage represents trap generation at and near IL. Higher annealing temperature further drives Al atoms toward IL, which contributes to a degraded SILC for 800°C annealed sample despite less charge formation in the bulk of dielectrics as represented by the stress-induced flat-band voltage shift characteristics.

Both V_{FB} shift and SILC evolution followed a power law function with stress time as observed from Figure 1.13(a,d). Table 1.3 lists the power exponents (n) for ΔV_{FB} and $\Delta J_g/J_{g0}$ with stress duration for dielectrics with different Al contents for Lot A (A_1, A_2, and A_3), Lot B (B_1 and B_2), and the control device, C.

The observed stress-induced flat-band voltage shift (Figure 1.13(a) and Table 1.3) suggests that, for Lot A dielectrics, devices with ~2% Al/(Hf+Al)% (sample: A_1) have 55% reduction in the rate of stress-induced charge formation as compared to the control sample with no Al content (sample: C), whereas only 11% reduction was observed for devices with ~7% Al/(Hf+Al)% (sample: A_3). For Lot B dielectrics, we observed a significant improvement for B_1 with 10 Cy $HfAlO_x$ (52% reduction). It can be inferred that a low percentage of Al incorporation helps to suppress the positive charge formation due to the applied stress, while an excess positive charge formation is possible if Al concentration increases beyond a certain percentage. Evaluation of

Table 1.3 Power exponent (n) for ΔV_{FB} and $\Delta J_g/J_{g0}$ and $\Delta D_{it}/D_{it0}$ comparison for different dielectrics

Dielectric	Annealing T[°C]	Power exponent (n) for ΔV_{FB}	Power exponent (n) for $\Delta J_g/J_{g0}$
C – HfO_2	800	0.084	1.287
A_1 – 10 Cy $HfAlO_x$	800	0.037	0.417
A_2 – 20 Cy $HfAlO_x$	680	0.071	0.441
	700	0.054	0.506
	800	0.039	0.67
A_3 – 30 Cy $HfAlO_x$	800	0.074	0.758
B_1 – 10 Cy $HfAlO_x$	800	0.04	0.314
B_2 – 20 Cy $HfAlO_x$	800	0.052	0.432

Table 1.4 Impact of stress on interface state density, D_{it}, in the Si mid-gap level for different dielectrics

Dielectric	C	A_1		A_2		A_3	B_1	B_2
Annealing T (°C)	800	800	680	700	800	800	800	800
D_{it} [$\times 10^{10}$ cm^{-2} eV^{-1}]	2.66	25.2	18.7	25.6	31.3	3.59	6.98	65.6
$\Delta D_{it}/D_{it0}$	3.58	1.85	0.33	0.52	4.75	18.5	6.83	13.6

power exponent (n) for $\Delta J_g/J_{g0}$ with stress time also showed an improvement for dielectrics with <2% Al/(Hf + Al) (Figure 1.13(b) and Table 1.3). The lowest n value (80% lower than C) was observed for B_1 with ~0.6% Al, while the highest n value (41% lower than C) was observed for A_3 with ~7% Al.

1.6.4.5 Interface state density, D_{it}

Table 1.4 shows the comparison of the interface state density, D_{it} in the Si mid-gap level estimated by conductance method. With the addition of Al, all devices showed a moderate increase in the mid-gap D_{it} level. Except A_3, an increase in the Al concentration showed a corresponding increase in the D_{it} for all device types from both lots. In addition, comparison for A_2 and B_2 showed that a decrease in the distance of $HfAlO_x$ layer from the Si/SiO_2 interface also increases the mid-gap D_{it} (Table 1.4). Therefore, the observed increase in the mid-gap D_{it} can be attributed to the Al diffusion from $HfAlO_x$ through SiO_2 to the Si/SiO_2 interface after annealing [66]. Also, dielectrics with 20 Cy $HfAlO_x$ (A_2) annealed at 680°C, 700°C, and 800°C showed a subsequent increase in the mid-gap D_{it} (Table 1.4), which further confirms the fact that excess Al presence at Si/SiO_2 interface is detrimental as it contributes to a reduction in the carrier mobility. On the other hand, A_3 showed comparatively higher dielectric thickness due to less crystallization (Table 1.2) and hence, showed a comparable D_{it} value with the control device [58].

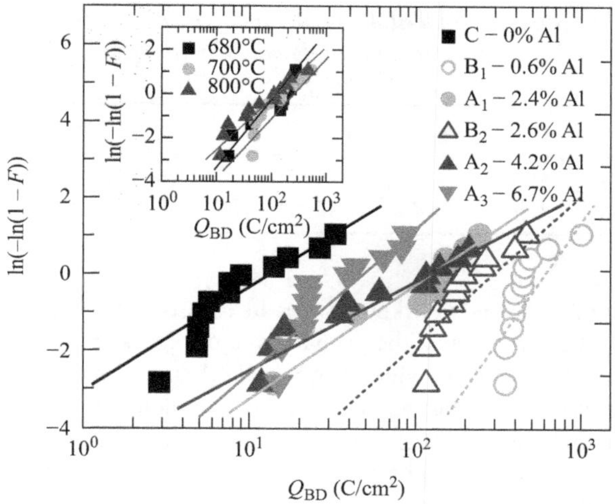

Figure 1.14 Weibull plot of charge to breakdown, Q_{BD} for devices with Al incorporation in HfO_2. Inset shows Weibull plot for A_2 with different annealing temperatures

When devices were stressed in the gate injection mode, devices with lower (<4%) Al concentration (A_1, A_2, and B_1) showed relatively higher resistance to stress-induced interface state generation in the mid-gap level (Table 1.4). A_3 and B_2 on the other hand showed more degradation due to the applied stress. As discussed earlier, an increase in the $HfAlO_x$ layer thickness or reduction of the distance from the IL was found to increase the stress-induced interface state generation in the mid-gap level (Figure 1.11(a) and Table 1.4). In addition, when dielectrics were compared for different annealing temperatures, the one with PDA at 800°C showed a significantly higher interface state generation ($\Delta D_{it}/D_{it0} \sim 4$). This can be attributed to more Al diffusion down to the IL as compared to dielectrics annealed at 680°C ($\Delta D_{it}/D_{it0} \sim 0.3$) and at 700°C ($\Delta D_{it}/D_{it0} \sim 0.5$). This clearly suggests that an excess aluminum presence at the interface increases the D_{it} that contributes to the SILC (Figure 1.13(b,d)) as observed earlier. Therefore, an optimized Al concentration in HfO_2 and at the interface can improve the gate stack quality.

1.6.4.6 TDDB characteristics

Figure 1.14 shows the Weibull plots extracted for Lot A (A_1, A_2, and A_3), Lot B (B_1 and B_2), and the control device, C, without any aluminum. For each device type, 12 devices were stressed in the gate injection mode. The stress voltage did not exceed −3 V for any device type during stress. It is observed that Al incorporation enhances the TDDB characteristics as the charge to breakdown, Q_{BD} increases significantly for both Lot A and Lot B devices as compared to the control device, C. It is also observed that for both Lot A and Lot B the Weibull plot shifts toward a lower Q_{BD} value with an increase in the Al content, which is consistent with their SILC characteristics

Table 1.5 Weibull slope, β for Lot A (A_1, A_2, and A_3), Lot B (B_1 and B_2) with $HfAlO_x$ layer, and the control device, C with HfO_2 only

Dielectric	**C**	**A_1**		**A_2**		**A_3**	**B_1**	**B_2**
Annealing T (°C)	800	800	680	700	800	800	800	800
Weibull slope β	1.40	1.43	1.44	1.24	0.97	1.67	2.70	2.15

shown in Figure 3(b). It is well known that in case of time-dependent breakdown for high-κ/SiO_2 dielectric stack, the primary role of high-κ layer is to determine the dominating current component which degrades both high-κ and ILs. The IL initiates the breakdown process, and subsequently the whole dielectric stack collapses when a percolation path creates through the entire dielectric [25]. Therefore, the highest charge to breakdown or time to breakdown was observed for sample B_1 having the lowest Al available to the IL.

Table 1.5 shows the variation in the Weibull slope, β, for different dielectrics. It is observed that samples B_1 and B_2 from Lot B have the superior breakdown characteristics in terms of trap distribution in the dielectrics, as they have higher Weibull slope. Except A_3, all device types from Lot A and Lot B showed a reduction in β with addition of Al (Figure 1.14 and Table 1.5). An increase in the Al content in the dielectrics increases the trap distribution due to the newly generated traps in the dielectrics due to stress, as the reduction in Weibull slope is related to the trap generation rate rather than initial trap concentration [79]. On the other hand, A_3 has higher dielectric thickness and IL thickness than A_1 and A_2 (Table 1.2), and therefore, showed a moderate increase in β as thinner oxides require few traps to form a conductive breakdown path and consequently they have lower value of β due to larger statistical spread on the average density to form such conductive path as compared to thicker oxides [64]. Similar increase in Weibull slope ($\beta = 1.44$) was also observed for A_2 with 680°C PDA temperature, and at 700°C ($\beta = 1.24$) because of having higher dielectric thickness as compared to the dielectrics annealed at 800°C ($\beta = 0.97$) as shown in inset of Figure 1.14 and Table 1.5.

From the above observations, an increase in the Al/(Hf + Al)% contributes to the less crystallization of the dielectrics even after annealing at 800°C. This characteristic modifies the EOT for these dielectrics. If the crystallization is higher, the dielectric constant increases and the EOT goes down. Higher crystallization also leads to the higher flat-band voltage and leakage current by increasing grain boundary–induced charge [56]. In addition, an increased Al incorporation and a higher crystallization increase interface state density due to Al diffusion to the interface [68]. Furthermore, the SILC and the interface state density behavior in B_1 and B_2 devices clearly suggest that when $HfAlO_x$ layer is closer to the interface, the D_{it} is degraded as the Al concentration is increased. For a low aluminum concentration ($<2\%$) bulk and interface trap creation showed significant improvement as evidenced by stress-induced flat-band voltage shift, SILC, ΔD_{it}, and TDDB. This can be attributed to the level of crystallization and availability of aluminum.

1.7 Conclusion

In this research, electrical characterization and reliability study have been done for Zr- and Al-incorporated Hf-based high-κ dielectrics with advanced processing conditions. MOS capacitors with p-Si substrate and TiN metal gate were fabricated with ALD $Hf_{1-x}Zr_xO_x$ and $HfAlO_x$ as high-κ dielectrics. A cyclic plasma treatment with SPA Ar plasma (DSDS process) was applied during the deposition process of ALD $Hf_{1-x}Zr_xO_2$. DSDS $Hf_{1-x}Zr_xO_2$ ($x = 0$, 0.31, and 0.8) was compared with the control sample deposited with standard As-Dep process. Extremely low percentage of Al (Al/(Hf+Al) in the range <1% to ~7%) were incorporated in HfO_2 by depositing ALD $HfAlO_x$ layer along with HfO_2 layer in a multi-layered gate stack. Chemically grown SiO_2 IL has been used for Al-incorporated dielectrics, while SiON formed by RFN of chemically grown SiO_2 has been considered for Zr-incorporated dielectrics.

The impact of cyclic plasma treatment with SPA Ar plasma and cyclic annealing on the EOT, the flat-band voltage (V_{FB}), the gate leakage current density (J_g), and the interface state density (D_{it}) have been observed for ALD $Hf_{1-x}Zr_xO_2$ with different Zr contents. Devices were subjected to a constant voltage stress in the gate injection mode to study the reliability of these dielectrics. Stress-induced flat-band voltage shift, ΔV_{FB}, SILC, $\Delta J_g/J_{g0}$, and stress-induced interface state generation, ΔD_{it}, were compared for these dielectrics. A moderate increase in the D_{it} was observed due to an increased Zr incorporation into HfO_2, while DSDS processed dielectrics showed a reduced mid-gap D_{it}. When devices were subjected to a constant voltage stress in the gate injection mode, the Zr addition and the SPA plasma treatment showed to result a suppressed stress-induced interface state generation. In addition, all devices were subjected to TDDB stress in the gate injection mode. It was observed that DSDS $Hf_{0.2}Zr_{0.8}O_2$ with a SiON IL exhibits better reliability as compared to other dielectrics. Zr addition in HfO_2 helps the EOT downscaling for DSDS and As-Dep $Hf_{0.2}Zr_{0.8}O_2$. The suppression of oxide trap formation due to the cyclic SPA plasma exposure is believed to contribute to a superior $Hf_{0.2}Zr_{0.8}O_2$, EOT downscaling ability, and good reliability performance. Breakdown characteristics suggest that the devices go through electron trapping, the SBD, the PBD, and subsequently the HBD. The Weibull characteristic and the Weibull slope suggest that the variation in electron affinity for HfO_2 and ZrO_2 contributes to the improvement in DSDS $Hf_{1-x}Zr_xO_2$ with $x = 0.8$ by suppressing the oxide trap formation.

A high-quality HfO_2-based gate stack by depositing ALD $HfAlO_x$ along with HfO_2 in a layered structure has been demonstrated in this research. Electrical characteristics of these dielectrics were found to be influenced by both Al/(Hf+Al)% and the position of the ALD $HfAlO_x$ layer in the stack. Increase in Al/(Hf+Al)% contributes to a less crystallization of the dielectrics even after annealing at 800°C. This characteristic modifies the EOT for these dielectrics. If the crystallization is higher, dielectric constant increases and the EOT goes down. An optimized Al incorporation (<2%) was found to be beneficial for both EOT downscaling and reliability enhancement. For low aluminum concentration (<2%) bulk and interface trap creation showed significant improvement as evidenced by stress-induced flat-band voltage shift, SILC, ΔD_{it}, and TDDB. This can be attributed to the level of crystallization and availability of aluminum. Also annealing temperature has significant influence on their electrical characteristics and reliability, as higher annealing temperature enhances the crystallization process and facilitates Al diffusion to the interfaces.

Acknowledgment

The authors would like to thank K. Tapily, R. D. Clark, S. Consiglio, C. S. Wajda, G. Nakamura, and G. J. Leusink of TEL Technology Center, America, Albany, NY, for supplying devices and helpful discussions.

References

[1] H. Iwai, "Future semiconductor manufacturing: Challenges and opportunities," in *Proc. IEDM Tech. Dig. 2004*, pp. 11–16.
[2] D. Misra, H. Iwai, and H. Wong, "High-k dielectrics," *Electrochem. Soc. Interface 2005*, pp. 30–34.
[3] http://www.intel.com/pressroom/kits/quickreffam.htm#atom.
[4] D.G. Schlom, S. Guha, and S. Datta, "Gate oxides beyond SiO_2," *MRS Bull. 2008*, vol. 33, pp. 1017–1025.
[5] J. Robertson, "Maximizing performance for higher-k dielectrics," *J. Appl. Phys. 2008*, vol. 104, pp. 124111-1–7.
[6] http://www.itrs.net/Links/2013ITRS/2013Chapters/2013PIDS.pdf.
[7] L. Wu, K.S. Yew, D.S. Ang, *et al.*, "A novel multi deposition multi room-temperature annealing technique via ultraviolet-ozone to improve high-K/metal (HfZrO/TiN) gate stack integrity for a gate-last process," in *IEDM Tech. Dig. 2010*, pp. 273–276.
[8] L. Wu, H.Y. Yu, K.S. Yew, J. Pan, W.J. Liu, and T.L. Duan, "Multideposition multiroom-temperature annealing via ultraviolet ozone for HfZrO high-k and integration with a TiN metal gate in a gate-last process," *IEEE Trans. Electron Dev. 2011*, vol. 58(7), pp. 2177–2181.
[9] R.D. Clark, S. Consiglio, G. Nakamura, Y. Trickett, and G.J. Leusink, "Optimizing ALD HfO_2 for advanced gate stacks with interspersed UV and thermal treatments – DADA and MDMA variations, combinations, and optimization," *ECS Trans. 2011*, vol. 41(2), pp. 79–88.
[10] M.H. Lin, C.H. Hou, J.Y. Wu, and T.B. Wu, "Thermal stability improvement via cyclic D_2O radical anneal interposed in atomic layer deposition process," *J. Electrochem. Soc. 2011*, vol. 158(3), pp. H221–H223.
[11] C.H. Tu, F.C. Chiu, C.H. Chen, *et al.*, "Reliability characteristics of D_2O-radical annealed ALD HfO_2 dielectric," *ECS Trans. 2010*, vol. 28(2), pp. 331–338.
[12] A. Delabie, M. Caymax, B. Brijs, *et al.*, "Scaling to sub-1 nm equivalent oxide thickness with hafnium oxide deposited by atomic layer deposition," *J. Electrochem. Soc. 2006*, vol. 153(8), pp. F180–F187.

[13] R.D. Clark, S. Aoyama, S. Consiglio, G. Nakamura, and G. Leusink, "Physical and electrical effects of the Dep-Anneal-Dep-Anneal (DADA) process for HfO_2 in high K/metal gate stacks," *ECS Trans. 2011*, vol. 35(4), pp. 815–834.

[14] K. Nagata, H. Akamatsu, D. Kosemura, *et al.*, "Improvement of CVD SiO_2 by post deposition microwave plasma treatment," *ECS Trans. 2009*, vol. 19(9), pp. 45–51.

[15] K. Kawase, A. Teramoto, H. Umeda, *et al.*, "Densification of chemical vapor deposition silicon dioxide film using oxygen radical oxidation," *J. Appl. Phys. 2012*, vol. 111, pp. 034101-1–7.

[16] T. Tanimura, Y. Watanabe, Y. Sato, Y. Kabe, and Y. Hirota, "Effect of microwave plasma treatment on silicon dioxide films grown by atomic layer deposition at low temperature," *J. Appl. Phys. 2013*, vol. 113, pp. 064102-1–5.

[17] C. Tian, T. Nozawa, K. Ishibasi, H. Kameyama, and T. Morimoto, "Characteristics of large-diameter plasma using a radial-line slot antenna," *J. Vac. Sci. Technol. A 2006,* vol. 24, pp. 1421–1424.

[18] R.I. Hegde, D.H. Triyoso, P.J. Tobin, *et al.*, "Microstructure modified HfO_2 using Zr addition with Ta_xC_y gate for improved device performance and reliability" in *IEDM Tech. Dig. 2005*, pp. 35–38.

[19] R.I. Hegde, D.H. Triyoso, S.B. Samavedam, and B.E. White, "Hafnium zirconate gate dielectric for advanced gate stack applications," *J. Appl. Phys.* 2007, vol. 101(7), pp. 074113-1–7.

[20] D.H. Triyoso, R.I. Hegde, and J.K. Schaeffer,"Characteristics of atomic-layer-deposited thin $Hf_xZr_{1-x}O_2$ gate dielectrics," *J. Vac. Sci. Technol. B 2007*, vol. 25(3), pp. 845–852.

[21] D.H. Triyoso, R.I. Hegde, J. Jiang, J.K. Schaeffer, and M.V. Raymond, "Improved electrical properties of ALD $Hf_xZr_{1-x}O_2$ dielectrics deposited on ultrathin PVD Zr underlayer," *IEEE Electron Dev. Lett. 2008,* vol. 29(1), pp. 57–59.

[22] C.K. Chiang, J.C. Chang, W.H. Liu, *et al.*, "A comparative study of gate stack material properties and reliability characterization in MOS transistors with optimal ALD zirconia addition for hafina gate dielectric," in *Proc. IEEE IRPS 2012*, pp. GD. 3.1–3.4.

[23] W.J. Zhu, T. Tamagawa, M. Gibson, T. Furukawa, and T.P. Ma, "Effect of Al inclusion in HfO_2 on the physical and electrical properties of the dielectrics," *IEEE Electron. Dev. Lett. 2007*, vol. 23(11), pp. 649–651.

[24] Y.-K. Chiou, C.-H. Chang, C.-C. Wang, *et al.*, "Effect of Al incorporation in the thermal stability of atomic-layer-deposited HfO_2 for gate dielectric applications," *J. Electrochem. Soc. 2007*, vol. 154(4), pp. G99–G102.

[25] K. Okada, H. Ota, A. Hirano, A. Ogawa, T. Nabatame, and A. Toriumi, "Roles of high-k and interfacial layers on TDDB reliability studied with $HfAlO_X/SiO_2$ stacked gate dielectrics," in *IEEE IRPS, 2008*, pp. 661–662.

[26] P.K. Park and S.W. Kang, "Enhancement of dielectric constant in HfO_2 thin films by the addition of Al_2O_3," *Appl. Phys. Lett. 2006*, vol. 89(19), pp. 192905-1-3.

[27] M.-Y. Ho, H. Gong, G.D. Wilk, *et al.*, "Suppressed crystallization of Hf-based gate dielectrics by controlled addition of Al_2O_3 using atomic layer deposition," *Appl. Phys. Lett. 2002*, vol. 81(22), pp. 4218–4220.
[28] H.Y. Yu, N. Wu, M.F. Li, *et al.*, "Thermal stability of $(HfO_2)_x(Al_2O_3)_{1-x}$ on Si," *Appl. Phys. Lett. 2002*, vol. 81(19), pp. 3618–3620.
[29] G.D. Wilk, M.L. Green, M.Y. Ho, *et al.*, "Improved film growth and flatband voltage (control of ALD HfO_2 and Hf–Al–O with n+ poly-Si gates using chemical oxides and optimized post-annealing," in *Proc. VLSI Tech. Dig. 2002*, pp. 88–89.
[30] V. Mikhelashvili, R. Brener, O. Kreinin, B. Meyler, J. Shneider, and G. Eisenstein, "Characteristics of metal–insulator–semiconductor capacitors based on high-k HfAlO dielectric films obtained by low-temperature electron-beam gun evaporation," *Appl. Phys. Lett. 2004*, vol. 85(24), pp. 5950–5952.
[31] T.J. Park, J.H. Kim, J.H. Jang, *et al.*, "Reduction of electrical defects in atomic layer deposited HfO_2 films by Al doping," *Chem. Mater. 2010*, vol. 22, pp. 4175–4184.
[32] K. Takeda, R. Yamada, T. Imai, T. Fujiwara, T. Hashimoto, and T. Ando, "Characteristic instabilities in HfAlO metal–insulator–metal capacitors under constant-voltage stress," *IEEE Trans. Electron. Dev. 2008*, vol. 55(6), pp. 1359–1365.
[33] M.-H. Cho, K.B. Chung, H.S. Chang, *et al.*, "Interfacial reaction depending on the stack structure of Al_2O_3 and HfO_2 during film growth and postannealing," *Appl. Phys. Lett. 2004*, vol. 85(18), pp. 4115–4117.
[34] T. Nishimura, T. Okazawa, Y. Hoshino, *et al.*, "Atomic scale characterization of HfO_2/Al_2O_3 thin films grown on nitride and oxidized Si substrates," *J. Appl. Phys.*, vol. 96(11), pp. 6113–6119.
[35] K. Tapily, S. Consiglio, R.D. Clark, *et al.*, "Higher-K formation in atomic layer deposited $Hf_{1-x}Al_xO_y$," *ECS Trans. 2014*, vol. 64(9), pp. 123–131.
[36] J. Robertson, "Interfaces and defects of high-K oxides on silicon," *Solid State Electron. 2005*, vol. 49(3), pp. 283–293.
[37] M. Houssa, S. DeGendt, G. Groeseneken, and M.M. Heyns, "Negative bias temperature instabilities in SiO_2/HfO_2-based hole channel FETs," *J. Electrochem. Soc. 2004*, vol. 151(12), pp. F288–F291.
[38] A. Kerber and P. Srinivasan, "Impact of stress mode on stochastic BTI in scaled MG/HK CMOS devices," *IEEE Electron. Dev. Lett. 2014*, vol. 35(4), pp. 431–433.
[39] M. Cho, B. Kaczer, T. Kauerauf, L.-A. Ragnarsson, and G. Groeseneken, "Improved NBTI reliability with sub-1-nanometer EOT ZrO_2 gate dielectric compared with HfO_2," *IEEE Electron. Dev. Lett. 2013*, vol. 34(5), pp. 593–595.
[40] A. Kerber and E. Cartier, "Reliability challenges for CMOS technology qualifications with hafnium oxide/titanium nitride gate stacks," *IEEE Trans. Dev. Mater. Reliab. 2009*, vol. 9(2), pp. 147–162.

[41] T.J. Ho, D.S. Ang, A.A. Boo, Z.Q. Teo, and K.C. Leong, "Are interface state generation and positive oxide charge trapping under negative-bias temperature stressing correlated or coupled?" *IEEE Trans. Electron. Dev. 2012*, vol. 59(4), pp. 1013–1022.

[42] H.D. Xiong, D. Heh, S. Yang, *et al.*, "Stress-induced defect generation in HfO_2/SiO_2 stacks observed by using charge pumping and low frequency noise measurements," in *Proc. IEEE IRPS 2008*, pp. 319–323.

[43] D.K. Schroder, "Negative bias temperature instability: What do we understand?" *Microelectron. Reliab. 2007*, vol. 47, pp. 841–852.

[44] J. Robertson, "Band offsets of high dielectric constant gate oxides on silicon," *J. Non-Cryst. Solids 2002*, vol. 303, pp. 94–100.

[45] K. Tapily, S. Consiglio, R.D. Clark, *et al.*, "Texturing and tetragonal phase stabilization of ALD $Hf_xZr_{1-x}O_2$ using a cyclical deposition and annealing scheme," *ECS Trans. 2012*, vol. 45(3), pp. 411–420.

[46] C.C. Yeo, B.J. Cho, M.S. Joo, *et al.*, "Improvement of electrical properties of MOCVD HfO_2 by multistep deposition," *Electrochem. Solid State Lett. 2003*, vol. 6(11), pp. F42–F44.

[47] D. Ishikawa, S. Kamiyama, E. Kurosawa, T. Aoyama, and Y. Nara, "Extended scalability of HfON/SiON gate stack down to 0.57 nm equivalent oxide thickness with high carrier mobility by post-deposition annealing," *Jpn. J. Appl. Phys. 2009*, vol. 48(4S), pp. 04C004-1–5.

[48] T. Nabatame, K. Iwamoto, H. Ota, *et al.*, "Design and proof of high quality $HfAlO_x$/film formation for MOSCAPs and nMOSFETs through layer-by-layer deposition and annealing process," in *Symp. VLSI Tech. Dig. 2003*, pp. 25–26.

[49] M. Kobayashi, G. Thareja, M. Ishibashi, *et al.*, "Radical oxidation of germanium for interface gate dielectric GeO_2 formation in metal–insulator–semiconductor gate stack," *J. Appl. Phys. 2009*, vol. 106(10), pp. 104117-1–7.

[50] M.N. Bhuyian, D. Misra, K. Tapily, *et al.*, "Cyclic plasma treatment during ALD $Hf_{1-x}Zr_xO_2$ deposition," *ECS J. Solid State Sci. Technol. 2014*, vol. 3(5), pp. N83–N88.

[51] A. Stesmans and V.V. Afanas'ev, "Paramagnetic defects in annealed ultrathin layers of SiO_x, Al_2O_3, and ZrO_2 on (1 0 0)Si," *Appl. Phys. Lett. 2004*, vol. 85(17), pp. 3792–3794.

[52] S. Ferrari and G. Scarel, "Oxygen diffusion in atomic layer deposited ZrO_2 and HfO_2 thin films on Si(1 0 0)," *J. Appl. Phys. 2004*, vol. 96(1), pp. 144–149.

[53] G. Chang *et al.*, United states Patent Publication No: US 2010/0075507 A1, Publication Date, March 25, 2010.

[54] H.-S. Jung, S.-A. Lee, S.-H. Rha, *et al.*, "Impacts of Zr composition in $Hf_{1-x}Zr_xO_y$ gate dielectrics on their crystallization behavior and bias-temperature-instability characteristics," *IEEE Trans. Electron Dev. 2011*, vol. 58(7), pp. 2094–2103.

[55] J.R. Hauser and K. Ahmed, "Characterization of ultra-thin oxides using electrical C–V and I–V measurements," in *Proc. AIP Conf. 1998*, vol. 449, pp. 235–239.

[56] E.H. Nicollian and A. Goetzberger, "The Si–SiO interface—Electrical properties as determined by the metal–insulator–silicon conductance technique," *Bell Syst. Tech. J. 1967*, vol. 46, pp. 1055–1133.
[57] W. Zheng, K.H. Bowen, Jr., J. Li, I. Daübkowska, and M. Gutowski, "Electronic structure differences in ZrO_2 vs HfO_2," *J. Phys. Chem. A 2005*, vol. 109(50), pp. 11521–11525.
[58] P.T. Chen, B.B. Triplett, J.J. Chambers, L. Colombo, P.C. McIntyre, and Y. Nishi, "Analysis of electrically biased paramagnetic defect centers in HfO_2 and $Hf_xSi_{1-x}O_2$/(1 0 0)Si interfaces," *J. Appl. Phys. 2008*, vol. 104(1), pp. 014106-1–7.
[59] E. Cartier and A. Kerber, "Stress-induced leakage current and defect generation in nFETs with HfO_2/TiN gate stacks during positive-bias temperature stress," in *Proc. IEEE IRPS 2009*, pp. 486–492.
[60] N. Rahim and D. Misra, "TiN/HfO_2/SiO_2/Si gate stack breakdown: Contribution of HfO_2 and interfacial SiO_2 layer," *J. Electrochem. Soc. 2008*, vol. 155(10), pp. G194–G198.
[61] M. Houssa, "Defect generation under electrical stress: Experimental characterization and modeling," in *High-κ Gate Dielectrics*, Ed. M. Houssa, Bristol and Philadelphia: Institute of Physics Publishing, 2004, p. 467.
[62] Y. Kabe, R. Hasunuma, and K. Yamabe, "Oxidation of silicon utilizing a microwave plasma system: Electric-stress hardening of SiO_2 films by controlling the surface and interface roughness," *Jpn. J. Appl. Phys. 2012*, vol. 151, pp. 041104-1–5.
[63] F. Mondon and S. Blonkowski, "Electrical characterization and reliability of HfO_2 and Al_2O_3–HfO_2 MIM capacitors," *Microelctron. Reliab. 2003*, vol. 43(8), pp. 1259–1266.
[64] R. Degraeve, G. Groeseneken, R. Bellens, M. Depas, and H.E. Maes, "A consistent model for the thickness dependence of intrinsic breakdown in ultra-thin oxides," in *Proc. IEDM Tech. Dig. 1995*, pp. 863–866.
[65] M.L. Green, E.P. Gusev, R. Degraeve, and E.L. Garfunkel, "Ultrathin (<4 nm) SiO_2 and Si–O–N gate dielectric layers for silicon microelectronics: Understanding the processing, structure, and physical and electrical limits," *J. Appl. Phys. 2001*, vol. 90(5), pp. 2057–2121.
[66] W. Wang, K. Akiyama, W. Mizubayashi, T. Nabatame, H. Ota, and A. Toriumi, "Effect of Al-diffusion-induced positive flatband voltage shift on the electrical characteristics of Al-incorporated high-k metal-oxide-semiconductor field-effective transistor," *J. Appl. Phys. 2009*, vol. 105(6), pp. 064108-1–6.
[67] K. Kita, L.Q. Zhu, T. Nishimura, K. Nagashio, and A. Toriumi, "Formation of dipole layers at oxide interfaces in high-k gate stacks," *ECS Trans. 2010*, vol. 33(6), pp. 463–477.
[68] S. Hibino, T. Nishimura, K. Nagashio, K. Kita, and A. Toriumi, "Interface dipole cancellation in SiO_2/high-k/SiO_2/Si gate stacks," *ECS Trans. 2012*, vol. 50(4), pp. 159–163.

[69] H. Ota, A. Ogawa, M. Kadoshima, *et al.*, "Significance of nitrogen and aluminum depth profile control in HfAlON gate insulators," *ECS Trans*. 2006, vol. 3(3), pp. 41–47.

[70] P. Samanta, C.-L. Cheng, Y.-J. Lee, and M. Chan, "Electrical stress-induced charge carrier generation/trapping related degradation of HfAlO/SiO_2 and HfO_2/SiO_2 gate dielectric stacks," *J. Appl. Phys. 2009*, vol. 105(12), pp. 124507-1–8.

[71] K. Iwamoto, A. Ogawa, Y. Kamimuta, *et al.*, "Re-examination of flat-band voltage shift for high-k MOS devices," in *Proc. VLSI Tech. Dig. Symp.*, pp. 70–71, 2007.

[72] P. Samanta, C.-L. Cheng, and Y.-J. Lee, "Charge trapping related degradation of thin HfAlO/SiO_2 gate dielectric stack during constant-voltage stress," *J. Electrochem. Soc. 2009*, vol. 156(8), pp. H661–H668.

[73] J.J. Kim, M. Kim, U. Jung, *et al.*, "Intrinsic time zero dielectric breakdown characteristics of HfAlO alloys," *IEEE Trans. Electron. Dev. 2013*, vol. 60(11), pp. 3683–3689.

[74] J. Robertson, "Band offsets of wide-band-gap oxides and implications for future electronic devices," *J. Vac. Sci. Technol. B 2000*, vol. 18(3), pp. 1785–1791.

[75] W.M. Tang, U. Aboudi, J. Provine, R.T. Howe, and H.-S. Wong, "Improved performance of bottom-contact organic thin-film transistor using Al doped HfO_2 gate dielectric," *IEEE Trans. Electron Dev. 2014*, vol. 61(7), pp. 2398–2403.

[76] Q. Li, K.M. Koo, W.M. Lau, *et al.*, "Effects of Al addition on the native defects in hafnia," *Appl. Phys. Lett. 2006*, vol. 88(18), pp. 182903-1–3.

[77] X.F. Wang, Q. Li, R.F. Egerton, *et al.*, "Effect of Al addition on the microstructure and electronic structure of HfO_2 film," *J. Appl. Phys. 2007*, vol. 101(1), pp. 013514-1–5.

[78] M. Houssa, "Defect generation under electrical stress: Experimental characterization and modeling," in *High-κ Gate Dielectrics*, Ed. M. Houssa, Bristol and Philadelphia: Institute of Physics Publishing, 2004, p. 310.

[79] D. Ielmini, A.S. Spinelli, A. Rigamonti, and A.L. Lacaita, "Modeling of SILC based on electron and hole tunneling—Part II: Steady-state," *IEEE Trans. Electron Dev. 2000*, vol. 47(6), pp. 1266–1272.

[80] P.E. Nicollian, M. Rodder, D.T. Grider, P. Chen, R.M. Wallace, and S.V. Hattangady, "Low voltage stress-induced-leakage-current in ultrathin gate oxides," in *Proc. IEEE IRPS 1999*, vol. 37, pp. 400–404.

[81] K. Okada, W. Mizubayashi, N. Yasuda, *et al.*, "Degradation mechanism of $HfAlO_X$/SiO_2 stacked gate dielectrics studied by transient and steady-state leakage current analysis," *J. Appl. Phys. 2005*, vol. 97(7), pp. 074505-1–7.

Chapter 2

High mobility n and p channels on gallium arsenide and silicon substrates using interfacial misfit dislocation arrays

E. J. Renteria[1], *S. Addamane*[1], *D. Shima*[1], *and G. Balakrishnan*[1]

For technology scaling issues, the use of high-K as dielectric has been adopted as device-level solution. The use of graphene in transistor structure is being evaluated. The previous chapter focused on these aspects. This chapter also discusses a material level solution of technology scaling and focuses on the use of compound semiconductor with Si to solve the challenges of technology scaling.

2.1 Introduction

The performance of silicon microchips has consistently improved in the past 50 years in accordance with Moore's law. The ability to achieve more transistors per unit area in a microprocessor is credited to the use of the complementary metal-oxide semiconductor (CMOS) technology. Advanced lithography techniques have allowed the reduction in the gate length of the field effect transistors resulting in the improvement of their performance. As the transistor count increases, each individual device has been predicted to become smaller, faster, and cheaper. Today, the most advanced transistors face challenges such as excessive leakage currents, degradation of carrier mobility, rapid device breakdown, and the increasing variability between individual devices. Despite significant achievements in shrinking gate lengths, the technology will soon reach a point where integration of Si-CMOS materials with compound semiconductors will be required to allow for the progress of Moore's law. The integration of compound semiconductor with Si is an attractive solution to such challenges. The most researched compound semiconductor for such applications are the III-V alloys containing elements of column III (Al, Ga, and In) and column V (N, P, As, Sb, and Bi) of the periodic table. The case for III-V semiconductors may be made by highlighting the extraordinary carrier mobility of these materials. For InAs, e.g., the carrier mobility has been reported to be over an order of magnitude higher than that

[1]Center for High Technology Materials, University of New Mexico, Albuquerque, NM, USA

of silicon. Also, transistors fabricated from these materials have resulted in very high switching frequencies of between 600 GHz and 1 THz. Therefore, the integration of III-V semiconductors with silicon will enable further development in the performance of silicon-based microchips.

The integration of III-V compound semiconductors with silicon can be achieved through monolithic integration or through a wafer-bonding process. In the case of monolithic integration, the III-V alloys are grown on a silicon wafer using an epitaxial process such as molecular beam epitaxy (MBE), or metal-organic chemical vapor deposition (MOCVD). However, III-V materials are highly lattice-mismatched when grown on silicon. Lattice mismatch in crystal growth leads to strain and in most cases leads to dislocations, which are highly detrimental to the performance and lifetime of devices. Wafer bonding also involves the growth of an epitaxial structure. However, this is not done directly on silicon but rather on a III-V substrate such as GaAs or InP. After growth, the epitaxial layer is bonded to the silicon substrate and the III-V substrate is either removed or lifted off. Both techniques have their advantages and disadvantages. However, from the point of view of economic feasibility, a continuous crystalline III-V epilayer on silicon would perhaps be the optimal solution.

The early body of research on the integration of III-Vs with Silicon involved the growth of GaAs on silicon.[1] GaAs and Si have a 4.2% lattice mismatch that can produce threading dislocations that can propagate through the crystal. In addition to the lattice mismatch issue, the thermal expansion coefficient mismatch between silicon and GaAs can result in strain in the structure when the sample is cooled from growth temperature (usually in the range of 450°C–600°C) to room temperature. Furthermore, the growth of polar material on a nonpolar substrate gives rise to antiphase domains (APDs). While semiconductor lasers operating continuously at room temperature have been reported, there have only been a few isolated cases of successful fabrication of such lasers.[2,3,4] The primary limitation is that the defect density is extremely high in these materials.

The work on the topic of III-V/Si integration began in the early 80s and lasted for over a decade. During this time, several key problems were addressed by researchers such as lowering threading dislocation density (TDD) in the material, overcoming the issue of thermal expansion coefficient mismatch between silicon and GaAs, and annihilating APDs in the material grown. The effort for optimizing the GaAs material involved the optimization of the substrate preparation process, growth time parameters, and postgrowth treatment.[5] Of considerable importance was the study of silicon substrate orientation and buffer layers on the quality of the III-V grown, which helped achieve single-domain material. The complete absence of antiphase boundaries (APBs) and APDs was achieved by using vicinal double-stepped substrates.

Alternate technologies to improve threading dislocation densities were also tried including the growth of the GaAs in small isolated regions using an oxide/nitride layer as a mask with exposed Si regions where the growth occurs. This mode of growth is called "selective area epitaxy" (SAE). Yet another approach was used by Motorola using strontium titanate interfacial layers. However, even this breakthrough did not help GaAs on Si to mature to be a commercially feasible technology.[6]

The defect density in GaAs grown on silicon was drastically reduced through the use of low temperature nucleation. At these temperatures, GaAs grows planar from the very early stages of growth, following the homogeneous nucleation of 3D GaAs islands, resulting in the complete elimination of planar faults. Nevertheless, accurate TEM dislocation counts indicate a dislocation density in the low $10^8/cm^2$ range. It is concluded that by either increasing the GaAs epilayer thickness or the sample temperature, one produces a residual compressive stress that forces the threading dislocations to slip, thus reducing their density by reactions that become more probable with proximity. The residual dislocation density of about $10^8/cm^2$ is attributed partly to threading dislocation generation during the early stages of epitaxy and only partly to generation from tensile thermal stress during cooling. This low temperature growth on vicinal GaAs with low defect density and single domain represents one of the best results for GaAs monolithically achieved on Si to date. The use of an MBE grown homoepitaxial Si buffer layer resulted in an addition improvement to the quality of the GaAs layer. This growth method was successfully undertaken by Motorola, as demonstrated by similar mobility and RF performance for power amplifiers fabricated on Si and GaAs, respectively.[7] Lifetime for these majority carrier devices seemed reasonable: 800 hours of usage at 200°C resulted in only 1.2% degradation. A series of buffer layers (SiO_2, $SrTiO_3$, GaAs) allows mechanical decoupling between the device layer and the substrate.

Step-graded metamorphic buffers have been put to great use in various mismatched systems.[8] The fact that germanium has the same lattice constant as GaAs allows the use of a step-graded buffer from Si to GaAs via the SiGe step-graded buffer. GaAs/AlGaAs quantum-well devices have been demonstrated via organometallic chemical vapor deposition on relaxed graded Ge/GeSi virtual substrates on Si. A number of GaAs/Ge/Si integration issues including Ge autodoping behavior in GaAs and reduced critical thickness due to thermal expansion mismatch. Despite these issues, surface threading dislocation densities for GaAs/AlGaAs lasers on Si substrates was as low as 2×10^6 defects/cm^2 permitting the realization of electronic and optoelectronic devices. There have been many studies of growth approaches (MBE, MOCVD, etc.) as well as different procedures used within each general deposition approach to minimize defect densities even further. Despite these many studies, GaAs heteroepitaxy on silicon does not generally provide a sufficiently high-quality material for fabrication of commercial electronic devices.

In this chapter, we explore the growth of III-Sb–based compound semiconductors on silicon as a potential embodiment of III-V/silicon integration. III-Sb alloys, such as GaSb, AlSb, and InGaSb, are typically grown on GaSb substrates. High mobility n and p channels such as InAs and InGaSb can be grown on GaSb. Therefore, III-Sb–based semiconductors are an interesting candidate for development of high electron mobility transistors. This chapter describes a novel growth mode for the realization of large area growth of GaSb on silicon involving the formation of interfacial misfit (IMF) dislocation arrays at the III-Sb/Si interface that allows for the realization of large area growth of III-Sb alloys without the need for specialized growth processes like SAE.

2.2 IMF versus pseudomorphic growth

One of the main constraints on achieving new and unique devices with the present semiconductor technology is that imposed by lattice mismatch. In the late 1940s Frank and Van der Merwe established the foundation for future studies in mismatched yet crystalline growth. These studies were further advanced in the 1970s by Matthews and Blakeslee,[9] who established that mismatched epitaxy resulted in coherent strain and not polycrystalline or amorphous incoherent growth. Lattice-matched epitaxy implies layer-by-layer growth of material where the epilayer and the substrate have the same crystallographic type and the same lattice constant. This results in a coherent strain-free growth. However, if the lattice constant of the epilayer is not the same as the substrate, it results in strained growth. Depending on whether the epilayer's in-plane lattice constant is larger or smaller than the substrate, the strain is classified as compressive or tensile. In case of compressive strain, the larger epilayer conforms to the smaller substrate to achieve a one-to-one correspondence with the substrate. This results in the cell expanding along the direction perpendicular to the growth surface. Tensile strain has the opposite effect. These instances of strain are called *"pseudomorphic"* because the epilayer takes on the morphology or the lattice constant of the underlying substrate, and the distortion of the material grown is termed as *"tertragonal distortion."*

As the growth of the strained material continues, the strain energy increases. However, if this strain energy in the crystal becomes relatively large due to either a very large mismatch to begin with or a smaller mismatch but very thick epilayer, it is relieved through a network of dislocations. These dislocations are initially in the form of misfit dislocations. The misfit dislocations act as sources for threading dislocations, which as the name suggests refers to defects that thread though the material-breaking bonds as they propagate. These dislocations either terminate by propagating to one of the crystal surfaces or annihilate themselves through interactions with another threading dislocation.

The extent to which a mismatched epilayer can be realized without the formation of dislocations is called the critical thickness. A classic equation for the critical thickness, which takes into consideration only misfit dislocations, is provided below

$$h_c = \frac{a_0\left(1 - \frac{\nu_{PR}}{4}\right)\left[\ln\left(\frac{h_c\sqrt{2}}{a_0}\right) + 1\right]}{2\sqrt{2}\pi|f|(1 + \nu_{PR})} \tag{2.1}$$

where, a_0 = Substrate lattice constant; f = mismatch; ν_{PR} = Poisson's ratio (~1/3 for most semiconductors).

When the epitaxial process exceeds this critical thickness the epilayer starts to develop dislocations which relieves the strain and results in *relaxation*. While the theory of strained growth states that a critical thickness has to be achieved prior to the onset of misfit dislocations, in certain materials systems such as GaSb on GaAs, a plastic relaxation facilitated by two-dimensional (2D) array of misfit dislocations is present at the interface of the GaSb on GaAs growth.[10] This is a fundamentally different growth mode that results in lower defect epilayer (compared to growth that proceeds through a pseudomorphic phase) in which strain energy is solely relieved by laterally propagating network of 90° misfit dislocations (also known as Lomer dislocations or pure edge dislocations) confined to the episubstrate interface.[11]

As mentioned above, the IMF array does not proceed through the critical thickness route, but instead the arrangement of large Sb atoms on the GaAs substrate results in spontaneously relaxation. There is typically some residual strain in the layer but this is attributed to the slight thermal expansion coefficient mismatch between GaSb and GaAs. The IMF growth mode could also be employed to integrate III-Sb alloys with silicon; however, there are certain key differences when compared to GaSb on GaAs IMF; these will be discussed in the following sections.

2.3 III-Sb on GaAs substrates[19]

The growth of GaSb on GaAs is of much interest due to lower substrate cost and larger wafer sizes of GaAs compared to GaSb. Moreover, GaAs has semi-insulating properties and is suitable to form excellent Ohmic contacts.[12] However, growing GaSb on GaAs introduces a 7.8% lattice mismatch, giving rise to a high density of defects.[13] Several methods have been tried to mitigate these defects caused by the lattice mismatch between GaSb and GaAs including strained-layer superlattices, graded metamorphic buffers, and the introduction of an IMF array.[14,15,16] Although all the above-mentioned techniques help in reducing threading dislocations, the IMF technique is fundamentally different and has proven to be the most effective method to obtain high-quality buffer-free GaSb on GaAs.[17] Consequently, several antimonide-based devices have been successfully grown on GaAs using the IMF technique.[18,19]

The growth of GaSb on GaAs has been shown to start as islands and later coalesce into a uniform layer. At the interface, the strain energy is relieved by both 90° and 60° misfit dislocations. While most of the strain is believed to be relieved by the laterally propagating 90° misfits, the minority 60° dislocations end up causing threading dislocations in the GaSb bulk layer. The root cause of the formation of the 60° dislocations has been traced back to the coalescence of the GaSb islands. The fundamental approach behind the IMF technique is to induce a highly periodic array of 90° misfit dislocations along both [110] and [1–10] directions which would solely relieve all the strain energy at the interface.[16] The following paragraph describes the procedure for the growth of GaSb on GaAs by MBE using the IMF technique.

The IMF-grown bulk GaSb is sensitive to the growth temperature which affects relaxation, surface morphology, and threading dislocation densities in the epilayer. The evidence for the formation of the IMF array is seen in transmission electron microscopy (TEM) analysis of the interface. High-resolution XRD measurements also verify the relaxation of the GaSb epilayer at all growth temperatures.

The typical growth of GaSb on GaAs involves removal of the oxide on the substrates at 630°C for 20 minutes before a 200-nm GaAs smoothing layer is grown at 580°C. At this point, the As cracker is valved off to initiate the desorption of arsenic from the surface. The reflection high-energy electron diffraction (RHEED) pattern transforms from a (2×4) As-stabilized GaAs surface to a (4×2) Ga-rich surface. Then the sample is exposed to Sb flux and the RHEED pattern changes to a (2×8) reconstruction. As shown by Huang *et al.*,[20] this interaction of Sb with a Ga-rich surface is critical to forming the IMF array instead of causing tetragonal distortion in the subsequent GaSb layer. Once the (2×8) reconstruction is observed, the substrate temperature is brought down to the growth temperature of GaSb under Sb overpressure and the growth is initiated. However, if the conditions for the formation of the IMF are not fully optimized, the GaSb will still turn out to have a high density of threading dislocations. Therefore, it is critical to understand the effect of parameters such as nucleation temperature, growth sequence, and III:V ratio. Among these parameters, it has been conclusively shown that the effectiveness of the IMF growth mode in reducing TDD is dependent on the GaSb growth temperature following the formation of the misfit dislocation array.

Figure 2.1(a–c) shows the 10 μm × 10 μm AFM images of the GaSb bulk surfaces grown at different temperatures. These images clearly show that the GaSb

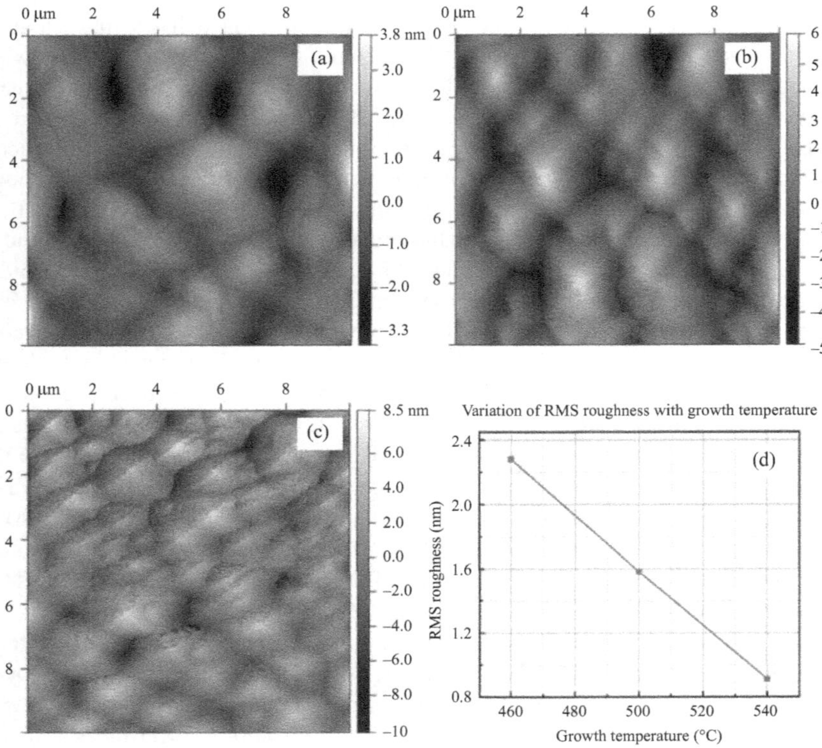

Figure 2.1 AFM images of 2 μm thick IMF-grown GaSb epilayers on GaAs at different growth temperatures. (a) 540°C, (b) 500°C, (c) 460°C, (d) graph showing the variation of surface roughness with growth temperature, RMS_{sq} = 2.28 nm at 460°C, 1.582 nm at 500°C, and 0.912 nm at 540°C

growth proceeds in a step-flow growth mode originating at the misfit dislocations. On closer observation, it is found that the RMS roughness of the surfaces increases with decrease in temperature as shown in Figure 2.1(d). Brown *et al.* observed a similar trend in surface roughness when GaSb is grown on GaAs without trying to induce the formation of a periodic IMF array.[21] In order to explain the variation in surface roughness at different growth temperatures, the widths of the steps formed during the step-flow growth mode are to be considered at each temperature. It is found that the terrace widths in the step-flow growth mode increase with the increase in temperature. It is suspected that this increase in terrace width manifests as a decrease in the surface roughness in the AFM images.

The bright-field cross-sectional TEM image along [110] direction of the 2 μm thick IMF-grown GaSb epilayers at different growth temperatures (460, 500, and 540°C) are shown in Figure 2.2(a–c). At a higher magnification (Figure 2.2(d)), all samples show a highly periodic array of misfits at the GaAs/GaSb interface confirming the formation of the IMF. It can be seen that at a higher growth temperature (540°C),

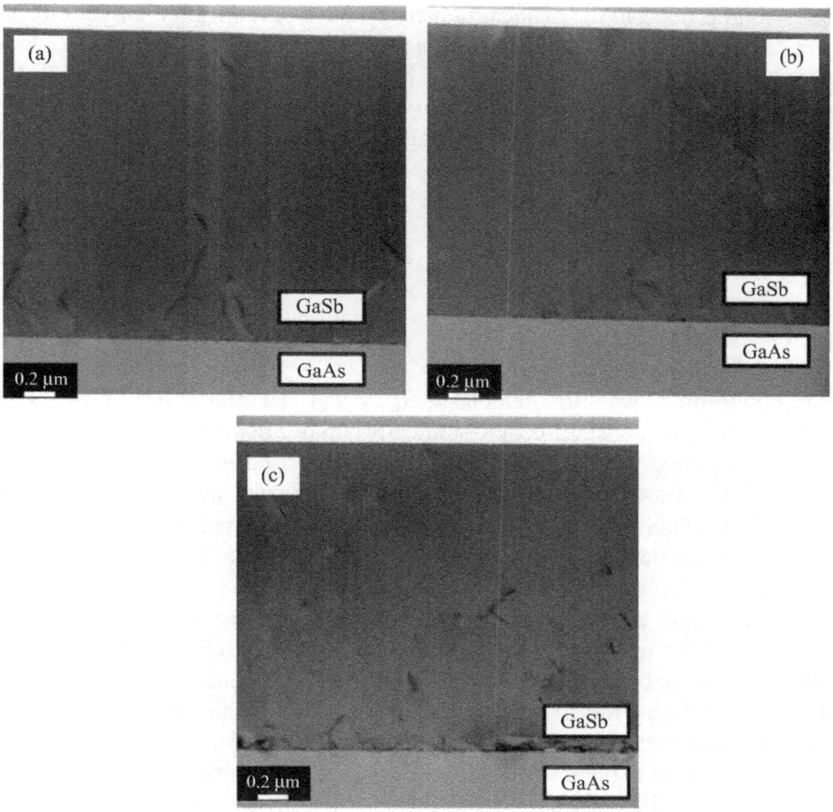

Figure 2.2 Cross-sectional TEM images of 2 μm thick IMF-grown GaSb epilayers at (a) 540°C, (b) 500°C, and (c) 460°C

the initial growth conditions for GaSb are not optimized and the TDD formation at the interface is higher. As demonstrated before,[22] it can be seen that the TDD dramatically decreases with increase in the film thickness due to mutual interactions between dislocations at all growth temperatures. Using the projected length of dislocation lines in a TEM, the TDD can be estimated by:[23]

$$N_D = \frac{4}{\pi} \frac{l'}{At} \tag{2.2}$$

where, t is the thickness of the sample and A is the area over which the projected length l' is estimated. From Figure 2.2(a–c), the TDD is estimated using (2) at different temperatures within the first 0.5 μm GaSb layer. The projected length l' is measured using the appropriate scale. While the dislocation density is estimated to be $\sim 1 \times 10^9$ cm^{-2} for the sample grown at 540°C, there is a decrease in the TDD to 4×10^8 cm^{-2} at a growth temperature of 460°C. Plan-view TEM images (Figure 2.3) of the GaSb surfaces grown at different temperatures (540, 500, and 460°C) are analyzed to verify the TDD obtained. As observed by cross-sectional TEM, a similar trend in the dislocation density can be seen from plan-view TEM in Figure 2.3(d) with the sample grown at a higher growth temperature showing a higher TDD. Analyzing a typical 3-μm^2 surface by directly counting the number of dislocations, the TDD from the plan-view images could be estimated to be 9×10^8 cm^{-2} for GaSb surface grown at 540°C and again decreases to $\sim 3 \times 10^8$ cm^{-2} for the sample grown at 460°C. These results clearly show that the formation of the IMF array to reduce TDD in GaSb grown on GaAs is highly dependent on the growth conditions.

2.4 III-Sb on silicon substrates

III-Sb growth on Si was first demonstrated by Van Der Ziel *et al.*[24] This research culminated in the demonstration of RT optically pumped double heterojunction lasers. These devices were grown using the same formula used in the LT GaAs growth, but instead of a GaAs nucleation layer, an AlSb layer was used. The active region comprised an AlGaSb/GaSb/AlGaSb active region that emitted at 1.6–1.8 μm. While they had extensive threading dislocations (around 5×10^7/cm^2), they were among one of the first III-V devices to lase on Si.

AlSb is a very promising material for mismatched growth on both III-V and Si substrates.[25] As previously mentioned, the growth of high crystal quality III-V materials on Si substrates is hindered by lattice mismatch, thermal expansion mismatch, and formation of APDs. The IMF growth mode allows the growth of highly mismatched and yet low-defect density materials. An AlSb nucleation layer is helpful to overcome the thermal expansion mismatch problem as AlSb has a thermal expansion coefficient very close to the one of Si. The focus of this section is to describe the growth of III-Sb on Si substrates by using an AlSb nucleation layer and the IMF growth mode along with the approaches to decrease the formation of APDs.

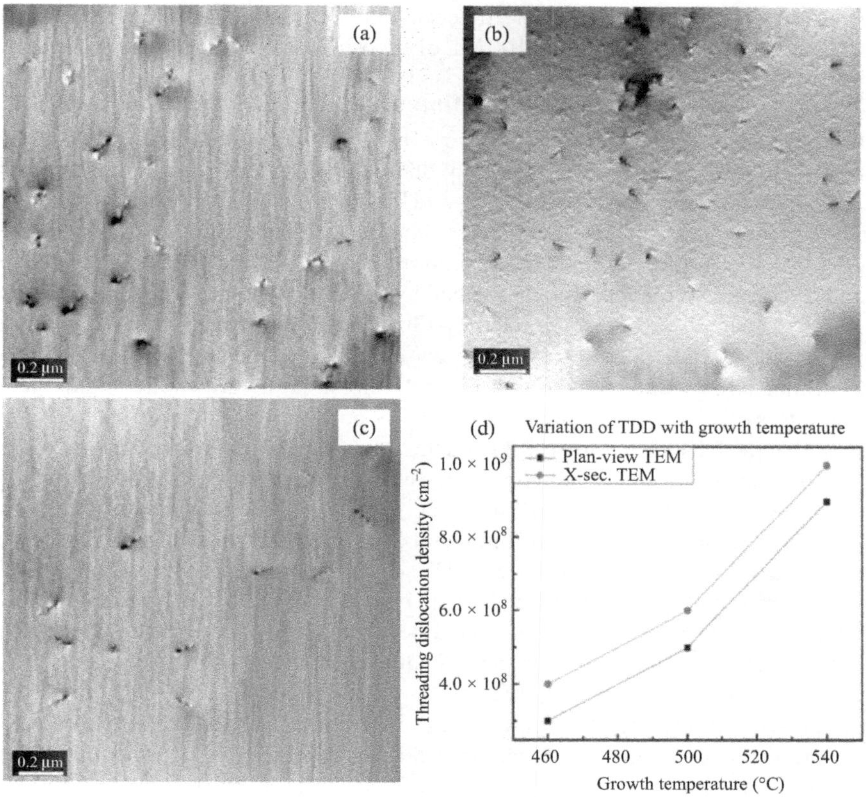

Figure 2.3 *Plan-view TEM of 2 μm thick IMF-grown GaSb epilayers on GaAs at (a) 460°C, (b) 500°C, and (c) 540°C, (d) graph showing the variation of TDD with growth temperature. TDD estimated from cross-sectional TEM is calculated considering the first 0.5 μm of GaSb grown after IMF formation*

2.4.1 Lattice mismatch solution: IMF layer

If one of the constituents of the epitaxial layer has a sufficiently large atomic size in comparison to Si, the IMF layer would form providing us with a platform to grow the III-V. The obvious choice would be antimony. With Sb, an interfacial array of 90° misfit dislocations can be observed at growth temperatures as high as ~540°C. Since the 90° misfit dislocation array depends strongly on the ability to have a self-assembled layer of ad-atoms, the larger surface residence time, and the high surface mobility of the Sb atoms would lead to improved growth results for the growth of antimonides on Si.

During the first few monolayers (MLs) of AlSb growth on Si, highly crystalline quantum dots (QDs) form. With continued deposition, the islands coalesce into a planar material with no detectable defects. The QD nucleation layer facilitates a completely relaxed AlSb within ~100 ML of deposition according to x-ray diffraction.

2.4.1.1 Growth and characterization of AlSb on silicon[26,27]

Prior to growth, the Si substrate surface is hydrogen-passivated through an hydrofluoric acid (HF) etch. The HF is usually diluted to a 1:10 ratio, and the Si wafer is dipped in it. The HF reacts with the SiO_2 and leaves behind a clean Si surface with the dangling bonds passivated by hydrogen atoms. The hydrogen passivation is removed by heating the substrate at 500°C in vacuum. A thermal cycle at 800°C ensures the removal of oxide remnants. This is verified by RHEED. The substrate temperature is reduced and stabilized at 500°C, and the AlSb is then grown at the same temperature.

The substrate preparation is extremely critical in achieving a good III-V surface. Since an Si homoepitaxial smoothing layer cannot be grown, the atomic level smoothness of the Si wafer achieved during its polishing has to be preserved. This means that the HF etch has to be brief and the material transferred into vacuum within 5–10 minutes of the etch. The longer a wafer is exposed to the atmosphere prior to growth the higher the chances of the oxide forming. While a variety of other etch and wash procedures have been recommended for Si wafers prior to the epitaxial process, a single immersion of the wafer in HF for a few seconds is sufficient to remove the oxide.[28] Initial growth of AlSb (3 MLs) results in the RHEED pattern switching from a 2×1 to a chevron-like reconstruction shown in the inset of Figure 2.4(a). The connected chevron pattern is characteristic of islands with an abrupt truncated pyramidal shape that are completely relaxed, as would be the case if IMF dislocations were present in the material. Figure 2.4(c)'s inset shows the 3×1 pattern associated with the approaching planar growth after 54 ML deposition. Corresponding AFM images of the AlSb/Si surface are shown in Figure 2.4(a–c) after 3, 18, and 54 MLs of deposition. At 3 MLs, the QD density is 10^{11} QDs/cm^2 with dot height and diameter of 1–3 nm and 20 nm, respectively. Continued deposition causes the individual islands to coalesce but remain crystallographic in contrast to InAs/GaAs QD growth where island coalescence leads to large defective islands. Figure 2.4(b), at 18 MLs, indicates a crystallographic preference of the coalescence along the [110] direction. Figure 2.4(c) shows continued coalescence towards planar growth with 54 MLs deposition. The growth of the AlSb past the 54 MLs results in very smooth and high-quality surfaces as has been shown in Figure 2.4(d). Figure 2.5 shows an HR-TEM image of the (110) crystal plane at the AlSb/Si interface. The image shows three distinct regions on the Si substrate labeled (i), (ii), and (iii). Region (i) is the bright region along the AlSb and Si interface. The deteriorated resolution in this region compared to the surrounding material makes analysis of this region difficult. The white appearance in contrast to the surrounding material is due to a changed density of atoms. This could possibly be due to very closely arranged IMF dislocations or possibly due to a twisted lattice causing the region to have a higher atomic density compared to the imaged (110) plane. Artificially induced twist-bonded substrates have been known to accommodate considerable interfacial mismatch as has been shown by Ejeckam *et al.*[29,30,31] Region (ii), about 5 ML in thickness, represents the nucleation layer formed by QD growth and coalescence. This material has a low-defect density and shows a planar

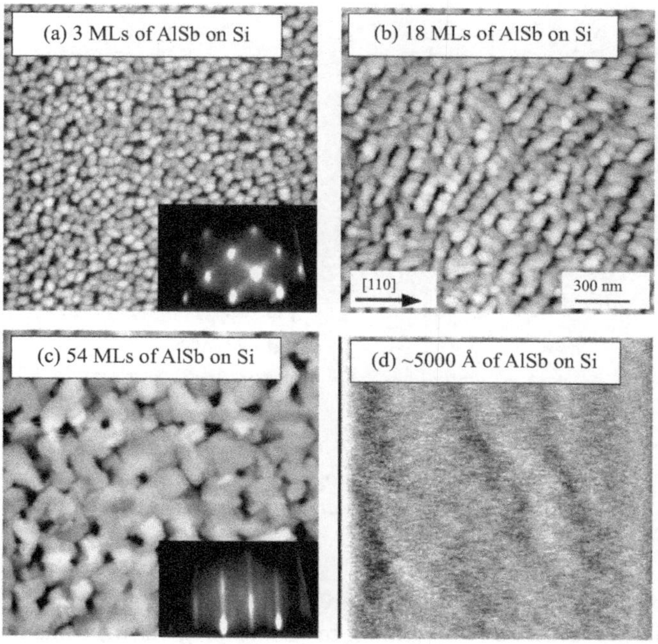

Figure 2.4 *AFM images showing surface structure after (a) 3 ML, (b) 18 MLs, and (c) 54 MLs of AlSb deposition on Si. (a) and (c) also show the RHEED image for the corresponding growths. (d) Shows the surface after 500 nm of growth, and this is an extremely smooth surface*

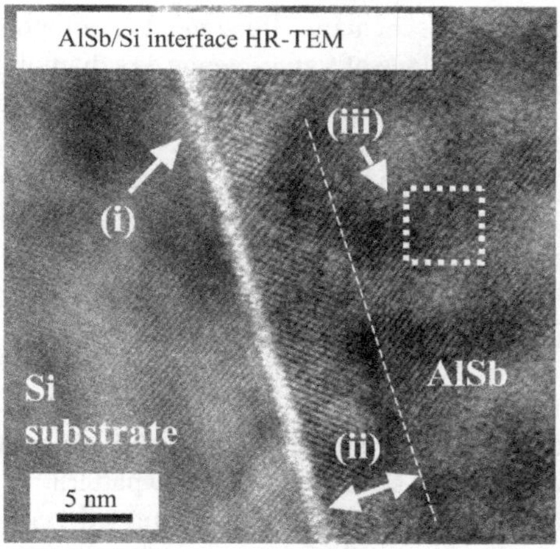

Figure 2.5 *Cross-sectional HR-TEM image of the AlSb/Si interface, showing the (110) plane with changes in crystallographic orientation*

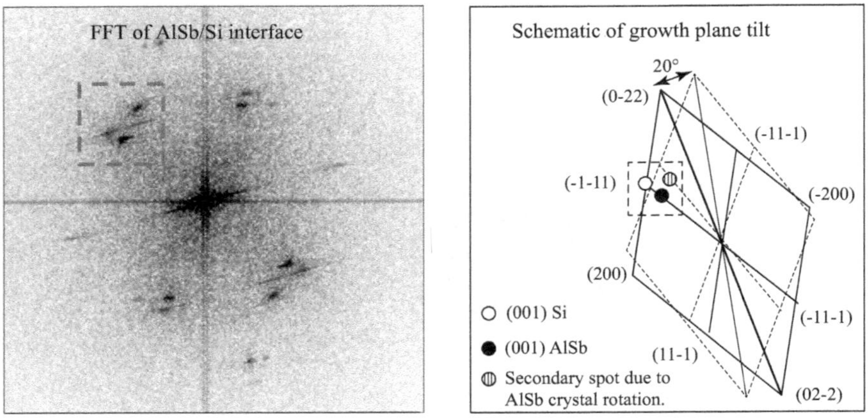

(a) (b)

Figure 2.6 Cross-sectional HR-TEM image of the AlSb/Si interface, showing the (110) plane with changes in crystallographic orientation

homogeneous zinc-blende crystal structure in which the arrangement of the atoms is in the form of consecutive (100) planes. In contrast, region (iii) contains undulating bulk material denoted by a measurable rotation of the zinc-blende crystal lattice.

The crystallographic undulations or bending lead to misfit dislocations that propagate parallel to the substrate. In other material systems such as arsenides, misfit dislocations lead to vertical propagating defects, such as threading or screw dislocations. However, the AlSb does not propagate these vertical defects perhaps due to the strong AlSb bond at these growth temperatures.

Figure 2.6 shows the reciprocal space analysis of Figure 2.5(a) and includes regions (i), (ii), and (iii). The schematic illustrates the components associated with the reciprocal view of a [110] plane of a zinc-blende or a diamond lattice. In analyzing the (-1-11) component, three spots are indicated. One spot indicates the Si (100) lattice and another that is closer to the (000) point, corresponds to the AlSb (100). A third spot corresponds to the AlSb point on a rhombus rotated clockwise by 20°. This indicates a completely different plane of growth that is rotated clockwise from the (100) plane. A variety of rotations, both clockwise and anticlockwise are measured at other locations within this sample. The crystallographic rotations measured in the reciprocal image are indicative of undulations in the AlSb bulk and verify the real-space TEM analysis of Figure 2.5. The undulations are ~10 nm wide and 1 nm high. With continued growth, the surface undulations merge, become shallower, and considerably broader until they can no longer be detected by TEM after ~1 μm. However, evidence of the undulation is still visible in the RHEED pattern for AlSb layer thickness ~10 μm. These undulations can however be suppressed very effectively by keeping the AlSb nucleation layer to ~100 Å and immediately following it with a GaSb layer. The suppression of the undulations can be seen in HR-XRD studies shown in Figure 2.7. Part (a) shows the XRD spectrum of AlSb grown on Si directly and part (b)

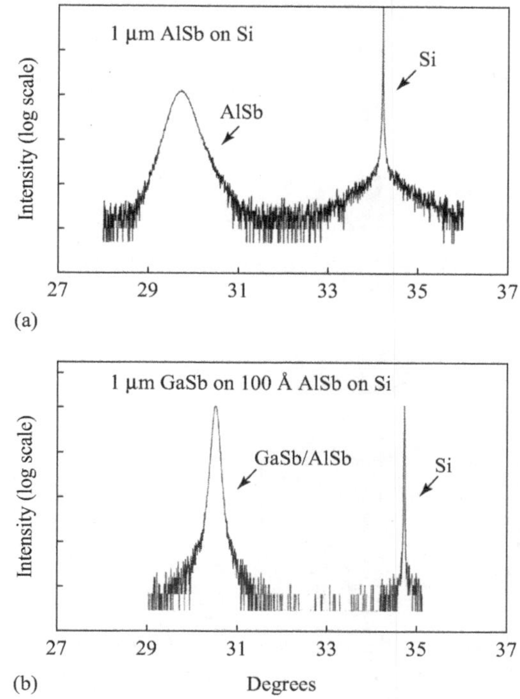

Figure 2.7 *(a) HR-XRD (004) scan of AlSb on Si and (b) GaSb on AlSb on Si. The reduction in the full width for the same growth thicknesses indicates that the GaSb layer is successful in suppressing the undulations in the AlSb*

shows that of GaSb growth on a 100 Å AlSb nucleation layer. The FWHM in part (a) is ~1700 arc seconds and that in part (b) is ~320 arc seconds. Undulations in semiconductors due to misfit dislocations have been noted and modeled by other researchers.[32]

2.4.2 *Antiphase domains (APDs)*

The diamond structure consists of two interpenetrating face-centered cubic lattices. The two sublattices differ from each other only in the spatial orientation of the four tetrahedral bonds that connect each atom to its four nearest neighbors, which are on the other sublattice. For example in Figure 2.8, the atoms with bond orientations indicated as "A" and "B" belong to sublattices "A" and "B." There is no distinction between the sublattices otherwise. Both are occupied by the same atomic species. In the zinc-blende structure in which both GaAs and AlSb crystallize, one of the sublattices is occupied by the group III and the other by the group V. In a crystal without antiphase disorder, the sublattice allocation is the same throughout the crystal. If this allocation changes somewhere inside the crystal, the interface between

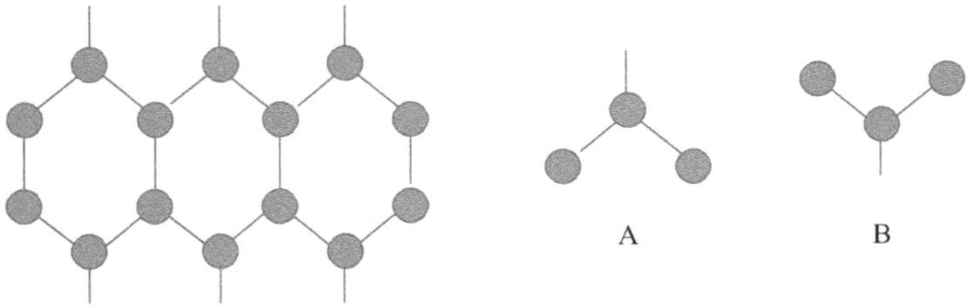

Figure 2.8 Interpenetrating sublattices

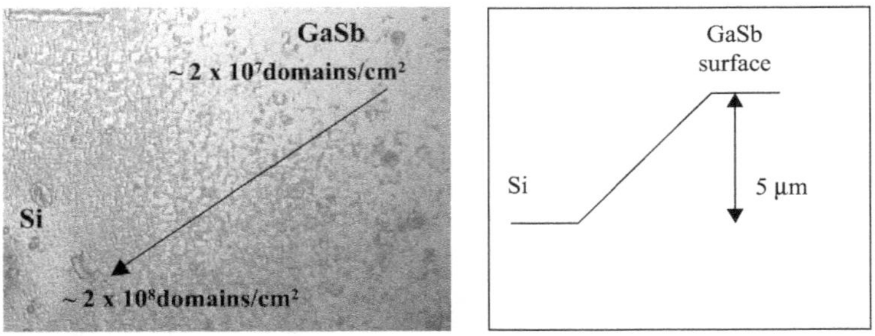

Figure 2.9 An etch gradient showing changes in the domain density from the Si surface to the top of a 2-μm epilayer. The schematic depicts the nature of the gradient

the domains with opposing sublattice allocation forms a two-dimensional structural defect called APB, and the domains themselves are called APDs.

Ideally if the silicon substrate could be manufactured without a single step in it, the problem of APBs and APDs would not exist. But since this is impossible to achieve and the problem is further accentuated by lack of Si homoepitaxy (resulting in rougher surfaces), extensive domain formation is observed in the growth of all III-Vs (and polar) alloys on Si.

The growth of AlSb on Si (100) results in domain formation. The schematic for the study as well as a Nomarski image of the end product is shown in Figure 2.9.

The image shows us that the domain density is $\sim 2 \times 10^7/\text{cm}^2$ at the surface and increases to $\sim 2 \times 10^8/\text{cm}^2$ at the growth interface. This reduction in the domain density with growth thickness is due to a process called domain annihilation, where two domains run into each other and one of them survives and the other does not.

Figure 2.10 shows a high-resolution AFM image of an AlSb/Si surface with domain formation. It also shows us that the domains themselves are extremely smooth surfaces.

The domains can however be eliminated by using a simple procedure. The key factor is having a substrate that contains only double steps. This can be achieved through the use of offcut Si substrates. When the growth is performed on Si (100), 2.5°–4°, a high density of double steps are formed in the Si surface. However, there

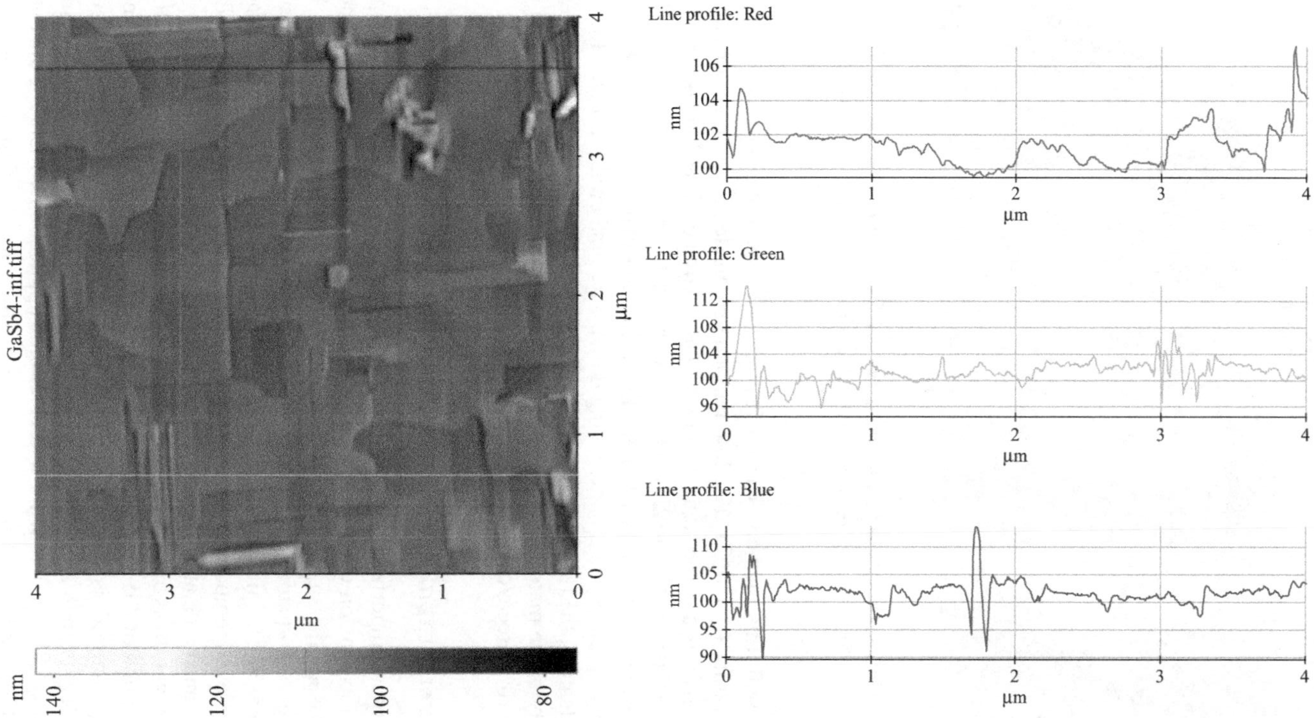

Figure 2.10 HR-AFM of domained GaSb/AlSb/Si surface

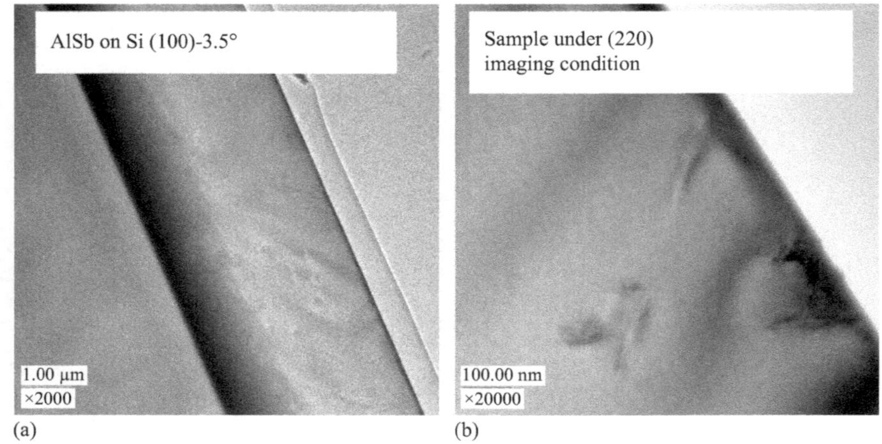

Figure 2.11 Dark-field (220) image of 3 μm of AlSb on Si (100), 3.5°, with no domains visible in the picture, but the TDD is 6 × 10⁸ defects/cm²

is one issue with this method. The nucleation has to be done on this substrate at a low temperature (~350°C). The low temperature nucleation allows for superior domain annihilation, and subsequent epilayers have extremely low domain densities.[33] The results are shown in Figure 2.11. The figure shows the growth of AlSb on Si (100), 4° substrate, which has resulted in practically no domains in the material. The problem however is that in the process of achieving domain-free material through a low temperature nucleation, the IMF array is not formed properly. This results in an increased defect density in the material. The best results for single-domain AlSb that we have seen are 6×10^8 defects/cm^2.

2.4.3 Thermal expansion coefficient

One of the most important problems in the growth of III-Vs on Si has been the huge thermal expansion coefficient mismatch between virtually every III-V and Si. Phosphide-based alloys such as GaP that are practically lattice matched to Si have not resulted in high-quality devices due to the fact that GaP has a thermal expansion coefficient that is 2.75 times that of Si. While an IMF layer may guarantee lattice matching at growth temperature, a mismatch in expansion coefficient may result in a severe lattice mismatch at room temperature.

Also, noteworthy is the fact the Si has a lower expansion coefficient than all III-Vs with the exception of AlSb, implying a net tensile strain in the material when cooled to room temperature. It is a known fact that mild tensile strain often results in highly detrimental results such as microcracks and threading dislocations resulting in the material quality being too poor for optoelectronic applications. Table 2.1 documents the various III-Vs and their thermal expansion coefficients. The first column contains thermal expansion coefficients of the alloys calculated through plasmon energies

Table 2.1 Thermal expansion coefficients for Si and selected III-V materials. Column 1 shows the values calculated by Kumar et al. and columns 2 and 3 show the experimental value at 300 K

	Calculated (10^{-6} K^{-1})	**Experimental[30] (10^{-6} K^{-1}) (RT)**	**Experimental[31] (10^{-6} K^{-1}) (RT)**
Silicon	2.58	2.5	
Ge	5.8		
AlP	5.161		
AlAs	3.756		3.5
AlSb	2.551	4.5	4.2
GaP	6.899	6.1	5.3
GaAs	6.928	7.2	5.4
GaSb	8.866	6.5	6.1

by Kumar *et al.*,[34] and the second and third columns contain experimental data by Neumann *et al.*[35] and Miauchi *et al.*[36] Kumar *et al.*'s calculations show that AlSb is the only III-V that comes close to being thermally matched to Si. Furthermore, AlSb has a smaller expansion coefficient than Si. This would result in a small compressive strain in the AlSb layer at room temperature not tensile. The high-bond strength of AlSb and the miniscule mismatch therefore ensure that the material is lattice matched at both growth temperatures as well as at room temperature.

2.5 GaSb membranes

An alternative way to integrate III-Sb materials with silicon substrates is by wafer bonding. For this the III-Sb layer must be isolated from its original substrate and then transfer to Si. The isolation of epitaxial layers from their substrate can be achieved by substrate removal or by epitaxial liftoff (ELO).

2.5.1 Substrate removal technique

The substrate removal technique involves the complete dissolution of the substrate with a selective etchant that does not react with the next layer called an etch stop layer. Subsequently, the etch stop layer must be removed with etchant solution that has a higher affinity for the etch stop layer than for the layers above.

The isolation of GaSb epilayers grown lattice matched to a GaSb substrate is rather difficult due to the lack of highly selective etchants in the GaSb system. GaSb can be selectively etched using an $InAs_{0.91}Sb_{0.09}$ etch stop layer with a CrO_3:HF:H_2O solution that has a selectivity of 100:1 for GaSb over the $InAs_{0.91}Sb_{0.09}$ etch stop layer.[37] However, CrO_3:HF:H_2O gives a nonuniform etch causing the GaSb substrate to etch faster in some areas than others.[38] This and the low selectivity of the etchant can lead to the possibility of etching through the etch stop layer and damage the

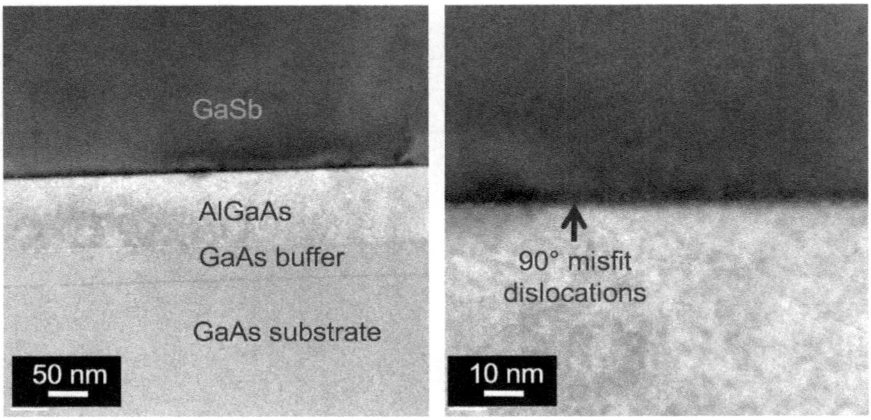

(a) (b)

Figure 2.12 Transmission electron microscopy (TEM) images: (a) Low-resolution TEM of structure epitaxial layers and (b) TEM image of the GaSb/AlGaAs interface

top membrane. To minimize this nonuniform etch problem, GaSb substrates are mechanically thin down to about 100 μm before an etchant solution is used. However, the selectivity of the etchant used to remove the etch stop layer is more critical than the selectivity of the etchant used to remove the substrate to avoid any etching in the layer of interest. In the GaSb system, the $InAs_{0.91}Sb_{0.09}$ etch stop layer can be removed with a $C_6H_8O_7$:H_2O_2 with a maximum selectivity of 127.[39] These selectivities for III-Sb compound semiconductors are low when compared with the very high selectivity of the etchants for GaAs systems.

The isolation of GaAs epilayers from their GaAs substrates can be achieved without difficulty. GaAs substrates can be removed with a variety of highly selective etchants using an AlGaAs or AlAs etch stop layer.[40,41,42,43] In addition, the etch stop layer can be removed with diluted hydrofluoric acid that has an extremely high selectivity of $\sim 10^7$:1 for AlAs over GaAs.

GaSb epitaxial structures can be isolated with fewer complications by growing them metamorphically on GaAs substrates and then using the highly effective GaAs-based etch chemistry.[44] To test such idea, three epitaxial structures had been studied. The first structure has an AlGaAs etch stop layer grown on the GaAs substrate followed by a thin layer of GaAs where the GaSb epilayer is grown. The second structure does not have the GaAs epilayer and GaSb was grown directly on the AlGaAs etch stop layer. The third structure has GaSb directly grown on the GaAs substrate without an etch stop layer. The IMF array was formed in all structures even in the second structure, where GaSb was directly grown on the AlGaAs etch stop layer as shown on Figure 2.12. The GaAs substrate is etched with a NH_4OH:H_2O_2 solution with a volume ratio of 1:33. Then, the AlGaAs layer is removed by dipping the sample in a NH_4F:HF 6:1 solution for 30 seconds. After substrate removal, the first structure is

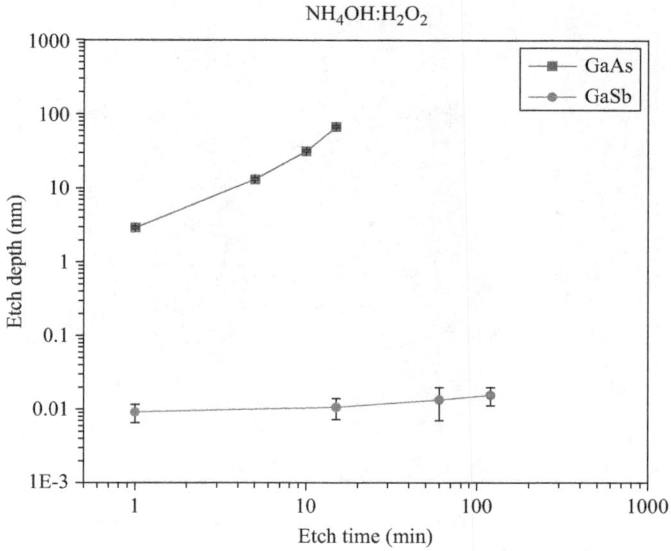

Figure 2.13 Etch depth measurements for GaAs and GaSb substrates as a function of etch time

left with a GaSb/GaAs membrane, and the second and third structures with a GaSb membrane. The isolation of the GaSb membrane from the third structure without an etch stop layer was possible due to significant $NH_4OH:H_2O_2$ etch rate differential between GaSb and GaAs as shown on Figure 2.13.[45]

2.5.2 ELO technique

The ELO process involves the lateral etch of a sacrificial layer with a highly selective etchant that enables the separation of the device from the substrate with minimal damage either. ELO studies have been restricted to GaAs and InP material systems. In the case of GaAs, the successful application of the ELO process is due to the extreme selectivity of dilute hydrofluoric acid between AlAs and GaAs.[46] In GaSb lattice-matched systems, the lack of highly selective etchants for antimonide-based materials makes it very difficult to achieve ELO. However, the ELO of GaSb membranes is still possible if grown metamorphically on GaAs substrates. The IMF growth mode and the extreme selectivity of HF for AlGaAs allow the liftoff of GaSb films from typical GaAs ELO structures. ELO of GaSb membranes from GaAs substrates is possible using the first and second structure described above. The first structure gives GaSb/GaAs membranes (Figure 2.14) where GaSb can be completely isolated by etching the GaAs layer with the $NH_4OH:H_2O_2$ etchant solution that has a minimum effect on GaSb material. The second structure gives a GaSb membrane. However, its surface morphology shows some oxidation due to the reaction of GaSb with hydrofluoric acid.[47]

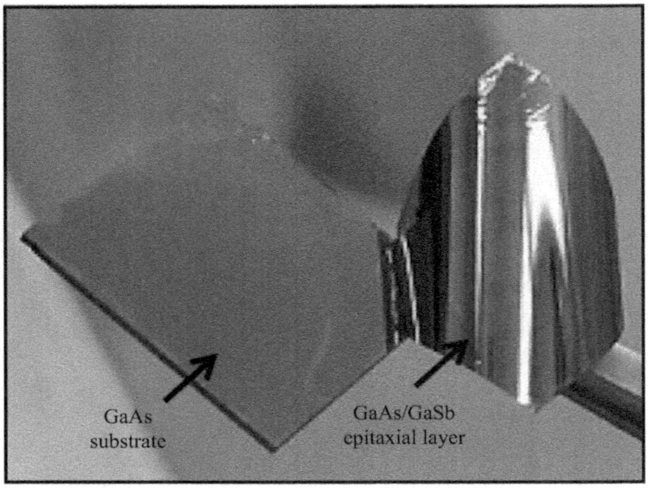

Figure 2.14 Image of the GaSb/GaAs membrane after etching the AlGaAs sacrificial layer

2.6 InAs and InGaSb channels on GaAs

The two materials of choice for creating channels on GaSb are InAs and pseudomorphic InGaSb for n and p channels, respectively. The biaxial strain in the pseudomorphic InGaSb channel layer is known to split the light and heavy hole subbands and causes the light hole band to rise above the heavy hole by thus significantly reducing the in-plane hole effective mass. This was first demonstrated in p-channel InGaAs/(Al)GaAs quantum wells.[48,49] Recently, this concept has been adapted in strained Ge layers on relaxed SiGe buffer layers and room-temperature mobilities greater than 2000 cm^2/V·s have been reported.[50,51] Hole mobilities of about 1230 cm^2/V·s have previously been achieved with 40-nm gate-length InSb HFETs.[52] The binary compounds GaSb and InSb are the only two III-V materials with higher hole mobility than Si, and therefore, the $In_xGa_{1-x}Sb$ alloy system is of interest. Furthermore, $In_xGa_{1-x}Sb$ could be used with AlSb barriers since the valence band offset will provide confinement for the holes, and varying the composition of InGaSb could control the compressive strain. P-channels based on $In_xGa_{1-x}Sb$/AlGaSb quantum wells have been demonstrated and excellent hole mobility as high as 1500 cm^2/V·s with a carrier concentration of 0.7×10^{12} cm^{-2} have been achieved.[53,54] These heterostructures are based on GaAs and involve growth of thick buffer layers to accommodate the 8% lattice mismatch between AlGaSb and GaAs. The use of the IMF technique does not involve growth of thick buffer-layers and hence could be used to fabricate thinner devices.

Pure InAs n-channels with nearly lattice matched GaSb or Al(Ga)Sb barriers have many desirable properties including high-electron mobility (30,000 cm^2/V·s at 300 K) and velocity.[55] The GaSb/InAs/GaSb quantum-well system is particularly interesting

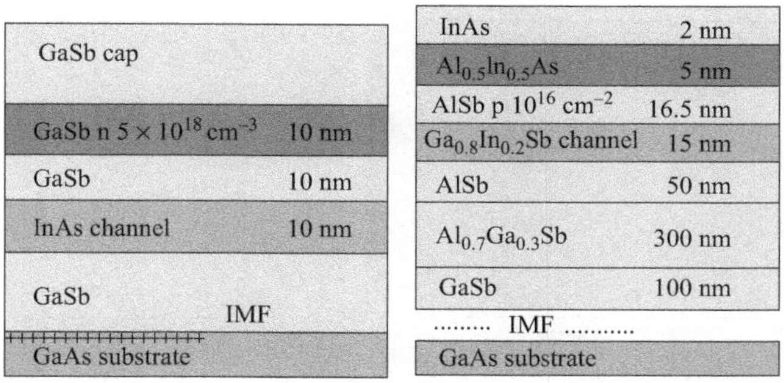

Figure 2.15 Growth structures for InAs based n-channel and InGaSb based p-channel on GaAs substrates

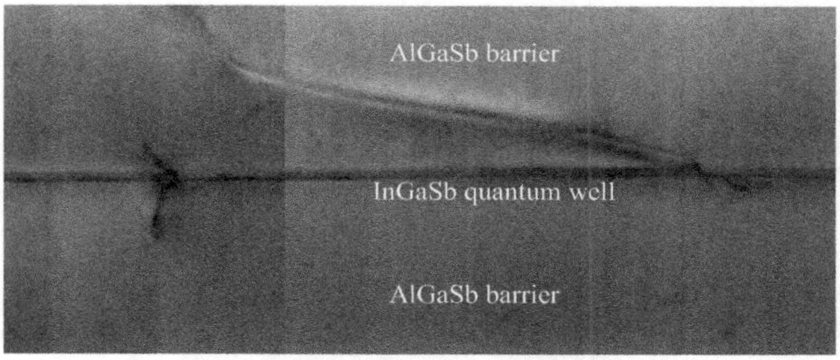

Figure 2.16 Cross-section TEM image of the InGaSb channels in GaSb barrier. The image shows the presence of residual threading dislocations in and around the channel

because of the band lineup between these two materials. Under the right conditions, the valence electrons in GaSb are transferred to the InAs layer. This would mean that the electrons and holes are spatially separated and coexist as two-dimensional gases in two separate layers. This type II band provides excellent confinement to the electrons and has been shown to exhibit high-electron mobilities (in excess of 10^5 $cm^2/V{\cdot}s$) at low temperatures.[56] This system could be incorporated on GaAs or Si using the IMF array.[57] Room temperature mobility of 13,900 $cm^2/V{\cdot}s$ and a peak mobility of 25,200 $cm^2/V{\cdot}s$ have been measured.

Therefore, InAs and InGaSb materials are ideal for the formation of n and p channels, respectively. Figure 2.15 shows InAs and InGaSb channel designs. In both designs, the channel is within 500 nm of the highly mismatched GaSb/GaAs interface. While the IMF growth method mitigates the effect of the mismatch and results in reduced TDD, there are still substantial residual threading dislocations in the channel (1×10^7–1×10^8 defects/cm^2).

In Figure 2.16, we can observe the InGaSb channel with several threading dislocations intersecting the layer. A higher resolution image captures a clear section of the channel showing the precise interfaces. The TDD in both samples with InAs

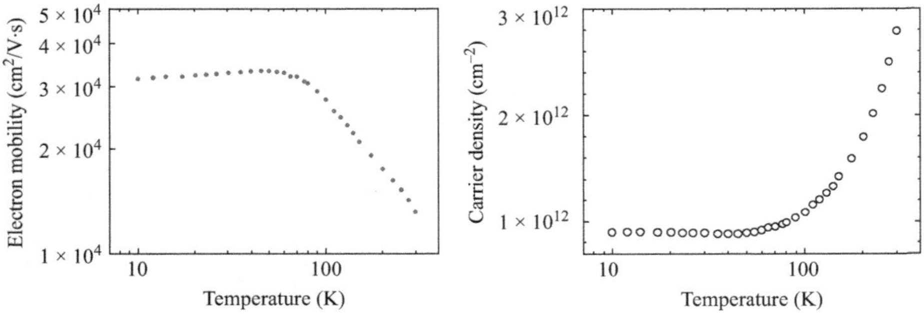

Figure 2.17 Plots showing electron mobility and the carrier density as a function of temperature for the InAs n-type channel

Table 2.2 Mobility and carrier density for n-type InAs channel

T (K)	Carrier density (cm^{-2})	Mobility ($cm^2/V \cdot s$)
300	2.79×10^{12}	13,230
77	9.74×10^{11}	31,130

Table 2.3 Mobility and carrier density for p-type InGaSb channel

T (K)	Carrier density (cm^{-2})	Mobility ($cm^2/V \cdot s$)
300	1.5×10^{12}	595
77	8×10^{11}	2850

and InGaSb channels are in the range of 1×10^7 to 1×10^8 dislocations per cm^2. The dislocation density is dependent on the quality of the IMF layer, and this can vary depending upon the growth parameters used.

The channel mobility is measured using Hall measurement using the Van der Pauw method. The samples have Indium contacts that are annealed at 280°C in nitrogen ambient. The sample size in both cases is 1 cm × 1 cm. The results for the InAs channel is shown in Figure 2.17. Figure 2.17 shows the electron mobility and the carrier concentration for the InAs channels as a function of temperature. The mobility for the electrons peaks at 77 K at a value of $\sim$32,000 $cm^2/V \cdot s$. The room temperature mobility of the channel is $\sim$13,000 $cm^2/V \cdot s$. These results can further be improved with the optimization of the delta-doped layer grown above the channel.

A comparison of the values for the InAs and the InGaSb channels is shown in Tables 2.2 and 2.3. As expected the n-type channel has significantly higher electron mobility at both 77 and 300 K compared to the p-type InGaSb channel. These are expected values for InGaSb and can also be improved by increasing the strain in the channel. The channel mentioned has 20% Indium; however, this can be increased to as much as 45% without significant deterioration of the semiconductor. The increased pseudomorphic compressive strain also leads to the splitting of the heavy hole–light hole bands and results in increased hole mobility.

2.7 Conclusions

III-Sb alloys represent an excellent choice for high mobility n and p channels on GaAs and Silicon substrates. High-quality low-defect density growth of antimonides on silicon can be achieved using an AlSb nucleation layer. A very thin AlSb layer (100 Å) nucleated on Si, which relieves almost the entire strain caused by the 13% lattice mismatch via a 2-D array of 90° misfit dislocations. These dislocations form at the III-V epi/silicon interface, propagate within that plane, and do not thread vertically into the material. This growth mode produces relaxed very low-defect density material as indicated by x-ray diffraction, TEM and etch pitch density measurements.

References

[1] A. Georgakilas *et al.*, 'Achievements and limitations in optimized GaAs films grown on Si by molecular-beam epitaxy'. *J. Appl. Phys.* **71**(6), 2679–2701 (1992).

[2] D. G. Deppe, R. F. Naresh Chand, J. P. van der Ziel, and G. J. Zydzik, 'Al_xGa_{1-x} As-GaAs vertical-cavity surface-emitting laser grown on Si substrate'. *Appl. Phys. Lett.* **56**, 740 (1990).

[3] D. G. Deppe, N. Holonyak Jr., D. W. Nam, *et al.*, 'Room-temperature continuous operation of p-n Al_xGa_{1-x} As-GaAs quantum well heterostructure lasers grown on Si'. *Appl. Phys. Lett.* **51**, 637 (1987).

[4] T. H. Windhorn, G. M. Metze, B-Y. Tsaur, and John C. C. Fan, 'AlGaAs double-heterostructure diode lasers fabricated on a monolithic GaAs/Si substrate'. *Appl. Phys. Lett.* **45**, 309 (1984).

[5] R. F. Naresh Chand, A. T. Macrander, J. P. van der Ziel, *et al.*, 'GaAs-on-Si: Improved growth conditions, properties of undoped GaAs, high mobility, and fabrication of high-performance AlGaAs/GaAs selectively doped heterostructure transistors and ring oscillators'. *J. Appl. Phys.* **67**, 2343 (1990).

[6] C. C. Phua, T. W. Chong, W. S. Lau, *et al.*, 'Application of semiconducting low temperature grown GaAs to improve laser diodes grown on Si substrates'. *Jpn. J. Appl. Phys.* **36**, 1888–1891 (1997).

[7] http://www.motorola.com/mot/doc/0/284_MotDoc.pdf.

[8] M. E. Groenert, C. W. Leitz, A. J. Pitera, *et al.*, 'Monolithic integration of room-temperature cw GaAs/AlGaAs lasers on Si substrates via relaxed graded GeSi buffer layers'. *J. Appl. Phys.* **93**, 362 (2003).

[9] J. W. Matthews and A. E. Blakeslee, 'Defects in epitaxial multilayers: I. Misfit dislocations'. *J. Cryst. Growth* **27**, 118 (1974).

[10] A. Yu Babkevich, R. A. Cowley, N. J. Mason, and A. Stunault, 'X-ray scattering, dislocations and orthorhombic GaSb'. *J. Phys. Condens. Matter.* **12**, 4747 (2000).

[11] A. Rocher, 'Interfacial dislocations in the GaSb/GaAs (001) heterostructure'. *Solid State Phenom.* **19/20**, 563 (1991).

[12] A. G. Baca Baca, F. Ren, J. C. Zolper, R. D. Briggs, and S. J. Pearton, 'A survey of ohmic contacts to III-V compound semiconductors'. *Thin Solid Films* **308**, 599 (1997).

[13] B. Berinder and D. Leonard, 'Spiral growth of GaSb on (001) GaAs using molecular beam epitaxy'. *Appl. Phys. Lett.* **66**, 463 (1995).

[14] H. Ruiting, S. Deng, L. Shen, *et al.*, 'Molecular beam epitaxy of GaSb on GaAs substrates with AlSb/GaSb compound buffer layers'. *Thin Solid Films* **519**, 228 (2010).

[15] E. A. Pease, L. R. Dawson, L. G. Vaughn, P. Rotella, and L. F. Lester, '2.5–3.5 μm optically pumped GaInSb/AlGaInSb multiple quantum well lasers grown on AlInSb metamorphic buffer layers'. *J. Appl. Phys.* **93**, 3177 (2003).

[16] S. H. Huang, G. Balakrishnan, A. Khoshakhlagh, A. Jallipalli, L. R. Dawson, and D. L. Huffaker, 'Strain relief by periodic misfit arrays for low defect density GaSb on GaAs'. *Appl. Phys. Lett.* **88**, 131911 (2006).

[17] A. Jallipalli, G. Balakrishnan, S. H. Huang, *et al.*, 'Structural analysis of highly relaxed GaSb grown on GaAs substrates with periodic interfacial array of 90° misfit dislocations'. *Nanoscale Res. Lett.* **4**, 1458 (2009).

[18] J. Tatebayashi, A. Jallipalli, M. N. Kutty, *et al.*, *Appl. Phys. Lett.* **91**, 141101 (2007).

[19] E. Plis, J. B. Rodriguez, G. Balakrishnan, *et al.*, 'Mid-infrared InAs/GaSb strained layer superlattice detectors with nBn design grown on a GaAs substrate'. *Semicond. Sci. Technol.* **25**, 085010 (2010).

[20] S. Huang, G. Balakrishnan, and D. L. Huffaker, 'Interfacial misfit array formation for GaSb growth on GaAs'. *J. Appl. Phys.* **105**, 103104 (2009).

[21] S. J. Brown, M. P. Grimshaw, D. A. Ritchie, and G. A. C. Jone, 'Variation of surface morphology with substrate temperature for molecular beam epitaxially grown GaSb(100) on GaAs(100)'. *Appl. Phys. Lett.* **69**, 1468 (1996).

[22] W. Qian, M. Skowronski, and R. Kaspi, 'Dislocation density reduction in GaSb films grown on GaAs substrates by molecular beam epitaxy'. *J. Electrochem. Soc.* **144**, 1430 (1997).

[23] T. W. Butler (No. USNA-E-69-1), Department of Engineering, Naval Academy, Annapolis, MD (1969).

[24] J. P. van der Ziel, R. J. Malik, J. F. Walker, and R. M. Mikulyak, 'Optically pumped laser oscillation in the 1.6–1.8 μm region from $Al_{0.4}Ga_{0.6}Sb/GaSb/Al_{0.4}Ga_{0.6}Sb$ double heterostructures grown by molecular beam heteroepitaxy on Si'. *Appl. Phys. Lett.* **48**, 454 (1986).

[25] C. A. Chang, H. Takaoka, L. L. Chang, and L. Esaki, 'Molecular beam epitaxy of AlSb'. *Appl. Phys. Lett.* **40**, 983 (1982).

[26] G. Balakrishnan, S. Huang, L. R. Dawson, Y. C. Xin, P. Conlin, and D. L. Huffaker, 'Growth mechanisms of highly mismatched AlSb on a Si substrate'. *Appl. Phys. Lett.* **86**, 034105 (2005).

[27] G. Balakrishnan, S. Huang, A. Khoshakhlagh, *et al.*, 'High quality AlSb bulk material on Si substrates using a monolithic self-assembled quantum dot nucleation layer'. *J. Vac. Sci. Technol.* B **23**, 1010 (2005).

[28] S. M. Koch, S. J. Rosner, Darrell Schlom, and J. S. Harris, 'The growth of GaAs on Si by molecular beam epitaxy'. *MRS Proceedings*, **67**, 37 (1986).

[29] F. E. Ejeckam, M. L. Seaford, Y. H. Lo, H. Q. Hou, and B. E. Hammons, 'Dislocation-free InSb grown on GaAs compliant universal substrates'. *Appl. Phys. Lett.* **71**, 776 (1997).

[30] M. L. Seaford, P. J. Hesse, D. H. Tomich, and K. G. Eyink, 'Strain relief by surface undulations of dislocation free MBE grown antimonides on compliant universal GaAs substrates'. *J. Electron. Mater.* **28**, 878 (1999).

[31] F. E. Ejeckam, C. L. Chua, Z. H. Zhu, Y. H. Lo, M. Hong, and R. Bhat, 'High-performance InGaAs photodetectors on Si and GaAs substrates'. *Appl. Phys. Lett.* **67**, 3936 (1995).

[32] F. Jonsdottier, 'Computation of equilibrium surface fluctuations in strained epitaxial films due to interface misfit dislocations'. *Mater. Sci. Eng.* **3**, 503 (1995).

[33] J. W. Lee, 'MBE growth of low dislocation and high mobility GaAs-on-Si'. *Mat. Res. Soc. Proc.* **67**, 29 (1986).

[34] V. Kumar and B. S. R. Sastry, 'Thermal expansion coefficient of binary semiconductors'. *Cryst. Res. Technol.* **36**, 6 (2001).

[35] H. Neumann, 'Trends in the thermal expansion coefficients of the $A^IB^{III}C_2^{VI}$ and $A^{II}B^{IV}C_2^V$ chalcopyrite compounds'. *Krist. Tech.* **15**, 849 (1980).

[36] N. Yamamoto, H. Horinaka, and T. Miyauchi, 'Temperature dependence of tetragonal distortion and crystal field splitting in $CuGaS_2$'. *Jpn. J. Appl. Phys.* **18**(2), 225 (1979).

[37] J. P. Perez, A. Laurain, L. Cerutti, I. Sagnes, and A. Garnache, 'Technologies for thermal management of mid-IR Sb-based surface emitting lasers'. *Semicond. Sci. Technol.* **25**, 45021 (2010).

[38] P. Y. Delaunay, B. M. Nguyen, D. Hofman, and M. Razeghi, 'Substrate removal for high quantum efficiency back side illuminated type-II InAs/GaSb photodetectors'. *Appl. Phys. Lett.* **91**, 231106 (2007).

[39] G. C. DeSalvo, R. Kaspi, and C. A. Bozada, 'Citric Acid Etching of $GaAs_{1-x}Sb_x$, $Al_{0.5}Ga_{0.5}Sb$, and InAs for Heterostructure Device Fabrication'. *J. Electrochem. Soc.* **141**, 3526 (1994).

[40] K. Kenefick, 'Selective etching characteristics of peroxide/ammonium-hydroxide solutions for $GaAs/Al_{0.16}Ga_{0.84}As$'. *J. Electrochem. Soc.* **129**, 2380 (1982).

[41] J. Novak, M. Morvic, J. Betko, A. Förster, and P. Kordoš, 'Wet chemical separation of low-temperature GaAs layers from their GaAs substrates'. *Mater. Sci. Eng.* B **40**, 58 (1996).

[42] H. J. Yeh and J. S. Smith, 'Integration of GaAs vertical-cavity surface emitting laser on Si by substrate removal'. *Appl. Phys. Lett.* **64**, 1466 (1994).

[43] M. Konagai, M. Sugimoto, and K. Takahashi, 'High efficiency GaAs thin film solar cells by peeled film technology'. *J. Cryst. Growth* **45**, 277 (1978).

[44] E. J. Renteria, P. Ahirwar, S. P. R. Clark, *et al.*, 'Isolation and characterization of large-area GaSb membranes grown on GaAs substrates'. *IEEE 39th Photovoltaic Specialists Conference (PVSC), 2013*, 2459 (2013).

[45] E. J. Renteria, A. J. Muniz, S. J. Addamane, D. M. Shima, C. P. Hains, and G. Balakrishnan, 'Isolating GaSb membranes grown metamorphically on GaAs substrates using highly selective substrate removal etch processes'. *J. Electron. Mater.* **44**, 1327 (2015).

[46] E. Yablonovitch, T. Gmitter, J. P. Harbison, and R. Bhat, 'Extreme selectivity in the lift-off of epitaxial GaAs films'. *Appl. Phys. Lett.* **51**, 2222 (1987).

[47] J. Fastenau, G. Tuttle, and F. Laabs, 'Epitaxial lift-off of thin InAs layers'. *J. Electron. Mater.* **24**, 757 (1995).

[48] G. C. Osbourn, 'Electron and hole effective masses for two-dimensional transport in strained-layer superlattices'. *Superlattices Microstruct.* **1**, 223 (1985).

[49] M. Jaffe, J. Oh, J. Pampulapati, P. Bhattacharya, and J. Singh, 'Experimental and theoretical-studies of carrier mass in pseudomorphic N-type and P-type type MOSFET with excess indium in the active channel'. *Inst. Phys. Conf. Ser.* **96**, 255 (1989).

[50] T. Irisawa, S. Tokumitsu, T. Hattori, K. Nakagawa, S. Koh, and Y. Shiraki, 'Ultrahigh room-temperature hole Hall and effective mobility in $Si_{0.3}Ge_{0.7}$/Ge/$Si_{0.3}Ge_{0.7}$ heterostructures'. *Appl. Phys. Lett.* **81**, 847 (2002).

[51] M. L. Lee, E. A. Fitzgerald, M. T. Bulsara, M. T. Currie, and A. Lochtefeld, 'Strained Si, SiGe, and Ge channels for high-mobility metal-oxide-semiconductor field-effect transistors'. *J. Appl. Phys.* **97**, 011101 (2005).

[52] M. Radosavljevic, T. Ashley, A. Andreev, *et al.*, 'High-performance 40 nm gate length InSb p-channel compressively strained quantum well field effect transistors for low-power (VCC = 0.5 V) logic applications'. *IEDM Tech. Dig.* 1–4 (2008).

[53] B. R. Bennett, M. G. Ancona, J. B. Boos, and B. V. Shanabrook, 'Mobility enhancement in strained *p*-InGaSb quantum wells'. *Appl. Phys. Lett.* **91**, 042104 (2007).

[54] H.-C. Ho, Z.-Y. Gao, H.-K. Lin, P.-C. Chiu, Y.-M. Hsin, and J.-I. Chyi, 'Device characteristics of InGaSb/AlSb high-hole-mobility FETs'. *IEEE Elect. Dev. Lett.* **33**, 964–966 (2012).

[55] Z. Dobrovolskis, K. Grigoras, and A. Krotkus, 'Measurement of the hot-electron conductivity in semiconductors using ultrafast electric pulses'. *Appl. Phys. A Mater.* **48**, 245–249 (1989).

[56] E. E. Mendez, L. Esaki, and L. L. Chang, 'Quantum Hall effect in a two-dimensional electron-hole gas'. *Phys. Rev. Lett.* **55**, 2216 (1985).

[57] D. Shima and G. Balakrishnan, *IEEE Computer Society Annual Symposium on VLSI*, 374–379 (2014).

Chapter 3
Anodic metal-insulator-metal (MIM) capacitors

D. Kannadassan[1], *Partha S. Mallick*[1], *and Maryam Shojaei Baghini*[2]

Metal-insulator-metal (MIM) capacitor is an important passive component in RF, analog and mixed signal (RF-AMS) circuits. It takes a large circuit area of integrated circuits (ICs) compared to other passive and active components. So miniaturization of MIM capacitors, along with transistors, has become essential in design and fabrication of future ICs. This has made a trend to design high capacitance density MIM capacitors with novel dielectric materials. In this regard, many works were carried out in fabrication of various nanostructured high-*k* dielectric MIM capacitors over the last decade. However, many of them suffered with structural defects, interface traps, and poor polarization process due to limitations of fabrication processes. The anodization process is an electrochemical oxidation of metals which had been demonstrated for the preparation of high-*k* dielectrics with improved crystalline properties, low structural defects, and improved ionic polarization. In this chapter, the fabrication and characterization of nanostructured anodic high-*k* MIM capacitors are presented. Many of these capacitors are meeting the requirements of International Technology Roadmap for Semiconductor with crystalline properties and improved ionic polarization.

3.1 Introduction

"Capacitor" is a significant and useful passive element in various system applications such as radio-frequency (RF), digital, analog and mixed signal (AMS) integrated circuits (ICs). Capacitors are often used for DC isolation, coupling, decoupling, and bypass in analog circuits. Few RF-AMS applications are shown in Figure 3.1(a–d). In these circuits, various sizes of capacitors take a large portion of IC area equal to that of transistors. In this regard, miniaturization of capacitors with high capacitance and low leakage current has become a challenge in IC fabrication. Between 1960 and 2000, many researchers developed various planar capacitors for wireless radio circuits. Mostly MOS capacitor were developed since the MOS fabrication methods

[1]VIT University, Vellore, India
[2]Indian Institute of Technology Bombay, India

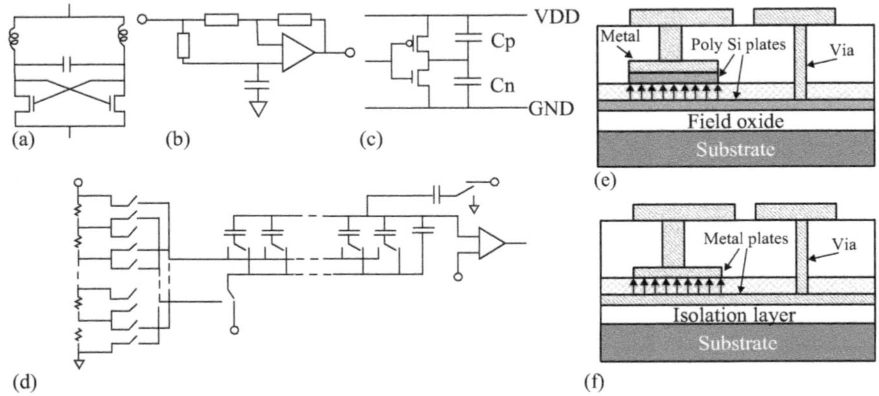

Figure 3.1 RF-AMS applications of capacitors: (a) cross-coupled LC oscillator, (b) phase-shift circuit, (c) decoupling capacitors, (d) analog–digital converters. Schematic cross-section views of (e) PIP and (f) MIM capacitors

were largely optimized and became the most successful technology. High-*k* materials, such as Si_3N_4, Al_2O_3, TiO_2, and Ta_2O_5, were predominately used in capacitors as insulators to increase the capacitance density.

In 1990s, the nanostructured thin film technologies were evolved in fabrication of single chip RFICs [1, 2]. On those days, MOS capacitor was the only candidate for AMS IC technologies. However, it had undesirable variations in capacitance with voltage while considering RF-AMS applications. This is due to the poor interface quality with high-*k* materials with polysilicon and semiconductor. These variations are tolerable in some applications, such as dynamic random-access memory (DRAM) and few analog applications. However, it cannot be acceptable in high precision circuits which have higher resolution of more than 10 bits, like digital signal processors, analog-to-digital (AD) and digital-to-analog (DA) converters [1, 2]. The precision is not about the value of capacitance, but it is a measure of sensitivity of capacitance with voltage, frequency, and temperature. This precision is measured using parameter "Voltage coefficient of capacitance" (VCC). VCC highly depends on the interface property of metal/insulator or polysilicon/insulator, thickness of dielectric layer, and quality of dielectric material. It is worth to note that these properties of dielectric thin film are largely affected by the fabrication method for various materials.

Polysilicon–insulator–polysilicon (PIP) capacitors replaced MOS capacitors in 1999 [3]. PIP capacitors were realized using low-pressure chemical vapor deposition (LPCVD) grown polysilicon layers and with high temperature deposition of dielectric layer, such as thermal oxidation. Capacitance density of $>5\ fF/\mu m^2$ and high breakdown field of $>2\ MV/cm^2$ were achieved with 45-nm thick SiO_2 dielectrics. However, PIP capacitors also suffered variation of capacitance with voltage due to the depletion effect at polysilicon–substrate interface and associated parasitic capacitance.

Onge *et al.* had reported VCC of $<2000\,ppm/V^2$ while MOS capacitors showed upto $30{,}000\,ppm/V^2$ [2]. With careful reduction of depletion effect, PIP capacitors had shown lower voltage linearity than MOS capacitors. However, the in situ doping of polysilicon and additional mask for etching had increased the manufacturing cost compared to CMOS technology. Though PIP capacitor technology was well established, it showed low quality factor (<50) with poor RF compatibility at very high frequencies. This is due to resistive losses at polysilicon plates, intraelement parasitic capacitance and lossy silicon substrate were the dominant weaknesses [2, 4].

Metal electrodes were used to replace the polysilicon in top and bottom contacts, particularly Pt or TiN [4], so they were called as "MIM" capacitors. It was constructed over top of metal lines to avoid series resistance and cross talk between silicon substrate. Cross-section views of typical PIP and MIM capacitors are shown in Figure 3.1(e) and (f), respectively. In Reference 1, the PIP capacitor had a thin dielectric film which was deposited between heavily doped polysilicon. These layers were placed over a thick field oxide (SiO_2). It was observed that increase in doping of polysilicon improved the VCC [4]. This is due to reduction of depletion at polysilicon–insulator interfaces and more metalic nature of heavily doped polysilicon. On the other hand, MIM capacitor with high-*k* dielectric layer is shown in Figure 3.1(f) which exhibit a very low dependence with voltage and frequency [5]. This is due to improved metal–insulator interface and field distribution of MIM capacitors lies within two metal electrodes.

3.2 MIM capacitor

MIM capacitors are usually fabricated as follows. First of all, the wafer or substrate will be cleaned using RCA technique. An insulating layer of SiO_2 of higher thickness is deposited using thermal oxidation. A bottom electrode of metal or metal-alloy thin film will be deposited using PVD or thermal evaporation. Nanostructured dielectric thin film will be deposited by unique deposition tools. High temperature annealing will be carried out to crystallize and reduce the oxygen vacancies in bulk dielectrics. After cleaning thoroughly using deionized water, top metal electrode will be deposited as similar to bottom electrode. Lithography and etching process will be carried out to pattern the top electrode area to form capacitor and to reach the bottom electrode. The selection of electrode metal, metal deposition technology, dielectric material, dielectric deposition technology, and thickness of each layer are considered based on the applications.

MIM capacitors were largely employed with conventional SiO_2 ($k = 3.9$) and Si_3N_4 ($k = 7$). Those capacitors have shown capacitance density of $\leq 2fF/\mu m^2$ with high VCC of $>100\,ppm/V^2$ and very low leakage current density of $<1\,nA/cm^2$ [2]. Although MIM capacitors show a stable characteristics compared to MOS capacitors, they occupy a large IC area in many RF-AMS applications to get high capacitance since they posses low capacitance density. At the same time, this area increases about five times for memory applications. This may lead to increase in noise, IC size, and fabrication cost. So miniaturization of MIM capacitors has become essential in IC technology which was recognized by International Technology Roadmap for Semiconductor (ITRS).

ITRS is a nonprofit organization which predicts the future scopes on the advancement of IC fabrication. It draws a map of future requirements in IC manufacturing and modeling. For AMS processing [6], MIM capacitors should hold a high capacitance density of $>5\ \mathrm{fF}/\mu\mathrm{m}^2$, low voltage linearity of $<100\ \mathrm{ppm}/\mathrm{V}^2$ and low leakage current density of $<10\ \mathrm{nA}/\mathrm{cm}^2$ in consideration of future requirements. On the other hand, ITRS restricts the maximum temperature of dielectric processing up to 400°C to make compatibility with backend fabrication processes [6]. This opens a challenge to IC designers and material scientists to develop low temperature deposition tools for nanostrucutred dielectrics and semiconductors. Many works were carried out on various high-*k* dielectrics over the last decade. Along with high capacitance density, many studies were dedicated to achieve low voltage linearity, low leakage current density, and improved reliability. However, many of them are facing problems with structural defects, interface traps, and poor polarization process due to limitations of fabrication process.

DRAM and AMS ICs need simple, low temperature, and low cost dielectric deposition technique. Various fabrication methods have been proposed in MIM capacitor technology to meet such requirements, such as atomic layer deposition (ALD), sol-gel, sputtering, thermal oxidation, anodic oxidation, and physical/chemical vapor deposition (PVD/CVD). Using these technologies, various dielectric structures, such as single layer, bilayer, and multilayer, were developed for MIM capacitors for the past 10 years. Most popular oxides, such as Al_2O_3, TiO_2, Ta_2O_3, and HfO_2, are extensively investigated. Some of the rare earth and ferroelectric dielectric materials are also receiving attention for high density MIM capacitors.

3.3 Anodization for nanoelectronics

Anodic oxidation or anodization is an electrochemical oxidation process which results in growth of low defect metal oxides. It can be performed using an anodizing cell which consists of a container with an electrolyte solution, anode, cathode, and DC/AC power supply. Since the nature of electrolyte, temperature, and voltage are highly influencing the anodization process, the necessary measurement and control setup can be added further. A typical anodization cell is shown in Figure 3.2(a). Anode is the metal to be oxidized and cathode should be a nonreacting metal or alloy. Cathode should be larger or equal as anode and insoluble in electrolyte. Anode and cathode are closely placed to form uniform field. Once the power supply is turned on, the electric field between anode and cathode initiated the electron transport from surface of cathode to anode via electrolyte medium. During this time, the oxygen ions (O^{2-}) migrate into anode metal and forms metal oxides.

After the development of electron microscopy, the porous and barrier type anodic structures are classified. While anodization, the barrier type oxide is formed if the resultant anodic oxide is insoluble in the electrolyte [7]. However, if resulting anodic oxide is soluble in electrolyte, it results in porous anodic structures due to nonuniform

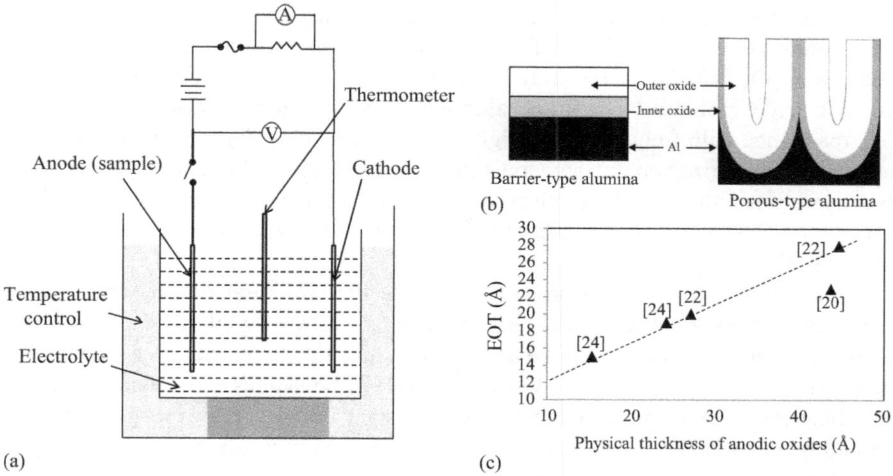

Figure 3.2 (a) Anodization cell and (b) schematic cross-section view of barrier type (left) and porous type (right) anodic oxides, (c) Achieved EOT (Å) in MOS capacitors with anodic oxides [20–23]

oxidation of metal. Cross-section schematic of these structures are shown in Figure 3.2(b). Barrier type and porous type anodic oxides have been intensively investigated for coloring of metal layers [8], surface protection [9], corrosion resistance [10], gas sensors [11], and thin film capacitors [12, 13]. In 1965, C. G. Thornton had reported the various new possibilities in microelectronics fabrication, among which anodization was promising for thin metal-oxide growth, particularly Al_2O_3 and Ta_2O_5 [14]. Later, few authors have reported fabrication and electrical properties of anodic oxides capacitors for microelectronics [13, 15–18]. Jawalekar *et al.* and Wyatt *et al.* were demonstrated the fabrication of Al_2O_3 and Ta_2O_5 capacitors using anodization, respectively [16, 19]. Those thin film capacitors showed less sensitivity of capacitance with frequency and temperature. It was reported that the anodization controls the thickness of dielectric layer precisely with low defect [16]. However, the limited number of materials can only be anodized. Also electrical breakdown of oxide during anodization was mentioned as a limiting factor [16].

A detailed study on incorporating anodic alumina in MOSFET fabrication was reported by M. B. Das *et al.* in 1976 [24]. It was concluded that anodization is compatible with existing fabrication methods such as thermal oxidation and CVD. Alternatively, Hwu *et al.* have demonstrated the importance of anodization/reoxidation in various dielectric material processing for microelectronics [21–23, 25]. They demonstrated that anodization can be used in MOS fabrication process for the deposition of ultra thin metal-oxides with EOT of 15–28Å as shown in Figure 3.2(c) [20–23]. These anodic alumina MOS capacitors show improved leakage characteristics compared to SiO_2 of same EOT. This is due to improved ionic polarization of anodic alumina which reduced the ionic and thermionic conduction of dielectrics. Few authors have reported the fabrication of MIM capacitors with anodic oxides, such as Al_2O_3 [18, 26] and Ta_2O_5 [27]. Hourdakis *et al.* were anodized Al using sulfuric acid which results in porous alumina. This capacitor shows a high sensitivity of capacitance with frequency and high VCC ($>$1000 ppm/V^2). This is due to instability in oxide formation during anodization which results in high defect density and noncrystalline alumina. Wet anodization of Ta_2O_5 results in barrier type oxide which shows a low leakage current density of $<$10 nA/cm^2 for 5 V and high breakdown field of $>$4.3 MV/cm^2 [27].

Annealing is a common process to reduce the structural defects and oxygen vacancies in the dielectrics. However, it may reduce the quality of metal electrodes or substrate. Anodic oxidation can be used to reoxidize the dielectric layer prepared by other techniques, such as ALD, PVD, and CVD. The breakages and structural defects in dielectrics can be removed at low temperature itself. However, the metal or substrate should be chemically independent of anodization, which needs suitable electrolyte. Anodization is one of the nonlithography techniques, which utilizes the masking cum chemical etching processes for micrometer level fabrications. This practice largely reduces the time and cost of IC fabrication.

Dielectric polarization processes involved in formation of capacitance for barrier type anodic oxides were reported by Kosjuk *et al.* [28]. It is understood that deformation and ionic polarizations are dominant in anodic oxides of Al, Ta, and Nb. It is also reported that anodic alumina shows a less sensitivity of capacitance with frequency compared to anodic Ta_2O_5 and Nb_2O_5 [28]. In this chapter, the anodization was utilized to prepare barrier type Al_2O_3, TiO_2 and bilayer of TiO_2/Al_2O_3 for the fabrication of MIM capacitors. Various studies, such as voltage linearity, leakage characteristics, and reliability are reported in detail. It was observed that anodization has large potential to solve many problems in nanoelectronics.

3.4 Anodic alumina MIM capacitors

Al_2O_3 is one of the attractive dielectric materials with wide bandgap of 8.3 eV. Its dielectric constant varies from 8 to 10 based on the fabrication process. Al_2O_3 MOS structure using ALD shows more than 10 years of lifetime at low voltage operation [29] with high breakdown field of 30 MV/cm [30]. Dielectric properties of Al_2O_3, such as leakage, dielectric relaxation, and reliability, were investigated by K. Allers *et al.* [31]. In which, Al_2O_3 shows more reliable and optimum performance compared to SiO_2 and Ta_2O_5, with low leakage current density [31]. ALD and thermal evaporation techniques were successfully demonstrated for Al_2O_3 MIM capacitors [5, 32]. Porous type anodization also have been used in fabrication of MIM capacitor which results in a high capacitance density of $>5\,fF/\mu m^2$ [17]. However, it shows ~40% reduction in capacitance value in the frequency range of 1 kHz to 1 MHz. Moreover, the capacitance–voltage variation coefficients are highly sensitive with frequency and temperature. This is the sign of thermal and frequency instability due to charge traps available at the metal–insulator interface.

The barrier type anodic oxide is a better solution to improve the capacitor's performance because of its crystalline and low defects [20]. Barrier type anodic γ-Al_2O_3 was obtained using various aqueous electrolytes, such as ammonium pentaborate (APB) dissolved in H_2O (bor-H_2O), sulfuric acid, APB dissolved in ethyline glycol (bor-gly), and citric acid by many authors [18, 24, 33]. It was observed that bor-gly solution results in low leakage and high effective barrier height compared to bor-H_2O [33]. Sato *et al.* have studied the effect of electrolyte on the crystalline properties of anodic alumina [34]. Also the bor-gly solution yields improved crystalline oxide than that of other electrolytes. In this work, we have fabricated the MIM capacitor with anodization on bor-gly electrolytes based on suitable approach.

3.4.1 Fabrication process flow and crystalline properties

Initially, the Si wafers (100) were cleaned thoroughly by regular RCA cleaning method. An isolation layer of SiO_2 was grown over Si substrate using wet oxidation. Over that an *Al* (99.99% pure) thin film of 300 nm was deposited by thermal deposition using tungsten filament at pressure of 2.5×10^{-5} Torr. At approximately 0°C, surrounded by ice bath, the Al film was anodized in a solution of APB dissolved in ethylene glycol (20 gl^{-1}) by platinum cathode of equal size as Pt anode in a constant current density of 0.5 mA/cm^2. The solution was prepared by adding 17 g of APB (99% pure) for every 100 ml of ethylene glycol [24]. To avoid the etching for bottom electrode, three quarters of the sample area was dipped in the electrolyte at constant voltage of V_A. Once cleaned thoroughly by deionized water, the 50 nm thick Al top electrode was deposited using thermal deposition with the shadow mask area of ~0.6 mm^2.

Anodization was performed for various anodization voltages (V_A) over anodization time (T_A). The thickness of anodic dielectric layer was measured using elipsometry test. The measured thickness of anodic Al_2O_3 thin film for various V_A and T_A for anodization current density of 0.5 mA/cm^2, shown in Figure 3.3(a). Rate of growth of anodic Al_2O_3 was found as 1.4 nm/V per minute. This rate is increased to 2.1 nm/V per minute for current density of 1 mA/cm^2. Figure 3.3(b) shows the SEM cross-sectional view of anodization region which confirms "non-porous" or "barrier type" anodic Al_2O_3. Figure 3.3(c) shows the X-ray diffraction (XRD) spectra as a function of scattering angle (2θ) of sample anodized at $V_A = 30$ V at current density of 0.5 mA/cm^2 [35]. The crystalline peaks at 46.2° and 67.7° are observed which confirms that the formed oxide is γ-Al_2O_3. The delamination or removal of oxide layer from metal is observed at higher voltage (>40 V), which affects the surface of dielectric layer and later deposition of top electrode.

3.4.2 Capacitance and voltage linearity

The capacitance and leakage current have been measured using HP4155C semiconductor parameter analyzer. Figure 3.3(d) shows measured capacitance–voltage (C-V) characteristics of MIM capacitors for various dielectric thicknesses (inset). It is found that the stability of capacitance with applied voltage improves with dielectric

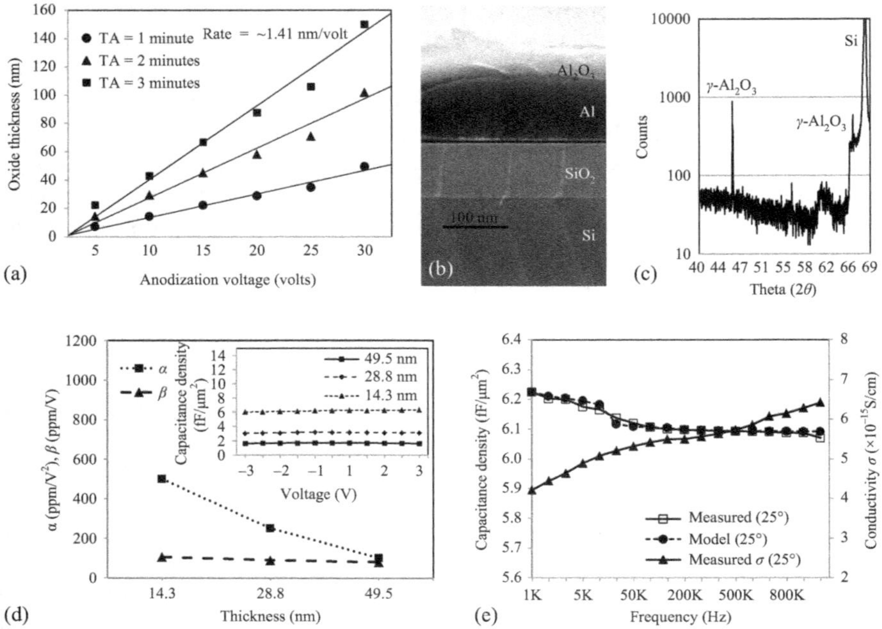

Figure 3.3 (a) Measured thickness of anodic Al_2O_3 thin film for various V_A and T_A, (b) SEM images of anodized samples—cross-sectional view of sample anodized at 30 V for 1 minute at 0.5 mA/cm^2, (c) XRD spectra of sample anodized at $V_A = 30$ V at current density of 0.5 mA/cm^2, (d) calculated voltage coefficients of capacitance from the measured C-V characteristics (inset) of MIM capacitors for various dielectric thicknesses, and (e) frequency dependence of capacitance and conductance of 14.3 nm thick sample at 25°C [35]

thickness. From the measured capacitance for various oxide thicknesses, it is observed that the linear relation between capacitance and thickness, $C = \varepsilon_0 \varepsilon_r A/d$, is valid up to 25 nm. At higher thickness, the stabilization of amorphous layer reduces the effective dielectric constant of anodic Al_2O_3. This is due to the immigration of boron ions into oxide layer near to top-electrode interface [33]. Variation of capacitance due to applied voltage and temperature were estimated by calculating the VCC and temperature coefficient of capacitance (TCC) [2].

$$\mathrm{VCC} = \left[\frac{C(V) - C_0}{C_0}\right] \times 10^6 \tag{3.1}$$

$$\mathrm{TCC} = \left[\frac{C(T) - C_0}{C_0}\right] \times 10^6 \tag{3.2}$$

In general, the VCC values are lower at higher thicknesses. The parabolic nature of VCC is often described by linear (β) and quadratic coefficients (β) of capacitance using the following equation,

$$C(V) = C_0 \left(\alpha V^2 + \beta V + 1\right) \tag{3.3}$$

Figure 3.3(d) shows the calculated α (ppm/V^2) and β (ppm/V) for various thicknesses at 100 kHz. It shows that the value of α decreases from 605 to 102 ppm/V^2 as thickness increases; however, β does not change significantly. This is due to the reduction of field intensity across the dipoles of dielectric layer at higher thickness. Figure 3.3(e) shows the frequency dependence of capacitance of 14.3 nm thick sample at 25°C and 125°C. The results show that the sensitivity to frequency variation is low compared to the earlier reports at 25°C [5, 17]. The stable nature of capacitance with voltage and frequency is due to the low defect density available at the bulk and near to the metal–insulator interface.

Beaumont and Jacobs developed the electrode polarization model which explains the dispersion of capacitance with input signal frequency [36]. This model is helpful to understand the dielectric polarization process and nature of induced and intrinsic defects. In many high-*k* oxides, oxygen vacancies are considered as dominant intrinsic defects and the density of defects depends on the oxide growth processes. These vacancies lead to localized conduction by hoping of electrons. When an AC signal is applied, the mobile charges of oxide form a double layer near to bottom electrodes. This double-layer is considered as injected free electrons from electrode or oxygen vacancies near interface [37]. For applied bias, the mobile charges are accumulated at a distance L_d from the electrode, called Debye length. This modulation of space charge region under the AC field is referred as "electrode polarization." According to this model, the capacitance is [37],

$$C = C_m \left[1 + \frac{A_c}{\omega^{2n} \tau^{2n}}\right] \tag{3.4}$$

where C_m is the capacitance for no electrode polarization (at zero bias voltage), expressed as $C_m = \varepsilon_0 \varepsilon_r S/L$, with top electrode area S and oxide thickness L. In (3.4), the slowly varying quadratic second term has $(\omega\tau)^{2n}$, called Jonscher response, with $0 < n < 1$. ω and τ are angular frequency of AC signal and relaxation time of oxide respectively. Parameters A_c and τ are expressed as,

$$A_c = \frac{2}{(2+\rho)^2} \frac{L}{L_d}, \tag{3.5}$$

$$\tau = \tau_0 \frac{1}{(2+\rho)} \frac{L}{L_d}. \tag{3.6}$$

In these (3.5) and (3.6), $\tau_0 = \varepsilon_0 \varepsilon_r / \sigma$ is intrinsic relaxation time and $L_d = (\varepsilon_0 \varepsilon_r k_B T / N_t q^2)^{1/2}$ is Debye length, where N_t is density of intrinsic defects and σ is conductivity of dielectric. ρ is called "blocking parameter" which is a measure of the electrode transparency. It is defined as $\rho = \alpha \nu (L/D) \exp(-E_i/k_B T)$, where α and

ν are hopping distance and hopping frequency normal to the interface, respectively, and D is the bulk diffusion coefficient. For strongly injecting contacts, like ohmic contacts, ρ tends to infinity which further gives $A_c = 0$ and $C \approx C_m$. This indicates that space charge is not formed at the metal–dielectric interface. In contrast, when the contact is not injecting any charges, A_c is very large and ρ is very small. This describes importance of the effect of space charge [37]. From References 36 and 37, a = 0.5 nm, $\nu = 10^{12}$ Hz and interfacial energy barrier for Al/Al_2O_3 $E_i = 0.98$ eV. The measured conductivity of the anodic oxide is shown in Figure 3.3(e). For this model, the best fit has been obtained by considering $N_t = 3.2 \times 10^{15}/cm^3$ and $n = 0.072 \pm 0.001$ at 25°C which yields $L_d = 1.1$ nm and $\rho = 3.4 \pm 2$. The model fits at $n = 0.09 \pm 0.001$ for 125°C for the same defect density [35].

Electrode polarization model gives a confirmation of low defect density ($\sim 10^{15}/cm^3$) at the bulk insulator. This is due to the anodization process which alters the atomic structure of oxide during oxidation process and makes denser and low defect dielectric layer. Such barrier oxide layer results stable frequency and temperature response of capacitance. The second term of the model refers the contribution of relaxation polarization during formation of capacitance. It is slowly varying in the order of $n = 0.072 \pm 0.001$ as frequency increases. This slow variation of capacitance indicates that the polarization process is dominated by ionic or displacement polarization rather than the relaxation polarization. This agrees to the results reported by L. M. Kosjuk *et al.* [28].

3.4.3 Leakage characteristics and conduction mechanisms

The measured leakage current density for three samples is shown in Figure 3.4(a). The sample of 49.5 nm thick anodic alumina shows a leakage current density of ~ 1 nA/cm^2 for applied voltages up to 5 V which is much lower than ITRS recommendation. The breakdown voltage obtained from the leakage characteristics is ~ 12.5 V for thickness of 14.3 nm. It is worth to note that this breakdown voltage value is close to that of Al_2O_3 MIM capacitor prepared using ALD [39]. According to Reference 40, this breakdown field of resulting oxide is due to the low defect density at the bulk.

The conduction in this barrier type anodic alumina is analyzed based on Schottky emission (SE), Poole–Frenkel (PF) emission, and trap-assisted tunneling (TAT) mechanisms. In Figure 3.4(a), the higher slope at very low voltage indicates the Schottky thermionic emission of electrons to the unoccupied defect or trap states near metal–insulator interface. Low-field current density is dominated by TAT mechanism of electrons, which depends on temperature, defect density, and trap well depth. High-field region of leakage characteristics is dominated by PFT which accounts the trapped electron enhanced from defect states to conduction states of the dielectrics. Transition from SE to TAT is observed by "1st knee" point. The knee point varies in magnitude with thickness of the oxide layer. This is due to the change in barrier height for various oxide thickness [41]. According to Morgan *et al.*, the barrier height of Al/Al_2O_3 interface is expressed as $\phi = q^2N_tL/2\varepsilon_0\varepsilon_r$. The effective barrier height of the bottom electrode decreases as the thickness increases. This indicates that the trap wall of oxygen vacancies near metal–insulator interface is deep. On the other

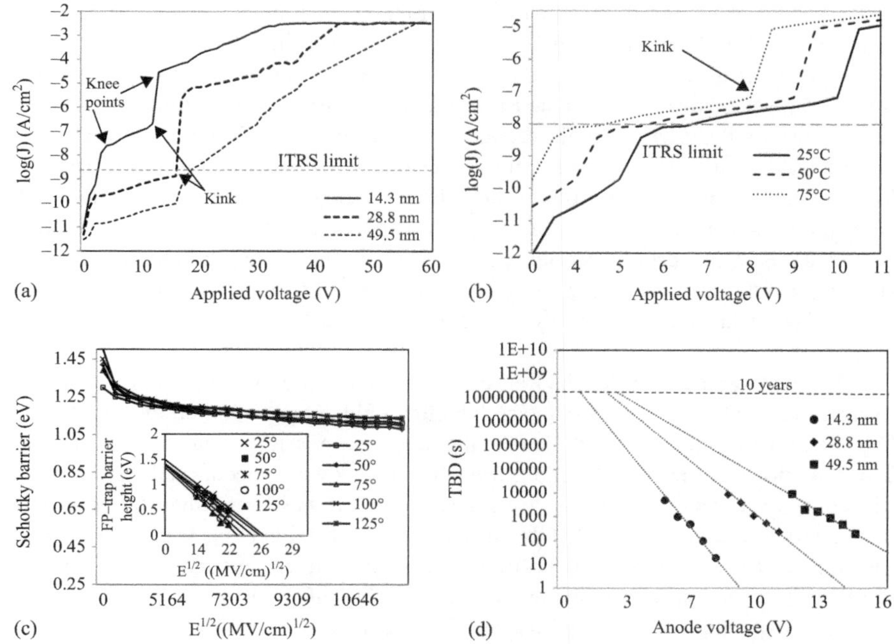

Figure 3.4 *(a) Measured leakage current density for three samples, (b) measured leakage current density for various temperatures (c) extracted barrier height and trap barrier height at various temperatures, and (d) measured time-to-breakdown for various dielectric thicknesses at room temperature [35, 38]*

hand, the "kink" point can be observed in transition of TAT to PFT mechanisms. The low-field region between the kink point from 1st knee point is almost a straight line with similar slope at all thicknesses. This indicates the uniform trap density and deep trap energy over bulk oxide. This is a useful feature for tunnel barrier structure.

The high fields are dominated by PF emission which is a hopping conduction of the trapped charges between trap potential wells. This hopping rate is further increased by applied voltage and temperature. The SE and PF emission current densities are expressed as References 42 and 43,

$$J_{SE} = A_R T^2 \exp\left\{-\frac{1}{k_B T}\left(q\phi_B - \beta_{SE}\sqrt{E}\right)\right\} \tag{3.7}$$

$$J_{PF} = CE \exp\left\{-\frac{1}{\xi k_B T}\left(q\phi_{PF} - \beta_{PF}\sqrt{E}\right)\right\} \tag{3.8}$$

where $\beta_{SE} = \left(q^3/4\pi\varepsilon_0\varepsilon_r\right)^{1/2}$ and $\beta_{PF} = \left(q^3/\pi\varepsilon_0\varepsilon_r\right)^{1/2}$ and C is pre-exponential factor. ϕ_B and ϕ_{PF} are PF trap energy and barrier height of dielectrics. It is observed that the best fit occurs at $\phi_{PF} = 1.47$ eV for the dynamic relative permittivity of Al_2O_3 ($\varepsilon_r = 3.25$ [42]) at higher fields. Trap barrier height is reduced nearly to zero for voltages at or above 2nd knee point, thus the charged (Coulombic) traps have no effect on the carriers [44]. This PF saturation dominates PFT at higher thickness which ensures the barrier height reduction of the metal–oxide interface for thicker anodic oxides. It's clear from the tunneling mechanisms that bulk oxide has very low and nonuniform defect profile. The Schottky emission at the very low field indicates the higher deep trap states or oxygen vacancies near metal–insulator interface. This ensures that the bulk barrier anodic oxide is highly crystalline, whereas the surface or outer layer is amorphous. The insolubility of inner layer and slight solubility of outer layer with electrolyte lead to such defect profile across the dielectric layer.

The leakage current density for various temperatures from 25°C to 75°C was measured and shown in Figure 3.4(b). The low fields are dominated by TAT and the high fields are dominated by PFT as per the model proposed by Atanassova *et al.* [45]. The transition from TAT to PFT is observed by the kink which occurs at different values of field strength as temperature varies. After 75°C, it was found that the knee point disappears due to dominant PFT for a wide range of fields [35]. It gives a physical meaning that the defects are deep trap energy and highly sensitive to temperature. Similar observations were made by others on ALD Al_2O_3 [39].

These speculations can be ensured from extraction of Schottky barrier height and trap barrier height of dielectrics from (3.7) and (3.8). Barrier height (ϕ_B) and trap barrier height (ϕ_{PF}) are extracted and shown in Figure 3.4(c) and its inset, respectively. Here the dielectric constant ε_r is assumed as 9. It is observed that the Schottky barrier height is $\sim$1.25 eV while extrapolating the calculated barrier height at high fields to zero field. This high barrier height is responsible for low leakage current density at low fields. Extracted trap barrier heights at high fields for various temperature are shown in inset of Figure 3.4(c). The intrinsic trap barrier height is obtained by extrapolating the linear fit to zero field which yields $\phi_{PF} = 1.47$ eV. This agrees with earlier results [46]. Moreover, the trap height reduces for increase in temperature with a rate of 0.063 eV/°C. The stable characteristics of Schottky and trap barrier heights with temperature is due to improved lattice arrangement during anodic polarization. This enhances the breakdown field strength of Al_2O_3.

Constant voltage stress (CVS) measurement is a useful tool to study the reliability behavior of devices. It can be carried out by measuring leakage current density as a function of stress time at a constant applied voltage. During this condition, the stress-induced traps create a platform for carrier transport. This leads to increase in current density of several decade compared to initial which causes electrical breakdown [47]. The time taken for this breakdown is called time-to-breakdown (TBD) which is one of the important parameters to assess the reliability and lifetime of capacitors. Measured CVS results of Al_2O_3 MIM capacitor are reported in Reference 38. TBD is measured at various CVS for different thicknesses of anodic alumina at room temperature and plotted in Figure 3.4(d). By extrapolating these measured values of TBD to 10 years line, it is found that anodic Al_2O_3 MIM capacitors can operate up to 10 years for the continuous stress of 2 V.

Table 3.1 A performance comparison of Al_2O_3 MIM capacitors formed using various dielectric deposition techniques (~15 nm)

Al_2O_3 MIM capacitors	**Thermal oxidation [5]**	**Atomic layer deposition [48]**	**Porous anodization [17]**	**Barrier type anodization [35]**
Capacitance density (fF/μm^2)	5	6.05	5.1	6.01
Leakage current density at 1 V (A/cm^2)	10^{-8}		10^{-9}	10^{-11}
Leakage current density at 2 V (A/cm^2)	10^{-7}	10^{-8}	10^{-9}	10^{-10}
Breakdown filed (MV/cm)		8.61	3.6	8.77
VCC (ppm/V)	>1000	795	>1000	400
Variation of capacitance (%) with frequency	10		40	6
TBD (s)		120		122

Table 3.1 specifies the performance of Al_2O_3 MIM capacitors using various dielectric deposition techniques. Thermal oxidation of Al shows low capacitance density and high leakage current due to high oxygen vacancies and incomplete oxidation of Al even at 400°C [5]. Also the fabricated porous anodic oxide MIM capacitor results in low breakdown field and high sensitivity to the frequency. This indicates the sign of high defect/trap density at the interface and bulk because of solubility in the electrolyte during anodization [17]. ALD and current work on barrier type anodic MIM capacitors exhibit excellent reliability and high capacitance density. However, the anodization provides best quality oxides at the lowest fabrication cost.

3.5 Anodic titania MIM capacitors

Titanium oxide or Titania (TiO_2) is used for variety of applications such as gas sensors, photovoltaic devices, and capacitors [49–51]. TiO_2 has three crystalline phases namely rutile, anatase and brookite, with dielectric constant of 40 to 170 and energy band gap of ~3.0 eV [52]. TiO_2 MIM capacitors have been fabricated using DC magnetron sputtering [53] and thermal oxidation [51]. Although these fabrication methods results a very high capacitance density, they suffer with high leakage current density ($>10^{-4}$ A/cm^2) and capacitance variation $\Delta C/C_0$ of more than 10^4 ppm [51, 53]. This is due to several reasons. The low band gap of TiO_2 gives a very low effective thickness between metal electrode which leads to a high rate of tunneling. Also the structural defects/traps available in bulk oxide and metal–insulator interface yield slow relaxation time of oxide, which show a fast reduction of dielectric constant with increase in signal frequency. In this section, the fabrication and characterization of the anodic TiO_2 MIM capacitors are explained. The leakage mechanisms, frequency dependency of capacitance, and structural properties of these capacitors are studied in detail.

3.5.1 Fabrication process, oxide formation, and crystallization

Rutile and anatase crystalline phases in barrier type anodic titania were obtained and studied by many authors using various aqueous electrolytes, such as APB (bor-H_2O), sodium tetraborate [54, 55]. Anodization of titanium using APB in ethylene glycol (bor-gly) electrolyte were demonstrated by a few authors [56, 57]. However, many authors have worked on anodization of aluminum with bor-gly electrolyte which results crystalline barrier oxide with low defect [33, 35]. We have already observed that bor-gly solution results low leakage and high effective barrier height compared to bor-H_2O [33]. This indicates that bor-gly electrolyte is a promising candidate for preparation of barrier type anodic oxides. In this work, the anodization of titanium using APB dissolved in ethylene glycol is demonstrated from MIM capacitors. However, controlling the oxidizing rate of TiO2 is challenging because the barrier height is low and highly leaky.

The Al/TiO_2/Al MIM structures were fabricated in the following manner. On the Si wafer (100), an isolation layer of 100 nm silicon dioxide was grown using thermal oxidation. Over that a Ti/Al bilayer of thickness 15/100 was deposited using electron-beam evaporator with tungsten filament at a pressure of 8×10^{-5} mBar. Here, bottom Al acts as bottom electrode and also controls the thickness of TiO_2 during anodization. Ti film was potentiostatically anodized in a nonaqueous solution of APB dissolved in ethylene glycol ($20\,gl^{-1}$) by the same size of platinum cathode. Preparation of the electrolyte was reported elsewhere [58]. Anodization was done for anodization voltages of 10, 15, and 20 V till the current density reduced to 1 $\mu A/cm^2$. The resulting TiO_2/Al samples were named as T1, T2, and T3, respectively. Only a three quarters of sample area was dipped in the electrolyte to avoid etching of bottom electrode. These samples were cleaned thoroughly by deionised water and dried. A 50 nm thick Al top electrode was deposited on the anodized samples using thermal evaporator with the shadow mask diameter of ~1 mm. Figure 3.5(a–c) shows the SEM cross-section view of all the three samples before top electrode deposition.

From SEM images, it is clear that the formed oxides are barrier type and uniformly thick. It is observed that the bottom Al electrode is also anodized at higher anodization voltages (15 and 20 V). This forms a thin amorphous layer of ~2 nm AlTiO (an alloy of both TiO_2 and Al_2O_3) which stops the further migration of oxygen ions into Al electrode. Figure 3.5(d) shows the depth profile of the T1 sample using secondary ion mass spectrometry (SIMS) in positive mode with 1 kVCs. This shows the distribution of Ti, Al, O, Si, Ti-O, and Al-O ions. It is observed that the count of Al-O and Al is significant near TiO_2/Al interface which forms the AlTiO composite layer in T2 and T3 samples (not shown, reported elsewhere [58]). The formation of AlTiO composite layer is formed due to outward migration of Al ions into TiO_2 region. However, Al ions rapidly decrease into TiO_2 region since it migrates slower than Ti ions [54]. This composite layer reduces the effective thickness of TiO_2

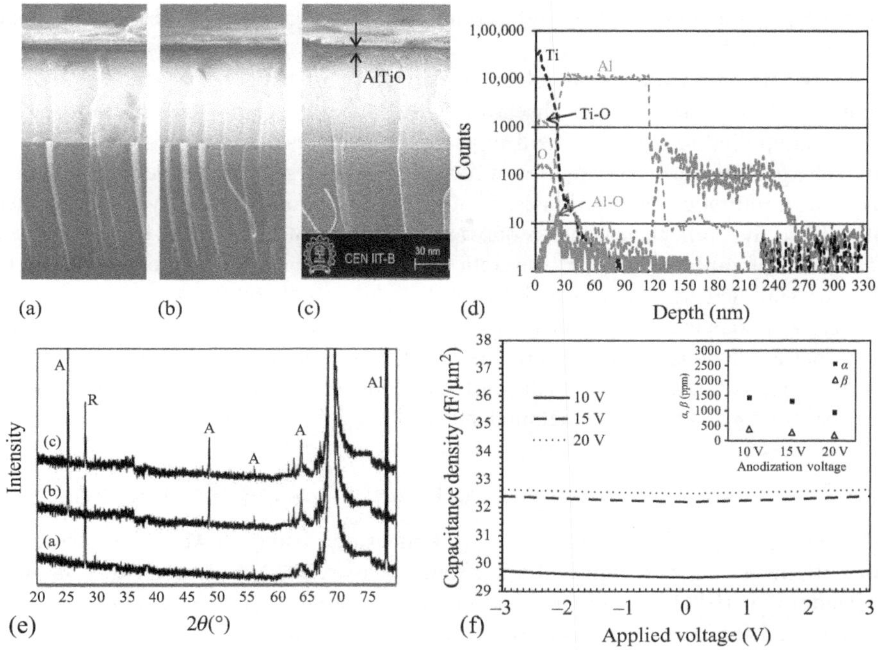

Figure 3.5 SEM cross-section image of anodized region before top electrode deposition, (a) AV = 10 V (b) AV = 15 V and (c) AV = 20 V, (d) SIMS depth profile of T1 sample, (e) XRD spectra of all samples (A: Anatase, R: Rutile), (f) Measured C-V characteristics of all TiO_2 capacitors at room temperature [58]. ©2013 Elsevier. Reprinted with permission from Reference 58

region. The XRD spectra of samples prepared at various anodization voltages are shown in Figure 3.5(e). The spectra shows that the prepared TiO_2 at lower anodization voltage has crystalline phases of dominantly rutile with anatase and partially amorphous. At higher anodization voltages, the amorphous state is transferred to crystalline/quasicrystalline state (anatase). Outer layer of anodic film has amorphous structure (~30%) which has been stabilized by "electrolyte-derived species" and a crystalline layer is available near bottom electrode [54].

Crystallization of anodic titania has been studied by many authors. Some of them reported that amorphous to crystalline transition occurs at low voltages (<10 V) [17, 55, 59]. H. Habazaki *et al.* observed that during anodization the amorphous phase has been transformed to anatase [17]. According to Vasil'eva *et al.*, borate and fluoride electrolytes have generally result in titania with rutile and anatase modifications, respectively [55]. Felshe *et al.* have reported that anodic titania shows rutile phases at low anodization voltages [59]. Unfortunately, there are inconsistencies found in the experimental conditions for TiO_2 crystallization such as electrolyte temperature, electrolyte composition, and applied anodization voltage [33, 54–57, 59].

Crystallization process involved in formation of barrier type anodic TiO_2 is addressed here based on the observations of Habazaki *et al.* and Piyus Kar [54, 60]. The applied anodization field impacts the near insulator–metal interface with high energy. This forms a thin layer of rutile titania with bulk defect/bubbles ($\sim$1 nm) by inward migration of O^{2-} ions [54, 60]. An amorphous titania layer was formed just above the rutile layer due to outward migration of boron ions [54]. After the formation of rutile, the inward migration of O^{2-} ion was suppressed by the high-field crystalline (high ionic resistivity) and slightly anodized the Al regions. At this case, amorphous to anatase phase transition occurs at defect sites above rutile region at higher anodization voltages. The lower activation energy of bulk defects aids the heterogeneous nucleation of anatase titania [60]. Readers are recommended to see References 54 and 60 for the detailed study of crystallization of titania.

3.5.2 Capacitance, voltage linearity, and leakage characteristics

The capacitance and leakage current were measured using HP4155C semiconductor parameter analyzer. Figure 3.5(f) shows measured C-V characteristics of MIM capacitors for 10 kHz at room temperature. It is observed that the capacitance density is increased about 2 $fF/\mu m^2$ for higher anodization voltage. This is due to improved crystalline property at bulk TiO_2. The thin AlTiO layer acts as interfacial layer between TiO_2 and Al bottom electrode which reduces the effective thickness of dielectric layer. This helps in formation of higher capacitance and reduction of leakage current [61]. Extracted α (ppm/V^2) and β (ppm/V) for various anodization voltages are shown in the inset of Figure 3.5(f). Value of α reduces from 1431 to 938 ppm/V^2 as anodization voltage increases from 10 to 20 V. High capacitance density and low VCC at high anodization voltage ensure the reduction of traps/defects at the bulk and metal–dielectric interface. The obtained α values are comparable to the earlier reports [17, 33, 35, 49, 62–64]. It is observed that VCC is inversely proportional to square of oxide thickness of MIM capacitor [65]. Therefore, one can reduce VCC and leakage current density as per recommendations of ITRS by increasing the thickness of anodic TiO_2 with capacitance density of >5 $fF/\mu m^2$.

Figure 3.6(a) shows the frequency-dependent capacitance for various anodization voltages at 25°C. It is found that the capacitance is less sensitive to frequencies after 100 kHz at higher anodization voltages; this indicates the stronger dipolar polarization is formed. Beaumont and Jacobs's electrode polarization model is used to explain the dispersion of capacitance with frequency which is described in earlier section. Parameters of modified Beaumont and Jacobs model [37] are obtained from measured capacitance C and considering $\alpha = 0.5$ nm, $\nu = 10^{12}$ Hz, L = 15 nm, $L_d \approx 0.9$ nm and interfacial energy barrier $E_i \approx 0.94$ eV for Al/TiO_2. The values of τ, N_t , and n are extracted for the best fit with measured capacitance of three capacitors. These values are presented in Table 3.2. It is observed that the defect density N_t and factor n are decreased as anodization voltage increases. Hence, sensitivity of the TiO_2 MIM capacitor to frequency reduces. Figure 3.6(a) illustrates the compatibility of the model and measured results.

Table 3.2 Measured and extracted parameters of anodic TiO_2 at various anodization voltages [58]. ©2013 Elsevier. Reprinted with permission from Reference 58.

AV* (volts) Sample name and L_{eff}**	Measured C ($fF/\mu m^2$) at 1 V and 10 kHz	Measured σ ($\times 10^{-12}$ S/cm) at 1 V and 1 MHz	Extracted model parameters				
			ρ	τ(s)	A	N_t cm^{-3}	n
10 (T1), ~17 nm	29.6	2.2	4.4 ± 2	3.72×10^{-5}	0.79 ± 0.2	6.6×10^{18}	0.2 ± 0.1
15 (T2), ~15 nm	32.2	0.97	6.1 ± 2	1.17×10^{-5}	0.57 ± 0.2	7.2×10^{17}	0.1 ± 0.1
20 (T3), ~14 nm	32.5	0.90	8.6 ± 2	1.01×10^{-5}	0.32 ± 0.2	3.1×10^{17}	0.1 ± 0.1

*AV – anodization voltage;
**L_{eff} – effective thickness of TiO_2

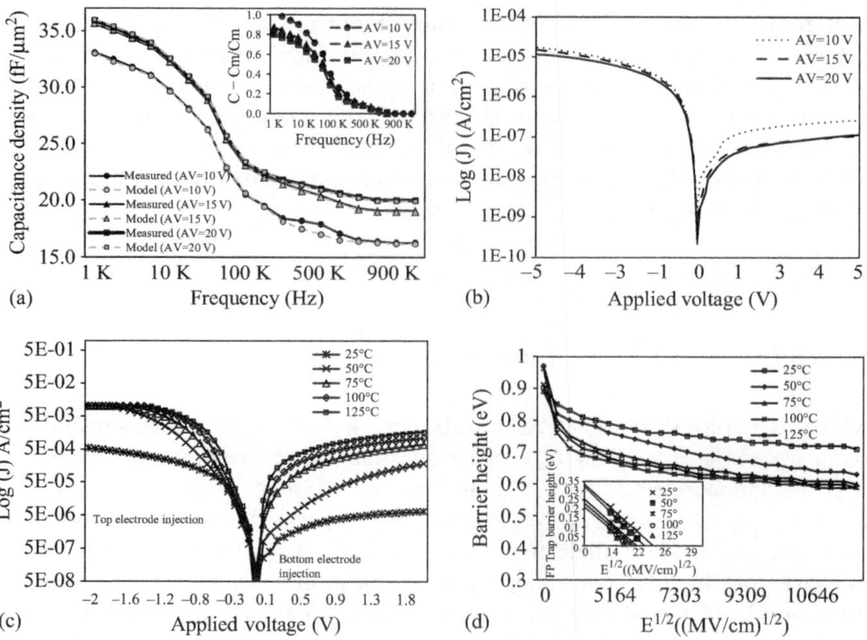

Figure 3.6 (a) Frequency-dependent capacitance for various anodization voltages at 25°C, (b) measured leakage current density of all the samples at room temperature, (c) measured leakage current density of T1 sample at various temperatures, (d) extracted barrier height and trap barrier height of T3 sample at various temperature [58]. ©2013 Elsevier. Reprinted with permission from Reference 58

According to Kosjuk *et al.* [28], the second term of the model in (3.4) refers the contribution of relaxation polarization in formation of capacitance. This term can be extracted using $(C - C_m)/C_m$ which is shown in the inset of Figure 3.6(a). This slowly varying term decreases in the order from 0.2 to 0.1. At lower anodization voltages, the relaxation polarization is dominant due to the presence of large defects and slow relaxation time. But the ionic polarization improves as amorphous state transferred to crystalline with low defect density at higher anodization voltages.

The measured leakage characteristics for forward and reverse biases are shown in Figure 3.6(b) for all the samples at room temperature. Figure 3.6(c) shows the measured leakage current density of sample T3 at various temperatures for different applied voltages. It is found that the I-V characteristics obtained is asymmetric which is due to nonuniform crystalline structure and anodization of Al (AlTiO). This crystalline profile results in different effective barrier heights at the top Al/TiO_2 and bottom TiO_2/Al interfaces. At the same time, the leakage current is more sensitive to temperature in forward bias than reverse bias. These properties of anodic TiO_2 are explained by conduction mechanisms such as SE and PF emission.

In Figure 3.6(b), a higher slope from 0 to 0.3 V indicates the SE tunnelling of electrons to the unoccupied defect or trap states near metal–insulator interface. Between 0.3 and 1 V, moderate fields are dominated by PF tunnelling emission mechanism, where the trapped electrons enhance from defect states to conduction states of the dielectric. Above 1 V, PF saturation is observed. This is because the trap barrier height is reduced to zero at higher fields, thus the charged (coulombic) traps have no effect on the carriers [44]. The expression for leakage current density due to SE and PF mechanisms are expressed in Reference 43 as,

$$J_{SE} = A_R T^2 \exp\left\{-\frac{1}{kT}\left(q\phi_B - \beta_{SE}\sqrt{E}\right)\right\} \tag{3.9}$$

$$J_{PF} = CE \exp\left\{-\frac{1}{\xi kT}\left(q\phi_{PF} - \beta_{PF}\sqrt{E}\right)\right\} \tag{3.10}$$

where Richardson's constant $A_R = 1200\,cm^{-2}\,k^{-2}$. C is the proportionality constant. ϕ_B and ϕ_{PF} are Schottky barrier and trap barrier heights, respectively, and E is the applied electric field. β_{SE} and β_{PF} are constants which are expressed as $\beta_{SE} = (q^3/4\pi\varepsilon_0\varepsilon_r)^{1/2}$ and $\beta_{PF} = (q^3/\pi\varepsilon_0\varepsilon_r)^{1/2}$, respectively, where ε_0 is permittivity of the free space and ε_r is dielectric constant of the insulator. The dielectric constant, ε_r can vary from 40 to 120 depending on crystalline property and structural defects. We have considered it as 60 due to the presence of amorphous layer and rapid degradation of capacitance with frequency. Barrier height ϕ_B is extracted as a function of applied electric field E for various temperatures using (3.9). Figure 3.6(d) shows the extracted barrier height for T3 sample (anodized at 20 V) at forward bias. As expected, a small difference of 0.05 eV in barrier height at bottom Al/TiO_2 and top Al/TiO_2 interface is observed at low fields. This confirms the formation of AlTiO at bottom–electrode interface.

While considering reverse bias, barrier height shows a large variation with increase in temperature at forward bias. This indicates that the crystalline TiO_2 region and interface is sensitive to temperature. This similar observation is reported for ALD of TiO_2 in MIM capacitor [66]. According to Hickmott, the effective tunnelling barrier of the empty traps has largely decreased due to increase in temperature [26]. PF tunnelling mechanism is used to extract the trap barrier heights at various temperatures. From the temperature-dependent leakage characteristics shown in Figure 3.6(c), the *ln* (J/E) versus 1/T plot is obtained which is used to extract the trap barrier height at specified fields. This is shown in inset of Figure 3.6(d). Now, the intrinsic trap barrier height is obtained by extrapolating the obtained curves to zero field [46]. Similar procedure is followed to extract trap barrier height in reverse bias (not shown, reported elsewhere [58]). It is observed from that intrinsic trap barrier height (at $E^{1/2} = 0$) of crystalline region (near bottom electrode) is largely degraded with increase in temperature. At reverse bias, the amorphous region shows a trap barrier height of ~0.15 eV which is stable with temperature. The partial crystalline structure of anodic TiO_2 and interfacial layer cause asymmetric leakage/breakdown at positive and negative half cycle. In a big wafer level anodization, there is a possibility of nonuniform oxide thickness due to unequal cathode size with wafer and/or nonuniform anodic field distribution. This may lead to undesirable effects in circuit level; however, it is purely a statistical impact. One can study the process variations in detail with respect to anodization voltage, electrolyte composition, and temperature.

3.6 Anodic bilayer MIM capacitors

The band gap of dielectric materials is inversely proportional to dielectric constant [67, 68]. Nobuyuki Mise *et al.* have reported in study of high-*k* MIM structures that the effective barrier thickness decreases with increment in permittivity of dielectric material [68]. This reduction in barrier height and effective thickness of dielectric layer will further lead to high leakage and poor reliability. Stack engineering of high-*k* dielectrics has emerged in fabrication of MIM capacitor to solve these issues. In general, the stack is made of a thin layers of large band gap dielectric material such as Al_2O_3, SiO_2 and very high dielectric constant material (ZrO_2, TiO_2, and HfO_2). Many authors have reported on various combination of bilayer stack MIM capacitors such as HfO_2/SiO_2, $HfTiO/Y_2O_3$, TiO_2/SiO_2, and $SrTa_2O_7/SrTiO_3$ [69–72].

Laminated HfO_2/Al_2O_3 stack MIM capacitors were reported by Ding *et al.* for RF applications [42]. These capacitors exhibit a capacitance density of more than $>4\ fF/\mu m^2$ with acceptable VCC of 200 ppm/V^2 at 1 MHz. It was found that the thickness of Al_2O_3 significantly reduces the VCC and sensitivity of capacitance with frequency. Along with HfO_2, many high-*k* oxides, such as $HfO_xC_yN_z$ [73], $LaAlO_3$ [74], were stacked. Of note, 3 nm Al_2O_3 on 40 nm Ta_2O_5 as dielectric stack in MIM capacitor was reported by Ishikawa *et al.* which showed a capacitance density of $>4.4\ fF/\mu m^2$ and VCC of 400 ppm/V^2 [75]. However, the capacitance density increased to $>9.2\ fF/\mu m^2$ with VCC of 3580 ppm/V^2 while the Ta_2O_5 thickness was reduced to 16 nm [75]. Here Al_2O_3 acts like a barrier layer which reduces the leakage current density. The Al_2O_3/Ta_2O_5 show a low leakage current density of $10^{-8}A/cm^2$ compared to HfO_2/Ta_2O_5. This is due to large band gap of Al_2O_3 compared to HfO_2.

Although, fabrication of multilayer stack of dielectric materials is regular practice in nanoelectronics, anodization of multilayer is not common. Few possible approaches can be adopted: (1) anodization of individual layer after deposition of each metal layer and (2) simultaneous anodization of thin film layers of metals. The former one is easier but time consuming and may not yield the expected stack. When the second metal layer is deposited on anodized first layer, the metal ion may migrate into anodized region which degrades the formation of stack. In Case 2, the electrolyte should anodize all metal layers simultaneously. Since, the top metal layer is oxidized first, it will block the migration of oxygen to second metal layer. However, on both the cases, rarely reports are available.

Anodic oxidation of superimposed multilayer metals was studied in detail by J. Perriere *et al.* about three decades ago [76–78]. In those works, Ta-Nb, Al-Ta, and Al-Nb couples were anodized and observed that the bottom metal-oxides are amorphous. Later, Thompson *et al.* have reported the anodization Al-Zr bilayer using 0.1 M APB solution at 25°C [79]. Yao Lei *et al.* have reported the fabrication procedure for TiO_2/Al_2O_3 dielectric stack using combined sol-gel and anodization processes [80]. Sol-gel-coated TiO_2 on aluminum foil has been anodized using aqueous ammonium adipate solution in 25°C. It is observed that at higher annealing temperatures (600°C) the crystalline transformation from amorphous to anatase in TiO_2 was observed [80]. This section presents the fabrication and characterization of a bilayer TiO_2/Al_2O_3 MIM capacitor using anodization process. These bilayer capacitors show a high capacitance density, low leakage current density, and low VCC which are close to ITRS recommendations for the year 2015. The formation of bilayer, crystalline properties, and conduction mechanisms are studied in detail. It is observed that the crystalline oxides and excellent polarization properties in dielectric stacks have improved the performance of MIM capacitors.

3.6.1 Fabrication process flow

A 100 nm SiO_2 was grown on two-inch Si (100) wafer. This sample was thoroughly cleaned by deionized water and dried with nitrogen gun. On SiO_2 isolation layer, a two thin layers of 15 nm Ti on 100 nm Al was deposited using electron beam evaporator with tungsten filament at a pressure of 8×10^{-5} mBar. This Ti/Al bilayer film was anodized using nonaqueous solution of APB dissolved in ethylene glycol (20 gl^{-1}), potentiostatically, by the same size of Pt cathode. Oxidation was performed for various anodization voltages of 15, 20, 25, and 30 V till the anodization current density reduces to 1 $\mu A/cm^2$. A three-quarters of sample area was dipped into electrolyte, so that we may avoid etching for bottom electrode. It was observed that a barrier type anodic TiO_2 and bilayer of TiO_2/Al_2O_3 were formed at low and high anodization voltages, respectively. After cleaned thoroughly by deionized water and dried, a 50 nm thick Al top electrode was deposited on the samples using thermal evaporation. A circular shadow mask area of $\sim$0.61 mm^2 was used to make the area of capacitors. Samples AT1 and AT2 are referring the single layer TiO_2 MIM capacitors whereas samples AT3 and AT4 refer bilayer TiO_2/Al_2O_3 MIM capacitors.

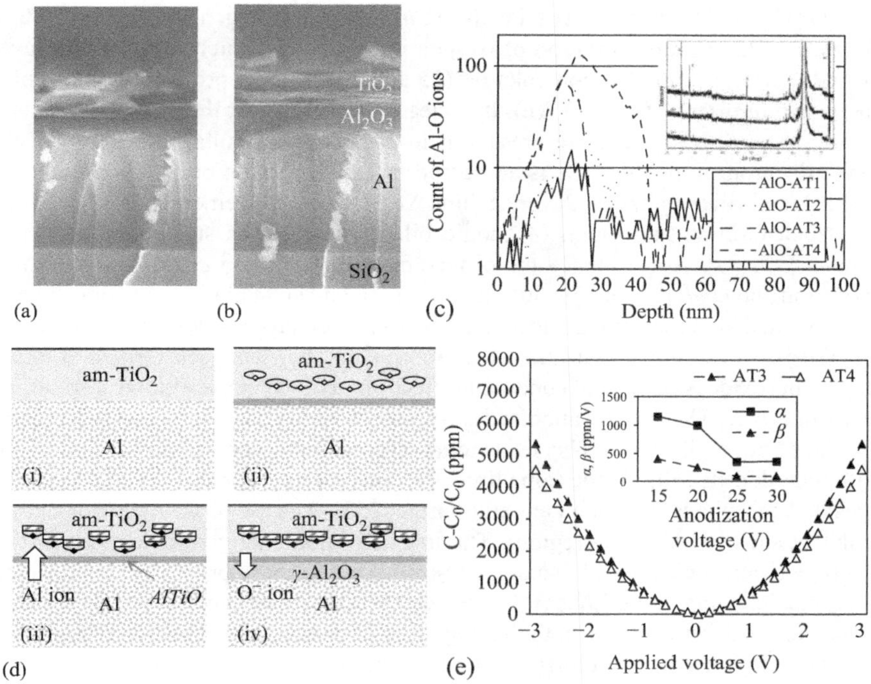

Figure 3.7 SEM cross-section images of anodized region of (a) AT3 and (b) AT4 samples before top-electrode deposition, (c) a comparison of Al-O ion distribution profile of all samples using SIMS and XRD spectra of all samples (inset), (d) schematic view of crystallization processes during anodization of Ti/Al layers, and (e) extracted $(C–C_0)/C_0$ plot from measured CV characteristics of bilayer MIM capacitors and calculated voltage nonlinearity coefficients (inset) [81]

3.6.2 Formation of bilayer and crystallization

SEM cross-section view of anodized regions of AT3 and AT4 samples are shown in Figure 3.7(a,b). The top Ti layer is anodized fully at low anodization voltages ($\leq$20 V) with a thin (<2 nm) interfacial layer of AlTiO. The bottom Al was also anodized at higher anodization voltages and formed Al_2O_3. Anodization voltage was restricted to 30 V due to delamination of TiO_2 from bottom Al_2O_3 at higher anodization voltage (AV > 30 V, not shown). The depth profile of all sample was measured using SIMS in positive mode with 1 kVCs and reported in Reference 81. Figure 3.7(c) shows the Al-O ion distribution profile from surface. It can be clearly observed that Al bottom electrode is slightly anodized at low anodization voltages (15 and 20 V). The outward migration of Al ion takes place at low anodization voltages and forms such a thin interface layer of AlTiO. Notably, the composite layer reduces the effective thickness of TiO_2 which shall help in formation of higher capacitance and reduction of leakage current. The AlTiO layer reduces the inward migration of oxygen ions into Al bottom electrode. The inward migration of oxygen ion increases which forms a thin layer of Al_2O_3 at higher anodization voltages (25 and 30 V). XRD profiles of all samples are shown in inset of Figure 3.7(c). It is clearly observed that the crystalline phases of TiO_2 anatase and rutile are present at low anodization voltages (<20 V) itself. Crystallization of anodic TiO_2 is addressed in earlier section on anodic titania. At higher anodization voltages, the crystalline Al_2O_3 (γ-Al_2O_3) emerges at $2\theta = 65.5°$.

Nucleation/crystallization of anodic bilayer oxides was studied by very few authors [79, 80]. According to the observations of Habazaki *et al.*, Kar, Shimizu *et al.*, and Yao *et al.* [54, 60, 79, 80], the formation and crystallation of anodic bilayer of titania and alumina are explained here. Step-by-step process flow is shown in Figure 3.7(d). During the anodization, the amorphous TiO_2 has been formed at initial stage. This oxide is capable of conducting electrons and oxygen evolution at solution–oxide interface. The applied anodization voltage/field impacts at the metal–insulator interface with high energy. This leads to transformation of amorphous to rutile TiO_2 with bulk defects [60]. At the same time, the outward migration of boron ions stabilizes the amorphous TiO_2 at outer oxide surface [54]. Anatase phase transition occurs at defect sites (above rutile region). The low activation energy at defects helps this heterogeneous nucleation of anatase TiO_2[60]. During these processes, anodization of Al results a thin amorphous AlTiO which reduce further evolution of oxygen. At higher anodization voltages, both the migration of oxygen and evolution of electrons into Al region increase and forms a thin layer of crystalline Al_2O_3 near TiO_2/Al_2O_3 interface. For voltages greater than 30 V, the density of defect sites at TiO_2/Al_2O_3 interface increases rapidly and forms local cavities. This ruptures the oxide due to pressure within defects and delaminate TiO_2 from surface of Al_2O_3.

3.6.3 Capacitance, voltage linearity, and leakage characteristics

The capacitance and leakage current density were measured using semiconductor parameter analyzer (HP4155C). Figure 3.7(e) shows the variation in capacitance from the zero bias case from the measured C-V characteristics. It was observed the capacitance density of AT3 is $>10.32\ fF/\mu m^2$ [81] which is three times lower compared to TiO_2 MIM capacitors ($>30\ fF/\mu m^2$). Formation of bilayer TiO_2/Al_2O_3 at higher anodization voltages reduces the capacitance density. Inset of Figure 3.7(e) shows the extracted α (ppm/V^2) and β (ppm/V). As anodization voltage increases from 15 to 30 V, α reduces from 995 to 150 ppm/V^2. This is due to low defect and low dielectric constant alumina which plays a dominant role in formation of capacitance.

The leakage characteristics of bilayer MIM structure was measured by injection of electrons from top and bottom electrodes. Figure 3.8(a) shows the measured leakage current density as a function of applied voltage AT3 and AT4 samples at room temperature. It is observed that both samples show a high degree of asymmetry at forward and reverse biases. While comparing the leakage current density of TiO2 capacitors, AT3 and AT4 have several decade lower leakage due to formation of AlTiO interfacial layer and TiO_2/Al_2O_3 stack. Conduction mechanism of MIM structure is analyzed using SE, PFE, and TAT which are discussed in the previous section.

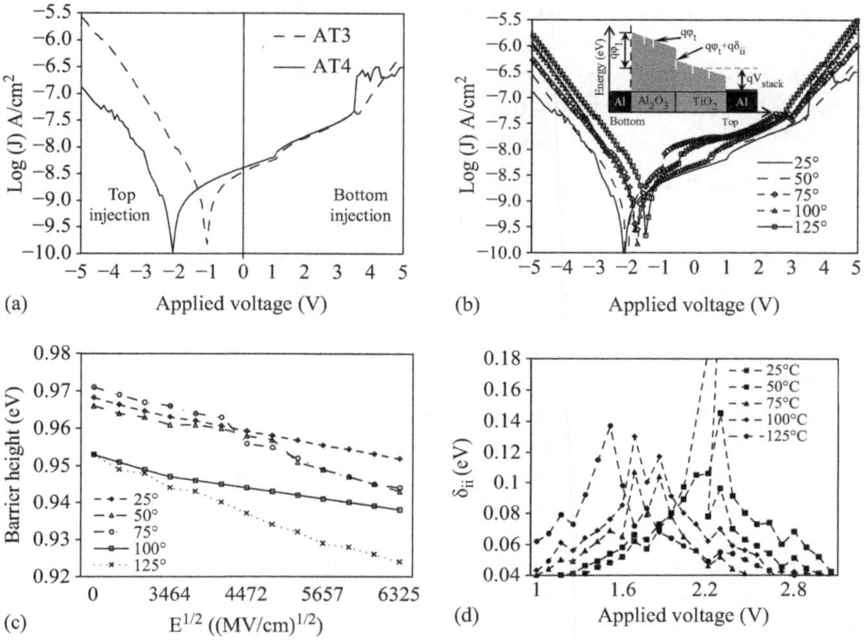

Figure 3.8 (a) Measured leakage current density of all samples at room temperature, (b) measured leakage current density of AT4 sample at various temperature, (c) extracted barrier height at various temperatures of AT2 in forward bias, and (d) extracted interface trap barrier heights of AT4 sample, from difference of extracted trap heights in forward and reverse biases [81]

Figure 3.8(b) shows the measured current density of the sample AT4 as a function of applied voltage at various temperatures. At low field ($-1\,V < V_{bias} < 1\,V$), the current density is less than 10 nA/cm^2, and no significant variation is observed due to change in temperature. This shows that the emission of electron is blocked by large effective barrier height which is independent of temperature. Each J-V characteristic in sample AT4 exhibits a sharp transition or kink at low fields ($-1\,V < V_{bias} < 2.5\,V$). It is also observed that the kink is occurring at lower bias voltages for higher temperature. This shows the presence of positive traps at the interface of TiO_2/Al_2O_3. At high fields ($|V_{bias}| > 2.5\,V$), the leakage increases rapidly and varies with temperature.

Figure 3.8(c) shows the extracted Schottky barrier height for AT4 at forward bias. A difference of ~0.25 eV is observed in AT4 compared to TiO_2 samples that of AT2. This is due to outward migration of Al into TiO_2 region. Variation in extracted barrier height at forward bias is less significant with temperature compared to reverse bias. This is due to the presence of barrier type anodic Al_2O_3 which shows stable barrier height and traps which are insensitive to temperature. Inset of Figure 3.8(b) shows the model band diagram to explain these barrier and trap barrier heights at dielectrics.

The kink at low fields due to positive traps at the interface of TiO_2/Al_2O_3 is explored using TAT mechanism. Houssa *et al.* used TAT model which has not included the insulator–insulator interface trap [82]. In this study, a term δ_{ii} has been introduced to specify the trap barrier height of insulator–insulator interface traps. Therefore the modified TAT model can be expressed as,

$$J_{TAT} = A_T N_t \exp\left[\left(qV_{stack} - \phi_1 + \phi_2 + \phi_t - \delta_{ii}\right)/k_B T\right] \quad (3.11)$$

Here ϕ_1 and ϕ_2 are barrier height at Al/TiO_2 and Al_2O_3/TiO_2 interface, respectively. ϕ_t is trap barrier height at TiO_2 region. The new term δ_{ii} specifies the positive traps at insulator–insulator interface with negative sign. If ϕ_1 and ϕ_2 are known, the ϕ_t and δ_{ii} can be extracted using (3.11) from leakage characteristics of AT4 (Figure 3.8(b)). The barrier heights used are $\phi_1 = 3.29$ eV and $\phi_2 = 2.29$ eV [26]. Since δ_{ii} has no effect in forward bias (Figure 5.10(a)), it is considered as zero. This results the barrier height of traps (ϕ_t) available at anodic TiO_2 region. In reverse bias, the extracted trap barrier height shows a large difference at kink points. From the difference of trap barrier heights in forward and reverse bias, the relative trap barrier height δ_{ii} values are plotted as a function of voltage and shown in Figure 3.8(d). It is clear that relative barrier depth of traps at Al_2O_3/TiO_2 interface decreases with temperature. Figure 3.9(a,b) compare the leakage characteristics and VCC of various stacked MIM capacitors in terms of capacitance density with ITRS recommendations. It is observed that the samples AT3 and AT4 show high capacitance density and low VCC comparable with other bilayer MIM capacitors [6, 69–72]. The structures with SiO_2, such as HfO_2/SiO_2 and TiO_2/SiO_2 show a very low VCC (<100 ppm/V) due to the canceling effect.

3.7 Modeling of high-*k* MIM capacitors

Capacitance–voltage characteristics and voltage linearity of MIM capacitor are important performance parameters for design of AMS ICs. Achieving low VCC is still a challenge to meet ITRS recommendations. However, the origin of such nonlinear behavior of capacitance with voltage is not clearly understood. Some authors attempted to model the voltage nonlinearity using orientation polarization [61], electrode polarization [37], electrostriction [83], and ionic polarization [84]. However, most of them are either complex or biased to particular materials. For instance, Phung *et al.* demonstrated the modeling of negative VCC for SiO_2 MIM capacitors using orientation polarization of polar dielectrics [61]. This model cannot be used for nonpolar dielectrics, such as Al_2O_3 and Ta_2O_5. Also the dipole moment of O-Si-O bond was calculated using complex techniques. Kim *et al.* reported that the VCC is a linear function of temperature [70]. However, the models mentioned above did not considered the effect of temperature.

In this section, a generalized model of voltage nonlinearity for MIM capacitors is presented using microscopic and macroscopic ionic polarizations. The model was verified with fabricated MIM capacitors with low and high dielectric constant materials such as Al_2O_3 and TiO_2, respectively, at various temperatures. The model maps

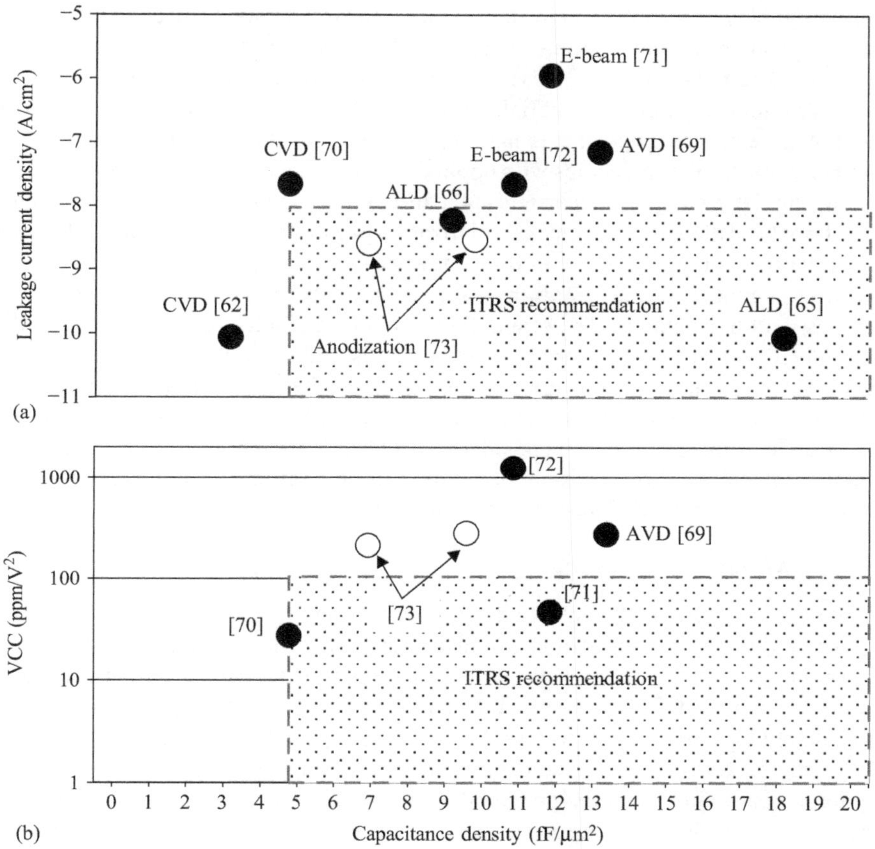

Figure 3.9 Comparing performances of various dielectrics stack MIM capacitors with ITRS recommendations. (a) Leakage current density at 1 V and (b) voltage linearity at 100 kHz

the dependable elements of voltage linearity, such as dielectric thickness and dielectric constant, to meet the recommendations of International Technology Roadmap for Semiconductor (ITRS) [6].

3.7.1 Modeling the voltage linearity

Dielectric materials are largely influenced by their polarization properties with applied electric field. In MIM capacitors, the formation of capacitance is due to various polarization mechanisms, namely electronic polarization (P_e), ionic polarization (P_i), orientation polarization (P_o), and space charge polarization (P_{sc}). The total polarization of dielectric layer can be expressed as $P = P_e + P_i + P_o + P_{sc}$. Among these, ionic and electronic polarization are almost independent of applied field. Electronic polarization is due to deformation of electrons in molecules which lead to dipole formation. For the applied field, the induced shift of cations (metal) of molecule with respect to neighbor atoms (oxide) of dielectric layer causes induced dipole moment. The polarization of these induced dipoles are called ionic polarization. It occurs for ionically bonded materials (Al_2O_3, HfO_2, and Ta_2O_3), sometimes called as nonpolar dielectrics. Dipolar/orientation polarization is due to polarization of inherent/permanent polar molecules with applied dielectric field. For the applied AC field, the accumulation of charges/carriers at the insulating boundaries (interface and interface traps) is called as space charge polarization or interfacial polarization. Electrode polarization is highly sensitive to frequency because of its time dependency.

In this model, the susceptibility for ionic and electric polarizations are considered as field independent. The orientation of induced dipoles for the applied field is considered as macroscopic case where the field is uniform throughout the dielectric layer. However, according to Lorentz approach, the field cannot be uniform within the dipole molecule or a cluster of dipole molecules. For this case, the internal field or local field inside dipole is considered which orients the electronics charges of dipole molecules. This electronic orientation of charges are treated as microscopic case which is known as Clausius–Mossotti model.

3.7.2 Macroscopic model

In dielectric materials, the voltage dependency of dielectric constant is based on the ionic polarization of induced dipoles and orientation polarization of permanent dipole. The polar dielectric materials have permanent dipoles which offer the polarization due to the applied field. The paraelectric materials are also called nonpolar oxides since they do not have permanent dipoles. Most common paraelectric materials are Al_2O_3, TiO_2, and Ta_2O_3. The metal and oxygen atoms of paraelectric materials are coupled by ionic bond. This metal-oxygen bond is distorted by the applied field which alters the inter-atomic distance between metal and oxygen atoms. This distortion lead to the formation of induced dipole in such dielectrics [85]. Distortion of metal ions (cations) in Al_2O_3 and induced dipole for the applied field are shown in Figure 3.10(a).

In macroscopic scale, the polarization of nonpolar dielectrics can be modeled using the orientation of induced dipole for the applied field. If angle between the induced dipole $\mu = \alpha_{ie}$ E and the applied electric field E is θ, then the average dipole moment can be expressed as $M = N_{di}\, \alpha_{ie}\, \overline{\cos^2\theta}\, E_{loc}$ [86]. Here α_{ie} is the internal electronic polarizability which is a microscopic quantity. Local electric field E_{loc} within the dipole sphere changes with respect to the position in dielectric layer from electrode. This is expressed as $E_{loc} = \lambda E$, for the applied external field E and field correction factor λ. According to Onsager's model, the field correction factor was derived as $\lambda = 3\varepsilon_r / [2\varepsilon_r + 1]$ based on the interaction of neighboring molecules which are affecting the polarization [86]. The average value of $\cos^2\theta$ can be obtained using Boltzmann statistics as,

$$\overline{\cos^2\theta} = \frac{\int_0^\pi \exp\left(\frac{\alpha_{ie}\cos^2\theta\, E_{loc}}{k_B T}\right)\cos^2\theta\, \sin\theta\, d\theta}{\int_0^\pi \exp\left(\frac{\alpha_{ie}\cos^2\theta\, E_{loc}}{k_B T}\right)\sin\theta\, d\theta} \tag{3.12}$$

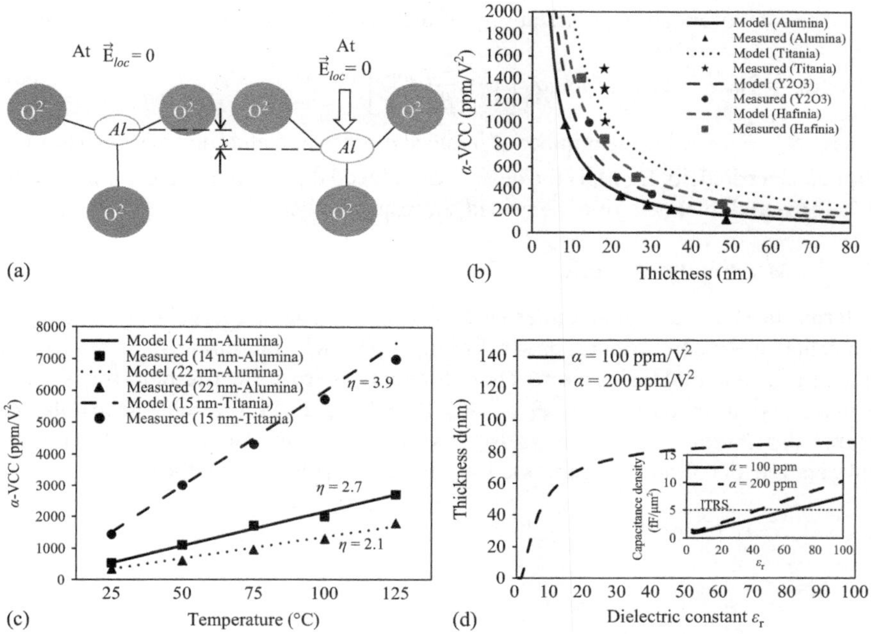

Figure 3.10 (a) Al_2O_3 molecule at equilibrium and distortion for the applied field, (b) measured and modeled quadratic coefficient of capacitance α of various MIM capacitors [35, 58, 83], (c) measured and modeled quadratic coefficient of capacitance α of various temperature, and (d) required thickness of dielectric to meet the ITRS recommendations for various dielectric constants [88]. ©2014 Elsevier. Reprinted with permission from Reference 88

This can be reduced to 2nd order Langevin function with axial symmetry [87],

$$\overline{\cos^2\theta} = \mathrm{L}_2(\beta) = 1 - \frac{2}{\beta}\mathrm{L}(\beta) \tag{3.13}$$

where,

$$\mathrm{L}(\beta) = \frac{e^{\beta} + e^{-\beta}}{e^{\beta} - e^{-\beta}} - \frac{1}{\beta} = \coth\beta - \frac{1}{\beta} \tag{3.14}$$

$\mathrm{L}(\beta)$ is referred as Langevin function [87] where $\beta = \frac{\mu \mathrm{E_{loc}}}{\mathrm{k_B T}}$. Therefore, the average dipole moment can be expressed as,

$$\overline{\mathrm{M}} = \mathrm{N_{di}}\alpha_e \mathrm{L}_2(\beta)\,\mathrm{E_{loc}} = \mathrm{N_{di}}\alpha_e\left[1 - \frac{2}{\beta}\mathrm{L}(\beta)\right]\mathrm{E_{loc}} \tag{3.15}$$

The total orientation polarizability of induced dipoles is $\alpha_{oi} = N_{di}\frac{\overline{M}}{E}$. Therefore,

$$\alpha_o = N_{di}\alpha_e L_2(\beta) = \lambda N_{di}\alpha_e\left[1 - \frac{2}{\beta}L(\beta)\right] \tag{3.16}$$

where, $N_{di} = n_{di}$ Ad is the number of induced dipoles in dielectric layer with uniform applied electric field $\overrightarrow{E}$, n_{di} is the density of induced dipoles per cubic volume, *A* and *d* are area and thickness of dielectric layer, respectively.

3.7.3 Microscopic model

External field and associated internal field within dipole are largely influencing the molecular electronic orientation. In Lorentz approach, the dipoles are modeled as sphere of a few molecules/atom. Then, the total internal electric field for isotropic materials is expressed as $E_{int} = E_{ext} + E_{pol}$ [85]. Here, E_{pol} is the field due to polarized charge distribution which is expressed as $E_{pol} = \frac{P}{3\varepsilon_0}$ with polarization P. The total polarization per dipole is expressed as, $P_{total} = \varepsilon_0\alpha\ E_{total}$ [85], therefore,

$$P_{ie} = \varepsilon_0\ N_{di}\ \alpha_{ie}\left(E + \frac{P_{ie}}{3\varepsilon_0}\right) \tag{3.17}$$

Also,

$$P_{ie} = \frac{\varepsilon_0\ N_{di}\ \alpha_{ie}\ E}{1 - N_{di}\alpha_{ie}/3} \tag{3.18}$$

where α_{ie} is internal/induced electronic polarizability of charges within the dipole sphere. Since $P = \varepsilon_0(\varepsilon_{r-1})E = \varepsilon_0\ \chi_{ie}\ E$, this equation can be written as,

$$\alpha_{ie} = \frac{3\varepsilon_0}{N_{di}}\left(\frac{\varepsilon_r - 1}{\varepsilon_r + 2}\right) \tag{3.19}$$

This is called Clausius–Mossotti equation. This equation relates the macroscopic element ε_r with microscopic element α_{ie}. Using this microscopic and macroscopic models of ionic polarization, the total polarization can be expressed as,

$$P = \left(\varepsilon_0\chi_e + N_{di}\left[\frac{3\varepsilon_0}{N_{di}}\left(\frac{\varepsilon_r - 1}{\varepsilon_r + 2}\right)\right]L_2(\beta)\right)E \tag{3.20}$$

The overall permittivity of dielectric layer of thickness d can be expressed in term of applied bias V (=Ed). The second-order Langevin function can be approximated to $L_2(\beta) \approx \frac{\beta^2}{15}$ for $\beta << 1$ or $\lambda\mu E << k_B T$ [87],

$$\varepsilon(V) = \varepsilon_0 + \varepsilon_0\chi_e + N_{di}\left[\frac{3\lambda\varepsilon_0}{N_{di}}\left(\frac{\varepsilon_r - 1}{\varepsilon_r + 2}\right)\right]^3 \frac{E^2}{15(k_B T)^2} \tag{3.21}$$

This equation can be compared with empirical relation, $C(V) = C_0(\alpha V^2 + \beta V + 1)$, where C_0 is the capacitance at zero bias. Therefore, the quadratic (α) coefficient of capacitance is,

$$\alpha \approx 1.8\left(\frac{\varepsilon_0}{N_{di}k_B T}\right)^2\left[\lambda\left(\frac{\varepsilon_r - 1}{\varepsilon_r + 2}\right)\right]^3 \frac{1}{d^2} \tag{3.22}$$

Here, N_{di} is the number of induced dipoles and ε_r is the static dielectric constant of material. The model shows that α is inversely proportional to square of the thickness which shows a good agreement with the model proposed by Wenger *et al.* and Phung *et al.* [61, 83]. It also shows a large dependence with ε_r, which indicates that higher dielectric constant materials shall lead to large α. Equation (6) shows that the α is inversely proportional to temperature, but it is observed that the α shows a linear relation with temperature T in many recent works [83, 70]. This is due to increase in induced dipole moment with temperature which can be introduced as $\overline{M}(T) = \overline{M}\eta T$. Here η is the increment factor of linear relation for the temperature T. This yields,

$$\alpha \approx 1.8\left(\frac{\varepsilon_0}{N_{di}\,k_B}\right)^2\left[\lambda\eta\left(\frac{\varepsilon_r - 1}{\varepsilon_r + 2}\right)\right]^3\frac{T}{d^2} \tag{3.23}$$

The polarization due to permanent dipoles of paraelectric materials can be modeled in accordance with Reference 61. Therefore, we can introduce the orientation polarization $P_o = N_{pd}\,\mu_d\,L(\beta)$ in equation 6.2.10 as described in Reference 61, where N_{pd} is the number of permanent dipoles and μ_{pd} is the dipole moment. It is observed that $L(\beta)$ converges to unity much faster than $L_2(\beta)$ [87] which indicates that the polar dielectric material show a high α than that of paraelectrics.

3.7.4 Model verification

Quadratic coefficient of capacitance α is extracted from measured C-V characteristics using empirical relation $C(V) = C_0(\alpha V^2 + \beta V + 1)$. For both TiO_2 and Al_2O_3 MIM capacitors, the modeled and extracted α are plotted as a function of dielectric thickness in Figure 3.10(b). Measured values of α of HfO_2 and Y_2O_3 MIM capacitors for various thicknesses are available in Reference 83, which is also included in Figure 3.10(b) to validate the model. Fitting parameters of the model for all materials are presented in Table 3.3. Al_2O_3 shows low α and good agreement with measured data. The crystalline state and strong ionic bond of anodic Al_2O_3 also support in reduction of α. It is observed that for the anodic TiO_2 capacitor, with same thickness of ~15 nm, the α decreases for higher anodization voltages due to transformation of amorphous to crystalline state which reduces the dependency of field. HfO_2 and Y_2O_3 MIM capacitors show a good fit with our model. Figure 3.10(c) shows the fitting compatibility of model with measured quadratic coefficient of capacitance at various temperatures. It is observed that titania MIM capacitor shows a strong dependence with temperature. This is due to week ionic bond which lead to large distortion

Table 3.3 Material parameters for modeling [88]. ©2014 Elsevier. Reprinted with permission from Reference 88.

Material	**Al_2O_3**	**TiO_2**	**HfO_2**	**Y_2O_3**
Dielectric constant (ε_r)	9	100	25	15
Induced dipole density (N_{di}) ($10^{22}/cm^3$)	6	10	4	4

of metal atoms. MIM capacitors with alumina shows a low dependence with temperature as thickness increases. It is due to decrease in field as thickness increases which intern reduces the local field. Also the ionic bond is strong and less sensitive to temperature.

The model can be used to explore the limitation of thickness and dielectric constant to meet the ITRS requirement. Figure 3.10(d) shows the required physical thickness of dielectric layer to meet $\alpha = 100$ ppm/V^2, assuming the average density of dipoles $N_{di} = 10 \times 10^{22}$ per cm^3. It is observed that required thickness d increases with dielectric constant ε_r. However, it saturates to ~100 nm after $\varepsilon_r \approx 30$. This indicates that the high dielectric constant MIM capacitors require >100 nm thickness of dielectric layer to meet the ITRS recommendation. Figure 3.10(d) also shows the required thickness if $\alpha = 200$ ppm/V^2 is acceptable, which shows 20% reduction in thickness. Inset of Figure 3.10(d) shows the maximum achievable capacitance density for the extracted thickness to meet $\alpha = 100$ ppm/V^2 and $\alpha = 200$ ppm/V^2. Figure 3.10(a,d) and the inset are highly useful to select the material thickness and electrode area as per the IC design requirement. For vertical or thickness miniaturization, one should go for low dielectric material with higher electrode area to achieve high capacitance. For horizontal or area reduction of ICs, the thicker and higher dielectric constant layer is preferable. However, the technology limitations such as oxidation rate, deposition rate, defect density, and ionic bonding of material play a major role in the performance of MIM capacitors.

3.8 Conclusion

In this chapter, the fabrication of high capacitance density single and bilayer MIM capacitors using anodization process is reported. It is observed that the performance of MIM capacitors is influenced by various fabrication conditions, such as anodization voltage, electrolyte, and temperature. These anodic dielectric oxide structures show crystalline, low defect, and improved ionic polarization which result higher capacitance density of more than >5 fF/μm^2 and low leakage current density of less than 10 nA/cm^2 with low VCC. These capacitors can be used for future AMS applications according to ITRS recommendations for the year 2015.

The low defect density and strong ionic polarization of anodic alumina provide a stable frequency-dependent capacitance characteristics. Leakage mechanism of anodic alumina is studied using SE and PF emission models. It is observed that large band gap and barrier height of Al_2O_3 yields low leakage current density. It is predicated that the defect/traps at bulk have deep energy depth of ~1.5 eV. The reliability and trap distribution are studied using CVS experiment and observed that the anodic alumina MIM capacitor can operate for more than 10 years continuously for the applied bias of 2 V. This is due to strong ionic bonds and low defect density of the anodic Al_2O_3. Anodic alumina MIM capacitors show stable and excellent performance, such as low VCC, low leakage current density, and high TBD, which are attractive features for AMS applications.

Very high capacitance density of $>30\,\mathrm{fF/\mu m^2}$ was achieved in barrier type anodic titania MIM capacitors. This high capacitance is achieved due to large dielectric constant of crystalline TiO_2 and a thin interfacial layer of AlTiO alloy. This thin interfacial layer is formed at higher anodization voltage which helps in formation of capacitance and reduction of leakage and VCC of MIM capacitor. However, the leakage current density and breakdown field are poor compared to alumina MIM capacitors. The high-defect density during crystallization, poor ionic polarization and large relaxation time exhibit a high sensitivity of capacitance with frequency and temperature. The asymmetry of leakage characteristics is observed which is due to the presence of thin interfacial oxide and nonuniform distribution of trap barrier height at metal–insulator interface. Such defect profile with amorphous region at top interface is the result of nucleation of oxide during anodization.

Physics and modeling of field dependent capacitance of MIM capacitors are useful to analyze the origin of nonlinearities, formation of capacitance, frequency and temperature dependence, and dielectric relaxation in MIM capacitors. The ionic polarization of metal–oxygen bond and bond distortions are statistically accounted in modeling of voltage nonlinearity coefficient α ($\mathrm{ppm/V^2}$) of high-k MIM capacitors. It is predicted that the bond distortion due to applied field is the origin of nonlinearities in many dielectrics. It also predicts that nonpolar dielectrics show low VCC compared to polar dielectric materials. The formula can be used to predict the required thickness of dielectric material to meet the ITRS requirement. It predicts that the thickness of dielectric layer should be $>100\,\mathrm{nm}$ for $\varepsilon_r > 30$ to achieve a low voltage linearity coefficient of $<100\,\mathrm{ppm/V^2}$. These observation are highly useful to inquire the effectiveness of fabrication/oxidation process to meet the ITRS requirements.

Anodized bilayers of TiO_2/Al_2O_3 shows polycrystalline and low defect which are favorable for many AMS and DRAM applications. These capacitor achieved a high capacitance density of $>7\,\mathrm{fF/\mu m^2}$ and low leakage current density of less than $9.1\,\mathrm{nA/cm^2}$ at 3 V. It is observed that these capacitors offers low VCC of $<200\,\mathrm{ppm/V^2}$ and quality factor of >50. These results are attractive for ITRS recommendation on AMS applications. The formation of bilayer metal-oxide and crystallization are discussed in detail with outward and inward migration of metal and oxygen ions. It is found that the bottom Al is anodized at higher anodization voltages of ≥ 25 V. The interface traps, Schottky barrier, and their temperature dependence are studied using various leakage models. It is observed that the anodic alumina acts like a barrier layer which reduces the leakage current density and VCC. Alumina's strong ionic bond improves the overall performance of dielectric stack. In path of these works, we developed the anodic trilayer $Al_2O_3/TiO_2/Al_2O_3$ MIM capacitors which exhibit improved performance. These all results are suggesting the anodization for future micro- and nanoelectronics fabrication. With careful integration of anodization with regular IC fabrication process, one can achieve compact and high performance ICs for various analog, digital, and mixed signal applications.

References

[1] S. G. T. Iida, M. Nakahara, and H. Akiba, "Precise capacitor structure suitable for submicron mixed analog/digital ASICs," *Custom Integration Circuits Conference*, 1990.

[2] S. Onge, S. G. Franz, A. F. Puttlitz, A. Kalinoski, B. E. Johnson, and B. El-Kareh, "Design of precision capacitors for analog applications," *IEEE Transactions on Components Hybrids and Manufacturing Technology*, vol. 15, p. 1064, 1992.

[3] M. J. Chen and C. S. Hou, "A novel cross-coupled interpoly-oxide capacitor for mixed-mode CMOS processes," *IEEE Electron Device Letters*, vol. 20, p. 360, 1999.

[4] C. H. Ng, C.-S. Ho, S.-F. S. Chu, and S.-C. Sun, "MIM capacitor integration for mixed-signal/RF applications," *Electron Devices, IEEE Transactions on*, vol. 52, pp. 1399–1409, 2005.

[5] S. Chen, C. Lai, A. Chin, J. Hsieh, and J. Liu, "High-density MIM capacitors using Al_2O_3 and AlTiOx dielectrics," *Electron Device Letters, IEEE*, vol. 23, pp. 185–187, 2002.

[6] ITRS, "International technology roadmap for semiconductor 2010, *Report on RF and Analog-Mixed Signal Design*," 2011.

[7] J. W. Diggle, T. C. Downie, and C. W. Goulding, "Anodic oxide films on aluminum," *Chemical Reviews*, vol. 69, no. 3, pp. 365–405, 1969.

[8] I. Tanabe, N. Yamaguti, J. Mizutani, T. Watanabe, and K. Itagaki, "Surface treatment about coloring of stainless steel and titanium using YVO 4 laser machine," *Nippon Kikai Gakkai Ronbunshu, C Hen/Transactions of the Japan Society of Mechanical Engineers, Part C*, vol. 69, no. 9, pp. 2470–2475, 2003.

[9] M. Thieme, R. Frenzel, V. Hein, and H. Worch, "Metal surfaces with ultrahydrophobic properties: Perspectives for corrosion protection and self-cleaning," *Journal of Corrosion Science and Engineering*, vol. 6, 2003.

[10] J. Hou and D. D. L. Chung, "Corrosion protection of aluminium-matrix aluminium nitride and silicon carbide composites by anodization," *Journal of Materials Science*, vol. 32, no. 12, pp. 3113–3121, 1997.

[11] Y. Kimura, S. Kimura, R. Kojima, M. Bitoh, M. Abe, and M. Niwano, "Microscaled hydrogen gas sensors with patterned anodic titanium oxide nanotube film," *Sensors and Actuators, B: Chemical*, vol. 177, pp. 1156–1160, 2013.

[12] F. Rosztoczy and M. Read, "Orientation dependence of the aluminum oxide dielectric in film capacitors," *Journal of the Electrochemical Society*, vol. 116, no. 12, pp. 1752–1756, 1969.

[13] R. K. Gupta, M. Katiyar, P. K. Ojha, K. N. Rai, and J. Narain, "Characterization and reliability study of Al_2O_3 as an alternative gate dielectric for ULSI technology," In: *Proceedings of SPIE—The International Society for Optical Engineering*, vol. 4746 I, pp. 659–663, 2002.

[14] C. Thornton, "New trends in microelectronics fabrication technology—1965," In: *IRE International Convention Record*, vol. 13, pp. 53–66, 1965.

[15] A. R. Morley and D. S. Campbell, "Electrolytic capacitors: their fabrication and the interpretation of their operational behaviour," *Radio and Electronic Engineer*, vol. 43, no. 7, pp. 421–429, 1973.

[16] C. N. Ajit and S. R. Jawalekar, "Thin film Al_2O_3 capacitors," *Thin Solid Films*, vol. 37, no. 1, pp. 85–89, 1976.

[17] E. Hourdakis and A. G. Nassiopoulou, "High-density MIM capacitors with porous anodic alumina dielectric," *IEEE Transactions on Electron Devices*, vol. 57, no. 10, pp. 2679–2683, 2010.

[18] E. Hourdakis and A. G. Nassiopoulou, "High performance MIM capacitor using anodic alumina dielectric," *Microelectronic Engineering*, vol. 90, pp. 12–14, 2012.

[19] P. W. Wyatt, "Dielectric anisotropy in amorphous Ta_2O_5 films," *Journal of the Electrochemical Society*, vol. 122, no. 12, pp. 1660–1666, 1975.

[20] S. Huang and J. Hwu, "Electrical characterization and process control of cost-effective high-k aluminum oxide gate dielectrics prepared by anodization followed by furnace annealing," *IEEE Transactions on Electron Devices*, vol. 50, no. 7, pp. 1658–1664, 2003.

[21] Y. Lin and J. Hwu, "Using anodization to oxidize ultrathin aluminum film for high-k gate dielectric application," *Journal of the Electrochemical Society*, vol. 150, no. 7, pp. G389–G394, 2003.

[22] Z. Chen, S. Huang, and J. Hwu, "Electrical characteristics of ultra-thin gate oxides (<3 nm) prepared by direct current superimposed with alternating-current anodization," *Solid-State Electronics*, vol. 48, no. 1, pp. 23–28, 2004.

[23] S. Huang and J. Hwu, "Lateral nonuniformity of effective oxide charges in mos capacitors with Al_2O_3 gate dielectrics," *IEEE Transactions on Electron Devices*, vol. 53, no. 7, pp. 1608–1614, 2006.

[24] R. K. Raymond and M. B. Das, "Fabrication and characteristics of MOSFET's incorporating anodic aluminum oxide in the gate structure," *Solid-State Electronics*, vol. 19, no. 3, pp. 181–184, 1976.

[25] C. Wang and J. Hwu, "Trench structure metal-oxide-semiconductor (MOS) solar cells with oxides prepared by anodization technique," In: *Technical Digest—15th OptoElectronics and Communications Conference, OECC2010*, pp. 366–367, 2010.

[26] T. W. Hickmott, "Temperature-dependent Fowler–Nordheim tunneling and a compensation effect in anodized Al–Al_2O_3–Au diodes," *Journal of Applied Physics*, vol. 97, no. 10, pp. 1–9, 2005.

[27] N. Sedghi, W. Davey, I. Z. Mitrovic, and S. Hall, "Reliability studies on Ta_2O_5 high-kappa dielectric metal-insulator-metal capacitors prepared by wet anodization," vol. 29, p. 01AB10, AVS, 2011.

[28] L. M. Kosjuk and L. L. Odynets, "Polarization processes in anodic oxide films," *Thin Solid Films*, vol. 302, no. 1–2, pp. 235–238, 1997.

[29] J. Lee, K. Koh, N. Lee, *et al.*, "Effect of polysilicon gate on the flatband voltage shift and mobility degradation for ALD – Al_2O_3 gate dielectric," in

Electron Devices Meeting, 2000. IEDM'00. Technical Digest. International, pp. 645–648, 2000.

[30] G. D. W. H. C. Lin and P. D. Ye, "Leakage current and breakdown electric-field studies on ultrathin atomic-layer-deposited Al_2O_3 on GaAs," *Applied Physics Letters*, vol. 87, p. 182904, 2005.

[31] K. Allers, P. Brenner, and M. Schrenk, "Dielectric reliability and material properties of Al_2O_3 in metal insulator metal capacitors (MIMCAP) for RF bipolar technologies in comparison to SiO_2, SiN and Ta_2O_5," in *Proceedings of the IEEE Bipolar/BiCMOS Circuits and Technology Meeting*, pp. 35–38, 2003.

[32] B. Miao, R. Mahapatra, R. Jenkins, J. Silvie, N. G. Wright, and A. B. Horsfall, "Radiation induced change in defect density in HfO_2-based MIM capacitors," *IEEE Transactions on Nuclear Science*, vol. 56, no. 5, pp. 2916–2924, 2009.

[33] T. W. Hickmott, "Electrolyte effects on charge, polarization, and conduction in thin anodic Al_2O_3 films. I. Initial charge and temperature-dependent polarization," *Journal of Applied Physics*, vol. 102, no. 9, p. 093706, 2007.

[34] H. A. Yoshiteru Sato and S. Ono, "Effect of electrolyte species on crystallinity and dielectric properties of anodic oxide films formed on aluminum," In: *Proceedings of the 12th International Conference*, 2010.

[35] D. Kannadassan, R. Karthik, M. Bhagini, and P. Mallick, "Nanostructured barrier type anodic oxide metal-insulator-metal capacitors," *Journal of Nanoelectronics and Optoelectronics*, vol. 7, no. 4, pp. 400–404, 2012.

[36] J. Beaumont and P. Jacobs, "Polarization in potassium chloride crystals," *Journal of Physics and Chemistry of Solids*, vol. 28, no. 4, pp. 657–667, 1967.

[37] P. Gonon and C. Vallae, "Modeling of nonlinearities in the capacitance-voltage characteristics of high-k metal-insulator-metal capacitors," *Applied Physics Letters*, vol. 90, no. 14, 2007.

[38] D. Kannadassan, R. Karthik, P. Mallick, and M. Baghini, "Temperature and stress dependent properties of barrier type anodic Al_2O_3 MIM capacitor," *International Conference on Emerging Electronics*, IIT Bombay, India, no. 6636238, 2012.

[39] M. Specht, M. Städele, S. Jakschik, and U. Schröder, "Transport mechanisms in atomic-layer-deposited Al_2O_3 dielectrics," *Applied Physics Letters*, vol. 84, no. 16, pp. 3076–3078, 2004. Cited By (since 1996): 52.

[40] C. Lhymn, P. Kosel, and R. Vaughan, "Thickness dependence of dielectric breakdown voltage," *Thin Solid Films*, vol. 145, no. 1, pp. 69–74, 1986.

[41] D. V. Morgan, A. E. Guile, and Y. Bektore, "Stored charge in anodic aluminium oxide films," *Journal of Physics D: Applied Physics*, vol. 13, no. 2, p. 307, 1980.

[42] S. Ding, H. Hu, C. Zhu, *et al.*, "Evidence and understanding of ALD – HfO_2–Al_2O_3 laminate mim capacitors outperforming sandwich counterparts," *IEEE Electron Device Letters*, vol. 25, no. 10, pp. 681–683, 2004.

[43] S. Chakraborty, M. K. Bera, S. Bhattacharya, and C. K. Maiti, "Current conduction mechanism in TiO_2 gate dielectrics," *Microelectronic Engineering*, vol. 81, no. 2–4, pp. 188–193, 2005.

[44] R. G. Southwick III, J. Reed, C. Buu, R. Butler, G. Bersuker, and W. B. Knowlton, "Limitations of Poole–Frenkel conduction in bilayer HfO_2/SiO_2MOS devices," *IEEE Transactions on Device and Materials Reliability*, vol. 10, no. 2, pp. 201–207, 2010.

[45] E. Atanassova, N. Stojadinovic, A. Paskaleva, D. Spassov, L. Vracar, and M. Georgieva, "Constant voltage stress induced current in Ta_2O_5 stacks and its dependence on a gate electrode," *Semiconductor Science and Technology*, vol. 23, no. 7, p. 075017, 2008.

[46] C. Yeh, T. P. Ma, N. Ramaswamy, *et al.*, "Frenkel–Poole trap energy extraction of atomic layer deposited Al_2O_3 and Hf_xAl_yO thin films," *Applied Physics Letters*, vol. 91, no. 11, p. 113521, 2007.

[47] T. Remmel, R. Ramprasad, and J. Walls, "Leakage behavior and reliability assessment of tantalum oxide dielectric MIM capacitors," in *Annual Proceedings – Reliability Physics (Symposium)*, pp. 277–281, 2003.

[48] S. Ding, Y. Huang, Y. Huang, S. Pan, W. Zhang, and L. Wang, "High density Al_2O_3/TaN-based metal-insulator-metal capacitors in application to radio frequency integrated circuits," *Chinese Physics*, vol. 16, no. 9, pp. 2803–2808, 2007. Cited By (since 1996): 2.

[49] A. Wisitsoraat, A. Tuantranont, E. Comini, G. Sberveglieri, and W. Wlodarski, "Characterization of n-type and p-type semiconductor gas sensors based on niox doped TiO_2 thin films," *Thin Solid Films*, vol. 517, no. 8, pp. 2775–2780, 2009.

[50] M. Gratzel, "Sol-gel processed TiO_2 films for photovoltaic applications," *Journal of Sol-Gel Science and Technology*, vol. 22, no. 1–2, pp. 7–13, 2001.

[51] K. C. Chiang, C. H. Lai, A. Chin, H. L. Kao, S. P. McAlister, and C. C. Chi, "Very high density RF MIM capacitor compatible with VLSI," *IEEE MTT-S International Microwave Symposium Digest*, vol. 2005, pp. 287–290, 2005.

[52] A. G. Mantzila and M. I. Prodromidis, "Performance of impedimetric biosensors based on anodically formed Ti/TiO_2 electrodes," *Electroanalysis*, vol. 17, no. 20, pp. 1878–1885, 2005.

[53] M. Stamate, G. Lazar, and I. Lazar, "Dimensional effects observed for the electrical, dielectrical and optical properties of TiO_2 DC magnetron thin films," *Journal of Materials Science: Materials in Electronics*, vol. 20, no. 2, pp. 117–122, 2009.

[54] H. Habazaki, M. Uozumi, H. Konno, K. Shimizu, P. Skeldon, and G. E. Thompson, "Crystallization of anodic titania on titanium and its alloys," *Corrosion Science*, vol. 45, no. 9, pp. 2063–2073, 2003.

[55] M. S. Vasil'eva, V. S. Rudnev, L. M. Tyrina, I. V. Lukiyanchuk, N. B. Kondrikov, and P. S. Gordienko, "Phase composition of coatings formed on titanium in borate electrolyte by microarch oxidation," *Russian Journal of Applied Chemistry*, vol. 75, no. 4, pp. 569–572, 2002.

[56] G. H. Gleaves, R. A. Collins, and G. Dearnaley, "A further investigation of the influence of implanted foreign ion species on the anodic oxidation of Ti," *Journal of Electroanalytical Chemistry*, vol. 137, no. 1, pp. 51–65, 1982.

[57] R. J. Soukup, "Observations of negative resistance in Ti–TiO_2–Au diodes," *Journal of Applied Physics*, vol. 43, no. 8, pp. 3431–3435, 1972. Cited By (since 1996): 2.

[58] D. Kannadassan, R. Karthik, M. Shojaei Baghini, and P. Mallick, "Nanostructured metal-insulator-metal capacitor with anodic titania," *Materials Science in Semiconductor Processing*, vol. 16, no. 2, pp. 274–281, 2013.

[59] A. Felske and W. J. Plieth, "Raman spectroscopy of titanium dioxide layers," *Electrochimica Acta*, vol. 34, no. 1, pp. 75–77, 1989.

[60] P. Kar, "Effect of anodization voltage on the formation of phase pure anatase nanotubes with doped carbon," *Inorganic Materials*, vol. 46, no. 4, pp. 377–382, 2010.

[61] T. H. Phung, P. Steinmann, R. Wise, Y. Yeo, and C. Zhu, "Modeling the negative quadratic VCC of SiO_2 in MIM capacitor," *IEEE Electron Device Letters*, vol. 32, no. 12, pp. 1671–1673, 2011.

[62] F. Mondon and S. Blonkowski, "Electrical characterisation and reliability of HfO_2 and Al_2O_3–HfO_2 MIM capacitors," *Microelectronics Reliability*, vol. 43, no. 8, pp. 1259–1266, 2003.

[63] S. Lee, H. Kim, P. C. McIntyre, K. C. Saraswat, and J. Byun, "Atomic layer deposition of ZrO_2 on W for metal-insulator-metal capacitor application," *Applied Physics Letters*, vol. 82, no. 17, pp. 2874–2876, 2003.

[64] C. H. Cheng, S. H. Lin, K. Y. Jhou, *et al.*, "High density and low leakage current in TiO_2 MIM capacitors processed at 300°C," *IEEE Electron Device Letters*, vol. 29, no. 8, pp. 845–847, 2008.

[65] J. Woo, Y. Chun, Y. Joo, and C. Kim, "Low leakage current in metal-insulator-metal capacitors of structural $Al_2O_3/TiO_2/Al_2O_3$ dielectrics," *Applied Physics Letters*, vol. 100, no. 8, 2012.

[66] M. Seo, S. H. Rha, S. K. Kim, *et al.*, "The mechanism for the suppression of leakage current in high dielectric TiO_2 thin films by adopting ultra-thin HfO_2 films for memory application," *Journal of Applied Physics*, vol. 110, no. 2, p. 024105, 2011.

[67] J. Robertson, "High dielectric constant oxides," *EPJ Applied Physics*, vol. 28, no. 3, pp. 265–291, 2004.

[68] N. Mise, A. Ogawa, O. Tonomura, *et al.*, "Theoretical screening of candidate materials for DRAM capacitors and experimental demonstration of a cubic-hafnia MIM capacitor," *IEEE Transactions on Electron Devices*, vol. 57, no. 9, pp. 2080–2086, 2010.

[69] C. B. Kaynak, M. Lukosius, I. Costina, *et al.*, "Enhanced leakage current behavior of $Sr_2Ta_2O_7/SrTiO_2$."

[70] S. J. Kim, B. J. Cho, M. Li, *et al.*, "Improvement of voltage linearity in high-k MIM capacitors using HfO_2–SiO_2 stacked dielectric," *IEEE Electron Device Letters*, vol. 25, no. 8, pp. 538–540, 2004.

[71] J. Wu, Y. Wu, C. Lin, W. Ou, M. Wu, and L. Chen, "Effect of nitrogen passivation on the performance of MIM capacitors with a crystalline-TiO_2/SiO_2 stacked insulator," *IEEE Electron Device Letters*, vol. 33, no. 6, pp. 878–880, 2012.

[72] B. Y. Tsui, H. H. Hsu, and C. H. Cheng, "High-performance metal-insulator-metal capacitors with HfTiO/Y_2O_3 stacked dielectric," *IEEE Electron Device Letters*, vol. 31, no. 8, pp. 875–877, 2010.

[73] J. M. Park, M. W. Song, H. K. Kang, *et al.*, "Mass production worthy MIM capacitor on gate polysilicon (MIM – COG) structure using $HfO_2/HfO_xC_yN_z/HfO_2$ dielectric for analog/RF/mixed signal application," *IEEE Electron Device Letter*, vol. 20, p. 993, 2007.

[74] D. S. H. C. L. Zhang, W. He, and B. J. Cho, "High-performance MIM capacitors using HfLaO-based dielectrics," *IEEE Electron Device Letter*, vol. 31, p. 17, 2010.

[75] T. Ishikawa, D. Kodama, Y. Matsui, M. Hiratani, T. Furusawa, and D. Hisamoto, "High-capacitance $Cu/Ta_2O_5/Cu$ MIM structure for SoC applications featuring a single-mask add-on process," In: *Technical Digest – International Electron Devices Meeting*, pp. 940–942, 2002.

[76] J. Perriere, S. Rigo, and J. Siejka, "Investigation of cation-transport processes during anodic oxidation of duplex layers of tantalum on niobium by the use of rutherford backscattering and nuclear microanalysis," *Journal of the Electrochemical Society*, vol. 125, no. 9, pp. 1549–1557, 1978.

[77] J. Perriere and J. Siejka, "Study of the anodization of niobium and tantalum superimposed layers by ^{18}O tracing techniques and nuclear microanalysis – part I. ^{18}O and cation depth profiles," *Journal of the Electrochemical Society*, vol. 130, no. 6, pp. 1260–1267, 1983.

[78] J. Perriere and J. Siejka, "Study of the anodization of niobium and tantalum superimposed layers by ^{18}O tracing techniques and nuclear microanalysis – part III. discussion," *Journal of the Electrochemical Society*, vol. 130, no. 6, pp. 1267–1273, 1983.

[79] K. Shimizu, K. Kobayashi, P. Skeldon, G. Thompson, and G. Wood, "Anodic oxidation of zirconium covered with a thin layer of aluminium," *Thin Solid Films*, vol. 295, no. 1–2, pp. 156–161, 1997.

[80] L. Yao, J. Hua Liu, M. Yu, S. Mei Li, and H. Wu, "Formation and capacitance properties of Ti–Al composite oxide film on aluminum," *Transactions of Nonferrous Metals Society of China*, vol. 20, no. 5, pp. 825–830, 2010.

[81] R. Karthik, D. Kannadassan, M. Baghini, and P. Mallick, "Nanostructured bilayer anodic TiO_2/Al_2O_3 metal-insulator-metal capacitor," *Journal of Nanoscience and Nanotechnology*, vol. 13, no. 10, pp. 6894–6899, 2013.

[82] M. Houssa, M. Tuominen, M. Naili, *et al.*, "Trap-assisted tunneling in high permittivity gate dielectric stacks," *Journal of Applied Physics*, vol. 87, no. 12, pp. 8615–8620, 2000.

[83] C. Wenger, G. Lupina, M. Lukosius, *et al.*, "Microscopic model for the nonlinear behavior of high-k metal-insulator-metal capacitors," *Journal of Applied Physics*, vol. 103, no. 10, 2008. Cited By (since 1996): 21.

[84] S. Bécu, S. Crémer, and J. Autran, "Capacitance non-linearity study in Al_2O_3 MIM capacitors using an ionic polarization model," *Microelectronic Engineering*, vol. 83, no. 11–12, pp. 2422–2426, 2006.

[85] E. Talebian and M. Talebian, "A general review on the derivation of Clausius–Mossotti relation," *Optik*, vol. 124, no. 16, pp. 2324–2326, 2013.

[86] E. Neumann, "Chemical electric field effects in biological macromolecules," *Progress in Biophysics and Molecular Biology*, vol. 47, no. 3, pp. 197–231, 1986.

[87] S. P. Gubin, *Magnetic Nanoparticles.* Wiley-VCH Verlag GmbH and Co. KGaA, Weinheim, 2009.

[88] D. Kannadassan, R. Karthik, M. Shojaei Baghini, and P. Mallick, "Modeling the voltage nonlinearity of high-k MIM capacitors," *Solid-State Electronics*, vol. 91, pp. 112–117, 2014.

Chapter 4
Graphene transistors—present and beyond

Ashok Srivastava[1] *and Yaser M. Banadaki*[1,2]

Graphene is being explored as a material to build scaled transistors for high speed operations (e.g., 10 s of GHz) of integrated circuits. This chapter discusses the state-of-art of the graphene-based transistors with a prediction for its future directions.

4.1 Introduction

Scaling down the CMOS transistors has enhanced the device performance and density, satisfying the prediction of Moore's law for decades [1]. However, electronic properties of silicon have imposed several challenges preventing further scaling in near future [2]. Novel materials such as carbon nanotube [3] and graphene [4] are potential candidates for alternative channel material in post-CMOS technology [5–7]. Graphene is a monolayer of carbon atoms in a two-dimensional (2D) honeycomb lattice, which was first discovered by Novoselov and Geim in 2004 [4]. Graphene shows exotic electronic properties. The carrier transport in graphene is similar to transport of massless particles, ballistic in nature and resulting in large mobility [8]. Two-dimensional electron gas of graphene provides high carrier velocity and concentration resulting in faster switching. While the bottleneck of scaling silicon channel is in heat removal of dissipated power, graphene has excellent thermal conductivity due to strong carbon–carbon bonding [9–11]. Atomically thin structure of monolayer graphene results in better gate control over the channel and its planar structure is compatible with current CMOS fabrication processes introducing the potential production of wafer-scale integrated circuits [12].

Despite many advantages, graphene is a semimetal with zero band gap, which limits its application as logic transistors [13, 14]. However, the band gap of several hundred meV can be opened by quantum confinement of graphene lattice in the form of nanoribbon with a few nanometers width [15]. While classical models [16, 17] like charge-collection equations [18] can be used to model graphene transistors with micrometer length and width, it is not suitable for modeling and simulation of graphene nanoribbon (GNR) field effect transistors (GNR FETs). The traditional

[1]Division of Electrical and Computer Engineering, Louisiana State University, Baton Rouge, LA
[2]College of Engineering, Southern University, Baton Rouge, LA

classical models usually focus on scattering effects inside the channel, whose length is much longer than the mean free path (mfp) of carriers. In GNR FET, the channel length is usually small and the gate electrostatic potential tunes the discrete energies of GNRs in the channel, leading to important effects of tunneling on carrier transport. Direct source-to-drain tunneling and band-to-band tunneling from drain to channel can be significant by scaling down the channel length and width of GNR. While semiclassical models [19–21] can be modified to incorporate tunneling current, the models cannot be used for GNR FET with channel length below 10 nm. Thus, by scaling down the channel length, the atomistic quantum-based models [22, 23] which can take into consideration tunneling effects in short channel GNR FET need to be employed in order to investigate the GNR FET performance and compare with the projection reported by the ITRS [2].

In the following, we begin with a brief overview of fabrication of atomic layer graphene in Section 4.2. In Section 4.3, properties of graphene and issues in using as a channel material in field effect transistors are described. A brief description on the modeling and simulation methods of graphene transistors is presented in Section 4.4. Finally, the investigation of GNR FET with channel length below 10 nm is presented in Section 4.5 followed by conclusion in Section 4.6.

4.2 Fabrication of graphene

Atomic layer graphene can be fabricated from various methods such as mechanical exfoliation of graphite, deposition of epitaxial graphene on SiC crystals, and chemical vapor deposition of graphene using metal catalyzer. The simplest method is mechanical exfoliation, in which graphene layers are peeled repeatedly using the popular scotch-tape technique [4] to achieve graphene monolayer that could then be transferred to an oxidized substrate. The number of graphene layers can easily be identified using optical microscopy due to contrast in a certain oxide layer thickness. However, the method cannot be used for the mass production. Epitaxial graphene can be formed directly on an insulating substrate like SiC by sublimating silicon atoms at high temperatures, preventing need for the transfer process [24]. The growth of large-area graphene is favorable for wafer-scaled lithography, which can be formed by chemical vapor deposition method. The carbon atoms are supplied by hydrocarbon molecules and dissolved on the metal surface like nickel and then transferred to an insulating wafer [25]. Raman spectroscopy of graphene can determine the qualities and the number of graphene layers. As can be seen from Raman spectroscopy of monolayer graphene in Figure 4.1(a), there are three peaks, 2D, G, and D at 2700/cm, 1580/cm, and 1350/cm, respectively, while peak D can be absent for defect-free graphene samples [26].

The fabrication of narrow strip of graphene has been demonstrated using e-beam lithography [27] and then the width of graphene ribbon can be further reduced by etching down to 4 nm [28] and 2 nm by chemical synthesis [29] with very smooth edge. Other lithography methods based on atomic force microscopy [30] and scanning tunneling microscopy (STM) [31] have been proposed for the fabrication of GNRs.

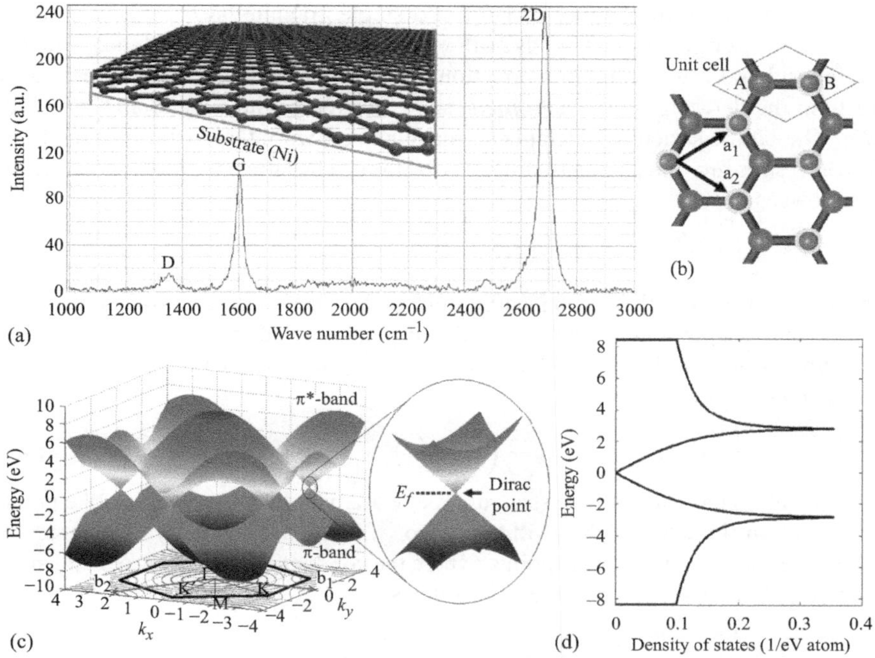

Figure 4.1 *(a) Raman spectroscopy of graphene on nickel substrate, (b) two-dimensional honeycomb lattice of graphene, which consists of two triangular sublattices A and B. Note that the unit cell and lattice vectors are also shown in the lattice. (c) Graphene band structure and first Brillouin zone in momentum space. Note that the position of Dirac points, K, K′, and reciprocal lattice vectors are also shown underneath of the graphene band structure. (d) DOS per unit cell as a function of energy (linear in energy near the Dirac point)*

GNRs of 1 nm width can be produced by unzipping carbon nanotubes with bottom-up chemical approach [32]. Mass production of GNRs can be made possible by using multiwalled CNTs (MWCNTs) as precursors such that the GNR widths can be controlled by controlling the size of the starting MWCNTs and the conditions of dry etching [33] or solution-based oxidative process [34].

4.3 Properties of graphene

4.3.1 Band structure

Graphene is a two dimensional (2D) material made of carbon atoms in a honeycomb-like hexagonal lattice, as shown inside Figure 4.1(a). The band structure of graphene was first calculated by Wallace in 1947 as a single graphite layer [35]. The carbon atoms form strong σ covalent bonds by three in-plane sp^2 hybridized orbitals, whereas the fourth bond is a π bond in z-direction [36]. The electron in this bond can move freely in the delocalized π-electronic system referred as the π-band and π*-bands [37]. The lattice structure of graphene made out of two interpenetrating triangular lattices results in a unit cell consisting of two atoms. The lattice vectors can be written as follows,

$$a_1 = \frac{a}{2}\left(3, \sqrt{3}\right), \quad a_2 = \frac{a}{2}\left(3, -\sqrt{3}\right) \tag{4.1}$$

where, $a = 1.42$ Å is the carbon–carbon distance. Since electronic transport can be two-dimensional in a graphene lattice, the dispersion relation for graphene has also two dimensions as shown in Figure 4.1(b). The reciprocal lattice vectors can be obtained as follows,

$$b_1 = \frac{2\pi}{3a}\left(1, \sqrt{3}\right), \quad b_2 = \frac{2\pi}{3a}\left(1, -\sqrt{3}\right) \tag{4.2}$$

Due to honey-comb lattice structure, there are two sets of three cone-like points K and K' on the edge of the Brillouin zone, which leads to valley degeneracy of $g_v = 2$. These are named Dirac points, where the conduction and valence bands meet each other in momentum space [36] as follows,

$$K = \left(\frac{2\pi}{3a}, \frac{2\pi}{3\sqrt{3}a}\right), \quad K' = \left(\frac{2\pi}{3a}, -\frac{2\pi}{3\sqrt{3}a}\right) \tag{4.3}$$

Theoretically, the valence π-band is completely empty and the conduction π*-band is full, which leads to the Fermi energy being located at the Dirac point, presenting the semimetallic behavior for graphene. The behavior of charge carriers near Dirac points resembles the Dirac spectrum for massless fermions [8] and can be described by the linear dispersion relation as follows,

$$E\left(\overrightarrow{k'}\right) = \pm\hbar\upsilon_F\left|\overrightarrow{k'}\right| \tag{4.4}$$

where k' is the momentum near the Dirac point, $\hbar$ is reduced Planck constant and υ_F is the Fermi velocity. The linear dispersion relation is contrary to most of materials as the solution of Schrodinger equation has second order in space and first order in time, leading to quadratic dispersion. Charge carriers near Dirac points behave like relativistic particles ideally transporting with Fermi velocity, which is theoretically 300 times smaller than the speed of light [8]. Thus, the Hamiltonian for electrons near Dirac Points in graphene can be calculated by Dirac equation with zero mass [26] as follows,

$$H = \hbar v_F \begin{pmatrix} 0 & k_x - ik_y \\ k_x + ik_y & 0 \end{pmatrix} = \hbar v_F \vec{\sigma} . \vec{k} \tag{4.5}$$

where $\vec{\sigma} = (\sigma_x, \sigma_y)$ is the 2D vector of the Pauli matrices, and $\vec{k}$ is the momentum of the quasi-particles in graphene. The massless chiral Dirac equation in this application was discussed by Semenoff in 1984 [38] to explain the low-energy band structure of graphene, where the term "graphene pseudospin" was used.

Assuming the first nearest neighbor interaction, the close form of dispersion relation near Dirac points can be obtained [36] as follows,

$$E\left(\vec{k}\right) = \pm t\sqrt{1 + 4\cos\frac{\sqrt{3}k_x a}{2}\cos\frac{k_y a}{2} + 4\cos^2\frac{k_y a}{2}} \tag{4.6}$$

where, minus and plus signs correspond to the conduction and valence bands, respectively. Graphene band structure including the 2D Brillouin zone of graphene in momentum space is shown in Figure 4.1(c). Figure 4.1(d) shows the density of state per unit cell of graphene, which can be approximated by the linear dispersion relation near the Dirac point as follows,

$$D(E) = \frac{g_v g_s}{2\pi}\frac{|E|}{(\hbar v_F)^2} \tag{4.7}$$

where, $g_v = 2$ is the valley degeneracy, $g_s = 2$ is the valley degeneracy corresponding to the K and K' points and $\upsilon_F = \sqrt{3}ta/2\hbar$ is Fermi velocity and t is the nearest neighbor hopping energy. The Fermi velocity can be as high as 3×10^6 m/s in suspended graphene [39] while it can be as low as $\sim 1 \times 10^6$ m/s when electron–electron interactions are weak [40]. Fermi velocity can be altered by the dielectric constant of the embedding environment because the electron self-energy is inversely decreased by increasing dielectric screening [41, 42].

4.3.2 Carrier density

The carrier density of 2D electron gas sheet in graphene can be calculated from,

$$n = \int_0^\infty D(E)f(E)dE \tag{4.8}$$

where $f(E) = (1 + \exp[(E - E_f)/k_B T])^{-1}$ is the Fermi–Dirac distribution function and $D(E)$ is the density of states (DOS) in (4.7), and E_f is the average Fermi level. The dominant carrier contribution in graphene carrier density induced by the gate voltage V_G is as follows,

$$n_G = p - n = -C_G(V_G - V_{Dirac})/q \tag{4.9}$$

where n_G is the induced carrier in graphene due to the gate voltage, C_G is the effective gate capacitance per unit area, and q is the electron charge. The gate induced carriers are negligible near Dirac point and the carrier density is determined by the electron and hole puddles carriers (n^*), the thermally generated carriers (n^{th}) as follows,

$$n_{Dirac} \approx [(n^*/2)^2 + n_{th}^2]^{1/2} \tag{4.10}$$

where n_{Dirac} is the carrier density in graphene at Dirac point under thermal equilibrium. The thermally generated carriers in 2D graphene can be obtained as follows [43],

$$n_{th} = \frac{\pi}{6}\left(\frac{k_B T}{\hbar v_F}\right)^2 \tag{4.11}$$

where k_B is Boltzmann constant, and T is absolute temperature on graphene. The residual charge puddle density n^* in graphene [44] has been modeled by assuming the spatial electrostatic potential as the periodic step function with equal size and amplitude $\pm\Delta$ as follows,

$$n^* = \int_{-\Delta}^{\infty} D(E+\Delta)f(E)dE + \int_{\Delta}^{\infty} D(E-\Delta)f(E)dE \tag{4.12}$$

By averaging the $\pm\Delta$ regions in the limit of $\Delta/k_BT \gg 1$, the equation can be simplified to $n^* \approx \Delta^2/\pi\hbar^2 v_F^2$ [45]. For graphene on SiO_2, $\Delta \approx 59\,\text{meV}$ has been measured using STM [46], which leads to $n^* \approx 2.6 \times 10^{11}/\text{cm}^2$ and Dirac voltage of 3.66 V. The total concentration of the electron and hole can be calculated by [46],

$$n,p \approx \frac{1}{2}\left[\pm n_g + \sqrt{n_g^2 + 4n_{Dirac}^2}\right] \tag{4.13}$$

where, upper and lower signs correspond to the electron and hole carriers. Due to thermally generated carriers, the carrier density versus gate voltage near Dirac points becomes nonlinear, and its range expands by increasing temperature, which makes the electron and hole puddles less important ($k_BT \gg \Delta$). The carrier density increases and mobility decreases with the temperature due to scattering mechanisms, which lead to the decrease in temperature dependence of conductivity ($\sigma(E_F) = en(E_F)\mu(E_F)$) near Dirac point [46].

4.3.3 Ambipolar field effect

Applying gate voltage (V_G) can induce a surface charge density and accordingly tunes the overall Fermi level. Increasing (decreasing) the gate voltage increases the electron (hole) carriers and correspondingly shifts the Fermi energy toward the conduction (valence) band. The graphene is in electron and hole regimes far from Dirac point. Carrier mobilities can be extracted from Drude model $\sigma(E_F) = en(E_F)\mu(E_F)$, where μ is the mobility, and n is the carrier concentration [4, 47] as shown in Figure 4.2(a). Since graphene is a gapless semiconductor, the gate voltage can tune the charge carriers continuously between electrons and holes. Thus, the graphene conductivity is due to the electron transport for $V_G > V_{Dirac}$, while it changes to hole transport regime for $V_G < V_{Dirac}$. In other words, the graphene displays ambipolar electric field effect when crossing the Dirac point. The graphene conductivity increases linearly by increasing $|V_G|$ away from Dirac voltage [48]. Graphene demonstrates anomalous non-zero minimum conductivity even when its carrier density vanishes at the Dirac point [49] due to the presence of inevitable disorders in graphene flakes [50, 51]. The residual conductivity can be diminished for improved samples moving closer to the quantum-limited unit $4e^2/h$, which is an intrinsic property of 2D Dirac fermions [52].

4.3.4 Conductivity

The experimental characteristic of graphene sheet is much different from an ideal theoretical graphene because many sources of disorders such as lattice imperfections [53],

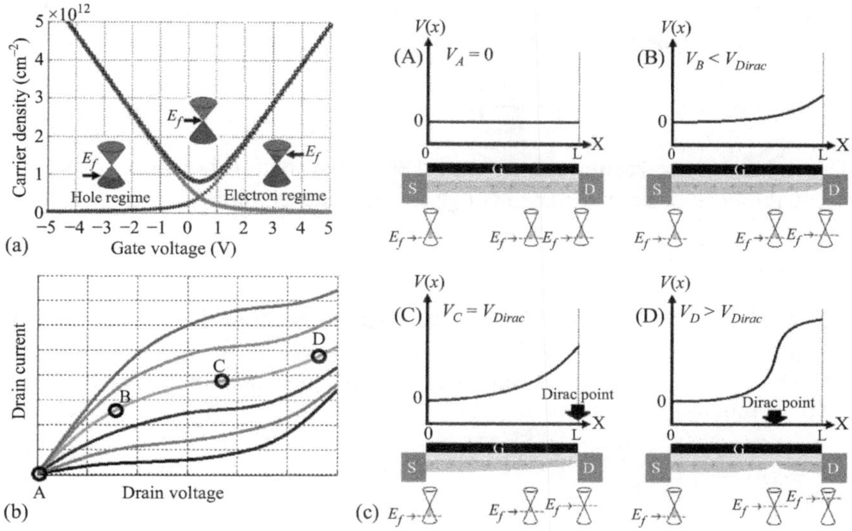

Figure 4.2 (a) Carrier density of graphene versus gate voltage and corresponding position of Fermi level in DOS, (b) and (c) conceptual description of ambipolar effects relating four points in I_D-V_D characteristic of GNRFET to the four points in I_{DS}-V_{DS} characteristic of GNRFET and corresponding Fermi level with respect to Dirac point under the channel

impurities [54], and phonons [55] can manipulate the carrier transport in graphene by increasing scattering mechanisms, such that the carrier mobility can reduce by two orders of magnitude from ~1,000,000 cm^2/Vs [56, 57] to ~10,000 cm^2/Vs [4]. The transport regime is determined based on the comparison between the graphene length L and the carrier mean free path (mfp) [58], which scales the strength of scattering mechanisms. In ballistic transport, Landauer formalism describes the transport since mfp is larger than the graphene length [59]. In this regime, the carrier can travel free of scattering at Fermi velocity (υ_F), and the conductivity can be calculated by [8],

$$\sigma_{bal} = \frac{L}{W}\frac{4e^2}{h}\sum_{n=1}^{\infty} T_n \tag{4.14}$$

the summation is over all available longitudinal transport modes, and T_n is the transmission probability of mode n. In a diffusive transport, the conductivity can be described as a random walk in two-dimension since graphene length is longer than the mfp of carriers. The carriers undergo elastic and inelastic collisions (scattering), and correspondingly the transport is incoherent. In literature, scattering mechanisms mostly discussed include short-range scattering (defects, adsorbates), long range scattering (Coulomb scattering by charged impurities), and electron-phonon scattering.

The semiclassical Boltzmann transport theory can treat the scattering mechanisms using their scattering time τ to calculate the conductivity as follows [60],

$$\sigma_{sc} = \frac{e^2}{2}\int D(\varepsilon)v_k^2\tau(\varepsilon)\left(-\frac{\partial f}{\partial \varepsilon}\right)d\varepsilon \tag{4.15}$$

where $v_{k'}$ and f are carrier velocity and Fermi distribution function, respectively. The equation can be approximated at low temperature as follows,

$$\sigma_{sc} = \frac{e^2 v_F^2}{2} D(E_F)\,\tau(E_F) \tag{4.16}$$

where $D(E_F)$ and $\tau(E_F)$ are DOS and scattering relaxation time at Fermi energy, respectively. By substituting (4.7) and Fermi energy with reference to K point as $E_F \approx \hbar v_F k_F \approx \hbar v_F \sqrt{\pi n}$, the conductivity can also be expressed in terms of carrier density as follows,

$$\sigma_{sc}(n) = \frac{e^2 v_F \tau}{\hbar}\sqrt{\frac{n}{\pi}} \tag{4.17}$$

Thus, the temperature and carrier density dependence of scattering mechanisms can be investigated to reveal the dominant scattering mechanism.

4.3.5 *Scattering mechanism*

4.3.5.1 Long- and short-range scattering

The phonon scattering is negligible at low temperatures, and the mobility is determined mostly by the dominant scattering mechanism of long-range scattering (Coulomb scattering) and short-range scattering. Lattice imperfections, edge roughness, and point defects are intrinsic sources of short-range scattering in graphene sheets. The Coulomb scattering is mostly because of trapped charges in the graphene-substrate interface [61]. These impurities cause long-range variations of the electrostatic potential, which are screened by the carrier transport in graphene sheet, leading to degradation of the mobility, shift of Dirac point, and increase in the minimum conductivity plateau width. A small carrier density, corresponding to small gate voltage, can contribute in long-range scattering mechanism while the short range scattering can be dominating in much higher carrier densities, leading to sublinear and eventually a constant conductivity [62]. In high-quality graphene, charge impurity concentration is small and consequently the short range disorder due to neutral impurity concentration can limit the conductivity in lower carrier densities, which shifts the sub-linear conductivity region to lower carrier density [60]. The short-range scattering is usually associated with vacancies in graphene flakes, which can produce mid-gap states in graphene [63]. The scattering time of short-range mechanism can be calculated [45, 61] and is proportional to $1/\sqrt{n}$, which leads to inverse proportionality of mobility to the carrier density. Chen *et al.* [64] deposited controlled potassium dopants on a clean graphene surface for the charged impurity and showed that the long-range Coulomb scattering is responsible for the linear dependence of conductivity on carrier density. Bolotin *et al.* [65] verified the importance of long-range Coulomb scattering by removing the impurity of the suspended graphene by current-induced heating which resulted in significant improvement of carrier mobility.

4.3.5.2 Acoustic phonon

Long- and short-range scatterings are extrinsic mechanism, which limit the intrinsic mobility of graphene samples at very low temperatures, while lattice vibrations are an intrinsic source of electron–phonon scattering and dominate the extrinsic scattering mechanisms at finite temperature and limit the carrier mobility in graphene. The acoustic and optical phonon scatterings can induce electronic transitions within a single valley (intravalley) or between different valleys (intervalley). In intravalley transitions, the low energy phonons contribute to elastic process of acoustic phonon scatterings while high-energy and low-momentum phonons contribute to optical phonon scatterings. In intervalley transitions, both the energy and momentum of acoustic or optical phonons are high, which can be a dominant scattering process at high temperatures [60]. In general, the dominant scattering mechanisms are changed by increasing temperature to acoustic phonons scattering and eventually optical phonons scattering at higher temperatures.

The acoustic phonons can contribute in quasi-elastic scattering since their energies are usually much less than the Fermi energy of electrons in graphene [55]. The graphene resistivity that is limited by the acoustic phonon can be investigated by defining two transport regimes based on the characteristic temperature of Bloch-Grüneisen [66] T_{BG} as follows,

$$k_B T_{BG} = 2k_F v_{ph} \tag{4.18}$$

where, k_B is the Boltzman constant and v_{ph} is sound velocity, which is 20 km/s for acoustic phonons in graphene [67]. The Bloch-Grüneisen temperature can be calculated as $54\sqrt{n}\,\mathrm{K}$ with density measured in units of $n = 10^{12}/\mathrm{cm}^2$ [68, 69]. In Bloch-Grüneisen regime ($T << T_{BG}$), phonons reduce exponentially with decreasing temperature and carrier density dependence of resistivity is $\rho_{BG} \approx T^4$ and $\rho_{BG} \approx n^{-3/2}$ while at temperature higher than T_{BG}, the resistivity of graphene is linear in temperature ($\rho_{BG} \approx T$) and independent of carrier density. In this regime, the low-field mobility of acoustic phonons is given by [69],

$$\mu_{ac} = \frac{e\rho_m \hbar v_F^2 v_{ph}^2}{\pi k_B} \frac{1}{D_{ac}^2 nT} \tag{4.19}$$

where, $\rho_m = 7.66 \times 10^{-11}\ \mathrm{kg/cm^2}$ is graphene mass density, and D_{ac} is a deformation potential, which has been reported between 10 and 30 eV in the literature.

4.3.5.3 Optical phonon

The optical phonons have relatively lower contribution on the low-field mobility since the acoustic phonons and impurity scatterings are dominant scattering mechanisms in graphene. The effects of intrinsic optical phonons are mostly due to inelastic scatterings in high-field and/or at high temperature, which is important in determining the velocity saturation in graphene [69]. The low-field mobility limited by intrinsic optical phonons in graphene can be obtained by [70],

$$\mu_{op} = \frac{e\rho_m v_F^2 \omega_{op}}{2\pi D_{op}^2} \frac{1}{nN_{op}} \tag{4.20}$$

where, D_{op} is the mode-specific optical deformation potential of graphene, ω_{op} is the optical phonon frequency, and N_{op} is the phonon occupation. The strongest intrinsic electron–phonon coupling in graphene is because of the zone-edge transverse optical mode, which has the energy around $\hbar\omega_{op} \approx 160\,\text{meV}$ and deformation potential $D_{op} = 25.6\,\text{eV/Å}$ [70].

4.3.5.4 Surface polar phonons

Another source of inelastic scatterings in graphene is surface polar phonons (SPPs), which is the coupling of electrons in graphene to thermally excited SPP phonons on the substrate [46]. The SPP effect in graphene FET is much more important than in a conventional MOS FET since the graphene layer has much smaller vertical dimension and consequently SPP phonons can induce higher electric field on the nearby sheet to manipulate the electron transport in graphene [71]. The strength of the dielectric polarization field depends on phonon frequencies ($\omega_{so,v}$) and dielectric constants in the substrate and gate materials, which is given by the Fröhlich coupling [72] as follows,

$$F_v^2 = \frac{\hbar\omega_{so,v}}{2\pi}\left(\frac{1}{\varepsilon_\infty + \varepsilon_{env}} - \frac{1}{\varepsilon_0 + \varepsilon_{env}}\right) \tag{4.21}$$

where, $\hbar\omega_{so,v}$ is surface phonon energy, ε_∞ and ε_0 are the dielectric constant of polar substrate in high and low frequencies, and ε_{env} is environment screening above the polar dielectric. The low-field mobility of graphene limited by SPP phonons can be calculated by [69],

$$\mu_{SPP}^{-1} = \sum_{v,n}\left(\sqrt{\frac{\beta}{\hbar\omega_v}}\frac{\hbar v_F}{e^2}\frac{ev_F}{F_v^2}\frac{\exp(k_0 z_0)}{N_{SPP,v}\sqrt{n}}\right)^{-1} \tag{4.22}$$

where, $k_0 \approx \sqrt{(2\omega_{so,v}/v_F)^2 + \alpha n}$. The parameters $\alpha \approx 10.5$ and $\beta \approx 0.153 \times 10^{-4}\,\text{eV}$ are fitting parameters, $N_{SPP,v}$ is SPP phonons occupation number, and z_0 is the van der Waal distance between the polar substrate and graphene sheet. The coupling of carriers in graphene to SPP phonons on SiO_2 substrate is much stronger than the intrinsic acoustic phonons of graphene since the van der Waal distance of the substrate is small $z_0 \approx 3.5\,\text{Å}$ [45].

4.3.6 *High-field transport*

By applying a low electric field (E) across the graphene flake, a carrier can achieve a velocity υ given by $\upsilon = \mu \times E$, where μ is the carrier mobility. In scaled FETs, the definition of constant mobility cannot describe the speed of carriers due to the existence of high-field across the channel region. The velocity of carriers saturates at high field, which can be six times higher than the saturation velocity in conventional semiconductors reaching as high as $\sim 6 \times 10^7$ cm/s for gapless large area graphene [73, 74]. The saturation in a field effect transistor is occurred when the carrier density decreases and correspondingly the voltage drop becomes high in a small area under the channel, where the induced high-field maximizes the carriers velocity to saturation velocity. In a conventional FET, the saturation region (pinch-off) is induced in drain-side of channel, which continuously increases by increasing the drain-source voltage. In a graphene FET, the minimum carrier density and correspondingly the maximum field can be achieved at Dirac point. Increasing the drain-source voltage decreases the carrier distribution toward the drain region and finally sets the Dirac point at drain-side of the channel as shown in Figure 4.2(b,c). By increasing drain-source voltage, however, the saturation region cannot extend and the Dirac point moves towards the source-side of the channel. The carriers type between the Dirac point and drain changes to electrons, leading to an increase in the carrier density again. Thus, the slope of the current in I-V curve of graphene decreases when the drain-source voltage becomes equal to the Dirac point, which corresponds to its appearance at drain-side of the channel. The current of the graphene FET enters the second linear regime (hole regime) when the drain-source voltage becomes more than Dirac voltage (D plot in Figure 4.2(c)). The optical phonon scattering contributes one order of magnitude higher than the acoustic phonons scattering at high-energy carrier transport [75]. The υ_{sat} can be expressed by inelastic emission of optical phonons at high-field as follows,

$$\upsilon_{sat}(n,T) = \frac{2}{\pi}\frac{\omega_{OP}}{\sqrt{\pi n}}\sqrt{1 - \frac{\omega_{OP}^2}{4\pi n v_f^2}\frac{1}{1+N_{OP}}} \tag{4.23}$$

where ω_{OP} is the effective frequency of the phonon responsible for the current saturation, $N_{OP} = 1/[\exp(\hbar\omega_{OP}/K_B T) - 1]$ is the phonon occupation, which applies the temperature dependence of the generated optical phonon scattering. It can be simplified to $\upsilon_{sat}(n) = (2/\pi)\omega_{OP}/(\pi n)^{1/2}$ at high carrier density, which corresponds to assuming only the contribution of carriers in the energy window $E_F \pm \hbar\omega_{OP}/2$ [69]. At low carrier density and low temperature, the saturation velocity is maximized, $\nu_{\max} = (2/\pi)\nu_F \approx 6.3 \times 10^7$ cm/s. For SiO_2 substrate, optical phonon and graphene zero-edge phonons have the energies of 55 and 160 meV, respectively, while the optimized equivalent energy of the optical phonons is equal to 81 meV [46] indicating the importance of substrate polar phonons in calculating the velocity saturation in graphene. The carrier density and temperature dependence of saturation velocity is shown in Figure 4.3(b). Shishir *et al.* [74] demonstrated that the drift velocity can increase initially by increasing electric field at low carrier density, leading to a velocity overshoot and a negative differential conductance. Meric *et al.* [76] showed that the transconductance of graphene FET is consistent with the velocity saturation and independent of channel length.

4.3.7 Low-field mobility

In diffusive regime, the low-field mobility of substrate-supported graphene can be degraded by the scattering mechanisms, such that the effective mobility of graphene

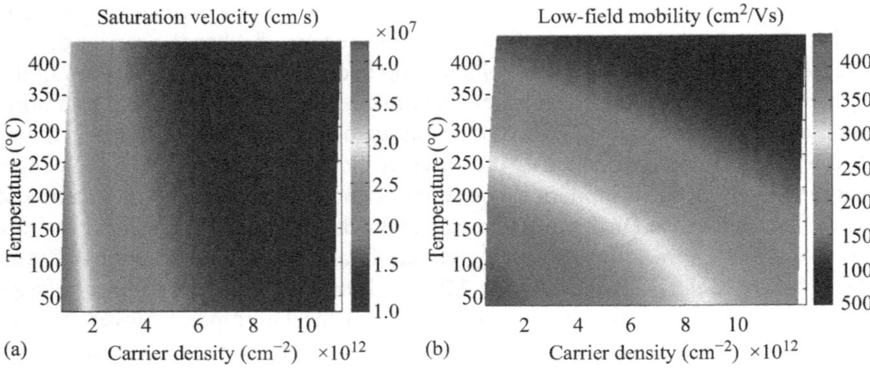

Figure 4.3 (a) Low field mobility and (b) saturation velocity of graphene as a function of carrier density and temperature

can be extracted by the Matthiessen's rule [69] by adding the inverse of the mobilities of scattering mechanisms as follows,

$$\mu_{eff}^{-1} = \mu_{ac}^{-1} + \mu_{op}^{-1} + \mu_{SPP}^{-1} \tag{4.24}$$

In suspended graphene, the extrinsic source of scattering such as charged impurities, substrate polar phonons and ripples can be eliminated, achieving the mobilities in excess of 200,000 cm^2/Vs for gapless large-area graphene [65], which is much higher than Si ($\sim$1000 cm^2/Vs). However, the suspended graphene is not a good choice for field effect transistors since the electrostatic attraction of charges in the detached gate allows only applying small gate voltages [77]. As the dominant scattering mechanism changes from Coulomb to phonon scattering by increasing carrier density and temperature, the mobility can be modeled by [46],

$$\mu(n, T) = \frac{\mu_0}{1 + (n/n_{ref})^{\alpha}} \times \frac{1}{1 + (T/T_{ref} - 1)^{\beta}} \tag{4.25}$$

where $\mu_0 = 4650\,cm^2/Vs$, $n_{ref} = 1.1 \times 10^{13}/cm^2$, $T_{ref} = 300\,K$, $\alpha = 2.2$, and $\beta = 3$. The carrier density and temperature dependence of low-field mobility is shown in Figure 4.3(a). In conventional semiconductors, hole mobility is significantly lower than the electron mobility, while it can be more than electron mobility in graphene. Low-field mobility of short-channel device decreases with the channel length since the transport regime changes from diffusive to quasi-ballistic [76].

4.3.8 Substrate and gate dielectrics

The roughness and impurities of the substrate can introduce additional scattering in graphene, urging the alternatives for SiO_2 substrates. The charged impurities in an isolating substrate generate inhomogeneous potential fluctuation, which can induce electron and hole puddles in graphene as a consequence of screening failure of insulator layer [78], causing impurity scattering as well. The screening strength of dielectric environment is described by the effective fine structure constant [60],

$$r_s = 4\pi e^2/(K_1 + K_2)hv_F \tag{4.26}$$

where, K_1 and K_2 are the dielectric constants of gate insulator and substrate on two sides of the graphene sheet. Increasing the dielectric constants $K_1 + K_2$ increases the short-range scattering of disorders in graphene [64] while it reduces the long-range scattering of charged impurities and the electron-phonon coupling at the high-symmetry point K [79], resulting in a possible increase of the carrier mobility [80]. Ponomarenko *et al.* [81] improved the mobility by covering graphene with high-dielectric liquids. When the impurity is the dominant mechanism, it can be reduced significantly by increasing the dielectric constant in the range of 1 to 47 while further increase cannot improve the mobility since the dominant limiting mechanism changes to the short-range scattering [62]. The positive charged impurities in graphene are reduced by increasing dielectric constant, which shifts Dirac point toward zero, decreases the minimum conductivity and narrows the minimum conductivity plateau. The high-K insulator has polar nature and soft energy bonds, which increase the coupling between remote phonons in the insulator to the plasmon in graphene known as interfacial plasmon–phonons (IPP) [82]. The electron coupling with remote phonons such as IPP and SPP phonons is another increasing source of scattering in high-K dielectrics, which counteracts any mobility improvement gained from charged impurity screening [77]. The phonons scatterings are predicted to change significantly with the insulator thickness, leading to the lower mobility in a thicker top oxide gate [83]. IBM has reported fabrication of the interfacial polymer layer in between high-K and graphene sheet, which can diminish the impact of remote phonon and charged impurity scatterings on carrier transport in graphene. Variety of gate dielectric have been studied for integration with graphene such as Al_2O_3 [84], AlN [85], Si_3N_4 [86], and HfO_2 [87].

Hexagonal boron nitride (h-BN) is a very promising candidate to integrate with graphene [88–90]. It has large surface optical phonon modes and consequently the lowest remote phonon scattering in thin insulators [82]. This increases the high-temperature and high-electric field performance of graphene on h-BN substrate. The h-BN has the same dielectric constant ($\varepsilon \cong 4$) and breakdown voltage ($V_B \cong 0.7$ V/nm) as SiO_2 insulator layer. However, it has larger band gap ($E_g \cong 5.97$ eV) and atomically smoother surface with similar lattice constant close to $\sim$1.7 per cent that is free of dangling bonds and charge traps [79, 91]. Epitaxial growth of graphene on SiC substrate is also promising as it allows the mass production and increases the effective van der Waal distance due to the existence of an intermediate dead layer between graphene and SiC substrate. However, the carrier motilities of epitaxial graphene on SiC substrate are smaller than exfoliated graphene on SiO_2 substrates [92] and is still the cost ineffective process [93].

4.3.9 Joule heating

The generated self-heating of graphene on the substrate at high bias can limit the saturation velocity [9] such that the experimental measurement of the saturation velocity in graphene at room temperature is significantly smaller than the theoretical value due to increase in scattering mechanisms by intrinsic phonon, SPP, acoustic phonon, and charged impurities. Perebeinos *et al.* [69] have modeled the generated Joule heating of the field (E) in Boltzmann transport calculation by the following equation,

$$T = T_{amb} + jE/r \tag{4.27}$$

where T_{amb} is the ambient temperature, $j = j(T)$ is the current density, which solves self-consistently to consider the Joule losses. The parameter $r = K/h$ models thermal conductance of dissipated heat by the substrate, where K and h are the thermal conductivity and thickness of the insulating substrate. The current density has been found to drop by approximately a factor of four due to the self-heating on SiO_2 substrate where $r \approx 0.47\,\mathrm{kW/(K\,cm^2)}$ for the insulator height $h = 300\,\mathrm{nm}$. The heat generated in a graphene channel has to cross the graphene/substrate interface, thereby experiencing the thermal contact resistance, known as Kapitza resistance. The high SPP scattering in SiO_2 substrate minimizes the effect of the thermal contact resistance. In contrast, the thermal contact resistance of SiC, h-BN, and HfO_2 substrates is much higher than substrate thermal conductance since the SPP scattering in graphene/substrate interface is lower than inside SiO_2 and the substrate thermal conductance is much higher than SiO_2 substrate. For instance, the graphene/substrate contact resistance of SiC substrate is more than a factor of ten larger than the SiO_2 substrate [13]. Thus, for increasing the saturation velocity, the heat dissipation limited by the graphene/substrate interface is an important factor to manipulate in substrate with high thermal conductance [94,95], while the thermal conductivity and thickness of the substrate are the dominant factors for the heat dissipation in SiO_2 substrate. Dorgan *et al.* [46] have modeled the average temperature of the graphene on SiO_2 substrate considering thermal resistances of the graphene/SiO_2 boundary (R_B), SiO_2 substrate (R_{ox}), and Si wafer (R_{Si}) as follows,

$$\Delta T = T - T_0 \approx P(R_B + R_{ox} + R_{Si}) \tag{4.28}$$

where P is the power delivered to graphene sheet, $R_B = 1/(hA)$, $R_{ox} = t_{ox}/K_{ox}A$, and $R_{Si} \approx 1/2K_{Si}A^{1/2}$. $h \approx 10^8\,\mathrm{Wm^{-2}\,K^{-1}}$ is thermal conductance of graphene/SiO_2 boundary, $A = LW$ is graphene channel area, K_{ox} and K_{Si} are thermal conductivities of SiO_2 and doped Si wafer. For graphene on SiO_2 substrate with thickness of 300 nm, the Joule heating contribution of the graphene/SiO_2 boundary, SiO_2 substrate, and Si wafer have been reported to be 4%, 84%, and 12%, respectively.

4.3.10 Contact resistance

In graphene transistors, contact resistance limits the performance as the graphene conductivity is much higher than in silicon MOSFET, and thereby lower contact resistance is required for realization of viable graphene material. For MOS FET, the contact resistance needs to contribute less than 10% of on-state resistance V_{DD}/I_{ON} [2]. For a junction, the screening length is given by,

$$\chi^{-1} = \sqrt{4\pi N(E_f)} \tag{4.29}$$

where $N(E_f)$ is the density of state at Fermi level. For a metal–metal junction, there is an abrupt change in the vacuum level without potential barrier at interface and thereby χ is very small. As graphene is a two-dimensional material and thereby its density of state is low, the Fermi level and screening length can be significantly changed by a small charge transfer, resulting in Fermi-level pinning and/or the dipole formation at the interface [96]. It can lead to the formation of p-n junction for positive gate voltages, resulting in asymmetric transfer characteristics of graphene field-effect transistors [97]. The potential barrier at the metal-graphene contact is presented by the photocurrent distribution [98]. The current crowding at the contact interface depends on the type of metal and contact length [99]. As the gate voltage can tune the DOS of graphene at the interface with contacts, the contact resistance has gate voltage dependence as well, such that it has been measured 300–500 Ωμm for Ti contacts at a charge density of $3 \times 10^{12}/\text{cm}^2$ [100]. The charge transfer of carrier from nickel contact to graphene is large due to the large work function difference between graphene and Ni, leading to contact resistivity range of 400–2000 Ωμm. It has been shown that the contact resistance can reduce to 100 Ωμm because nickel contact can make strong chemical bonds with zigzag-terminated graphene [101]. Grassi *et al.* [102] observed negative differential resistance due to the effect of contact-induced energy broadening and Dirac point in the source and drain regions.

4.3.11 Quantum capacitance

The quantum capacitance is an important quantity in operation of reduced-dimensional devices, which describes the response of the channel charge to the movement of the conduction and valence bands due to gate electrostatic potential. In a graphene sheet with a channel electrostatic potential V_S and the total charge density Q, the quantum capacitance is defined as $C_Q = dQ/dV_S$. Assuming a uniform channel potential, the quantum capacitance for 2D graphene can be obtained by [43],

$$C_Q = \frac{2q^2kT}{\pi(\hbar \upsilon_F)^2}\ln\left[2\left(1+\cosh\frac{qV_{ch}}{KT}\right)\right] \tag{4.30}$$

Xu *et al.* [103] showed that the above equation has excellent agreement with experimental results at large channel potential while the quantum capacitance deviate from theory and has a finite value due to residual carrier near Dirac point [104]. In general, the quantum capacitance of a clean channel at finite temperature is a function of its DOS $D(E)$ and can be determined as follows [105],

$$C_Q = q^2\int_{-\infty}^{+\infty} D(E)\left(-\frac{\partial f(E-E_f)}{\partial E}\right)dE \tag{4.31}$$

For GNR, it is confined in the transverse direction along with atomically thin in vertical direction, resulting in very low density of state. Thus, the corresponding quantum capacitance of GNR is also very low [20]. For armchair GNR, the edge state is small and thereby the charge distribution in the transverse direction is uniform, leading to a uniform voltage drop over the gate oxide [106]. Thus, the gate voltage can be obtained by summing the channel electrostatic potential and oxide voltage drop as $V_G = V_{ox} + V_S$, which leads to the following expression,

$$\frac{dV_G}{dQ} = \frac{dV_S}{dQ} + \frac{dV_{ox}}{dQ} \tag{4.32}$$

Thus, the total gate capacitance can be defined as series of two capacitances as follows,

$$\frac{1}{C_G} = \frac{1}{C_Q} + \frac{1}{C_E} \tag{4.33}$$

where $C_E = dQ/dV_{ox}$ is insulator capacitance and can be calculated as follows,

$$C_E = N_G \kappa \varepsilon_0 \left(\frac{W}{t_{ox}} + \alpha \right) \tag{4.34}$$

where N_G is the number of gates, κ is the relative dielectric constant of oxide layer, W is the channel width, t_{ox} is the gate insulator thickness, and $\alpha \cong 1$ is a dimensionless fitting parameter due to the electrostatic decay at the channel edge. In series combination of the quantum capacitance and the gate insulator capacitance, the smaller capacitance has a dominant effect in determining the gate capacitance [107].

4.4 Modeling and simulation

4.4.1 Classical transport

For the simulation of graphene transistors with the width and length of several microns, the carrier transport in graphene can be modeled using classical methods governed by Newtonian mechanics such as drift-diffusion [17, 108] and charge collection approach [9, 18]. The scattering effects are significant in classical transport calculation of graphene as the mfp of carriers is smaller than the channel length. In charge-collection model, the current in the graphene channel can be expressed by,

$$I_d = \frac{W}{L} q \int_0^L n(x) v_{drift}(x) dx \tag{4.35}$$

where, L and W are the channel length and width, $n(x)$ is the carrier density in graphene channel which can be find using (4.8). The carrier drift velocity v_{drift} is given by,

$$v_{drift}(x) = \frac{\mu E}{[1 + (\mu E / v_{sat})^{\gamma}]^{1/\gamma}} \tag{4.36}$$

where E is electric field, $\gamma \approx 2$ is the fitting parameter, μ and v_{sat} are the low-field mobility and saturation velocity of the carriers, which can be calculated using (2.24) and (2.23), respectively.

4.4.2 Semiclassical transport

4.4.2.1 Boltzmann transport equation

In semiclassical approach, the distribution function $f(r, k, t)$, which is the probability of finding a particle with momentum k at position r and time t, is calculated in order to extract the quantities such as charge density and current density. The basic equation in this regime is Boltzmann transport equation (BTE) where the transport of carrier is treated classically, while scattering mechanisms are modeled using quantum mechanical approach known as Fermi's Golden rule [109]. Boltzmann's transport equation is as follows,

$$\frac{\partial f}{\partial t} + \frac{\overrightarrow{F_{ext}}}{\hbar} \nabla_k f + \vec{v}\, \nabla_r f = \frac{\partial f}{\partial t}\Big|_{collision} \tag{4.37}$$

where f is the distribution function of carrier, $\overrightarrow{F_{ext}}$ is the external force on carriers due to electric field, $\vec{v}$ is the group velocity of a subband particular, and $\hbar$ is the reduced Planck's constant. $(\partial f/\partial t)_{|collision}$ is the collision term described as follows,

$$\begin{aligned}\frac{\partial f}{\partial t}\Big|_{collision} &= -\sum_{\vec{k'}} S\left(\vec{k}, \vec{k'}\right) f\left(\vec{k}\right)\left(1 - f\left(\vec{k'}\right)\right) \\ &\quad + \sum_{\vec{k'}} S\left(\vec{k'}, \vec{k}\right) f\left(\vec{k'}\right)\left(1 - f\left(\vec{k}\right)\right)\end{aligned} \tag{4.38}$$

where $S\left(\vec{k}, \vec{k'}\right)$ is the scattering rate for the transition of carriers from $\vec{k}$ state to $\vec{k'}$ state, which models the out-scattering and in-scattering of carriers in the first and second terms of the above equation. The classical distribution function $f(r, k, t)$ describes position and momentum simultaneously contrary to Heisenberg uncertainty principle and thereby it is required to be corrected to incorporate the quantum effects like tunneling phenomena. Chauhan *et al.* [70] simulated the BTE using Monte Carlo method in presence of optical phonon, acoustic phonon, and charge impurity scattering mechanisms and successfully verified the experimentally observed results in [18], showing that the saturation current scales as a function of the square root of the charge density.

4.4.2.2 Top-of-the-barrier approach

Another semiclassical model for simulating GNRFET with Ohmic contacts is top-of-the-barrier approach. The model is based on self-consistent calculation of carrier transport and electrostatic potential at the top of the potential barrier in the channel as shown in Figure 4.4(a), which has been used in simulating a variety of FET structures such as conventional silicon MOS FETs [110], CNT FETs [111], nanowire

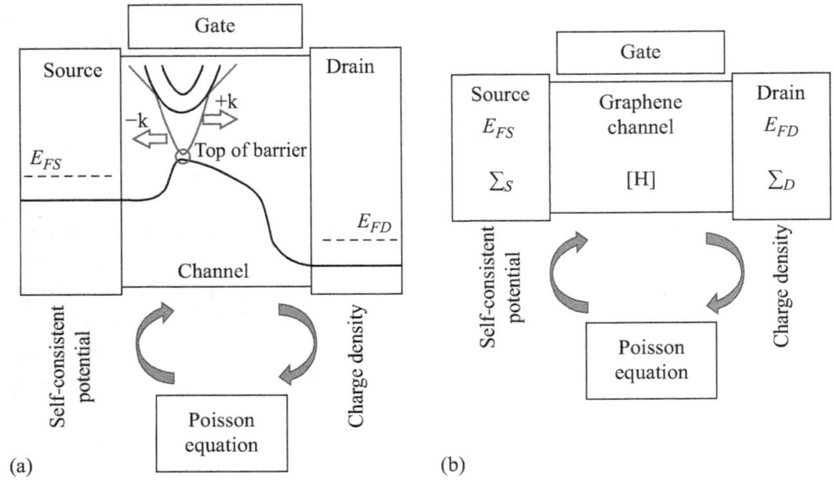

Figure 4.4 Schematic descriptions of self-consistent procedure for (a) top-of-the-barrier and (b) NEGF formalism. Note that the charge density based on transport equations and electrostatic potential based on Poisson equation are solved self-consistently. Once convergence is obtained, the drain current is calculated using transport equations

FETs [112], and GNR FETs [113]. In top-of-the-barrier model, the drain current can be calculated using Landauer equation as follows,

$$I = \frac{4e}{h}\int_{\varepsilon_{top}}^{\infty} T(E)[f_0(E - E_{FS}) - f_0(E - E_{FD})]\, dE \tag{4.39}$$

where h is the Planck's constant, $f(E)$ is Fermi function, E_{FD} and E_{FS} are the Fermi levels of the source and drain contacts. $T(E)$ is the probability of carriers to transport from one contact to another contact, which is "1" above channel barrier and "0" below the barrier for ballistic channel, providing the upper limit performance of GNR FETs. The energy at top of the potential barrier, ε_{top} is determined by the summation of Laplace potential U_L due to the applied bias at terminals and self-consistent potential U_P due to the carrier concentration in the channel [47]. For a bias condition, the device is not at equilibrium and channel states with positive and negative velocity are filled by the carrier injection from source and drain contacts corresponding to their Fermi levels. Thus, the charge density and the self-consistent potential at the top of the barrier must be solved together until full self-consistency is reached as shown in Figure 4.4(a). In Reference 113, the drain current at the ballistic limit has been obtained as follows,

$$I = \frac{2qk_BT}{h}\left[\ln\left(1 + e^{\frac{E_P - \varepsilon_{top}}{k_BT}}\right) - \ln\left(1 + e^{\frac{E_P - \varepsilon_{top} - qV_D}{k_BT}}\right)\right] \tag{4.40}$$

where k_B is Boltzmann constant, E_f is Fermi level at source terminal, and V_D is voltage at drain terminal. As the probability of carrier for reaching to other contact is decreased by scattering and the effect can be included by decreasing the transmission coefficient from unity to a factor determined by the mfp of carriers λ, optical phonon energy $\hbar\omega_{op}$, and channel length L_{ch} as follows,

$$T = \begin{cases} \lambda/(\lambda + L_{ch}) & if\ qV_D < \hbar\omega_{op} \\ \dfrac{\lambda}{\lambda + (\hbar\omega_{op}/qV_D)L_{ch}} & if\ qV_D > \hbar\omega_{op} \end{cases} \tag{4.41}$$

The above top-of-the-barrier approach cannot handle the tunneling effect, which is important for a FET structure with Schottky barrier at contact interfaces and for channel with small band gap material like graphene as band-to-band tunneling is increased by decreasing the band gap. While the transmission coefficient remains 1 for the thermionic emission of carriers above the barrier, it must be modified using Wentzel–Kramers–Brillouin approach [114] in order to incorporate tunneling probability of carriers [115, 116].

4.4.3 Quantum transport

Quantum simulation is the most computationally demanding approach as the quantum effects become more and more important by scaling down the channel length. The most accurate quantum-based method for bottom-up device simulation is non-equilibrium Green's function (NEGF) approach, where Schrodinger equation is solved under non-equilibrium condition. NEGF formalism provides the atomistic description of channel material as well as the effects of contacts and scattering on carriers transport in the channel. Figure 4.4(b) illustrates the basic principle of NEGF formalism to a generic transistor, where a channel is connected to the source and drain contacts and can be modulated by gate electrostatic. For the isolated channel, the Hamiltonian matrix H is obtained along with the self-energy matrices of source Σ_S and drain Σ_D contacts, which incorporate the effects of contacts on the channel subbands. The retarded Green's function is constructed as follows,

$$G = [EI - H - U - \Sigma_S - \Sigma_D - \Sigma_{Scattering}]^{-1} \tag{4.42}$$

where E is energy, I is identity matrix, U is the self-consistent potential, and $\Sigma_{Scattering}$ incorporates the effects of scattering mechanism on the channel. Then, the level broadening quantities Γ_S and Γ_D are calculated as follows,

$$\Gamma_{S/D} = i(\Sigma_{S/D} - \Sigma^{+}_{S/D}) \tag{4.43}$$

where $+$ and i refer to the Hermitian transpose operator and the imaginary unit, respectively. Then, in-scattering of electrons and holes from the source and drain contacts are calculated by,

$$\Sigma^{<}_{S/D}(E) = i\,\Gamma_{S/D} f_{S/D} \tag{4.44}$$

$$\Sigma^{>}_{S/D}(E) = i\,\Gamma_{S/D}[1 - f_{S/D}] \tag{4.45}$$

where $f_{S/D}$ is the Fermi function of source and drain contacts. Next, the electron and hole correlation functions can be calculated as follows,

$$G^{</>}(E) = G(E)[\Sigma_S^{</>}(E) + \Sigma_D^{</>}(E)]G^{+}(E) \tag{4.46}$$

The electron and hole numbers at atomic site (n,α) can be achieved by integration over energy and summation over all subbands as follows,

$$n_{n\alpha} = -2i\sum_b \left[|\varphi_{n\alpha}^b|^2 \int_{E_i^b(x)}^{\infty} \frac{1}{2\pi} G_b^{<}(n,n;E)dE \right] \tag{4.47}$$

$$p_{n\alpha} = 2i\sum_b \left[|\varphi_{n\alpha}^b|^2 \int_{-\infty}^{E_i^b(x)} \frac{1}{2\pi} G_b^{>}(n,n;E)dE \right] \tag{4.48}$$

The equations provide the charge density as an entry to Poisson equation in order to find a new potential energy in self-consistent iteration. Once the convergence condition is met, transmission coefficient $T(E)$ is determined as follows,

$$T(E) = Trace[\Gamma_S G \Gamma_D G^{+}] \tag{4.49}$$

Finally, the drain current can be calculated by integral over energy and summation over all subbands as follows,

$$I_{DS} = \frac{q^2}{h} \int_{-\infty}^{\infty} \sum_b \frac{4}{q} \Re\{H_b(n,n+1;E)G_b^{<}(n+1,n;E)\}dE \tag{4.50}$$

where symbol $\Re$ indicates real part, and h is the Planck constant. As the electron–phonon interaction is weak at room temperature [117], the mfp of carriers in GNR is around hundreds of nanometers. Thus, scattering term in (4.42) can be neglected for short channel GNR FETs, maintaining the accuracy of device simulation based on the ballistic transport. Assuming coherent transport, the above equation is reduced to Landauer equation as shown in (4.39). The discretization of device Hamiltonian (H) provides two alternative approaches for applying NEGF formalism: real space formulation [118] which can be used directly for any geometry and mode space formulation [119] which splits up the device simulation into a set of 1D problems over subbands. Mode-space approach can be applied for simulation of GNR FET by assuming smooth edges and negligible potential variation in transverse direction. It has been successfully applied for simulating a variety of nanometric channel materials such as carbon nanotube [120, 121], silicon MOS FET [122], and GNR [123, 124]. As power supply is scaled down at the same time with the scaling of channel, only a few lowest subbands participate in carrier transport and need to be taken into account, which leads to significant computational advantage. The energy-position-resolved local DOS of GNR (6,0) and (21,0) are shown in Figure 4.5(a,b), respectively, where the bandgap energies with low local DOS along with channel potential barrier are apparent for both GNR FETs. It can be seen that the band-to-band tunneling from drain to channel can be captured for GNR (21,0) with smaller bandgap.

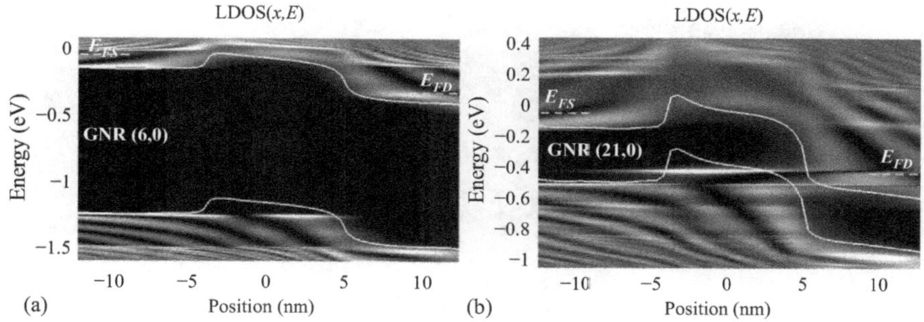

Figure 4.5 *Local DOS of GNRFET with the channel of (a) GNR (6,0) and (b) GNR (21,0). Note that the gate voltage $V_G = 0.5$ V and drain voltages of GNR (6,0) and GNR (21,0) are $V_D = 0.3$ and 0.4 V, respectively. The oxide thickness t_{ox} and the physical gate length L_G are 1 and 5 nm, respectively. The dielectric layer is assumed to be aluminum nitride (AlN) with the relative dielectric permittivity $k = 9$*

4.5 GNR FET

A full quantum transport model based on NEGF formulism is developed for the simulation of GNRFET [125, 126], where the energy-position dependent Hamiltonian is employed using non-parabolic effective mass model [127].

4.5.1 Graphene nanoribbon

Semimetallic nature of graphene with zero bandgap limits its application for logic transistors as it cannot fully switched off by tuning the Fermi level at the energy that conduction and valence bands cross each other. The required band gap of several hundred megaelectron volts can be introduced by quantum confinement of carriers in one-dimensional graphene, called GNR [14, 15]. Atomically smooth GNRs with width down to 1 nm has been already produced by unzipping of carbon nanotubes with bottom-up chemical approach [32]. The GNR is categorized in two typical types of armchair and zigzag depending on the ribbon edge since both the ribbon width and the direction of cutting graphene determine whether the GNR is metallic or semiconducting. The atomic view of armchair-edged GNR (N,0) is shown in Figure 4.6(a), where, N is called width index or ribbon index equal to the number of dimer lines in transverse direction. The atomic-level first principle calculation can obtain the electronic structure of GNR. It can be solved either by Dirac's equation of massless particles with an effective speed of light [128] or simple tight-binding (TB) approximation [129, 130].

TB is the state of the art for the calculation of GNR dispersion relation because the single-orbital Hamiltonian matrix can be constructed based on the nearest neighbor orthogonal P_z orbitals as basis functions equal to the number of atoms in a desired unit cell across width direction as shown in Figure 4.6(b). The TB calculation in a slab

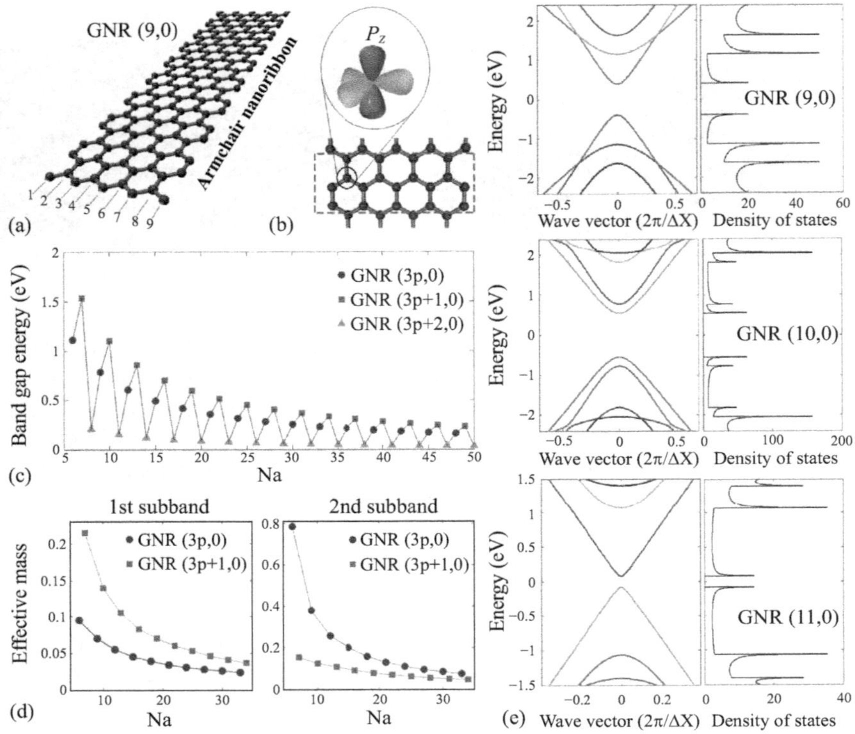

Figure 4.6 *(a) Structure of armchair-edge graphene nanoribbon, GNR (9,0), (b) slab of GNR (9,0) used in TB calculation as well as the outer p-orbitals of a carbon atom, (c) bandgap energy of three GNR families at the CNP versus GNR index, (d) effective mass of first and second subbands obtained by effective mass model for GNR (3p+1,0) and GNR (3p,0), and (e) band structure and DOS of GNR (9,0), GNR (10,0) and GNR (11,0)*

with 2N carbon atoms results in N conduction bands and N valence bands in energy dispersion relation of GNR. The elements of the Hamiltonian matrix between the αth atom within the nth slab and the βth atom within the mth slab is given by,

$$H_{n\alpha,m\beta} = H^0_{n\alpha,m\beta} + \delta_{n\alpha,m\beta} U_{n\alpha} \tag{4.51}$$

where $\delta_{n\alpha,m\beta}$ is the Kronecker delta and $U_{n\alpha}$ is the electrostatic potential energy at the (n,α) atom site. $H^0_{n\alpha,m\beta}$ is equal to the nearest neighbor hopping energy $t = -2.7$ eV if the atoms (n,α) and (m,β) are the first nearest neighbors, and equal to zero otherwise. All dangling bonds at the edge of the graphene lattice are assumed to be abruptly terminated and occupied by hydrogen atoms. The edge bond relaxation has been modeled by the modified hopping energy for atoms pairs along the edges equal to

$t(1+\gamma)$ with $\gamma = 0.12$ [129]. The first principle calculations shows identical results near the Fermi level of GNR [127]. It is shown in [131] that the edge bond relaxation has a significant effect on band gap energy and effective mass of GNR.

The zigzag-edge ribbons are always metallic, while the armchair-edge GNR can be semiconductor such that the corresponding bandgap energy decreases by increasing the ribbon index (width) as shown in Figure 4.6(c). The variation in band gap energy can be separated in three GNR families depending on ribbon index N, such that the ribbon with N = 3p and 3p+1, where p is an arbitrary integer, are semiconducting and 3p+2 is metallic with very small band gap energy. The band gap can be opened by quantum confinement of GNR in one dimension but it degrades the linearity of subbands near Dirac points. The reduction in electron velocity can be modeled by effective mass extraction [127, 132] as shown in Figure 4.6(d) for the first and second subbands. Figure 4.6(e) shows the calculated band structure and DOS of GNRs with N = 9, 10, and 11, resulting the corresponding bandgap energy $E_g = 0.78$, 1.1, and 0.15 eV, respectively.

Figure 4.7(a,b) shows the first conduction and valence bands of GNR (24,0) and (13,0), respectively and three Fermi levels in correspondence with three gate voltages. At positive gate voltage V_{GS1}, the Fermi level is in near or inside conduction band, only electrons contribute in carrier transport calculation while GNR (24,0) results in larger current as the smaller band gap can place Fermi level well inside conduction band. Decreasing the gate voltage shifts the Fermi level toward the valence band, the total carrier in the channel decreases and consequently minimizes at a gate voltage (V_{GS2}) corresponding to CNP, where the electron current is equal to the hole current. With regards to induced band gap energy and the Fermi distribution function, smaller carriers are in the conduction and valence bands to contribute in carrier transport of narrower GNR, e.g., GNR (13,0), and thereby the leakage current can be an order of magnitude smaller than the wider GNR. At negative gate voltage, V_{GS3}, the hole current increases as the Fermi level is near the valence band, showing the partial recovery of ambipolar transport with regards to subthreshold region as experimentally observed for the GNR with reduced impurity [133]. It can been seen that smaller leakage current is obtained at the expense of degrading band-linearity for narrower GNRs, which can be modeled using effective mass model with regard to the generated bandgap of nanoribbon [127].

While the gate insulator capacitance is smaller in conventional MOSFETs, the small quantum capacitance of GNR can be dominant, especially for a device with a thin t_{ox} and high-k dielectric constant. Thus, increasing insulator capacitance cannot make significant increase in equivalent gate capacitance of GNR FET at quantum capacitance limit and the DOS of a GNR is an important factor, which can alter the quantum capacitance as a function of gate voltage depending on subbands locations in energy as shown in Figure 4.7(c). For instance, while GNR (10,0) and GNR (6,0) have the same energy band gap, they have different quantum capacitance due to difference in the DOS. For GNR (3p+1,0), the second subband is close to the first subband and both subbands have larger effective masses than the first subband of GNR (3p,0), leading to a larger quantum capacitance for GNR (6,0) with sharper increase with increasing gate voltage [134].

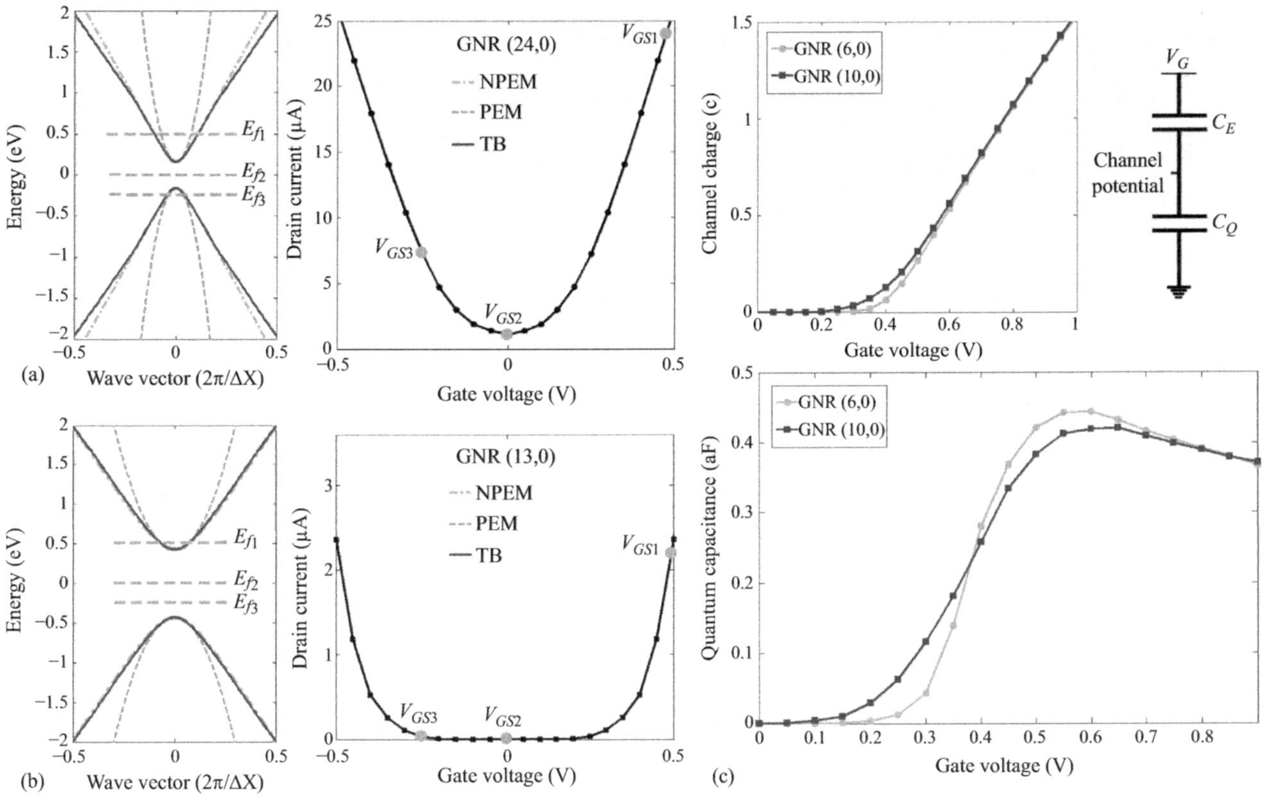

Figure 4.7 *First conduction and valence bands of (a) GNR (24,0) and (b) GNR (13,0) and corresponding drain current-gate voltage characteristics. Note that three Fermi levels and the corresponding gate voltages are also shown. (c) Channel charge and corresponding quantum capacitance versus gate voltage for GNR (6,0) and GNR (10,0). Note that the series configuration of electrostatic capacitance and quantum capacitance is also shown*

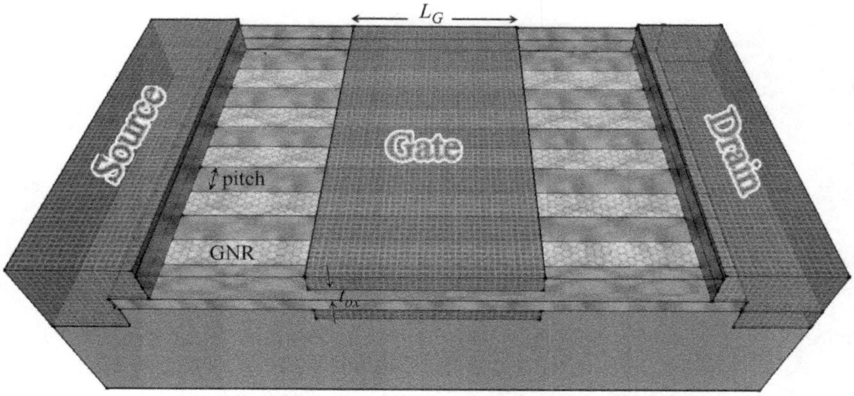

Figure 4.8 *3D schematic view of GNRFET structure with five parallel graphene nanoribbons in connection with two wide metallic contacts*

4.5.2 Device structure

As discussed earlier for Figure 4.7, the narrow GNR has the adequate band gap, which can decrease the off-state current and consequently increase the I_{ON}/I_{OFF} ratio as has been already demonstrated [135]. As the nanometer wide ribbon can have also smaller drive current, multiple parallel GNRs have been fabricated as the channels of a GNR FET [27, 136]. The GNR FET structure has been used in our simulation as shown in Figure 4.8, where GNRs are sandwiched between two thin insulator layers in a double-metal gate topology in order to maximize the electrostatic control of the gate electrostatic on the channel. The dielectric layer is assumed aluminum nitride (AlN) with the relative dielectric permittivity $k = 9$ as it reduces the phonon scattering in epitaxial graphene [137], and its thin film production is cost-efficient with good reproducibility and uniformity [138, 139]. The extensions of source and drain regions is heavily doped with n-type dopants in order to form Ohmic junctions between graphene and source (or drain) contacts minimizing contact resistances. The length of the metallic gates is assumed equal to the length of intrinsic GNR channel, and the pitch size between ribbons is kept equal to the width of ribbons. The oxide thickness t_{ox}, the physical gate length L_G, and the power supply voltage V_{DD} are variable parameters, which have been assigned based on the roadmap presented in ITRS report [2].

4.5.3 Device performance metrics

The I_{DS} versus V_{DS} for different V_{GS} of GNR FET is shown in Figure 4.9(a), where the channel is GNR (6,0) with the gate length of $L_G = 5$ nm. The strong saturation region indicates good MOSFET-type device behavior of GNR FET, where the saturation slope is significantly determined by GNR width rather than the gate length [140]. The transfer characteristics I_D-V_G for different drain voltages is shown in Figure 4.9(b). For a given drain voltage, the minimum current is obtained at CNP corresponding to

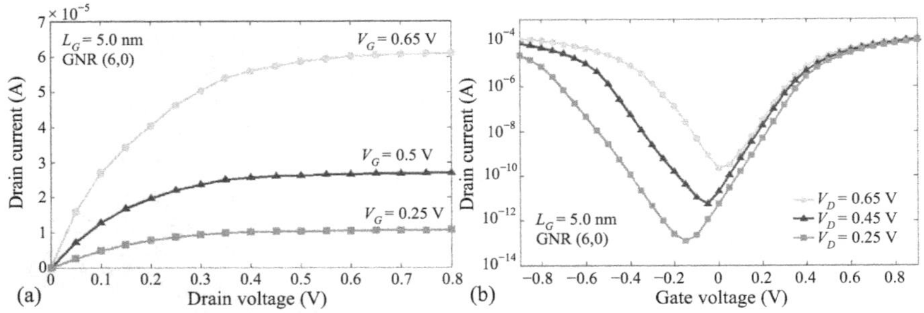

Figure 4.9 (a) Drain current versus drain voltage characteristics of GNRFET for varying gate voltages and (b) drain current versus gate voltage characteristics of GNRFET for varying drain voltages

a gate electrostatic potential at which the contribution of electron current becomes equal to hole current. The accumulation of holes in the channel is increased by increasing drain voltage due to increase in band-to-band tunneling from the source contact to channel, resulting in larger minimum current and shift of CNP to positive gate voltages.

The transfer characteristics of GNR FETs with different channel lengths are shown in Figure 4.10(a). In a short channel device, decreasing gate length results in significant decrease in both height and width of channel barrier, which can lead to the increase in thermionic current from over the channel barrier and direct tunneling current through the channel barrier, respectively [22, 141]. Increasing the drain voltage can lower the channel barrier at the beginning of the channel leading to degradation of gate electrostatic controllability on carrier transport in the short channel devices, known as drain-induced barrier lowering. Thus, the off-current density of GNR (6,0) is increased from 1.66×10^{-3} μA/μm to 0.36 μA/μm by scaling down the gate length from 10 to 5 nm as shown in Figure 4.10(b). It can be seen that GNR (6,0) with channel length larger than 6.5 nm can have smaller off-state current than that of silicon-based channels, 100 nA/μm for high-performance digital integrated circuit [2], while it remains more than two orders of magnitude larger than that of low-power criterion by scaling the channel length, and thus it is not a suitable channel material and structure. As band gap engineering of GNR FETs is possible, it can be designed for low power application with decreasing GNR width while high performance design is more desirable since graphene channel can have very high drive current at low voltage supply due to very large carrier injection velocity corresponding to the small effective mass of carriers. It can be seen in Figure 4.10(c) that GNR FET with 5-channels of GNR (6,0) shown in Figure 4.8 can have approximately two orders of magnitude larger on-current density than predicted by ITRS [2].

It is predicted [2] that channel material like graphene can continue the improvement of switching speed (I/CV) as the main criterion for logic scaling. The switching

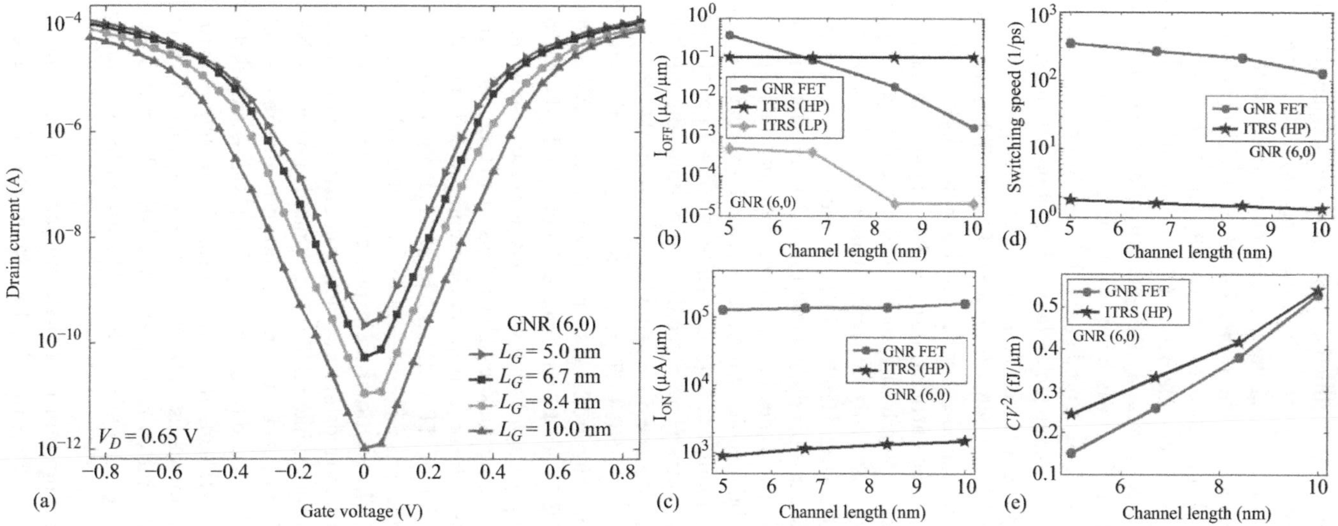

Figure 4.10 *(a) Transfer characteristics of GNRFET for different physical channel lengths, (b) off-current density, (c) on-current density, (d) switching speed, and (e) dynamic power. Note that the projections of ITRS are also shown in (a) to (e)*

speed of GNR FET can be much higher than conventional MOSFET as the carriers in graphene have very small effective mass and consequently can response to much higher clock frequencies. For GNR (6,0), the switching speed increases from 127 (1/*ps*) to 347 (1/*ps*) by scaling the channel length from 10 to 5 nm as shown in Figure 4.10(d), outperforming the projection of MOS FET by approximately two orders of magnitude. While the channel scaling increases the static power consumption, it decreases the dynamic power CV^2 as both the supply voltage and gate capacitance are decreased by scaling the gate length as shown in Figure 4.10(e). It can be seen that GNR FET shows better dynamic power performance by channel scaling than the MOS FET. It is predicted [2] that contrary to the tradeoff between power and speed in silicon-based technologies, high mobility materials can have a net improvement in total power consumption along with the operation speed.

By increasing the GNR width, off-current of GNR FET is increased and threshold voltage is decreased due to a smaller band gap and higher number of available conducting subbands in the carrier transport. The transfer characteristics I_D-V_G of four GNRs with different widths are shown in Figure 4.11(a). It can be seen that GNR (6,0) and GNR (10,0) with the similar bandgap of $E_g = 1.1$ eV have smaller off-current than GNR (12,0) and GNR (19,0) with the similar bandgap of $E_g = 0.6$ eV. Although they have the same bandgap, GNR (3p,0) has smaller effective mass, which leads to approximately two times higher off-current than GNR (3p+1,0). Both the leakage current and subthreshold swing are increased by decreasing the bandgap due to increase in the contribution of band-to-band-tunneling in total current. Although, the thermal emission of carriers over channel barrier limits the subthreshold swing of FET structure to 60 mV/decade, increasing bandgap from 0.6 to 1.1 eV can decrease the subthreshold swing of GNR FET from 118 mV/decade to 74 mV/decade. Comparing with the predicted subthreshold swing of 90 and 125 mV/decade for MOSFET and double-gate Fin-FET with 10-nm channel length [142], the narrow GNR allows the possibility of band gap engineering while the gate electrostatic have better control on GNR channel with atomic thickness. It can be seen in Figure 4.11(b,c) that increasing GNR width increases the on-current and switching speed of GNRFET, which can outperform the projection of ITRS for 5 nm channel length. Figure 4.11(d) shows the intrinsic gate-delay time versus I_{ON}/I_{OFF} ratio for four GNRs in the study. Narrower GNRs have higher I_{ON}/I_{OFF} ratio at the expense of higher intrinsic gate-delay time as smaller number of subbands are available for carrier transport, and their effective masses are higher than wider GNRs. The dynamic power of GNRs (6,0), (10,0), and (12,0) as a function of gate voltage are shown in Figure 4.11(e). The dynamic power of GNRFET is increased by decreasing GNR width, such that GNR (12,0) has dynamic power of 0.08 (fJ/μm) at the scaled supply voltage of 5 nm channel length $V_{DD} = 0.64$ V, while it increases to 0.16 and 0.29 fJ/μm for GNR (10,0) and GNR (6,0), respectively. Thus, the dynamic power of GNR (12,0) and GNR (10,0) can outperform the projected value of ITRS ~0.25 (fJ/μm) for high performance integrated circuit design.

The atomistic quantum transport modeling can be implemented in circuit simulators using look-up tables as in [143, 144] in order to fully capture the atomistic scale features like quantum effects. The technology exploration for GNR FET-based

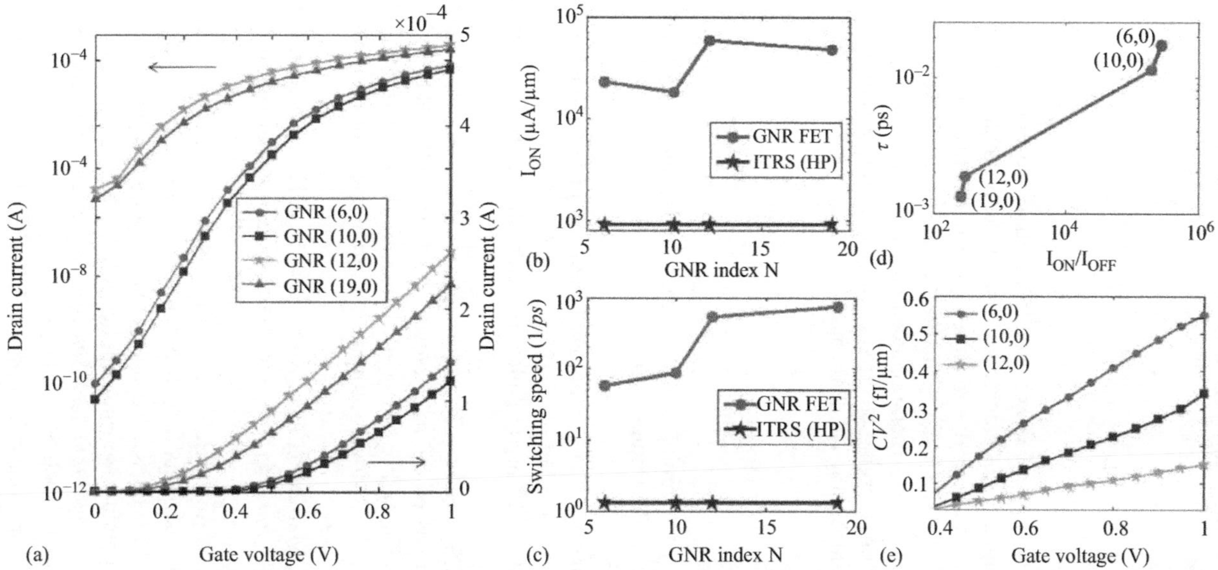

Figure 4.11 Transfer characteristics of GNRFET for different graphene nanoribbons, (b) on-current density versus GNR index, (c) switching speed versus GNR index and ITRS projection, (d) intrinsic gate-delay versus I_{ON}/I_{OFF} ratio, (e) dynamic power versus gate voltage for GNR (6,0), GNR (10,0), and GNR (12,0)

circuits can be performed in multiscale simulation framework [145, 146]. Simulations have shown that GNRFET-based integrated circuits can be designed with lower energy consumption and higher switching frequency than in scaled CMOS circuits [146] and comparable reliability.

4.6 Conclusion

Scaled-down CMOS technology is expected to encounter several challenging issues in near future [2], graphene has emerged as one of the promising alternative materials for post-CMOS technology due to large carrier motilities, high carrier velocity, high thermal conductivity, and planar structure. Novel circuits for varied applications can be designed from GNRs. Among reduced dimension materials, graphene holds tremendous promise for emerging electronics in post-CMOS electronics era though there are barriers and technological challenges.

References

[1] Moore G.E. 'Cramming more components onto integrated circuits'. Reprinted from Electronics, *IEEE Solid-State Circuits Society Newsletter*. 2006;38(8):33–5

[2] Wilson L. 'International Technology Roadmap for Semiconductors (ITRS)'. 2013. http://www.itrs.net/

[3] Iijima S. 'Helical microtubules of graphitic carbon'. *Nature*. 1991; 354(6348):56–8

[4] Novoselov K.S., Geim A.K., Morozov S., *et al.* 'Electric field effect in atomically thin carbon films'. *Science*. 2004;306(5696):666–9

[5] Neto A.H.C. 'The carbon new age'. *Materials Today*. 2010;13(3):12–17

[6] Van Noorden R. 'Moving towards a graphene world'. *Nature*. 2006; 442(7100):228–9

[7] Srivastava A., Marulanda J.M., Xu Y., Sharma A.K. *Carbon-Based Electronics: Transistors and Interconnects at the Nanoscale*. Pan Stanford Publishing, Singapore; 2015

[8] Cooper D.R., D'Anjou B., Ghattamaneni N., *et al.* 'Experimental review of graphene'. *International Scholarly Research Notes*. 2012;2012:1–56

[9] Banadaki Y., Mohsin K., Srivastava A. 'A graphene field effect transistor for high temperature sensing applications'. *Proc. SPIE 9060, Nanosensors, Biosensors, and Info-Tech Sensors and Systems*; San Diego, California, USA, 2014; 9060: 90600F-1-906003-7

[10] Mohsin K.M., Srivastava A., Sharma A.K., Mayberry C. 'A thermal model for carbon nanotube interconnects'. *Nanomaterials*. 2013;3(2):229–41

[11] Mohsin K., Srivastava A., Sharma A.K., Mayberry C. 'Characterization of MWCNT VLSI interconnect with self-heating induced scatterings'. *2014*

IEEE Computer Society Annual Symposium on VLSI (ISVLSI), Tampa, Florida, USA, July 2014, pp. 368–73

[12] Li X., Cai W., An J., *et al.* 'Large-area synthesis of high-quality and uniform graphene films on copper foils'. *Science*. 2009;324(5932):1312–4

[13] Schwierz F. 'Graphene transistors: status, prospects, and problems'. *Proceedings of IEEE*. 2013;101(7):1567–84

[14] Harada N., Sato S., Yokoyama N. 'Theoretical investigation of graphene nanoribbon field-effect transistors designed for digital applications'. *Japanese Journal of Applied Physics*. 2013;52(9R):094301

[15] Johari Z., Hamid F., Tan M.L.P., Ahmadi M.T., Harun F., Ismail R. 'Graphene nanoribbon field effect transistor logic gates performance projection'. *Journal of Computational and Theoretical Nanoscience*. 2013;10(5): 1164–70

[16] Chen M.T., Wu Y.R. 'Numerical study of scaling issues in graphene nanoribbon transistors'. *MRS Proceedings*, Cambridge University Press. 2011; 1344: mrss11-1344-y03-23

[17] Ancona M.G. 'Electron transport in graphene from a diffusion-drift perspective'. *IEEE Transactions on Electron Devices*. 2010;57(3):681–9

[18] Meric I., Han M.Y., Young A.F., Ozyilmaz B., Kim P., Shepard K.L. 'Current saturation in zero-bandgap, top-gated graphene field-effect transistors'. *Nature Nanotechnology*. 2008;3(11):654–9

[19] Liang G., Neophytou N., Nikonov D.E., Lundstrom M.S. 'Performance projections for ballistic graphene nanoribbon field-effect transistors'. *IEEE Transactions on Electron Devices*. 2007;54(4):677–82

[20] Ouyang Y., Yoon Y., Fodor J.K., Guo J. 'Comparison of performance limits for carbon nanoribbon and carbon nanotube transistors'. *Applied Physics Letters*. 2006;89(20):203107

[21] Rahman A., Guo J., Datta S., Lundstrom M.S. 'Theory of ballistic nanotransistors'. *IEEE Transactions on Electron Devices*. 2003;50(9): 1853–64

[22] Ouyang Y., Yoon Y., Guo J. 'Scaling behaviors of graphene nanoribbon FETs: A three-dimensional quantum simulation study'. *IEEE Transactions on Electron Devices*. 2007;54(9):2223–31

[23] Guan X., Zhang M., Liu Q., Yu Z. 'Simulation investigation of double-gate CNR-MOSFETs with a fully self-consistent NEGF and TB method'. *IEDM Tech Dig*. 2007;761(2007):764

[24] Emtsev K.V., Bostwick A., Horn K., *et al.* 'Towards wafer-size graphene layers by atmospheric pressure graphitization of silicon carbide'. *Nature Materials*. 2009;8(3):203–7

[25] Obraztsov A.N. 'Chemical vapour deposition: making graphene on a large scale'. *Nature Nanotechnology*. 2009;4(4):212–3

[26] Zhu Y., Murali S., Cai W., *et al.* 'Graphene and graphene oxide: synthesis, properties, and applications'. *Advanced Materials*. 2010;22(35):3906–24

[27] Han M.Y., Özyilmaz B., Zhang Y., Kim P. 'Energy band-gap engineering of graphene nanoribbons'. *Physical Review Letters*. 2007;98(20):206805

[28] Wang X., Dai H. 'Etching and narrowing of graphene from the edges'. *Nature Chemistry*. 2010;2(8):661–5
[29] Li X., Wang X., Zhang L., Lee S., Dai H. 'Chemically derived, ultrasmooth graphene nanoribbon semiconductors'. *Science*. 2008;319(5867): 1229–32
[30] Lu G., Zhou X., Li H., *et al.* 'Nanolithography of single-layer graphene oxide films by atomic force microscopy'. *Langmuir*. 2010;26(9):6164–6
[31] Tapasztó L., Dobrik G., Lambin P., Biró L.P. 'Tailoring the atomic structure of graphene nanoribbons by scanning tunnelling microscope lithography'. *Nature Nanotechnology*. 2008;3(7):397–401
[32] Xie L., Wang H., Jin C., *et al.* 'Graphene nanoribbons from unzipped carbon nanotubes: atomic structures, Raman spectroscopy, and electrical properties'. *Journal of the American Chemical Society*. 2011;133(27):10394–7
[33] Jiao L., Zhang L., Wang X., Diankov G., Dai H. 'Narrow graphene nanoribbons from carbon nanotubes'. *Nature*. 2009;458(7240):877–80
[34] Zhang Z., Sun Z., Yao J., Kosynkin D.V., Tour J.M. 'Transforming carbon nanotube devices into nanoribbon devices'. *Journal of the American Chemical Society*. 2009;131:13460–3
[35] Wallace P.R. 'The band theory of graphite'. *Physical Review*. 1947; 71(37): 622
[36] Neto A.C., Guinea F., Peres N., Novoselov K.S., Geim A.K. 'The electronic properties of graphene'. *Reviews of Modern Physics*. 2009;81(1):109
[37] Stampfer C., Fringes S., Güttinger J., *et al.* 'Transport in graphene nanostructures'. *Frontiers of Physics*. 2011;6(3): 271–93
[38] Semenoff G.W. 'Condensed-matter simulation of a three-dimensional anomaly'. *Physical Review Letters*. 1984;53(26):2449–52
[39] Li Z., Henriksen E.A., Jiang Z., *et al.* 'Dirac charge dynamics in graphene by infrared spectroscopy'. *Nature Physics*. 2008;4(7):532–5
[40] Trevisanutto P.E., Giorgetti C., Reining L., Ladisa M., Olevano V. 'Ab initio G W many-body effects in graphene'. *Physical Review Letters*. 2008;101(22): 226405
[41] González J., Guinea F., Vozmediano M. 'Unconventional quasiparticle lifetime in graphite'. *Physical Review Letters*. 1996;77(17):3589
[42] Hwang C., Siegel D.A., Mo S.K., *et al.* 'Fermi velocity engineering in graphene by substrate modification'. *Scientific Reports*. 2012;2
[43] Fang T., Konar A., Xing H., Jena D. 'Carrier statistics and quantum capacitance of graphene sheets and ribbons'. *Applied Physics Letters*. 2007;91(9):092109
[44] Adam S., Hwang E., Galitski V., Sarma S.D. 'A self-consistent theory for graphene transport'. *Proceedings of the National Academy of Sciences*. 2007;104(47):18392–7
[45] Zhu W., Perebeinos V., Freitag M., Avouris P. 'Carrier scattering, mobilities, and electrostatic potential in monolayer, bilayer, and trilayer graphene'. *Physical Review B*. 2009;80(23):235402
[46] Dorgan V.E., Bae M.H., Pop E. 'Mobility and saturation velocity in graphene on SiO_2'. *Applied Physics Letters*. 2010;97(8):082112-3

[47] Datta S. *Electronic Transport in Mesoscopic Systems*. Cambridge University Press, New York, NY; 1997

[48] Novoselov K., Geim A.K., Morozov S., *et al.* 'Two-dimensional gas of massless Dirac fermions in graphene'. *Nature*. 2005;438(7065):197–200

[49] Martin J., Akerman N., Ulbricht G., *et al.* 'Observation of electron–hole puddles in graphene using a scanning single-electron transistor'. *Nature Physics*. 2007;4(2):144–8

[50] Tworzydło J., Trauzettel B., Titov M., Rycerz A., Beenakker C.W. 'Sub-poissonian shot noise in graphene'. *Physical Review Letters*. 2006;96(24): 246802

[51] Tan Y.W., Zhang Y., Bolotin K., *et al.* 'Measurement of scattering rate and minimum conductivity in graphene'. *Physical Review Letters*. 2007;99(24): 246803

[52] Geim A.K., Novoselov K.S. 'The rise of graphene'. *Nature Materials*. 2007;6(3):183–91.

[53] Chen J.H., Cullen W., Jang C., Fuhrer M., Williams E. 'Defect scattering in graphene'. *Physical Review Letters*. 2009;102(23):236805

[54] Zhang Y., Brar V.W., Girit C., Zettl A., Crommie M.F. 'Origin of spatial charge inhomogeneity in graphene'. *Nature Physics*. 2009;5(10):722–6

[55] Hwang E., Sarma S.D. 'Acoustic phonon scattering limited carrier mobility in two-dimensional extrinsic graphene'. *Physical Review B*. 2008;77(11):115449

[56] Castro E.V., Ochoa H., Katsnelson M., *et al.* 'Limits on charge carrier mobility in suspended graphene due to flexural phonons'. *Physical Review Letters*. 2010;105(26):266601

[57] Mayorov A.S., Gorbachev R.V., Morozov S.V., *et al.* 'Micrometer-scale ballistic transport in encapsulated graphene at room temperature'. *Nano Letters*. 2011;11(6):2396–9

[58] Giannazzo F., Raineri V., Rimini E. 'Transport properties of graphene with nanoscale lateral resolution'. *Scanning Probe Microscopy in Nanoscience and Nanotechnology* 2: Springer; 2011. pp. 247–85

[59] Peres N.M. 'The transport properties of graphene'. *Journal of Physics Condensed Matter: An Institute of Physics Journal*. 2009;21(32):323201

[60] Sarma S.D., Adam S., Hwang E., Rossi E. 'Electronic transport in two-dimensional graphene'. *Reviews of Modern Physics*. 2011;83(2):407

[61] Farmer D.B., Perebeinos V., Lin Y.M., Dimitrakopoulos C., Avouris P. 'Charge trapping and scattering in epitaxial graphene'. *Physical Review B*. 2011;84(20):205417

[62] Chen F., Xia J., Ferry D.K., Tao N. 'Dielectric screening enhanced performance in graphene FET'. *Nano Letters*. 2009;9(7):2571–4

[63] Stauber T., Peres N., Guinea F. 'Electronic transport in graphene: a semiclassical approach including midgap states'. *Physical Review B*. 2007;76(20):205423

[64] Chen J.H., Jang C., Ishigami M., *et al.* 'Diffusive charge transport in graphene on SiO_2'. *Solid State Communications*. 2009;149(27):1080–6

[65] Bolotin K.I., Sikes K., Jiang Z., *et al.* 'Ultrahigh electron mobility in suspended graphene'. *Solid State Communications*. 2008;146(9):351–5

[66] Stormer H., Pfeiffer L., Baldwin K., West K. 'Observation of a Bloch–Grüneisen regime in two-dimensional electron transport'. *Physical Review B*. 1990;41(2):1278

[67] Fang T., Konar A., Xing H., Jena D. 'High-field transport in two-dimensional graphene'. *Physical Review B*. 2011;84(12):125450

[68] Hwang E., Sarma S.D. 'Acoustic phonon scattering limited carrier mobility in two-dimensional extrinsic graphene'. *Physical Review B*. 2008;77(11):115449

[69] Perebeinos V., Avouris P. 'Inelastic scattering and current saturation in graphene'. *Physical Review B*. 2010;81(19):195442

[70] Chauhan J., Guo J. 'High-field transport and velocity saturation in graphene'. *Applied Physics Letters*. 2009;95(2):023120

[71] Lin I.T., Liu J.M. 'Surface polar optical phonon scattering of carriers in graphene on various substrates'. *Applied Physics Letters*. 2013; 103(8):081606

[72] Wang S., Mahan G. 'Electron scattering from surface excitations'. *Physical Review B*. 1972;6(12):4517

[73] Bresciani M., Paussa A., Palestri P., Esseni D., Selmi L. 'Low-field mobility and high-field drift velocity in graphene nanoribbons and graphene bilayers'. *IEEE International Electron Devices Meeting (IEDM)*. 2010. pp. 32–1

[74] Shishir R., Ferry D. 'Velocity saturation in intrinsic graphene'. *Journal of Physics: Condensed Matter*. 2009;21(34):344201

[75] Fang T., Konar A., Xing H., Jena D. 'High field transport in graphene'. *Physical Review B*. 2011;84(12):125450

[76] Meric I., Dean C.R., Young A.F., *et al.* 'Channel length scaling in graphene field-effect transistors studied with pulsed current–voltage measurements'. *Nano Letters*. 2011;11(3): 1093–7

[77] Farmer D.B., Chiu H.Y., Lin Y.M., Jenkins K.A., Xia F., Avouris P. 'Utilization of a buffered dielectric to achieve high field-effect carrier mobility in graphene transistors'. *Nano Letters*. 2009;9(12):4474–8

[78] Du X., Skachko I., Barker A., Andrei E.Y. 'Approaching ballistic transport in suspended graphene'. *Nature Nanotechnology*. 2008;3(8):491–5

[79] Forster F., Molina-Sanchez A., Engels S., *et al.* 'Dielectric screening of the Kohn anomaly of graphene on hexagonal boron nitride'. *Physical Review B*. 2013;88(8):085419

[80] Liao L., Duan X. 'Graphene-dielectric integration for graphene transistors'. *Materials Science and Engineering*. 2010;70(3):354–70

[81] Ponomarenko L., Yang R., Mohiuddin T., *et al.* 'Effect of a high-κ environment on charge carrier mobility in graphene'. *Physical Review Letters*. 2009;102(20):206603

[82] Ong Z.Y., Fischetti M.V. 'Top oxide thickness dependence of remote phonon and charged impurity scattering in top-gated graphene'. *Applied Physics Letters*. 2013;102(18):183506

[83] Ong Z.Y., Fischetti M.V. 'Theory of remote phonon scattering in top-gated single-layer graphene'. *Physical Review B*. 2013;88(4):045405

[84] Fallahazad B., Lee K., Lian G., *et al.* 'Scaling of Al_2O_3 dielectric for graphene field-effect transistors'. *Applied Physics Letters*. 2012;100(9):093112

[85] Jin Z., Su Y., Chen J., Liu X., Wu D. 'Study of AlN dielectric film on graphene by Raman microscopy'. *Applied Physics Letters*. 2009;95(23):233110

[86] Habibpour O., Cherednichenko S., Vukusic J., Stake J. 'Mobility improvement and microwave characterization of a graphene field effect transistor with silicon nitride gate dielectrics'. *IEEE Electron Device Letters*. 2011;32(7):871–3

[87] Deen D.A., Champlain J.G., Koester S.J. 'Multilayer HfO_2/TiO_2 gate dielectric engineering of graphene field effect transistors'. *Applied Physics Letters*. 2013;103(7):073504

[88] Dean C., Young A., Meric I., *et al.* 'Boron nitride substrates for high-quality graphene electronics'. *Nature Nanotechnology*. 2010;5(10):722–6

[89] Gannett W., Regan W., Watanabe K., Taniguchi T., Crommie M., Zettl A. 'Boron nitride substrates for high mobility chemical vapor deposited graphene'. *Applied Physics Letters*. 2011;98(24):242105

[90] Levendorf M.P., Kim C.J., Brown L., *et al.* 'Graphene and boron nitride lateral heterostructures for atomically thin circuitry'. *Nature*. 2012;488(7413):627–32

[91] Meric I., Dean C.R., Young A.F., Hone J., Kim P., Shepard K.L. 'Graphene field-effect transistors based on boron nitride gate dielectrics'. *Proceeding of IEEE*. 2013;101(7):1609–19

[92] Berger C., Song Z., Li X., *et al.* 'Electronic confinement and coherence in patterned epitaxial graphene'. *Science*. 2006;312(5777):1191–6

[93] Ma P., Jin Z., Guo J., *et al.* 'Top-gated graphene field-effect transistors on SiC substrates'. *Chinese Science Bulletin*. 2012;57(19):2401–3

[94] Cai W., Moore A.L., Zhu Y., *et al.* 'Thermal transport in suspended and supported monolayer graphene grown by chemical vapor deposition'. *Nano Letters*. 2010;10(5):1645–51

[95] Mao R., Kong B.D., Kim K.W., Jayasekera T., Calzolari A., Nardelli M.B. 'Phonon engineering in nanostructures: controlling interfacial thermal resistance in multilayer-graphene/dielectric heterojunctions'. *Applied Physics Letters*. 2012;101(11):113111

[96] Nagashio K., Nishimura T., Kita K., Toriumi A. 'Metal/graphene contact as a performance killer of ultra-high mobility graphene analysis of intrinsic mobility and contact resistance'. *IEEE International Electron Devices Meeting (IEDM)*, 2009, pp. 1–4.

[97] Huard B., Stander N., Sulpizio J., Goldhaber G.D. 'Evidence of the role of contacts on the observed electron-hole asymmetry in graphene'. *Physical Review B*. 2008;78(12):121402

[98] Lee E.J., Balasubramanian K., Weitz R.T., Burghard M., Kern K. 'Contact and edge effects in graphene devices'. *Nature Nanotechnology*. 2008;3(8): 486–90

[99] Nagashio K., Nishimura T., Kita K., Toriumi A. 'Contact resistivity and current flow path at metal/graphene contact'. *Applied Physics Letters*. 2010;97(14):143514

[100] Blake P., Yang R., Morozov S., *et al*. 'Influence of metal contacts and charge inhomogeneity on transport properties of graphene near the neutrality point'. *Solid State Communications*. 2009;149(27):1068–71

[101] Leong W.S., Gong H., Thong J.T. 'Low-contact-resistance graphene devices with nickel-etched-graphene contacts'. *ACS Nano*. 2013;8(1):994–1001

[102] Grassi R., Low T., Gnudi A., Baccarani G. 'Contact-induced negative differential resistance in short-channel graphene FETs'. *IEEE Transactions on Electron Devices*. 2013;60(1):140–6

[103] Xu H., Zhang Z., Peng L.M. 'Measurements and microscopic model of quantum capacitance in graphene'. *Applied Physics Letters*. 2011;98(13): 133122

[104] Nagashio K., Nishimura T., Toriumi A. 'Estimation of residual carrier density near the Dirac point in graphene through quantum capacitance measurement'. *Applied Physics Letters*. 2013;102(17):173507

[105] Datta S. *Quantum Transport: Atom to Transistor*. Cambridge University Press, Cambridge; 2005

[106] Guo J., Yoon Y., Ouyang Y. 'Gate electrostatics and quantum capacitance of graphene nanoribbons'. *Nano Letters*. 2007;7(7):1935–40

[107] Chen Z., Appenzeller J. 'Mobility extraction and quantum capacitance impact in high performance graphene field-effect transistor devices'. *IEEE International Electron Devices Meeting (IEDM)*. 2008. pp. 1–4

[108] Ancona M.G. 'Electron transport in graphene from a diffusion-drift perspective'. *IEEE Transactions on Electron Devices*. 2010;57(3):681–9

[109] Lundstrom M. *Fundamentals of Carrier Transport*. Cambridge University Press, New York, NY; 2009

[110] Natori K. 'Ballistic metal-oxide-semiconductor field effect transistor'. *Journal of Applied Physics*. 1994;76(8):4879–90

[111] Wong H.S.P., Deng J., Hazeghi A., Krishnamohan T., Wan G.C. 'Carbon nanotube transistor circuits: models and tools for design and performance optimization'. *Proceedings of the IEEE/ACM International Conference on Computer-Aided Design*, 2006, pp. 651–4

[112] Szabo A., Luisier M. 'Under-the-barrier model: an extension of the top-of-the-barrier model to efficiently and accurately simulate ultrascaled nanowire transistors'. *IEEE Transactions on Electron Devices*. 2013;60(7): 2353–60

[113] Zhao P., Choudhury M., Mohanram K., Guo J. 'Computational model of edge effects in graphene nanoribbon transistors'. *Nano Research*. 2008;1(5): 395–402

[114] Frank N., Young L. 'Transmission of electrons through potential barriers'. *Physical Review*. 1931;38(1):80

[115] Jiménez D. 'A current–voltage model for Schottky-barrier graphene-based transistors'. *Nanotechnology*. 2008;19(34):345204

[116] Alam A.U., Holland K.D., Ahmed S., Kienle D., Vaidyanathan M. 'A modified top-of-the-barrier model for graphene and its application to predict RF linearity'. *IEEE International Conference on Simulation of Semiconductor Processes and Devices (SISPAD)*. 2013. p. 155–8

[117] Novoselov K.S., Fal V., Colombo L., Gellert P., Schwab M., Kim K. 'A roadmap for graphene'. *Nature*. 2012;490(7419):192–200

[118] Liang G., Neophytou N., Lundstrom M.S., Nikonov D.E. 'Ballistic graphene nanoribbon metal-oxide-semiconductor field-effect transistors: a full real-space quantum transport simulation'. *Journal of Applied Physics*. 2007;102(5):054307

[119] Grassi R., Gnudi A., Gnani E., Reggiani S., Baccarani G. 'Mode space approach for tight binding transport simulation in graphene nanoribbon FETs'. *IEEE Transactions on Nanotechnology*. 2011;10(3):371–8

[120] Javey A., Guo J., Wang Q., Lundstrom M., Dai H. 'Ballistic carbon nanotube field-effect transistors'. *Nature*. 2003;424(6949):654–7

[121] Marulanda J.M. 'Current transport modeling of carbon nanotube field effect transistors for analysis and design of integrated circuits'. Ph.D. (Electrical Engineering) Dissertation, Louisiana State University, Baton Rouge, USA, August 2008

[122] Martinez A., Bescond M., Barker J.R., *et al.* 'A self-consistent full 3-D real-space NEGF simulator for studying nonperturbative effects in nano-MOSFETs'. *IEEE Transactions on Electron Devices*. 2007;54(9):2213–22

[123] Fiori G., Iannaccone G. 'Simulation of graphene nanoribbon field-effect transistors'. *IEEE Electron Device Letters*. 2007;28(8):760–2

[124] Banadaki Y.M., Srivastava A. 'Investigation of the width-dependent static characteristics of graphene nanoribbon field effect transistors using non-parabolic quantum-based model'. *Solid-State Electronics*. 2015;111:80–90

[125] Banadaki Y.M., Srivastava A. 'A novel graphene nanoribbon field effect transistor for integrated circuit design'. *IEEE 56th International Midwest Symposium on Circuits and Systems (MWSCAS)*. 2013. pp. 924–7

[126] Srivastava A., Banadaki Y.M., Fahad M.S. '(Invited) dielectrics for graphene transistors for emerging integrated circuits'. *ECS Transactions*. 2014;61(2):351–61

[127] Grassi R., Poli S., Gnani E., Gnudi A., Reggiani S., Baccarani G. 'Tight-binding and effective mass modeling of armchair graphene nanoribbon FETs'. *Solid-State Electronics*. 2009;53(4):462–7

[128] Brey L., Fertig H. 'Electronic states of graphene nanoribbons studied with the Dirac equation'. *Physical Review B*. 2006;73(23):235411

[129] Son Y.W., Cohen M.L., Louie S.G. 'Energy gaps in graphene nanoribbons'. *Physical Review Letters*. 2006;97(21):216803

[130] Saito R., Dresselhaus G., Dresselhaus M.S. *Physical Properties of Carbon Nanotubes*, World Scientific Publishing, Singapore; 1998

[131] Sako R., Hosokawa H., Tsuchiya H. 'Computational study of edge configuration and quantum confinement effects on graphene nanoribbon transport'. *IEEE Electron Device Letters*. 2011;32(1):6–8

[132] Raza H., Kan E.C. 'Armchair graphene nanoribbons: electronic structure and electric-field modulation'. *Physical Review B*. 2008;77(24):245434

[133] Lin Y.M., Perebeinos V., Chen Z., Avouris P. 'Electrical observation of subband formation in graphene nanoribbons'. *Physical Review B*. 2008;78(16):161409

[134] Banadaki Y.M., Srivastava A., Sharifi S. 'Clocked adiabatic XOR and XNOR CMOS gates design based on graphene nanoribbon complementary field effect transistors'. *Proceedings of IEEE International Symposium on Nanoelectronic and Information Systems (iNIS)*. 2015: 13–17

[135] Wang X., Ouyang Y., Li X., Wang H., Guo J., Dai H. 'Room-temperature all-semiconducting sub-10-nm graphene nanoribbon field-effect transistors'. *Physical Review Letters*. 2008;100(20):206803

[136] Chen Z., Lin Y.M., Rooks M.J., Avouris P. 'Graphene nano-ribbon electronics'. *Physica E: Low-Dimensional Systems and Nanostructures*. 2007;40(2):228–32

[137] Konar A., Fang T., Jena D. 'Effect of high-K gate dielectrics on charge transport in graphene-based field effect transistors'. *Physical Review B*. 2010; 82(11):115452

[138] Owlia H., Keshavarzi P. 'Investigation of the novel attributes of a double-gate graphene nanoribbon FET with AlN high-κ dielectrics'. *Superlattices and Microstructures*. 2014;75:613–20

[139] Oh J.G., Hong S.K., Kim C.K., *et al.* 'High performance graphene field effect transistors on an aluminum nitride substrate with high surface phonon energy'. *Applied Physics Letters*. 2014;104(19):193112

[140] Imperiale I., Bonsignore S., Gnudi A., Gnani E., Reggiani S., Baccarani G. 'Computational study of graphene nanoribbon FETs for RF applications'. *IEEE International Electron Devices Meeting (IEDM)*. 2010. pp. 32–3

[141] Banadaki Y.M., Srivastava A. 'Scaling effects on static metrics and switching attributes of graphene nanoribbon FET for emerging technology'. *IEEE Transactions on Emerging Topics in Computing*. 2015;3(4):458–69

[142] Hasan S., Wang J., Lundstrom M. 'Device design and manufacturing issues for 10 nm-scale MOSFETs: a computational study'. *Solid-State Electronics*. 2004;48(6):867–75

[143] Sharifi M.J., Banadaki Y.M. 'A SPICE large signal model for resonant tunneling diode and its applications'. *International Conference on Computational Methods in Sciences and Engineering 2008 (ICCMSE 2008)*. AIP Publishing; 2009. p. 890–3

[144] Sharifi M.J., Banadaki Y.M. 'General SPICE models for memristor and application to circuit simulation of memristor-based synapses and

memory cells'. *Journal of Circuits, Systems, and Computers*. 2010;19(2): 407–24

[145] Choudhury M.R., Yoon Y., Guo J., Mohanram K. 'Graphene nanoribbon FETs: technology exploration for performance and reliability'. *IEEE Transactions on Nanotechnology*. 2011;10(4):727–36

[146] Choudhury M., Yoon Y., Guo J., Mohanram K. 'Technology exploration for graphene nanoribbon FETs'. *ACM Proceedings of the 45th Annual Design Automation Conference*. 2008. p. 272–7

Chapter 5

Junction and doping-free transistors for future computing

Chitrakant Sahu[1] *and Jawar Singh*[1]

Continued down-scaling of device dimensions poses severe challenges and difficulties for complementary metal-oxide semiconductor technology, particularly fabrication complexities, process variability, and short channel effects. These challenges mainly arise due to abrupt doping profile requirement at junctions and random dopant fluctuations (RDFs). Recently, the junctionless field-effect transistors (JLFETs), also known as gated resistors, have widely attracted attention, as they do not require formation of any metallurgical junctions (P-N, N^+-N, or P^+-P) and doping concentration gradient throughout the device. Thus, they relax abrupt doping profile requirements and greatly simplify the fabrication process. A key requirement for JLFETs is the formation of a semiconductor layer that should be thin and narrow enough to be depleted when the JLFET is in off-state. At the same time, semiconductor layer should be doped enough to achieve an adequate amount of drain current in on-state. Therefore, JLFETs are generally made of heavily doped silicon nanowires. The heavily doped nature of JLFETs causes certain problems, and to address them, the concept of doping-free (dopingless) JLFETs was recently proposed. In this chapter, detail study of both junction-free and doping-free transistors are presented based on 2-D device simulation with model calibrated to experimental data.

5.1 Introduction

Over the past few years, a tremendous attention has been paid for the advancement of metal-oxide semiconductor field-effect transistors (MOSFETs) that leads to miniaturization of device dimensions in the nanometer regime. This aggressive scaling of FETs has troubled the semiconductor industry specifically while formation of abrupt source and drain junctions in nano devices. Therefore, ultrafast annealing methods and the developments of novel doping techniques have been investigated which are complex as well as expensive. Among the many possible developments, multigate

[1]PDPM-Indian Institute of Information Technology Design and Manufacturing, Jabalpur, MP, India

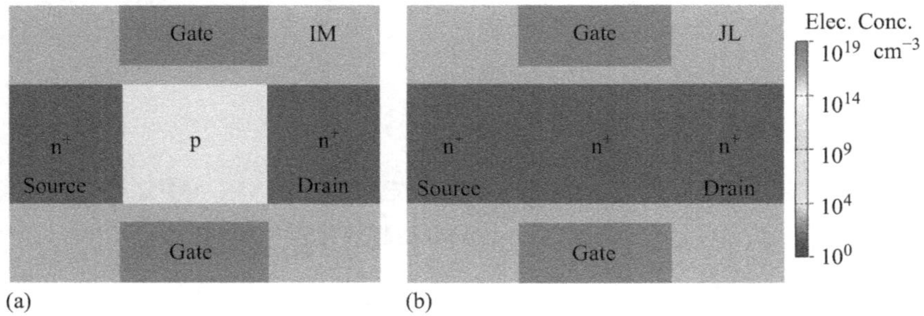

Figure 5.1 Schematic of (a) inversion mode and (b) JLFET

FETs or FinFETs are the most promising due to their superior gate controllability and better immunity towards short channel effects.

In addition, recently, junctionless (JL) field-effect transistor (JLFET) have also shown very good potential because of their simplified fabrication process, excellent gate controllability, and scalability. The JLFETs contains a single doping species of the order of 10^{19} atoms per cubic centimeter uniformly throughout its source, drain and channel regions; as a result, there is no metallurgical junctions (P-N, N^+-N, or P^+-P) and doping concentration gradient. Many reports on JLFETs are available in the literature based on Lilienfeld's first transistor architecture [1], Colinge *et al.* [2], Park *et al.* [3], and Jeon *et al.* [4] have successfully fabricated the multigate JL nanowire field-effect transistors. Different JLFET architectures include double-gate (DG) architecture [5–7], bulk planar architecture [8], trigate nanowire architectures with silicon-on-insulator (SOI) as well as bulk substrate [9–11], and gate all-around architecture [12, 13].

Figure 5.1 (a) shows the inversion mode (IM) DG n-channel FET having two p-n junctions: the source junction and drain junction. The source and drain regions are separated by the channel area with opposite doping polarity. The doping profile at these junctions must be abrupt (steep) to minimize the diffusion of source and drain doping atoms. Also, diffusion of source/drain carriers below gate region leads to shorter effective channel length (L_{eff}). Therefore, a novel device structure has recently been proposed referred as the JLFET, which exhibits an unconventional architecture that presents neither source/drain junction nor doping concentration gradients, as shown in Figure 5.1(b). It has a constant and heavy doping concentration (N-type in an n-MOS device and P-type in a p-MOS device) through source, channel, and drain regions. These inherent features automatically eliminate the impurity diffusion-related problems and abrupt (steep) doping profile requirements. Thus, JLFET is very different from conventional IM MOSFET in terms of operating principle. The current flow in JLFET is because of majority carrier instead of minority, and it flows in the volume instead of semiconductor dielectric interface. Unlike the conventional IM MOSFET, JLFET has heavily doped channel and is fully depleted under off-state. Thereby, an appropriate selection of gate metal which has a large work function difference to that of the Si-channel is highly desirable.

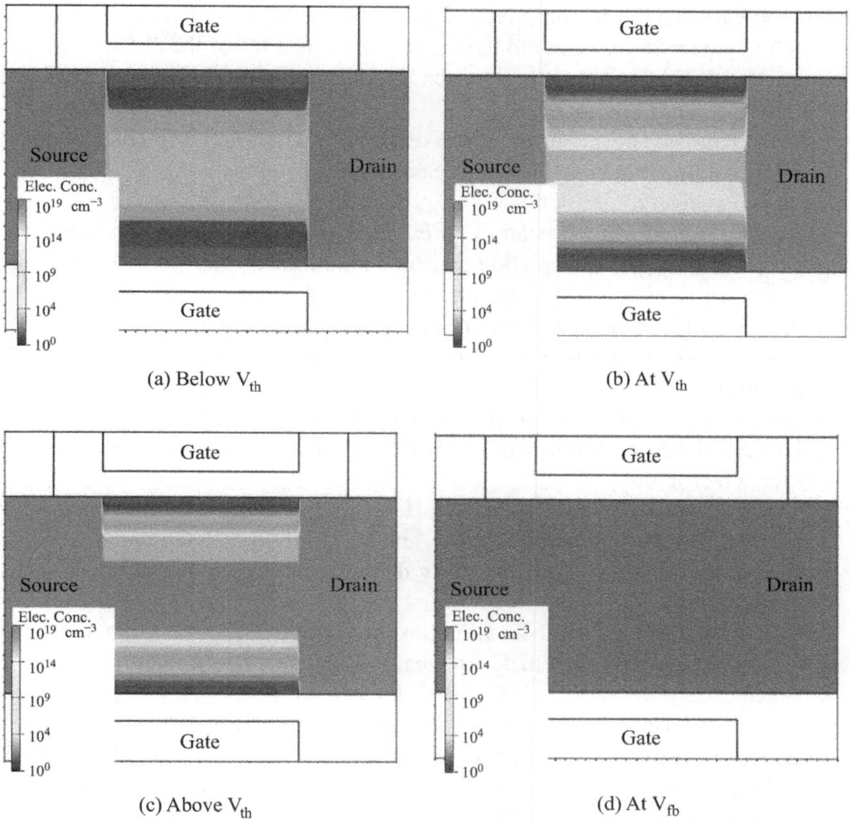

Figure 5.2 Carrier concentration contour plots in an n-type junctionless transistor with $V_d = 50\,mV$, when (a) $V_{gs} < V_{th}$; (b) $V_{gs} = V_{th}$; (c) $V_{gs} > V_{th}$; and (d) $V_{gs} = V_{fb}$

The highly doped and extremely thin channel can be fully depleted by appropriate gate metal work function as shown in Figure 5.2(a). Once the channel is fully depleted, the current between source and drain ideally becomes zero. As we increase the gate voltage, the JLFET enters into partially depletion region and current start flowing in the center of the nanowire when drain voltage is applied, as can be seen in Figure 5.2(b,c). At flat-band voltage, the depletion region is disappeared completely, as shown in Figure 5.2(d), and device behaves like a resistor; hence, it is also named as "gated resistor." In summary, JLFET offers following major advantages and differences in contrast with its counterpart inversion mode FET (IMFET):

- *Fabrication:* Fabrication complexity in JLFET is reduced due to the elimination of different polarity doping and steep concentration gradient requirements for p-n junction formation. This reduces the thermal budget and lowers overall manufacturing cost.

- *Gate metal:* In order to completely deplete the heavily doped channel, high work function gate metal is required to ensure proper turn-off of the JLFET. However, in case of IMFET, there is no as such special requirement for gate metal work function.
- *Volume:* To ensure proper turning OFF of the JLFET, it is important to have smaller channel volume that be depleted off easily and efficiently. But in case of IMFET, there is no such requirement.
- *Operating regime:* In ON state, IMFET mainly works in inversion regime, but JLFET works in partial depletion and accumulation regime.
- *Conduction mechanism:* In conventional IMFET, current flows at the interface of $Si-SiO_2$ layer due to inversion of carriers, while in JLFET, current flows mainly in bulk. There are several parallel current paths exist in the channel for conduction.
- *Vertical electric field:* The vertical electric is lower in JLFET during ON state due to flat-band condition, while IMFET (or MOSFET) experiences very high vertical electric field.
- *Mobility:* Due to highly doped nature JLFET, its mobility is limited by impurity scattering rather by phonon scattering, and its variation with temperature is much smaller than IM FETs. Also, mobility degradation is less due to lower vertical field in ON state.
- *SCEs:* The JLFET have near-ideal subthreshold slope, lower drain-induced barrier lowering (DIBL) and extremely low leakage currents due to better control over channel.
- *Capacitance:* The parasitic capacitances are low because JLFET mostly work in depletion regime; hence, effective capacitance will be a series combination of gate oxide and depletion capacitance at lower gate voltages. After flat-band condition, capacitance is similar for both types of devices because accumulation or inversion eliminates the effect of depletion.
- *Scalability:* In IMFET, effective channel length (L_{eff}) is always less than the physical gate length due to overlapping of doping profile of source/drain towards gate. However, in JLFET, L_{eff} is larger than the physical gate length due to depletion of carrier towards source/drain region. Hence, it shows better scalability due to larger L_{eff} and reduced SCEs.

Thus, the concept of junction-free JLFET with smaller channel volume and high doping concentration is very encouraging, however, it possess few severe challenges, such as:

- threshold voltage (V_{th}) variability due to RDFs,
- high leakage current due to band-to-band tunneling (BTBT),
- low drive current due to incomplete ionization of carriers below room temperature,
- higher gate work function (>5.5 eV) requirement to turn-off the device, and
- poor mobility due to heavy doping.

5.2 JLFET limitations

The JLFET offers many advantages, such as better scalability, simplified fabrication process, and less SCEs than its counterpart IMFET. However, some research groups have highlighted that the JLFETs are more sensitivity towards process variations [12]. As these JLFETs use similar type of lithography techniques as employed by the IMFETs, they will experience parametric variations similar to FinFETs or MOSFETs. Thus, physical dimensions such as gate length, silicon thickness, and oxide thickness variation may affect the device electrical characteristics. In addition, variability due to RDFs has been projected to be a serious concern when the devices are scaled to nanoscale dimensions [14]. More recently, the RDF impact on JLFET was found to be more significant as compared to an equivalent inversion-mode FET [15]. The RDFs mainly arise due to nonuniform placement of the dopant atoms in the channel during ion implantation, especially at nanoscale dimensions. Apart from random positioning, fluctuations may also occur in number of dopant atoms in a device. This may not be crucial in sufficiently large volumes, but it will become critical in nanoscale devices. These variability issues at device level can be encapsulate in threshold voltage (V_{TH}) to represent the electrical behavior of the device, as a function of variation in dopant atoms or physical dimensions.

Figure 5.3(a,b) shows the threshold voltage (V_{TH}) variation for a JL DG FET as a function of silicon (Si) thickness (T_{SI}), gate oxide thickness (T_{ox}), and for selected values of doping concentration (N_d). It can be observed that the V_{TH} decreases with thicker oxide or silicon layer, as well as higher impurity concentration. It is worth mention that a variation of 10% of T_{SI}, T_{ox}, and N_d lead to a 180 mV change in threshold voltage in worst case when all of the parameters deviate in the same direction. Furthermore, higher doping concentration (N_d) has dramatic effect on V_{TH} either it is varied with T_{SI} or T_{ox}, hence, causes device to fail even for minor variations in

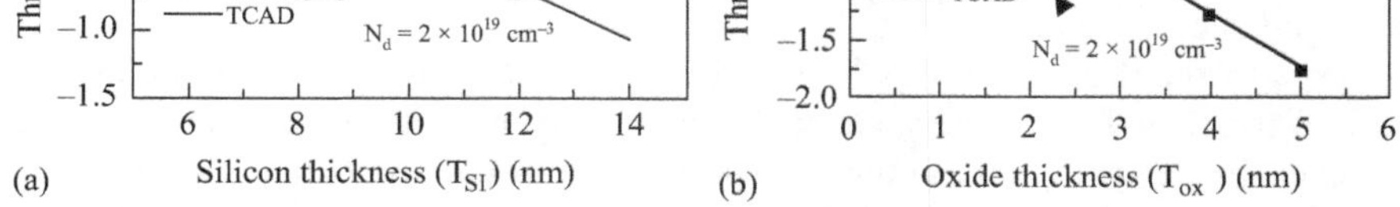

Figure 5.3 V_{th} fluctuation according to various (a) T_{SI} (b) T_{ox} in JL-DGFET as a parameter of doping concentration. V_{th} model data is taken from reference 10

Table 5.1 Standard deviation and its relative variation for various performance metrics [16]

Parameters	JL	IM
$\sigma V_{th}/V_{th}$	11.28%	0.48%
$\sigma SS/SS$	4.65%	0.77%
$\sigma I_{on}/I_{on}$	12.7%	8.65%
$\sigma I_{off}/I_{off}$	29.38%	5.38%

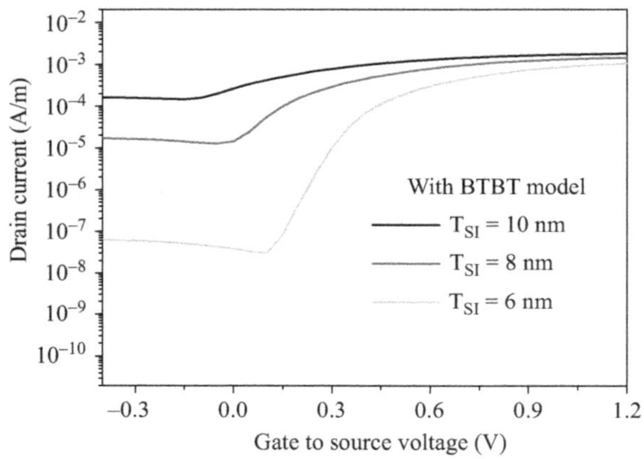

Figure 5.4 Effect of band to band tunneling on Id-Vgs characteristics of JL-DGFET at various Si thickness

these dimensions. A detailed comparison of different parameters as a function of RDFs for JLFET and IMFET is shown in Table 5.1. The impact of RDFs on different performance metrics estimated as the standard deviation of V_{TH}, subthreshold swing (SS), on-current (I_{on}) and off-current (I_{off}), normalized to their baseline values for both JLFET and IMFET from reference 16. From these observations, magnitude of variability in JLFET is quite large for all performance metrics as compared to IMFET with intrinsic doping. Hence, JLFET performance variability due to RDF is a major issue and need to be addressed.

Apart from these variability issues, JLFET also have larger I_{off} due to BTBT. To turn off the JLFET completely, channel volume must be fully depleted in OFF state; however, higher gate-drain potential than the gate-source potential due to applied drain bias causes band overlap between the valence band of channel and the conduction band of drain in lateral direction. As a result, higher BTBT probability from the valence band of the channel to the conduction band of the drain (for n-channel JLFET operation) makes OFF state current very high [17]. Figure 5.4 shows I_d-V_{gs} characteristics of a 20-nm gate length DG JLFET for different silicon (Si) thickness

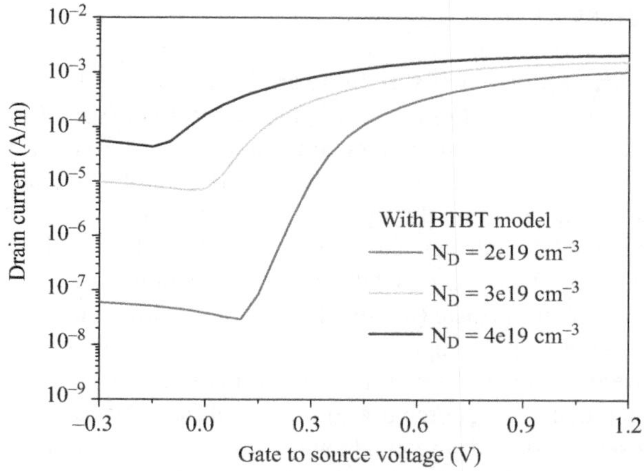

Figure 5.5 Effect of band to band tunneling on Id-Vgs characteristics of JL-DGFET at various channel doping

(T_{SI}) layers. The JLFET with thin T_{SI} has a higher threshold voltage and a lower OFF current compared to a device with thicker T_{SI}. The lower OFF current for thinner T_{SI} is due to reduction in lateral BTBT from channel to drain. It can also be observed that for a given channel doping, channel length and gate dielectric thickness, the OFF state currents are strongly dependent on the thickness of the silicon layer. Furthermore, we have observed the OFF state currents from the I_d-V_{gs} characteristics, as shown in Figure 5.5 for different channel doping concentrations (N_d). One can observe that a slight (double) variation in N_d significantly (almost 1000 times) increase the OFF state current. As fully depleting the channel would be difficult when there is a high channel doping; hence, BTBT is higher in these cases, leading to a high OFF state current. Therefore, a careful choice of T_{SI} and N_d should be made for an optimum value of I_{OFF}.

From the above discussion, it is clear that the JLFETs also have major limitations, and these limitations must be addressed before exploiting their full potential for future nano–complementary metal-oxide semiconductor (CMOS) computing applications. From above simulation results and analysis, it can be concluded that a major source of these limitations may be high doping concentration requirements for JLFET. For example, when we kept the doping of the order of 10^{15} atoms per cubic centimeter the V_{TH} variation with T_{OX} and T_{SI} was observed negligible, as shown in Figure 5.3. Intuitively, if we reduce the doping concentration in the JLFET throughout from source, channel, and drain region to keep JL concept intact, there will be chances to get rid-off from the requirement of high gate metal work function to deplete the channel, reduced OFF state current, and variability due to RDFs. However, reduction in doping concentration throughout in JLFET will reduce the drive (ON state) current significantly.

5.3 Dopingless FET

In JLFETs, the major problem of variability are from RDF and higher BTBT tunneling current which may be addressed by employing a thin intrinsic (undoped) silicon nanowire for the formation of source, channel, and drain regions, instead of heavily doped source, channel, and drain regions. As in JLFET, it is equally important to have high doping concentration throughout from source, channel, and drain regions to maintain sufficient ON state current. Therefore, to meet this requirement in doping-free (dopingless, DL) JLFET, the metal electrodes at source, gate, and drain regions are attached over a thin intrinsic (undoped/dopingless) silicon nanowire of different work functions. A uniform dopingless structure throughout from source, channel, and drain regions preserves the inherent merits of the JLFET, such as, JL device architecture (i.e., no p-n junctions at source/drain) and simplified fabrication process. As this device makes use of undoped silicon nanowire, thereby, no external ion implantation or annealing is required, hence, referred as a doping-free (or dopingless) junctionless field-effect transistor (DL-JLFET). Since, DL-JLFET has no free carriers under thermal equilibrium, it employs the reverse concept of gate metal work function engineering employed in JLFET for depleting the charge carriers in the channel region. In JLFET, a high work function gate metal was used to deplete the charge carriers; therefore, obverse of this concept was employed to accumulate the charge carriers of desired polarity and magnitude in the source and drain regions by appropriate selection of electrodes metal work function. This artificial injection of the doping concentration through work function engineering of metal electrodes was first time introduced for the P-N diode and referred as charge-plasma (CP) diode [18, 19]. However, there are certain prerequisites that need to be fulfilled before applicability of CP in any device.

In these devices, n-type or p-type region of desired doping concentration can be formed within an undoped silicon film using work function engineering of metal electrodes. The type of doping and its level of concentration introduced in different regions (source/drain) depend upon the work function difference between electrode metal employed for source/drain formation and the undoped silicon film. The artificial accumulation of electrons or holes is referred as the "charge-plasma" (electron or hole plasma) concept in literature. The work function difference of metal electrode and undoped silicon creates electric field in such a direction that it accumulate charge carrier (holes or electrons) concentration in undoped silicon. This concept of CP has been extended for realization of bipolar transistors [20] and dopingless tunnel FETs [21]. As this concept does not require external ion-implantation or annealing process, it minimizes the thermal budget requirements, while preserving the inherent merits of JLFET. The other issues and its (DL-JLFET) performance in contrast with JLFET have been explored in detailed in the next sections of this chapter.

For realization of any device (p-n diode, bipolar junction transistor (BJT) or FET) structure through CP, the accumulation of holes or electrons in an undoped silicon layer should have higher carrier concentration ($>10^{19}$ atoms/cm^3) in source/drain region. To see this phenomenon, we have considered two different thicknesses (1 μm and 15 nm thick) of SOI layers, and a metal of suitable work function is deposited over

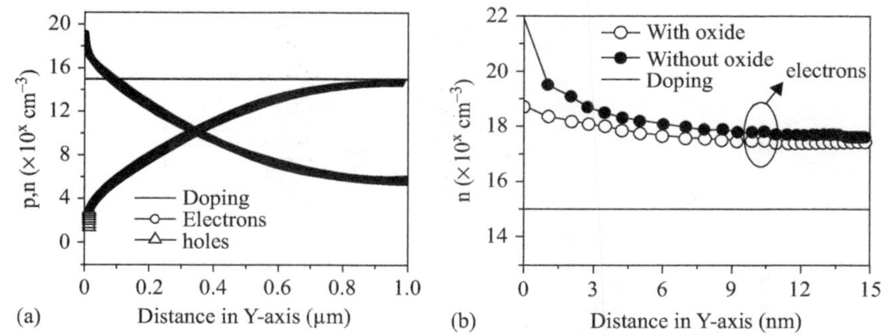

Figure 5.6 1-D vertical charge carrier distributions underneath an n-type metal contact, for two different SOI layer thicknesses: (a) 1 μm and (b) 15 nm. For (b), the distributions are shown for both a pure metal silicon junction (black symbols) and that for a gate with an oxide thickness of 2 nm (gray symbols)

it. For the simulation purpose, hafnium (HF) metal (work function = 3.9 eV) is considered, which creates a work function difference of approximately 0.7 eV (assuming undoped silicon work function = 4.6 eV). Through TCAD simulations, we observed the electron distribution along thickness of an undoped (10^{15}atoms/cm^3) silicon film (y-direction) for both samples (1 μm and 15 nm thick) under thermal equilibrium, as shown in Figure 5.6. For 1-μm thick sample, electron concentration degraded over the distance as shown in Figure 5.6(a). It is due to reduction in electric field intensity (caused due to work function difference between metal and undoped silicon) over a distance. However, for a 15-nm thick SOI layer, the electron concentration degradation is not significant, as shown in Figure 5.6(b).The electron concentration was observed under two circumstances, when (a) there is a direct contact of metal electrode with undoped silicon (or SOI layer), and (b) a thin insulating oxide layer was introduced between metal electrode with undoped silicon (or SOI layer). From these observations, it can be concluded that in a thinner SOI layer with electrode metal work function difference, a sufficient amount of carrier concentration can be induced. The induced carrier concentration in SOI layer is more uniform with oxide layer than without oxide layer, as can be seen in Figure 5.6(b). The electron concentration profile near interface region obtained using TCAD simulations is also verified by a model described in reference 22, which is given as $n_{yc} = n_i \exp\left(-\frac{\phi(y)}{v_t}\right)$, where, $\phi(y)$ is potential distribution along thickness of silicon film and n_i is intrinsic carrier concentration of undoped silicon.

Similarly, for accumulation of holes in a thinner SOI layer, platinum (work function = 5.65 eV) and gold (work function = 5.1 eV) electrodes were deposited over the SOI layer with a thin insulating oxide layer. Figure 5.7 shows the hole concentration along thickness of silicon film obtained from TCAD simulations which is consistent with the theoretical calculations. One can see that the platinum accumulates more holes than the gold due to larger work function difference between platinum and undoped silicon. Hence, majority carriers depend exponentially on the metal work

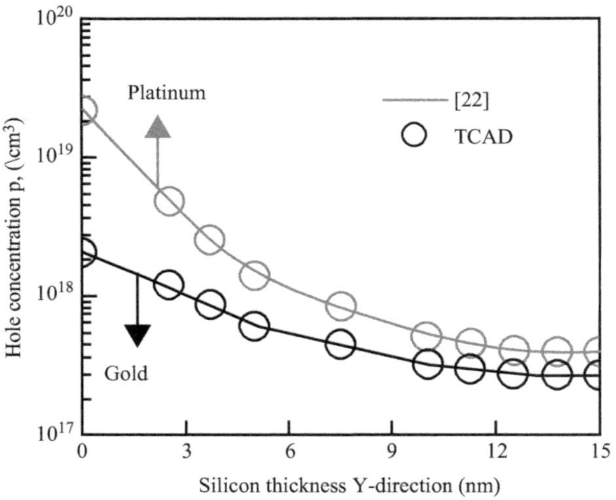

Figure 5.7 Hole concentration along thickness of silicon film (vertically) near S/D end at thermal equilibrium

function difference, but if work function difference between intrinsic silicon and metal is sufficiently large, then minority carrier can be neglected. In conclusion, the concept of CP can be exploited for a device without external doping requirement; however, functionality can only be ensured when following conditions are fulfilled:

- For inducing electron concentration (or electron plasma), the work function of source/drain (S/D) metal electrode should be less than the intrinsic (undoped) silicon, $\phi_m < \chi_{si} + (E_g/2q)$, where χ_{si} is the electron affinity of silicon ($\chi_{si} = 4.17$ eV) and E_g is the band gap of undoped silicon. Similarly to induce hole plasma, the work function of S/D metal electrode should be greater than the intrinsic silicon, i.e., $\phi_m > \chi_{si} + (E_g/2q)$. In general, the work function difference between metal and silicon ($\phi_m - \phi_{Si}$) must be greater than ± 0.5 eV for accumulation of electrons or holes.
- To ensure an uniform and sufficient amount of carrier concentration throughout the silicon thickness underneath S/D regions. The silicon film thickness (T_{SI}) should be less than the Debye length (L_D), i.e., $L_D = \sqrt{(\in_{si} v_t/qN)}$, where $\in_{si}$ is the dielectric constant of Si, v_t is the thermal voltage, N is the carrier concentration in the body, and q is the elementary charge.

5.4 Junction and doping-free FET

The absence of p-n junctions in the JLFETs make them scalable and greatly simplifies the fabrication process; however, heavy doping concentration requirements poses certain limitations. In order to get rid-off from these limitations and preserve the inherent merits of JLFET, the concept of CP was introduced and discussed in the previous section. It is evident that with CP, any type of solid-state device (p-n diode, BJT, field-effect transistor, and tunnel transistor) can be designed; therefore, major emphasis of this chapter will be on FET [23] and BJT [31].

5.4.1 Junction and doping-free DG FET

Figure 5.8 shows the cross-sectional views of n-type (a) JL and (b) doping-less (DL) as well as JL DGFETs. There are three major differences in the structure of DL-DGFET as compare to JL-DGFET: (a) doping level is reduced down to 10^{15} atoms/cm^3 instead of 10^{19} atoms/cm^3 throughout from source to drain, (b) HF metal is considered for source/drain (S/D) electrode instead of aluminium (AL), and (c) a thin insulating oxide layer is inserted between HF metal and undoped SOI layer, as well as sides of SOI layer in S/D region. The other device dimensions and simulation parameters are summarized in Table 5.2. For n-type JL-DGFET, some important parameters are: a uniformly doped 10-nm thick silicon (T_{SI}) having doping concentration (N_d) of 10^{19} atoms/cm^3, gate length (L_g) of 20 nm, and source and drain extension are 10 nm each. A high-k material (HfO_2) of dielectric constant 22 is considered and assumed

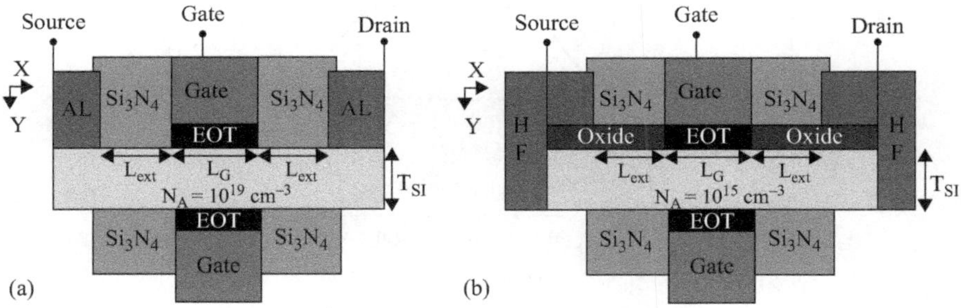

Figure 5.8 Cross sectional views of n-type (a) junctionless (JL) and (b) doping-less (DL) as well as JL (DGFETs)

Table 5.2 Device simulation parameters for DL- and JL-DGFET

Parameters	**Dopingless DGFET**	**Junctionless DGFET**
Silicon film thickness (T_{Si})	10 nm	10 nm
Effective oxide thickness (EOT)	1 nm	1 nm
Gate length (L_g)	20 nm	20 nm
Width (W)	1 μm	1 μm
Source/drain extension	10 nm	10 nm
Metal work function/doping for source/drain	3.9 eV(Hafnium)	10^{19}/cm^3
Metal work function for gate	4.66 eV (TiN)	5.25 eV(P^+ poly)
Doping	10^{15}/cm^3	10^{19}/cm^3

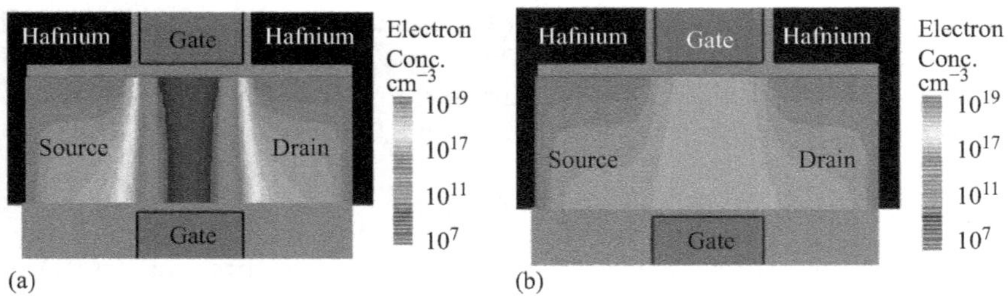

Figure 5.9 Electron contour inside the DL-DGFET at (a) thermal equilibrium ($V_{gs} = 0\,V$ and $V_{ds} = 0\,V$) and (b) ON-state ($V_{gs} = 1\,V$ and $V_{ds} = 50\,mV$)

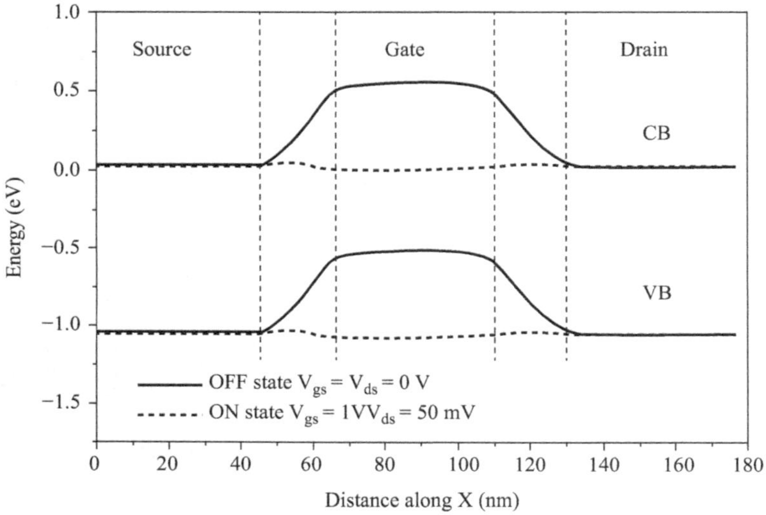

Figure 5.10 Energy band diagram along source to drain under zero bias and ON-state for DL-DGFET. Thermal equilibrium ($V_{ds} = 0$, $V_{gs} = 0\,V$), ON-state ($V_{ds} = 50\,mV$, $V_{gs} = 1\,V$)

that the gate leakage current is zero. The simulation parameters for DL-DGFET are also kept same except undoped (lowly) silicon body with carrier concentration $N_d = 10^{15}$ atoms/cm^3.

The behavior of DL-DGFET under thermal equilibrium ($V_{gs} = V_{ds} = 0$) and ON state ($V_{gs} = 1$ V, $V_{ds} = 50$ mV) was investigated for electron concentration and energy-band diagram. The DL-DGFET does not require any physical (external) doping for accumulation of carrier concentration due to HF metal (work function = 3.9 eV) electrode is considered for contact formation in source/drain region, as shown in Figure 5.9(a). One can observe that the DL-DGFET looks like an N^+-I-N^+ doped device structure at thermal equilibrium (without any external doping). It is due to work function engineering of source/drain metal electrodes as described in the previous section. The accumulation of electron is higher at metal contact edge of S/D

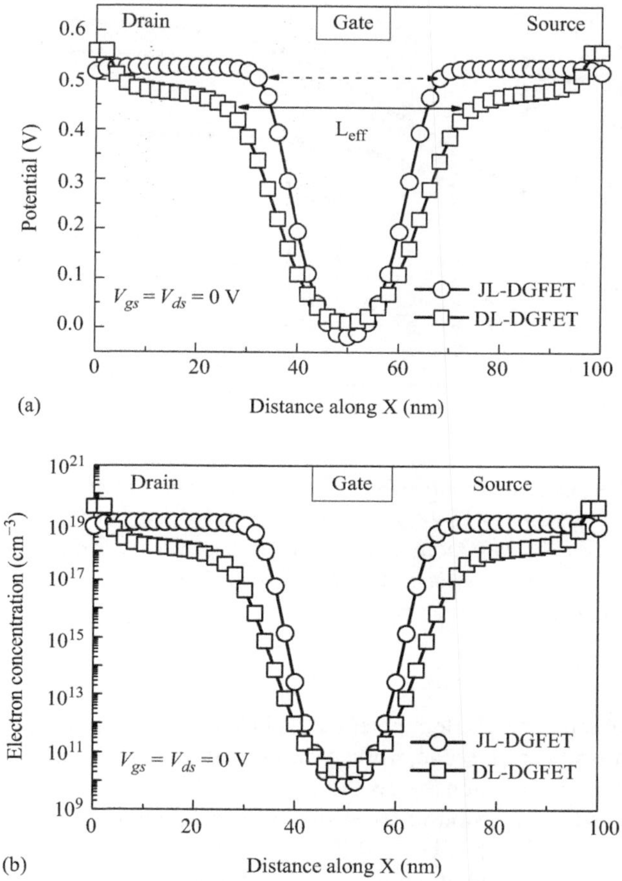

Figure 5.11 *(a) Potential, (b) electron concentration (n_e) along the channel direction (x) at thermal equilibrium for JL- and DL-DGFETs*

region, and it will degrade towards channel region. This can also be verified by OFF-state energy-band diagram as shown in Figure 5.10 where still a potential barrier exist that does not allow drift current to flow. The device is turned on by applying a positive gate voltage by accumulating electrons below gate as shown in Figure 5.9(b) and lower the barrier between source to drain as can be seen in Figure 5.10. It demonstrates the effective and successful formation of S/D region over the intrinsic silicon via CP, and behavior is consistent with the conventional JL-DGFET without heavily doped S/D region.

To further investigate the behavior of DL-DGFET and ensure the effectiveness of CP concept, the electrostatic potential (ϕ_c) and electron concentration (n_e) profile inside the device under thermal equilibrium were extracted along channel directional center of the silicon film, as can be seen in Figure 5.11(a,b). The gate extends from 40

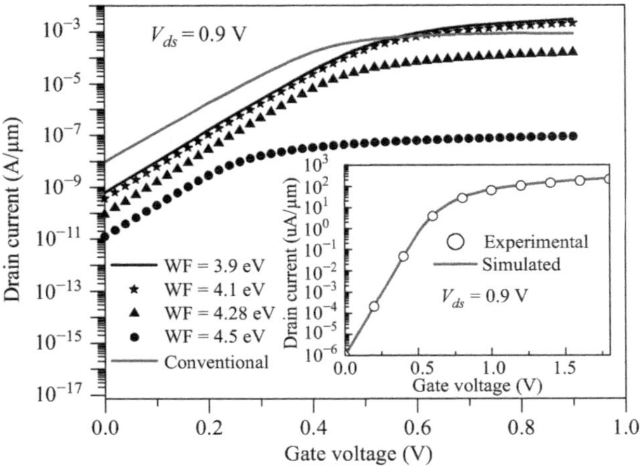

Figure 5.12 Transfer characteristics of conventional at S/D metal work function = 4.28 eV and dopingless JLFET for various work function metallic contact at S/D regions. V_{ds} = 0.9 V, inset shows model calibration against SOI-JLFET experimental $I_d - V_{gs}$ data from reference 24

to 60 nm (horizontal axis). Both potential and electron concentration of JL-DGFET decreases sharply near gate edge, while it varies gradually for DL-DGFET. The reduction of electron concentration (below the doping level in JL-DGFET) beyond the gate edge represents the extension of the depletion regions along the channel direction, i.e., unintentional underlap effect [60]. The length of depletion region extends beyond the gate (toward source and drain regions) depends on the doping level. This lateral extension of depletion region is significant for DL-DGFET due to dopingless architecture and signifies a longer effective gate length (L_{eff}) in off condition, as shown in Figure 5.11(a). The L_{eff} is 28 nm for JL-DGFET and 40 nm for DL-DGFET. The increase in L_{eff} minimizes the SCEs in DL-DGFET and makes it further scalable as compared to JL-DGFET. The study of energy band diagram, electrostatic potential (ϕ_c), and electron concentration (n_e) profile validate the working of CP concept for an n-type DL-DGFET.

To see the working of DL-DGFET, I-V characteristics are extracted through TCAD simulations. The transfer characteristics of the DL-DGFET are shown in Figure 5.12, and they follow a similar trend to that of a conventional JL-DGFET. The inset shows the calibration of our TCAD model parameters to the experimental SOI-JLFET data [24]. The results indicate that the drain current in JL-DGFET is higher to that of DL-DGFET due to very high doping (order of 10^{19} atoms/cm^3) at fixed S/D aluminium contact (work function = 4.28 eV). However, ON current in DL-DGFET device was improved by the use of lower work function material, such as HF (work function = 3.9 eV), as shown with solid black line in Figure 5.12. The HF metal

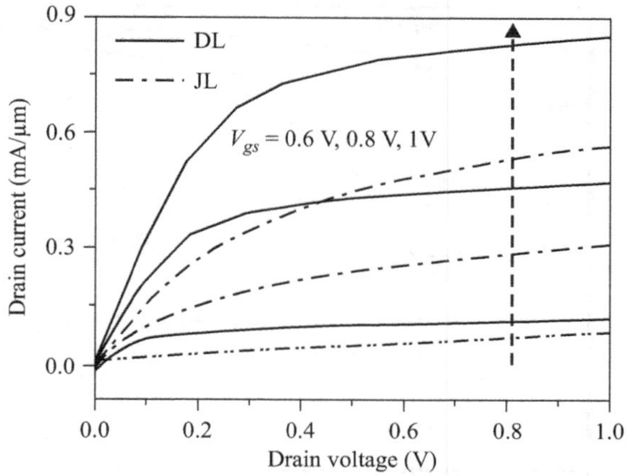

Figure 5.13 Drain current versus drain voltage for DL and JL DGFETs at different V_{gs}

electrodes will introduce higher work function difference, as a result, accumulation of higher carrier concentration near S/D contact and causes increase in overall drain current. Similarly, the output characteristics of DL- and JL-DGFETs are shown in Figure 5.13 for different V_{gs}. The drain current of DL-DGFET is higher and it saturates early at low drain voltages as compared to JL-DGFET. Furthermore, the drain current in saturation region is more flat in DL-DGFET as compared to JL-DGFET which shows its higher immunity towards SCEs. From these I-V characteristics, it can be concluded that the electrical characteristics of the junction and doping-free FET are at par with its counterpart JLFET. Subsequently, this analysis confirms the concept of CP and its applicability for FET, and it can be extended for other devices as well.

In JL-DGFET, gate-drain potential is higher than gate-source potential, due to applied drain bias. It causes band overlap between valence band of channel and conduction band of drain in the lateral direction. This results in higher probability of BTBT that leads to a significant leakage current as discussed in section of JLFET limitation (Figures 5.4 and 5.5). It means that the higher doping concentration of silicon triggers larger effect of BTBT. We analyzed an effective method to decline the influence of BTBT by reducing the channel doping concentration as in case of DL-DGFET. Hence, we observed low off current in DL-DGFET as compared to JL-DGFET as shown in Figure 5.14. It is due to reduction in band bending from the channel edge region to the drain region. Furthermore, in a DL-DGFET, larger depletion region in off state increase the tunneling width of the device; hence, it will reduce the possibility of the tunneling of electrons from the valence band of the channel to the conduction band of the drain. Furthermore, we analyzed $I_d - V_{gs}$ of JLFET for different silicon layer thicknesses (TSI) of 6, 8, and 10 nm, as shown in Figure 5.15. The JLFET with thin TSI has lower OFF current as compared to the

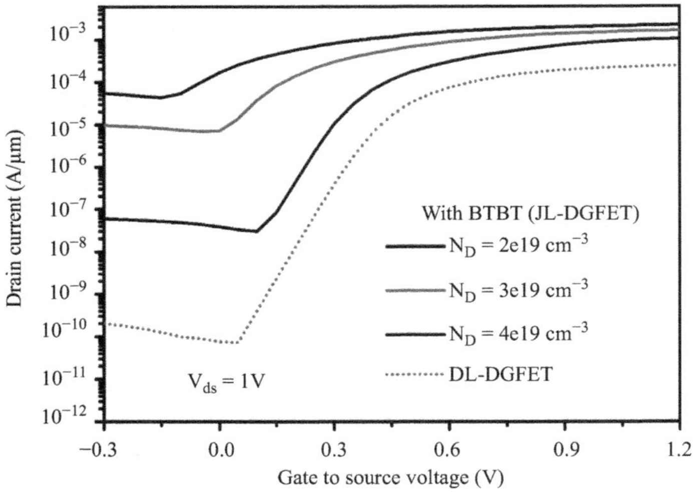

Figure 5.14 Effect of band to band tunneling on $I_d - V_{gs}$ characteristics of DL- and JL-DGFET at various channel doping

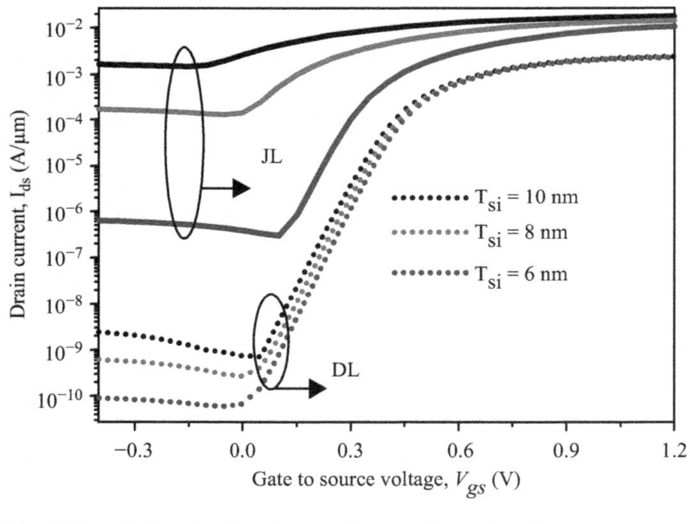

Figure 5.15 Effect of band to band tunneling on $I_d - V_{gs}$ characteristics of DL- and JL-DGFET at various Si thickness

device with thicker TSI. One can see that for all thickness, DL-DGFET shows lower OFF current than its counterpart.

In JLFET, it was observed that the variability in device dimensions is a challenging issue that needs to be addressed before exploiting its full benefits. Therefore, a detailed comparative study on different device dimensions were done for both JL and JL as

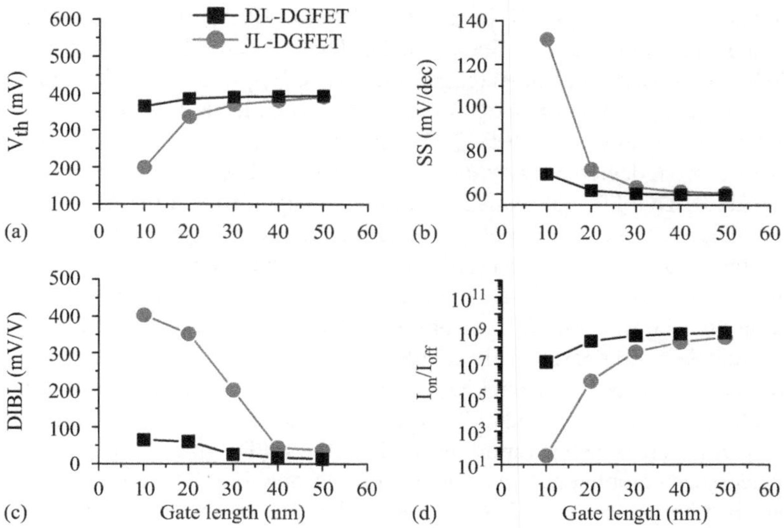

Figure 5.16 Comparison of JL and DL-DGFET for (a) V_{TH}, (b) SS, (c) DIBL, and (d) I_{on}/I_{off} as a function of gate length

well as dopingless devices. This study will also provide an incentive to gauge the robustness of both devices under process variations. The process variation at device level was analyzed for variation in different device dimensions, such as, gate length (L_g), silicon film thickness (T_{SI}), and equivalent oxide thickness (EOT). The variation in different device dimensions for both devices was observed as a change in electrical characteristics, such as threshold voltage (V_{th}), subthreshold swing (SS), DIBL, and ON to OFF current ratio (I_{on}/I_{off}). The large variations in these performance metrics as a function of device dimension (gate length) significantly hampers the overall performance of a device.

The conventional MOSFET scaling follows a set of predictable guidelines, where especially, gate length scaling is an important key factor for device miniaturization. To see the effect of variation in different electrical characteristics (performance metrics), the gate length scaling of both DL-DGFET and JL-DGFET is carried out from 50 to 10 nm keeping other parameter constant. Figure 5.16(a–d) shows the comparison of both devices on four different performance parameters, V_{TH}, SS, DIBL, and I_{on}/I_{off}, as a function of gate length. Initially, V_{TH} of both devices were adjusted to 400 mV by tuning the gate work functions for Lg = 50 nm. In Figure 5.16(a), the V_{TH} of both devices decrease with diminution of gate lengths and change of V_{TH} of DL-DGFET is significantly lower than JL-DGFET, which means that V_{TH} roll-off is less for DL-DGFET than the JL-DGFET. The SS for different L_g for both devices is shown in Figure 5.16(b). It can be seen that the SS increases with down scaling of L_g; however, down scaling has less influence on the SS of DL-DGFET than JL-DGFET. The DIBL for both devices is also compared for different L_g ranging from 50 to 10 nm. The L_g

scaling leads to increased DIBL in JL-DGFET but DIBL variation in DL-DGFET is less, as shown in Figure 5.16(c). Therefore, DL-DGFET can efficiently be turned off at $V_{gs} = 0$ V even at smaller channel lengths due to less V_{TH} roll-off, while highly doped JL-DGFET suffer from poor switch-off capability (i.e., lower I_{on}/I_{off}) at $L_g = 10$ nm, as can be seen in Figure 5.14(d). The gate length scaling has significantly lesser effect on different device performance parameters in DL-DGFET as compared to its counterpart JL-DGFET. Subsequently, it can be concluded that the DL-DGFET has reduced SCEs due to extension of the depletion regions along the channel direction, i.e., unintentional underlap effect. This lateral extension of depletion region is higher for lightly doped DL-DGFET that signifies a longer effective channel length (L_{eff}) in off condition. The increase in L_{eff} improves SCEs of the DL-DGFET and makes it further scalable as compared to JL-DGFET.

In nanowire-based FETs, thickness of nanowire is another most crucial parameter, specifically for JL heavily doped FETs because gate needs to deplete the entire channel region in OFF state. Smaller variations in Si thickness (T_{SI}) may or may not allow the gate to deplete it completely, hence, poor gate controllability. Therefore, effects of T_{SI} scaling for both devices (DL-DGFET and JL-DGFET) were observed for different performance metrics while keeping other device parameters constant, as can be seen in Figure 5.17. The threshold voltage (V_{TH}) variation increases with reduction in Si thickness in both devices, as shown in Figure 5.17(a). One can see that for lower doping concentration in DL-DGFET, V_{TH} is rather insensitive to Si thickness variation, but for JL-DGFET, variation is very high. In DL-DGFET, as T_{SI} scaled down from 9 to

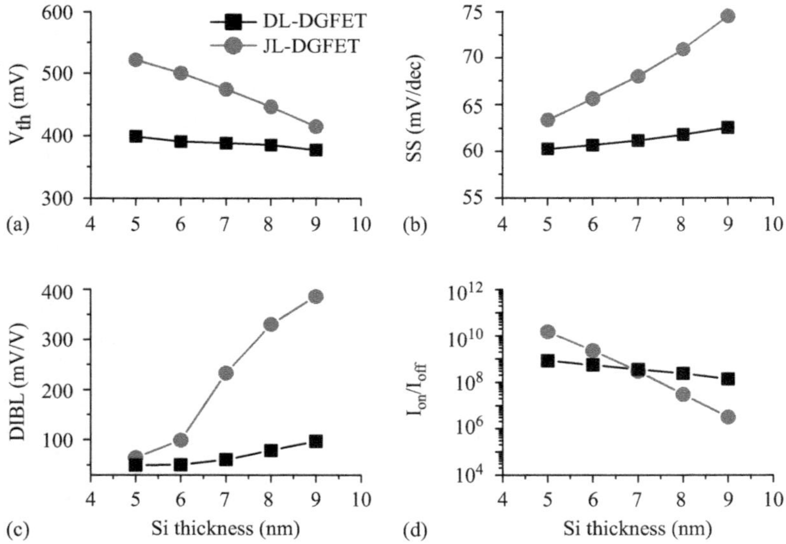

Figure 5.17 Comparison of JL and DL-DGFET for (a) V_{TH}, (b) SS, (c) DIBL, and (d) I_{on}/I_{off} as a function of Si thickness

5 nm, V_{TH} changes only 20 mV, but in JL-DGFET, V_{TH} shift occurs 110 mV for same T_{SI} scaling. The higher V_{TH} shift in JL-DGFET increases SCEs especially for larger T_{SI}. For JL-DGFET of $L_g = 20$ nm, $T_{SI} = 9$ nm, SS is 75 mV/dec, DIBL is 386 mV/V, and I_{on}/I_{off} is 2×10^6, while extracted values of DL-DGFET for SS, DIBL, and I_{on}/I_{off} are 96 mV/V, 62 mV/dec, and 2×10^8, respectively. In short, we can infer that the DL-DGFET is less susceptible to process induced variation in terms of T_{SI} scaling due to utilization of lightly doped silicon.

The equivalent oxide thickness (EOT) is an important parameter in any MOSFETs as it determines the gate capacitance C_{ox}. In conventional and accumulation mode MOSFETs, I_{on} is inversely proportional to EOT but in the JL-DGFET, I_{on} mainly depends on the channel doping concentration and moderately depends on the gate oxide capacitance [17]. As shown in Figure 5.18,the scaling of EOT for both JL-DGFET and DL-DGFET is not as critical as T_{SI}. In addition, it is not necessary to scale the EOT in a DL-DGFET as aggressively as in regular MOSFET to improve short-channel characteristics. Figure 5.18a shows the threshold voltage fluctuations due to EOT variation, which was observed about 60 mV/nm in both devices for a gate length of 20 nm. Even for EOT as thick as 1.2 nm, the SS is below 63 mV/dec and the DIBL is about 60 mV/V, but in classical MOSFETs, EOT requires to go down to 0.5 nm to attain devices with similar performances [25]. It is also shown that I_{on}/I_{off} ratio variation is lower in DL-DGFET with EOT. Hence, we infer that

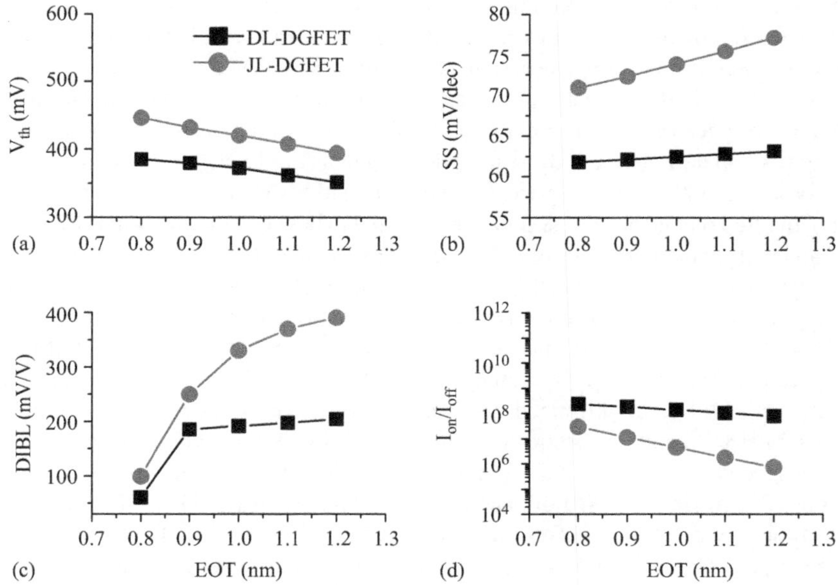

Figure 5.18 Comparison of JL and DL-DGFET for (a) V_{TH}, (b) SS, (c) DIBL, and (d) I_{on}/I_{off} as a function of EOT

the DL-DGFET shows moderate dependency on gate oxide capacitance similar to JL-DGFET and variation in SCEs is less due to lightly doped channel.

The heavily doped JLFETs have been considered as potential the candidates due to their good scalability and simplified fabrication process; however, heavy doping requirements poses certain limitations as discussed previously. Therefore, in that pursuit, the concept of junction and doping-free FETs was coined that preserves the inherent merits of JLFETs. Furthermore, to meet the ON-state current requirements for junction and doping-free transistors, the idea of CP was introduced that fulfills the doping requirements artificially without compromising over device performance. A comparative study of both devices reveals that the junction- and doping-free transistors (DL-JLFETs) not only offers competitive performance with its counterpart JLFET but also immunity towards RDFs and process variation, and lower SCEs make them highly scalable for nano regime. To see the transformation of device level benefits to the circuit level, a comparative study of standard six transistor (6T) static random access memory (SRAM) cell as a benchmark circuit is performed using JL- and DL-DGFET in comparison with IM and JL-DGFETs of identical dimensions.

In general, standard 6T SRAM cell is being considered as a standard benchmark circuit for assessment of any new technology and new technology success depends upon the successful realization of the SRAM cell. SRAM cells are widely used in electronic systems for cache memories, such as mobile phones, microcontrollers, and personal computers. In these devices a significant percentage of the total area and power dissipation are due to these SRAM cells. Also the SRAM leakage dominates leakage power and lowering the supply voltage (V_{dd}) for SRAMs may reduce the switching as well as leakage power dissipations. The SRAM cell must be designed in such a way that it should provide the non-destructive read operation and a reliable write operation. These two requirements impose certain constraints on SRAM cell sizing. The stability of a standard 6T SRAM cell is generally measured in terms of static noise margin (SNM) during read and write operations. As SRAM cell being most vulnerable to noise during read operations [26–28], read SNM as well as hold SNM (HSNM) for SRAM cells based on both devices were extracted. The read SNMs were extracted by square fitting method that is the largest square to be fitted in between the super imposed transfer and inverse characteristics of a SRAM cell using mixed-mode TCAD simulations.

The standard 6T SRAM cell (as shown in Figure 5.19) was designed and simulated using junctionless (JL-DGFET) and junctionless as well as dopingless (DL-DGFET) FETs with cell ratio and pull-up ratio equals to 1. The SNMs for these SRAM cells have been evaluated at a supply voltage (V_{dd}) of 0.8 V. The SNM comparison for both SRAM cells will yield optimized gate work function for both devices. The butterfly curves for HSNM and read static noise margin (RSNM) are shown in Figures 5.20 and 5.21, respectively. The DL-DGFET–based 6T SRAM cell shows an impressive HSNM of about 352 mV and RSNM of nearly 145 mV. The use of DL-DGFET reduces gate work function offset (0.4–0.2 eV) with respect to 4.74 eV for 6T SRAM cell. Hence, DL design not only achieves higher performance metrics and reduced parameter sensitivity but also relaxes the constraints on the selection of gate metal work function of JL devices. Figure 5.22 shows the different RSNM values for DL-DGFET 6TSRAM cell along with IM devices available in the literature [26–30]. The RSNM values

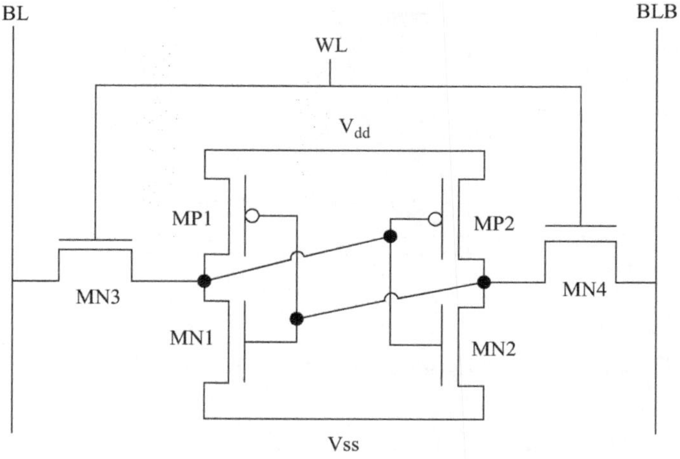

Figure 5.19 Schematic diagram of 6T-SRAM cell as a test circuit

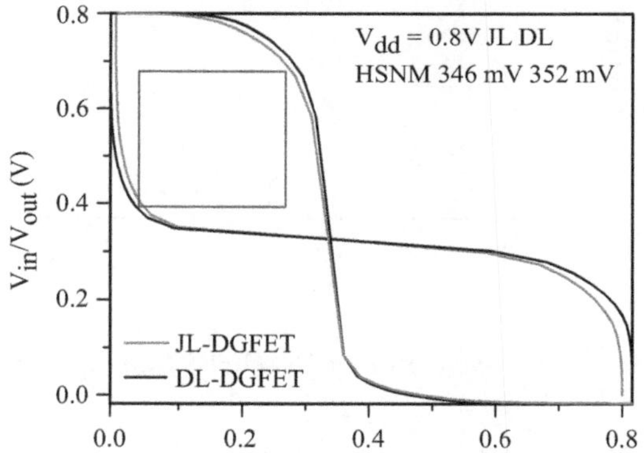

Figure 5.20 Butterfly curves in hold state of JL- and DL-DGFETs for 6T SRAM cell

(0.181 V_{dd}) for DL-DGFETs at gate length of 20 nm shows potential for low power digital applications.

5.4.2 Dopingless BJT

The BJTs have their own importance and figure-of-merits for mixed and high-speed radio frequency (RF) applications owning to their high switching speed and larger drive current. However, CMOS technology has its own merits and provides low power dissipation and larger packing density. Therefore, BiCMOS technology (integration of bipolar and CMOS devices on a single integrated circuit) is a promising candidate for future computing as it embraces the best of both worlds. The integration of both technologies makes BiCMOS fabrication process complex due to different types and

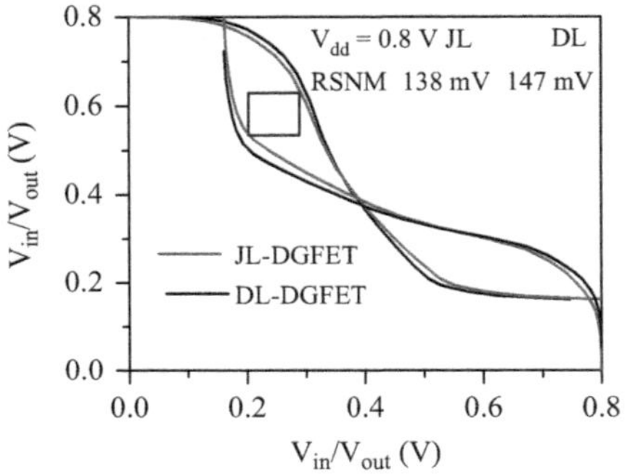

Figure 5.21 Butterfly curves in read state of JL- and DL-DGFETs for 6T SRAM cell

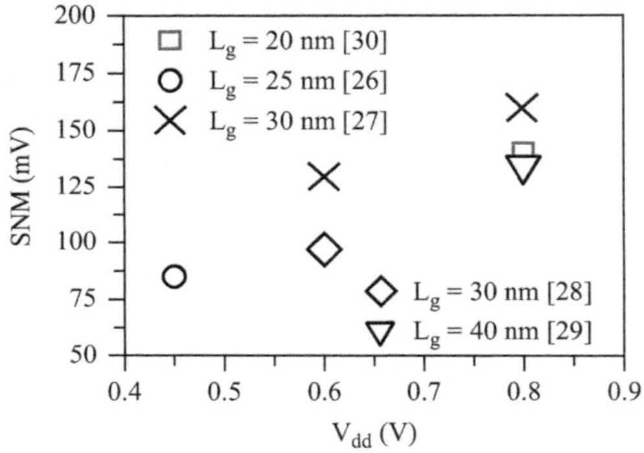

Figure 5.22 SNM values comparison of DL-DGFET and IM-FET 6T SRAM cells

levels of doping, isolation, and compatibility issues; subsequently, integration of both technologies becomes expensive. With the advent of SOI-CMOS technology, and junction-and doping-free FETs the challenges of integration and compatibility can be mitigated. An in-depth study of doping- and junction-free FETs (or DL-JLFETs) reveals a new horizon for simplified and inexpensive fabrication process for CMOS devices. Hence, with the same concept (i.e., CP), BJTs can also be realized with help of undoped SOI and yield the benefits of simplified process flow and low thermal budget requirements. This concept makes BiCMOS technology as a viable alternate for future computing with simplified and inexpensive fabrication process. Therefore, in this section, a detailed study on symmetric (where emitter and collector terminal are exchangeable similar to source and drain terminals in a MOSFET) BJTs without external doping is presented.

A symmetric BJT (where emitter and collector terminals are exchangeable) with CP concept on SOI referred as a symmetric bipolar charge-plasma transistor (BCPT) is shown in Figure 5.23(a). In this p-type symmetric BCPT, platinum metal electrodes of work function 5.65 eV are attached over undoped silicon film to form emitter and

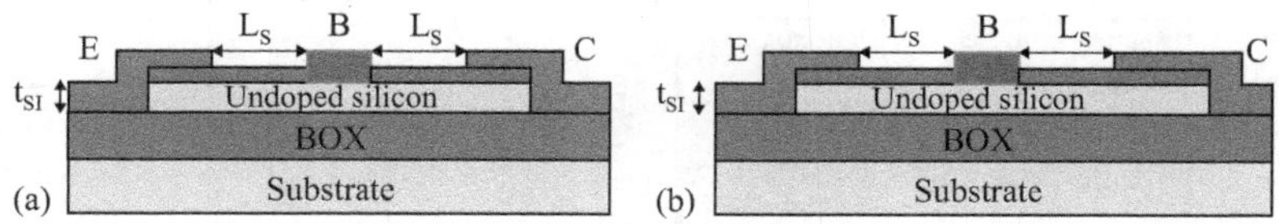

Figure 5.23 Schematic cross-sectional view of (a) symmetric BCPT; (b) asymmetric BCPT

collector regions of equal area [31]. However, in [20], two different metal electrodes were employed for the formation of emitter and collector regions, and the collector area is kept higher than emitter, as shown in Figure 5.23(b). The electrodes of different metal work function and uneven area for emitter and collector regions yield asymmetric BCPT; hence, terminals (emitter and collector) in this device cannot be exchangeable and device becomes asymmetric. The symmetric nature of a BJT yields following advantages:

- emitter and collector terminals are exchangeable, as a result, easy integration and complementary functionality can be achieved exactly in a similar fashion as in CMOS technology,
- reduced process complexity during the metal deposition process due to employment of same metal electrode for both emitter and collector regions, hence, reduced fabrication cost,
- reduced overall device area, i.e., higher integration density,

Furthermore, symmetric BCPT is free from statistical RDFs as external doping is not required for its implementation. It is also clear from the DL-JLFET study that the process variation–related issues that may arise during fabrication or lithographical process in the CP-based devices are very less as compared to conventionally doped devices. Also the CP-based devices can be processed at low temperature; hence, they reduces the thermal budget requirements significantly.

Through TCAD simulations, the transistor action of a symmetric BCPT was validated and compared with an asymmetric BCPT and conventional BJT. Simulation results illustrate that the symmetric BCPT exhibits almost similar behaviour as that of the asymmetric BCPT with a large current gain as compared with the conventional BJT. The parameters used during simulation for both BCPTs are: background doping of thin silicon film is $N_d = 10^{15}$ atoms/cm^3 (a dopingless silicon film), the buried-oxide layer thickness $t_{box} = 375$ nm, silicon film thickness $t_{SI} = 15$ nm, base length $= 0.1$ μm, intrinsic gap $= 0.1$ μm, and emitter/collector (E/C) length $= 0.2$ μm and collector length for an asymmetric BCPT is 0.4 μm. The silicon film thickness and metal work functions for different electrodes (emitter, base, and collector) were chosen such that they meet the basic requirements of CP concept as described in the previous section. To see the transistor action and appropriate induction of carrier concentration in the respective regions through CP, the carrier distribution in both equilibrium state ($V_{BE} = 0$ V, $V_{CE} = 0$ V) and forward active mode ($V_{EB} = 0.8$ V, $V_{EC} = 1$ V) for a symmetric p–n–p BCPT is observed in Figure 5.24(a,b), respectively.

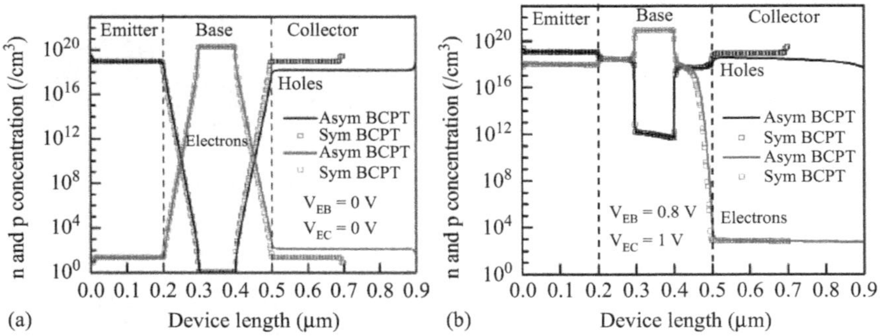

Figure 5.24 Carrier concentration (electron and hole) distribution in the asymmetric and symmetric BCPT (horizontally) under (a) thermal equilibrium (VBE = 0 V, VCE = 0 V), (b) active forward bias conditions (VEB = 0.8 V, VEC = 1 V)

Both hole (p) and electron (n) concentrations were extracted along the horizontal axis at a distance of 2 nm away from the silicon–oxide interface in the emitter, base, and collector regions for both symmetric and asymmetric BCPTs. The induced carrier concentration in asymmetric BCPT through CP in the emitter, base, and collector regions are $N_A = 10^{19}$ atoms/cm^3, $N_D = 10^{20}$ atoms/cm^3 and $N_A = 10^{18}$ atoms/cm^3, respectively. Since, asymmetric BCPT makes use of metal electrodes of different work functions and induces different hole and electron plasma in different regions; as a result, emitter and collector terminals cannot be exchangeable. Although the symmetric BCPT has uniform hole concentration of $N_A = 10^{19}$ atoms/cm^3 in collector and emitter regions that makes the device symmetric or bilateral, as shown in Figure 5.24(a). A clear difference in hole concentration in the symmetric and asymmetric BCPT is also observed near the collector region due to a difference in the collector electrode metal work functions of both devices.

A large work function difference between platinum electrode and undoped silicon in symmetric BCPT accumulates more holes than gold electrode used in asymmetric BCPT, which was verified through TCAD simulations. The symmetric BCPT has a smaller collector area and higher hole concentration in the collector due to large electrostatic field caused large work function difference. The hole distribution in forward active (ON state) state is shown in Figure 5.24(b), which is identical in both BCPTs due to the rearrangement of carriers with applied bias; as a result, current density does not affect the symmetric BCPT, as shown in Figure 5.25. The improvement in the collector hole concentration in the symmetric BCPT is compensated by a reduction in collector length, thereby, current density remains of the same order in the symmetric BCPT.

From the above carrier concentration analysis for BCPTs, it is clear that the CP concept works effectively and precisely to artificially induce the carrier concentration of certain types and levels in a designated region of an undoped silicon film. The working of a BJT based on CP is further studied with the Gummel plot, current

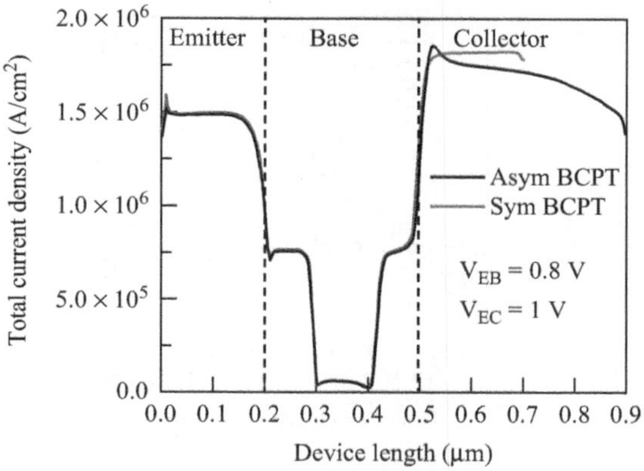

Figure 5.25 Total current density (horizontally) at distance of 2 nm away from the silicon-oxide interface

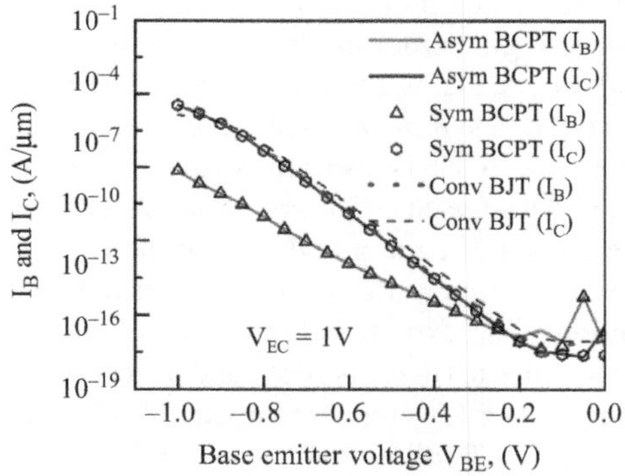

Figure 5.26 Gummel plot comparison of symmetric BCPT with BJT and asymmetric BCPT

gain, and cutoff frequency. The electrical characteristics for a symmetric BCPT are compared with that of the conventional SOI-BJT and asymmetric BCPT. The doping concentrations of the emitter, base, and collector regions of BJT are chosen in such a manner to have an equal neutral base width to BCPT structures. It is clear from Figure 5.26 that a small difference is present in the collector current between BCPT and conventional BJT, but the base current of the BCPT is significantly lower than the conventional BJT due to surface accumulation layer transistor (SALTran) effect [32]. The SALTran effect is generally observed in BCPT, where a lightly doped emitter and an emitter metal electrode contact with a work function higher than that of silicon will result in accumulation of hole at the metal–semiconductor (platinum–silicon) interface.

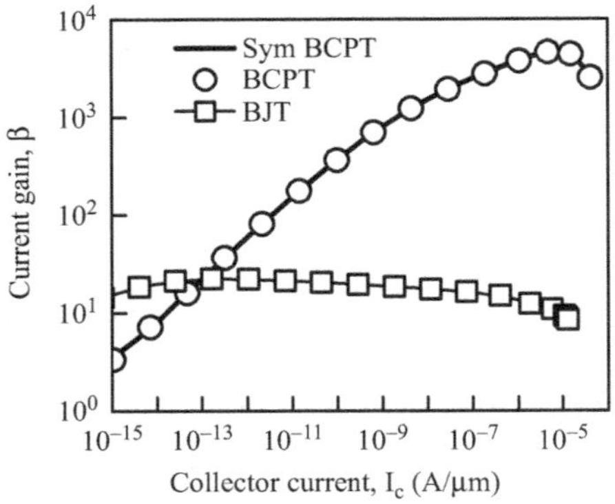

Figure 5.27 Current gain (β) comparison of symmetric BCPT with BJT and asymmetric BCPT

This results in a higher horizontal electric field that opposes the flow of holes entering into the emitter region from the base and acting as a reflecting boundary at the emitter contact, thus reducing the base current. The reduction in base current improves current gain (β) of a symmetric BCPT over the conventional BJT; however, current gain of a symmetric BCPT is also similar to the asymmetric BCPT, as shown in Figure 5.27. Owing to the large base Gummel number, as compared with the emitter Gummel number, the doping of emitter and base of the conventional BJT could not be chosen to be of the same order as that of the symmetric BCPT, because this would make β of the conventional BJT extremely low (<1) [20]. Furthermore, it is observed that the peak β of the symmetric BCPT is 1450, and the conventional BJT has its peak gain about 10. Assuming that BJT is made symmetrical, a considerable change in β is observed, but symmetric BCPT follows the same trend as asymmetric BCPT, as can be observed in Figure 5.27. A major challenge associated with the symmetric BCPT device is poor cutoff frequency (f_T) which reduces the operating speed of the device in comparison with the conventional BJT that can be seen in Figure 5.28. The peak f_T of both BCPT devices is 3.8 GHz, and for the conventional BJT, it is 7.9 GHz. Assuming the BJT is made symmetrical, f_T is affected severely but the symmetric BCPT follows similar behavior as the asymmetric BCPT. There are certain remedies that can be employed to improve the f_T such as (a) making the collector contact Schottky as discussed in [33], but asymmetry of the structure is still present and (b) replacement of SOI with selective buried SOI but it increases overall fabrication complexity [34].

As asymmetric and symmetric BCPTs have limited cutoff frequency, almost half of the conventional BJT which may hinder the applicability of these devices for RF and mixed signal applications. Therefore, an optimization of these devices for different device dimensions such as silicon film thickness (t_{SI}) and intrinsic gaps (L_s) was carried out to achieve higher cutoff frequency. Figure 5.29(a,b) shows f_T and β simulations performed for different devices for different L_s and t_{SI}. A significant change in f_T and β was observed while reducing L_s, because of reduction in L_s that

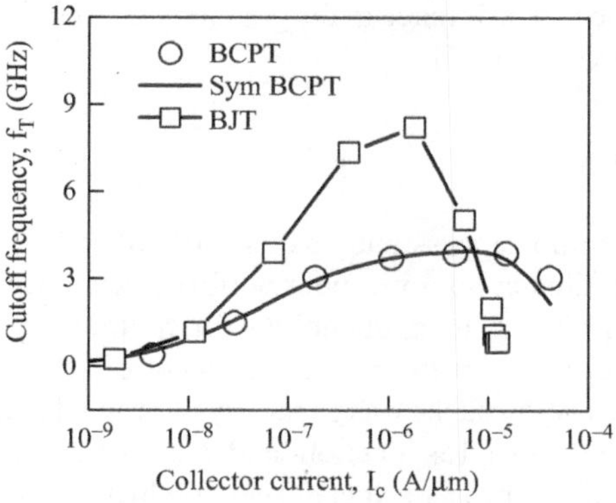

Figure 5.28 Cutoff frequency (f_T) comparison of symmetric BCPT with BJT and asymmetric BCPT

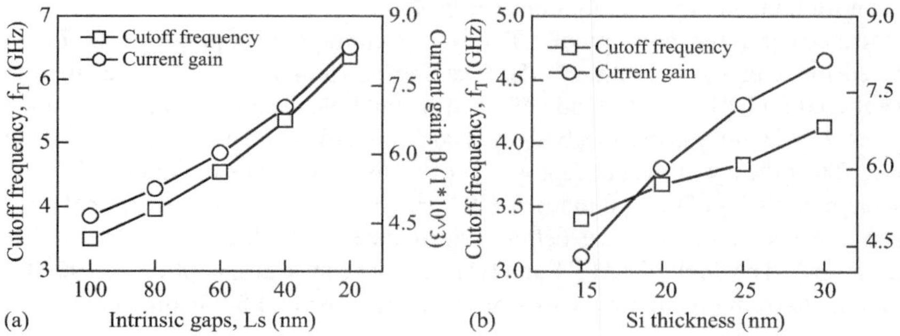

Figure 5.29 Effects on f_T and β on scaling of (a) intrinsic gaps and (b) silicon film thickness of symmetrical BCPT

minimizes the parasitic capacitances and also shortens the effective length between the emitter and collector. Hence, an improvement in collector current is obtained leading to enhanced f_T and β, as can be seen in Figure 5.29(a). The peak value of f_T is 6.5 GHz at $L_s = 20$ nm, which is still less than the f_T of the conventional BJT (7.9 GHz). Similarly, larger t_{SI} increases the overall collector current which results in improvement in β and f_T as shown in Figure 5.29(b). The peak value of f_T is 4.5 GHz at $t_{SI} = 30$ nm, which is still lower than the f_T of a conventional BJT. The optimization of both L_s and t_{SI} leads to $f_T = 7.9$ GHz (equivalent f_T of the conventional BJT) at $t_{SI} = 30$ nm and $L_s = 20$ nm. Thus, symmetric BCPT exhibits similar cutoff frequency as BJT, which is due to the reduction in the collector parasitic capacitance by optimization of L_s and t_{SI}. The consequence of scaling the device dimensions in BCPT is lower breakdown voltage (BV) which is a major tradeoff. Since BCPT exhibits lower collector BV due to a higher collector current when compared with the conventional BJT, but BCPT has negligible base width modulation due to a higher concentration of induced electrons in the base region [20]. Hence, BCPT can be made symmetrical due to a difference in the geometry as well as carrier concentration profiles when compared with BJT.

5.5 Conclusion

The simplified fabrication process and easy scaling of JLFET over IMFET makes JLFET a potential candidate; however, poor performance towards process variations and other limitations hinders its applicability for future CMOS technology–based computing. Therefore, an alternate device architecture that preserves the merits of JLFET and alleviates its drawbacks was exhaustively studied and investigated for digital applications, and is referred as junction- and doping-free FET (i.e., DL-DGFET). The DL-DGFET architecture has a metal–semiconductor interface in source/drain region, which required metal work function engineering instead of placing a simple ohmic contact. The benefits of DL-DGFET are large ON to OFF current ratio, lower V_{th} roll-off and minimized SCEs, at the same time, undoped silicon in this device makes it immune to process variations (or RDFs).

Furthermore, the concept of CP (i.e., junction- and doping-free FET) was explored for realizing a symmetric BJT having performance much better than the conventional BJT. The BCPT made of electron and hole plasmas on undoped silicon can be realized with less thermal budget as compared to doped transistors and CMOS devices. The efficacy of the concept was verified using TCAD simulations. The electrical characteristics of the symmetric BCPT are compared with a conventional doped bipolar transistor, and it exhibits better performance like high current gain than conventional BJT. The dopingless FET and BJT concept can be also applied for BiCMOS technology to overcome the drawback of low integration and fabrication complexity.

References

[1] Lilienfeld J.E., 'Method and apparatus for controlling electric current', U.S. Patent 1 745 175, 1925.

[2] Colinge J.P., Lee C.W., Afzalian A., *et al.*, 'Nanowire transistors without junctions', *Nat. Nanotechnol.*, 2010, 5(3), pp. 225–229.

[3] Park C.-H., Ko M.-D., Kim K.-H., *et al.*, 'Electrical characteristics of 20-nm junctionless Si nanowire transistors', *Solid State Electron.*, 2012, 73(7), p. 710.

[4] Jeon D.-Y., Park S.J., Mouis M., Barraud S., Kim G.-T., Ghibaudo G., 'A new method for the extraction of flat-band voltage and doping concentration in tri-gate junctionless transistors', *Solid State Electron.*, 2013, 81, pp. 113–118.

[5] Chen Z., Xiao Y., Tang M., *et al.*, 'Surface-potential-based drain current model for long-channel junctionless double-gate mosfets', *IEEE Trans. Electron Devices*, 2012, 59(12), pp. 3292–3298.

[6] Duarte J.P., Choi S.J., Moon D.I., Choi Y.K., 'Simple analytical bulk current model for long-channel double-gate junctionless transistors', *IEEE Trans. Electron Devices*, 2011, 32, pp. 704–706.

[7] Duarte J.P., Choi S.-J., Choi Y.-K., 'A full-range drain current model for double-gate junctionless transistors', *IEEE Trans. Electron Devices*, 2011, 32(6), pp. 704–706.

[8] Bulk planar junctionless transistor (BPJLT), 'An attractive device alternative for scaling', *IEEE Electron Device Lett.*, 2011, 32(3), pp. 261–263.

[9] Lee C.W., Afzalian A., Akhavan N.D., Yan R., Ferain I., Colinge J.P., 'Junctionless multigate field-effect transistor', *Appl. Phys. Lett.*, 2009, 94, pp. 1053511–1053511-2.

[10] Trevisoli R., Doria R., de Souza M., Das S., Ferain I., Pavanello M., 'Surface-potential-based drain current analytical model for triple-gate junctionless nanowire transistors', *IEEE Trans. Electron Devices*, 2012, 59(12), pp. 3510–3518.

[11] Han M.-H., Chang C.-Y., Chen H.-B., Cheng Y.-C., Wu Y.-C., 'Device and circuit performance estimation of junctionless bulk finFETs', *IEEE Trans. Electron Devices*, 2013, 60(6), pp. 1807–1813.

[12] Han M.-H., Chang C.-Y., Jhan Y.-R., *et al.*, 'Characteristic of p-type junctionless gate-all-around nanowire transistor and sensitivity analysis', *IEEE Electron Device Lett.*, 2013, 34(2), pp. 157–159.

[13] Su C.J., Tsai T.I., Liou Y.L., Lin Z.M., Lin H.C., Chao T.S., 'Gate-all-around junctionless transistors with heavily doped polysilicon nanowire channels', *Electron Device Lett.*, 2011, 32, pp. 521–523.

[14] Leung G., Chui C.O., 'Variability impact of random dopant fluctuation on nanoscale junctionless FinFETs', *IEEE Electron Device Lett.*, 2012, 33(6), pp. 767–769.

[15] Aldegunde M., Martinez A., and Barker J.R., 'Study of discrete doping induced variability in junctionless nanowire MOSFETs using dissipative quantum transport simulations', *IEEE Electron Device Lett.*, 2012, 33(2), pp. 194–196.

[16] Nawaz S.M., Dutta S., Chattopadhyay A., Mallik A., 'Comparison of random dopant and gate-metal work function variability between junctionless and conventional Fin-FETs', *IEEE Electron Device Lett.*, 2014, 35(6), pp. 663–665.

[17] Gundapaneni S., Bajaj M., Pandey R.K., 'Effect of band-to-band tunneling on junction-less transistors', *IEEE Trans. Electron Devices*, 2012, 59, pp. 1023–1029.

[18] Rajasekharan B., *et al.*, 'Fabrication and characterization of the charge-plasma diode', *IEEE Electron Device Lett.*, 2010, 31(6), pp. 528–530.

[19] Hueting R.J.E., Rajasekharan B., Salm C., *et al.*, 'Charge plasma P-N diode', *IEEE Electron Device Lett.*, 2008, 29(12), pp. 1367–1368.

[20] Kumar M.J., Nadda K., 'Bipolar charge-plasma transistor: a novel three terminal device', *IEEE Trans. Electron Devices*, 2012, 59(4), pp. 962–967.

[21] Kumar M.J., Janardhanan S., 'Dopingless tunnel field effect transistor: design and investigation', *IEEE Trans. Electron Devices*, 2013, 60(10), pp. 3285–3290.

[22] van Hemert T., Hueting R.J.E., Rajasekharan B., Salm C., Schmitz J., 'On the modelling and optimization of a novel Schottky based silicon rectifier',

In: *Proc. of ESSDERC*, Sevilla, Spain, September 2010, IEEE Conference, 2010, pp. 460–463.

[23] Sahu C., Singh J., 'Charge-plasma based process variation immune junctionless transistor', *IEEE Electron Device Lett.*, 2014, 35(3), pp. 411–413.

[24] Barraud S., *et al.*, 'Scaling of trigate junctionless nanowire MOSFET with gate length down to 13 nm', *IEEE Electron Device Lett.*, 2012, 33(9), pp. 1225–1227.

[25] International Technology Roadmaps for Semiconductor, www.itrs.net

[26] Collaert N., *et al.*, 'Low-voltage 6T FinFET SRAM cell with high SNM using HfSiON/TiN gate stack, fin widths down to 10 nm and 30 nm gate length', In: *Proc. Int. Conf. Integr. Circuit Design Technol. (ICICDT)*, June 2008, IEEE Conference 2008, pp. 59–62.

[27] Kawasaki H., *et al.*, 'Demonstration of highly scaled FinFET SRAM cells with high-K/metal gate and investigation of characteristic variability for the 32 nm node and beyond', In: *IEEE Int. Electron Device Meeting Tech. Dig.*, December 2008, IEEE Conference 2008, pp. 14–15.

[28] Mrelle T., *et al.*, 'First observation of FinFET specific mismatch behavior and optimization guidelines for SRAM scaling', In: *Int. Electron Device Meeting Tech. Dig.*, December 2008, IEEE Conference 2008, pp. 14–15.

[29] Wu C.C., *et al.*, 'High performance 22/20 nm FinFET CMOS devices with advanced high-K/metal gate scheme', In: *IEEE Int. Electron Device Meeting Tech. Dig.*, December 2010, IEEE Conference 2010, pp. 27(1)–127(4).

[30] Sahu C., Singh J., 'Potential benefits and sensitivity analysis of dopingless transistor for low power applications', *IEEE Trans. Electron Devices*, 2015, 62(3), pp.729–735.

[31] Sahu C., Ganguly A., Singh J., 'Design and performance projection of symmetric bipolar charge-plasma transistor on SOI', *Electronics Lett.*, 2014, 50(20), pp. 1461–1463.

[32] Kumar M.J., Parihar V., 'Surface accumulation layer transistor (SALTran): a new bipolar transistor for enhanced current gain and reduced hot-carrier degradation', *IEEE Trans. Device Mater. Rel.*, 2004, 4(3), pp. 509–515.

[33] Nadda K., Jagadesh Kumar M., 'Schottky collector bipolar transistor without impurity doped emitter and base: design and performance', *IEEE Trans. Electron Devices*, 2015, 60(9), pp. 2956–2959.

[34] Loan S.A., Bashir F., Rafat M., Alamoud A.R., Abbas S.A., 'A high performance charge plasma based lateral bipolar transistor on selective buried oxide', *Semiconductor Science and Technology*, 2014, 29(1), pp. 11–15.

Chapter 6
Nanoscale high-κ/metal-gate CMOS and FinFET based logic libraries*

Venkata P. Yanambaka[1], Saraju P. Mohanty[1], Elias Kougianos[1], and Dhruva Ghai[2]

During the last four decades, VLSI technology growth has been driven by miniaturization that reduces cost per transistor, power consumption per transistor, with higher packing density and reduced cost of operation. However, the small transistor size leads to very high electric fields across the gate oxide which causes the difficult problem of gate-oxide leakage. This problem is mitigated by high-κ/metal-gate (HKMG) technology, in which the gate material is copper (*going back to metal from polysilicon*), and the gate-oxide material is not silicon dioxide. At the same time, explosive growth of mobile portable electronics has been the driver for many scientific, engineering, and technological breakthroughs in the last few decades. Mobile electronics in particular, such as smart mobile phones, spend most of their operational time in waiting for a call or similar event. However, during these wait states, leakage power dissipation has been a major issue since it drains the battery continuously. The industry has explored various solutions to reduce the OFF-state leakage and multiple gate devices emerged as a solution to this problem. Double-gate FinFET technology is considered as a solution to reduce OFF-state leakage while having faster ON and OFF transitions and low-power (LP) dissipation. This chapter discusses these devices and presents logic libraries which can be used in the digital synthesis of large integrated circuits using such devices through electronic design automation (EDA) tools.

6.1 Introduction

The growth of VLSI technology is one of the fastest observed in human history. This has been made possible by several milestone inventions. Initially, metal-oxide semiconductor field-effect transistors (MOSFETs) had aluminum gates which were slow, large in area, unreliable, with high leakage currents. Then a self-aligned-gate process

*A preliminary conference version of this research was presented at [19, 39].

[1]University of North Texas, Denton, TX, USA
[2]Oriental University, Indore, India

arrived in which the gates of the transistors were made with polycrystalline silicon [60], *not a metal*. The scaling of CMOS technology has accelerated in recent years and will arguably continue toward the 8-nm regime [3]. The superior properties of SiO_2 permit the fabrication of properly functioning devices with SiO_2 layers as thin as 1.5 nm. Further scaling of the SiO_2 layer thickness leads to tunneling gate leakage [3, 37, 52]. In addition, the small transistor size leads to very high electric fields across the gate oxide which causes the difficult problem of gate-oxide leakage [41]. The use of ultra-low thickness gate oxide for short-channel transistors presents the problem of gate-oxide leakage in its ON, OFF, and transition states. These problems are mitigated by HKMG technology, in which the gate material is copper (*going back to metal from polysilicon*) and the gate oxide is not silicon dioxide [6]. An insulator with a higher dielectric constant κ than that of SiO_2 (= 3.9) is used. Statistical characterization of HKMG digital gates as a function of process parameter variation is needed for design. In this chapter, a methodology is presented for PVT-aware HKMG logic library characterization. The methodology considers the process variation effects of 15 device parameters. First, statistical models for gate-induced-drain leakage (GIDL) current ($\hat{I}_{GIDL}$), off-current ($\hat{I}_{OFF}$), and drive current ($\hat{I}_{ON}$) are presented at device level. This is followed by statistical characterization of the library logic cells at room temperature. Statistical results for sub-threshold current ($\hat{I}_{sub}$), GIDL current, dynamic current ($\hat{I}_{dyn}$), and delay are derived. This is followed by results for PVT-aware characterization of logic cells. The library can be used by circuit designers for digital synthesis and design exploration.

At the nanoscale, the short length of the channel affects the device characteristics and its operation dramatically. New devices are being explored to mitigate these short-channel effects (SCEs). One such device is the multi-gate transistor. The industry has been using triple-gate devices for high-performance low-leakage microprocessors. These multi-gate transistors (also known as FinFETs) show a promise of scaling beyond the conventional CMOS and also to overcome SCEs [49]. Ultra-thin specifications of device regions are maintained using the Silicon-On-Insulator (SOI) process. The electrical potential throughout the channel is accurately controlled by the gate voltage in these devices. Advantages of the FinFET over traditional CMOS are: the FinFET provides better area efficiency and the same manufacturing process can be used for both the conventional CMOS and FinFET. The OFF-state leakage current is reduced as the channel is surrounded by multiple gate surfaces. The ON-state drive current is increased, power dissipation is decreased, and the overall device performance is increased. Due to the compactness of multi-gate FETs and FinFETS, higher transistor density can be achieved.

Cross-sections of three different transistors are presented in Figure 6.1. The classic MOSFET is shown in Figure 6.1(a). In this transistor, a SiO_2 dielectric is used and the gate is deposited on top of it. This at nanoscale causes problems due to gate leakage and SCEs. Advanced circuit design exploration must go along with the process technology trends in order to solve design challenges posed by nanoscale CMOS. HKMG transistors are a promising alternative to traditional CMOS at nanoscale technologies [12]. The technology has been invented after extensive research and exploration of SiO_2/polysilicon, high-κ/polysilicon, and HKMG. It is determined that

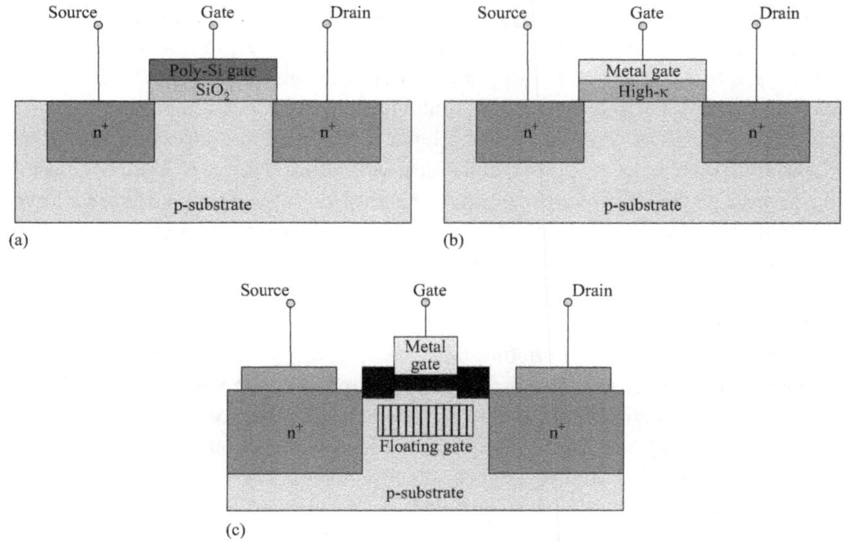

Figure 6.1 Nanoscale planar CMOS and DG-FinFET cross-sections [36, 26, 6, 39]. (a) Conventional SiO_2 gate dielectric, (b) HKMG dielectric, and (c) DG-FinFET structure

the HKMG is needed to have smooth mobility of the channel carriers [6]. Different high-κ dielectrics, including ZrO_2, TiO_2, Al_2O_3, SiON, Si_3N_4, and HfO_2, have been investigated [62, 31, 26]. However, use of HfO_2 with copper metal-gate seems to be promising as it has the highest dielectric constant. The high-κ-based MOSFET is shown in Figure 6.1(b). The SCEs of the bulk CMOS are compensated by introducing a Fin to the transistor. In the double-gate FinFET (DG-FinFET), the polysilicon gate straddles the Si-Fin which gives effective gate controlled characteristics compared to MOSFET. The Fin itself acts as channel, and it terminates on both sides of source and drain. The structure of a DG-FinFET is shown in Figure 6.1(c).

It is widely recognized that the statistical variability in device characteristics represents challenges to scaling and integration for present and next-generation nano-CMOS systems. This in turn demands revolutionary changes in the way in which future integrated circuits are designed. Strong links must be established between system design, circuit design, and fundamental device technology to allow circuits and systems to accommodate the wide variability. Major sources of variability are: process variation (P), supply voltage (V), and operating temperature (T) which may be due to self-heating effects, due to the environment, or a combination of the two. Process variations have profound effects on power and performance (e.g., oscillating frequency of a voltage-controlled oscillator) [36, 17, 18, 7]. Temperature is an important parameter due to its impact on leakage, performance (slowing down of devices and reduction of active current), reliability, and packaging and can lead to thermal breakdown [36, 10, 11]. Since reducing power consumption is increasingly becoming the most important goal for System-on-Chip (SoC) designs, especially for portable battery-driven embedded systems, it becomes essential to address the issue of reliable power estimation for these designs, in the face of PVT variability. PVT variability makes it hard to achieve "safe" integrated circuit designs in nanometer technologies. This is because PVT variability causes fluctuation in all important parameters of SoC designs [16, 7]. Till now several efforts have been made at addressing the effect of PVT variability on leakage, power, and delay estimation. However, the PVT variation in the context of HKMG nanoscale and DG-FinFET also needs additional investigation. This chapter presents a PVT-aware DG-FinFET-based statistical logic library and the device level characterization for its creation.

The notations used throughout the chapter are summarized in Table 6.1. The rest of the chapter is organized as follows: Section 6.2 discusses related research. The HKMG bulk MOSFET structure and its modeling are presented in Section 6.3. The DG-FinFET structure and its modeling are presented in Section 6.4. The proposed approach for logic library creation is discussed in Section 6.5 while the various sources of variability are highlighted. The power, leakage, and delay models used in this chapter are presented in Section 6.6. Section 6.7 summarizes the statistical characterization of HKMG NMOS and PMOS devices and DG-FinFET followed by the statistical characterization of the HKMG and DG-FinFET logic libraries at room temperature. PVT characterization results are presented in Section 6.8. The chapter concludes in Section 6.9.

6.2 Summary of this chapter

To minimize power consumption and maximize timing performance, designers require rapid library characterization and accurate modeling for specific operating environments [48]. Logic libraries are required for fast design exploration. In addition to nominal results, statistical characterization of logic gates as a function of process and environmental parameter variations is required for accuracy in the nanoscale domain. The *topics covered in this chapter* include the following:

1. A methodology for HKMG logic library development is presented.
2. A methodology for DG-FinFET logic library creation is presented.
3. The effect of process variations is taken into account during logic library creation.
4. Device level characterization of HKMG NMOS and PMOS transistors is presented.
5. Device level characterization of DG-FinFET is presented.
6. An HKMG logic library with statistical as well as nominal; characterization at room temperature (27°C) is presented.
7. A PVT-aware HKMG statistical logic library is developed.
8. A PVT-aware DG-FinFET-based statistical logic library is presented.

Highly scaled CMOS devices in the nanoscale regime inevitably exhibit probabilistic behavior due to process variations and other perturbations such as noise. Circuit design methodologies, which depend on the existence of deterministic and

Table 6.1 Notations used in this chapter

Notations	Definition of notations
A_{ch}	Area of the channel
$AGIDL$	The pre-exponential coefficient for GIDL
$BGIDL$	The exponential coefficient for GIDL
$CGIDL$	The parameter for body bias effect of GIDL
C'_{gate}	Capacitance per unit area
C_{eq}	Equivalent capacitance
C_{ins}	Insulator capacitance per unit length
C_L	Load capacitance in F
C_M	Effective Miller capacitance
$DG\text{-}FinFET$	Double-gate FinFET
$EGIDL$	The band bending parameter for GIDL
EOT	Equivalent oxide thickness
$GIDL$	Gate-induced-drain leakage
H_{fin}	Fin-height
$HKMG$	High-κ/metal-gate
$I_c(t)$	Current through the load capacitance C_L
I_d	Drain current
I_{dsat}	Saturated drain current
$\hat{I}_{dyn}$	Current associated with P_{dyn}
I_{GIDL}	GIDL current
IG mode	Independent-gate mode of DG-FinFET
$\hat{I}_{OFF}$	OFF-state current
$\hat{I}_{ON}$	ON-state current
$\hat{I}_{sub}$	Sub-threshold leakage current
L_{effp}	PMOS effective channel length (nm)
L_{effn}	NMOS effective channel length (nm)
L_{ext}	Extension length of DG-FinFET
L_{gate}	Length of the gate in DG-FinFET
L_{phy}	Geometrical channel length of DG-FinFET
LP mode	Low-power mode of DG-FinFET
N_{ch}	Channel doping
N_{chn}	NMOS channel doping concentration (cm^{-3})
N_{chp}	PMOS channel doping concentration (cm^{-3})
N_{gaten}	NMOS gate doping concentration (cm^{-3})
N_{gatep}	PMOS gate doping concentration (cm^{-3})
N_f	Number of fins in a transistor
N_{fin}	Number of fins of DG-FinFET
N_{sdn}	NMOS source/drain doping concentration (cm^{-3})
N_{sdp}	NMOS source/drain doping concentration (cm^{-3})
P_{dyn}	Dynamic power consumption
P_{ins}	Width of the channel in DG-FinFET
Q_M	Effective Miller charge
Q_{out}	Charge at output node
S	Activity factor
SG mode	Shorted gate mode of DG-FinFET
T_{fin}	Fin width of DG-FinFET
T_{gaten}	NMOS gate dielectric thickness (nm)
T_{gatep}	PMOS gate dielectric thickness (nm)
T_{ox}	Oxide thickness
T_{pd}	Propagation delay
T_{pdLH}	Time taken for low to high transition of the output
T_{pdHL}	Time taken for high to low transition of the output

(*Continues*)

Table 6.1 (Continued)

Notations	Definition of notations
T_{si}	Body thickness
V_{db}	Drain to body voltage
V_{dd}	Supply voltage (V)
V_{dsat}	Saturated drain voltage (V)
V_{ds}	Drain to source voltage (V)
V_{gf}	Potential difference between front gate and source
V_{gs}	Gate to source voltage (V)
V_{off}	Offset voltage
v_{therm}	Thermal voltage
V_{Thn}	NMOS threshold voltage (V)
V_{Thnf}	Threshold voltage of the front gate
V_{Thp}	PMOS threshold voltage (V)
W_{eq}	Equivalent width (nm)
W_{effcj}	Effective width of drain diffusions
W_{effn}	NMOS effective channel width (nm)
W_{effp}	PMOS effective channel width (nm)
W_{fin}	Fin width of DG-FinFET
W_{phy}	Geometrical channel width of DG-FinFET
$\hat{Y}$	Required response
α	Technology parameter
β	Technology-dependent constant
λ_w	Width correction factor
ε_{FB}	Bulk material Fermi Energy
ε_{FG}	Gate material Fermi Energy
κ_{gate}	Gate dielectric constant

uniform devices with no consideration for either power consumption or probabilistic systems, will no longer be sufficient to design robust circuits. This chapter provides statistical, input state-dependent characterization data for logic cells which can be used in making an RTL library [35, 40]. The issue of providing characterization data for systems built using these logic cells is not in the scope of this chapter. However, the data provided in this chapter will be useful at the system level when a probabilistic/statistical analysis is performed. In Reference 38, the authors characterize datapath components using a structural Hardware Description Language (HDL). The data presented in this chapter can also be useful for probabilistic-CMOS [1] where the analysis is done using stochastic metrics such as probability distribution functions (PDFs), instead of working with actual values. Once the probability of the occurrence of a particular state in a logic gate of the datapath component is determined, the effect of power, leakage, and delay in that state can be assessed.

6.3 HKMG bulk MOSFET

By the time silicon dioxide (SiO_2) dielectric transistors reached a size of 65 nm, problems started to arise [6, 41]. Figure 6.1(a) presents the cross-section of a conventional SiO_2 dielectric transistor. As a gate dielectric, the SiO_2 is used between the gate and

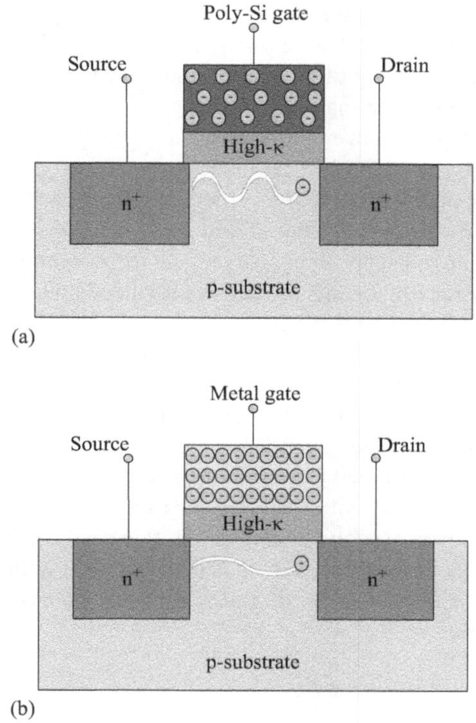

Figure 6.2 Electron movement in high-κ/poly-Si gate and HKMG [6]. (a) Electron movement in high-κ/polysilicon gate and (b) electron movement in HKMG

the substrate. In the 65-nm transistor, the size of the dielectric is four atoms thick [6]. This cannot be reduced further as the electrons were already leaking even in the OFF state of the transistor. So a new material had to be used in the place of SiO_2 in order to reduce the gate leakage current. Experiments have been done using different materials and materials having high dielectric constant are chosen to replace SiO_2. The use of HKMG reduces the gate leakage and improves the reliability of the gate [36, 6, 32]. In this chapter, this non-classical nano-CMOS structure is referred to as HKMG nano-CMOS.

6.3.1 HKMG device structure

SiO_2 in the transistor is replaced with a high dielectric material. Figure 6.2(a) shows the cross-section of a high-κ/poly-Si gate transistor. The dielectric materials used include ZrO_2, TiO_2, Al_2O_3, SiON, Si_3N_4. At first these alternative oxides were used with a polysilicon gate. This gave rise to new problems such as Fermi-Level pinning where the transistors needed higher voltage than usual to turn on and low charge carrier mobility where the transistor's switching speed was slowed down due to the sluggish movement of charge carriers. Figure 6.2(a) shows the movement of electrons in the HKMG gate transistor. The main reason for this behavior is because the dielectric is made up of dipoles and these dipoles lead to very strong vibrations in the semiconductor crystalline lattice (phonons). These phonons slow down the charge carriers in the channel affecting the switching times of the transistor. Hence a metal-gate replaced the traditional polysilicon gate material.

Figure 6.2(b) shows the movement of electrons through the channel of an HKMG transistor. The metal is packed with a many electrons compared to the polysilicon. This reduces the effects of phonons on the movement of charge carriers along the channel. The charge carriers move as they should and the transistor is fast compared to the high-κ/polysilicon gate transistor. This also reduces the minimum voltage required for turning on the transistor. Hence this model was finalized and commercially released into the market.

Initially, to deposit the dielectric layer, two techniques were used: reactive sputtering and metal organic chemical vapor deposition (CVD). These techniques produce a layer that was smooth but failed to fill pockets and gaps where the charges could get stuck. A new technique, Atomic Layer Deposition, was implemented. This technique allows the deposition of a single layer of atoms at a time. The surface of the wafer is exposed to a gas which reacts with the top layer of the wafer and deposits one layer of atoms. After one layer of atoms is deposited, there is no silicon wafer available to react with, so the gas is evacuated. This is then replaced with a second gas which will be able to deposit another layer of atoms, and this process can be repeated for multiple layers of atoms.

6.3.2 HKMG device modeling

An open boundary model based on the Non-Equilibrium Green's Function Formalism is used to model the HKMG FET in Reference 5. The use of non-equilibrium Green's functions allows for a full quantum mechanical treatment of conduction in the channel. Thermal equilibrium is assumed between the gate and the bulk regions, and they are characterized by Fermi energies ε_{FG} and ε_{FB}, respectively. The Green's functions are given by the following expression [5]:

$$\begin{aligned} G^R(r, r', \varepsilon) &= G^A(r, r', \varepsilon) \\ &= \left[\varepsilon I - H(r, r', \varepsilon) - \sum^R (r, r', \varepsilon) \right]^{-1}, \end{aligned} \tag{6.1}$$

where $H(r, r', \varepsilon)$ is the Hamiltonian of the system, and $\sum^R (r, r', \varepsilon)$ is the retarded self-energy. The carrier concentration and leakage current are determined by the following [5]:

$$\begin{aligned} n(r) &= -2i \int G(r, r', \varepsilon) \frac{d\varepsilon}{2\pi} \\ j(r) &= \frac{hq}{m*} \int \left[(\nabla - \nabla') \, G(r, r', \varepsilon) \right] \big|_{r'=r} \frac{d\varepsilon}{2\pi}. \end{aligned} \tag{6.2}$$

Although this approach can provide accurate atomistic simulations, it is not effective for circuit-level characterization of devices. Compact models are needed for use with standard analog simulators. For compact modeling of HKMG transistors using the BSIM4/5 model, two possible options can be considered [26, 39]:

1. Vary the model parameter in the model card that denotes relative permittivity (`EPSROX`).
2. Determine the equivalent oxide thickness (*EOT*) for a dielectric under consideration.

The first option may not be sufficient to model the behavior of non-classic nano-CMOS with non-SiO_2 dielectrics as it does not correctly account for the barrier height of such materials. In the second method, the *EOT* will be calculated so as to keep the ratio of relative permittivity over dielectric thickness constant. Both of these approaches ignore several aspects of the physics arising at the Si/dielectric interface. In the absence of available device data, the methodology presented will provide meaningful information of the various materials under consideration for EDA applications. We believe that along with the efforts in introducing high-κ gate dielectrics, future physical-aware LP synthesis methodologies should be developed in order to incorporate them into existing automatic design or synthesis flows.

The Predictive Technology Model (PTM) can be used for modeling HKMG transistors. The PTM provides a timely and effective analysis in the absence of published data and other device models [64]. With PTM, competitive circuit design and research can start even before the advanced semiconductor technology is fully developed. The simulation results obtained are of comparable accuracy to TCAD simulations. For high-κ dielectric modeling using PTM, two methods are used: (1) The SPICE model parameter for relative permittivity (`EPSROX`) is changed or (2) an *EOT* for the dielectric used is calculated. The *EOT* is calculated so as to keep the ratio of relative permittivity over dielectric thickness constant using the following expression [26]:

$$EOT = \left(\frac{\kappa_{SiO_2}}{\kappa_{gate}}\right) T_{gate}, \tag{6.3}$$

where T_{gate} is the thickness of the gate dielectric material, κ_{gate} is the relative permittivity, and κ_{SiO_2} is the dielectric constant of SiO_2 (= 3.9). As an example, $\kappa_{gate} = 21$ and $T_{gate} = 5$ nm to emulate a HfO_2-based dielectric. The *EOT* is calculated to be 0.9 nm for this specific example. For a 45-nm CMOS process, the BSIM 4.4 models provide a oxide thickness $T_{ox} = 1.4$ nm, threshold voltage $V_{Th} = 0.22$ V for the NMOS and $V_{Th} = -0.22$ V for the PMOS. The nominal power supply is $V_{DD} = 0.7$ V. These models are also scalable with respect to T_{ox} and channel length. The effect of varying oxide thickness (T_{ox}) was incorporated by varying `TOXE` in the SPICE model deck directly.

In the above, κ_{gate} is the relative permittivity (dimensionless) and T_{gate} is the thickness of the gate dielectric material (in m or Å) other than SiO_2, while κ_{SiO_2} is the relative permittivity of SiO_2 (= 3.9). A parametric study of device behavior versus κ was done, and the results are summarized in this section; however all real values of κ may not translate to a specific physical dielectric in nano-CMOS technology. The length of the device is proportionately changed to minimize the impact of higher dielectric thickness on device performance and to maintain the per width gate capacitance constant as per CMOS fabrication requirements [26, 39, 41, 57]. Hence the scaling ratio of channel length to gate thickness is maintained constant. In addition, the length and width of the transistors are chosen to maintain a (W/L) ratio of 4 : 1 for NMOS and 8 : 1 for PMOS to ensure equal flow of current through the devices and symmetric switching points. V_{DD} variation is achieved by running a parameter sweep in the simulator.

6.4 DG-FinFET device

The previous section presented the need for high-κ-based transistors. These devices reduce the gate leakage caused by the SiO_2 dielectric transistors but, when the device is reduced to a smaller size, new problems are introduced in the form of SCEs. GIDL also increases drastically as the length of the channel is decreased. This is due to the high electric field between the gate and drain. A study on GIDL has been presented in Reference 63. The DG-FinFET is introduced to minimize the problems of leakage. In this device, the source and the drain are extruded into the third dimension. The gate is then fabricated on the source and drain. In this structure of FinFET, the fins act as the channel. The length of the channel is equal to the width of the fin, and the width of the channel is equal to twice the height of the fins. In a DG-FinFET, there are two gates around the channel, the front gate and the back gate. These ensure that there is significant control over the charges that flow in the channel. Compared to the conventional MOSFET, the channel length is increased in the case of a FinFET. Hence the leakage current and the SCEs are reduced. Use of the high-κ as the dielectric allows the scaling of the transistor, as the length of the dielectric is almost equal to one atom thick. FinFET combined with high-κ allows scaling to sub-32-nm sizes. Research is being carried out to fabricate transistors of 10 nm. The DG-FinFET has many advantages besides scaling and leakage current compared to the conventional MOSFET and high-κ transistors. The ON-state drive current of the transistor is increased, power consumption is reduced and device performance is increased.

6.4.1 DG-FinFET device structure

There are two types of DG-FinFETs. The DG-FinFET with independent gates (IGs) and the DG-FinFET with unified gates. The unified gates are controlled by one voltage whereas the two IGs can be controlled by two voltages. SOI processes are used to fabricate the FinFET [36]. The SOI process is used to achieve ultra-thin specifications for the devices. Silicon nitride (SiN_3) and SiO_2 are deposited on the thin layer of SOI initially while fabricating the FinFET. Then the Fin is formed using Electron Beam Lithography. A heavily doped back gate serves as a ground plane in the DGFET which reduces electrostatic coupling. The two gates are formed using the conventional planar MOSFET manufacturing process. The drain–source channel is sandwiched between the gate oxide and the gate. The source and the drain are separated, and an insulator is formed. Thus the polysilicon straddles the fin structure to form perfectly aligned gates.

Structures of DG-FinFET are shown in Figure 6.3. Figure 6.3(a) presents the structure of a DG-FinFET with unified gate structure, and Figure 6.3(b) presents the structure of a DG-FinFET with IGs. An insulator is used to divide the front gate from

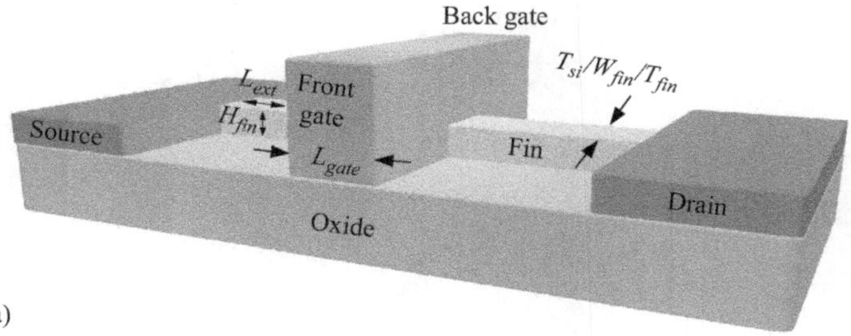

(a)

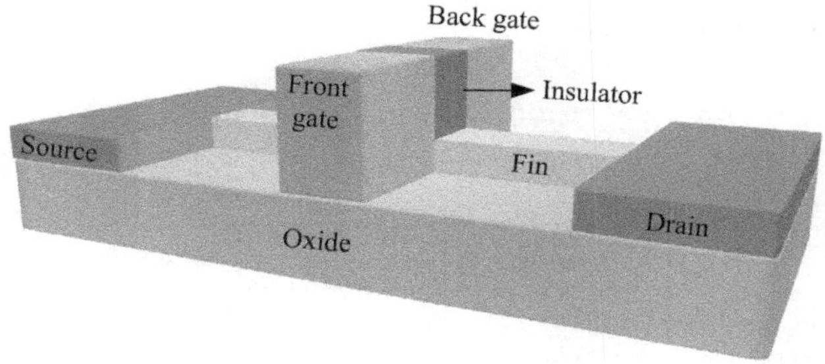

(b)

Figure 6.3 Structure of DG-FinFETs [36]. (a) DG-FinFET with a unified gate and (b) DG-FinFET with IGs

the back gate. Two IGs can take in two different voltages which helps in better control of the channel. In the unified gate structure, only one voltage is needed to control the channel. In this structure, the source and the drain of the transistor are extruded to the third dimension along with the fin. The metal-gate straddles the source and the drain. "Fin Pitch" is the term used to define the space between the source and the drain. L_{gate} is the length of the gate. As the gate is fabricated into the third dimension, the length of the channel is increased compared to conventional MOSFET. L_{ext} is the extension length. T_{fin} or W_{fin} is the width of the Fin. H_{fin} is the height of the fin. Typically, the height of the fin is more than that of the width of the fin. Fin height and fin width are related by the following expression:

$$H_{fin} = \left(\frac{W}{2}\right). \tag{6.4}$$

In the above expression, W is the Fin Pitch. The geometrical physical dimensions of the fin height and fin width are given by the following [22, 44]:

$$\text{Geometrical channel length } L_{phy} = L_{gate} + 2L_{ext}. \tag{6.5}$$

$$\text{Geometrical channel width } W_{phy} = T_{fin} + 2H_{fin}. \tag{6.6}$$

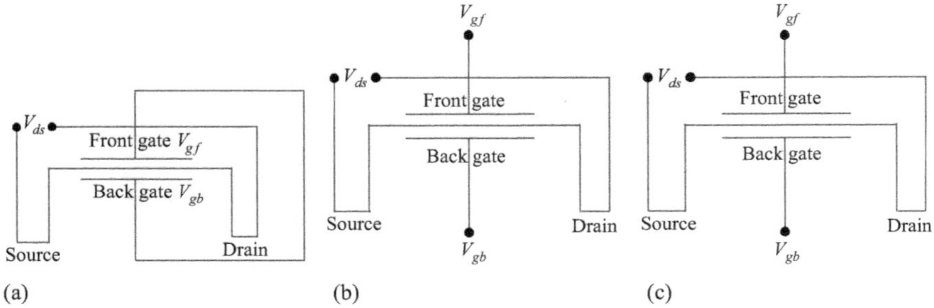

Figure 6.4 Different configurations for n-type DG-FinFET [20]. (a) SG, (b) IG, and (c) LP modes

Table 6.2 Parameter values of a 32-nm n-type DG-FinFET device [20]

Device parameter	Specific values
Oxide thickness T_{ox}	1.4 nm
Threshold voltage V_{Thn}	0.28 V
Threshold voltage V_{Thp}	−0.28 V
Channel doping N_{ch} (cm^{-3})	2×10^{16}
Fin-height H_{fin} (nm)	50 nm
Body thickness T_{Si}	8.6 nm

6.4.2 DG-FinFET device modeling

The FinFET is inherently an SOI transistor. The body thickness (T_{Si}) of a fin is analogous to the silicon channel thickness. Figure 6.4 shows the shorted-gate (SG), IG, and LP structures of an n-type FinFET. V_{gf} is the potential difference between the front gate and source. V_{gb} denotes the potential difference between the back gate and the source. In the SG mode, the front and back gates are tied together. In the IG mode, the top part of the gate is etched out for two IGs [27]. The LP mode applies a reverse-bias voltage to the back gate in order to reduce sub-threshold leakage. The SG mode has smallest delay, followed by IG and LP modes [13]. For power consumption, LP mode gives the lowest power consumption, followed by IG and SG modes.

In the typical FinFET process, the SOI thickness (T_{si}) is so thin that the silicon body is fully depleted. Two single-gate transistors have been used to capture the current conduction controlled by the front and back gates in a DG-FinFET transistor [56]. Each sub-transistor has its own definitions of gate voltage (V_g), threshold voltage (V_{Th}), and gate-oxide thickness (T_{ox}). The fully depleted SOI model of BSIM (BSIM FD SOI) is used for each sub-transistor. The key parameters for the FinFET model for the 32-nm node are shown in Table 6.2. For a DG-FinFET, each fin provides device width of 2 H_{fin}. For a good matching, a nominal size of $L = 100$ nm and $W = 500$ nm is assumed. The FinFET has $N_{fin} = 5$ fins which is determined as

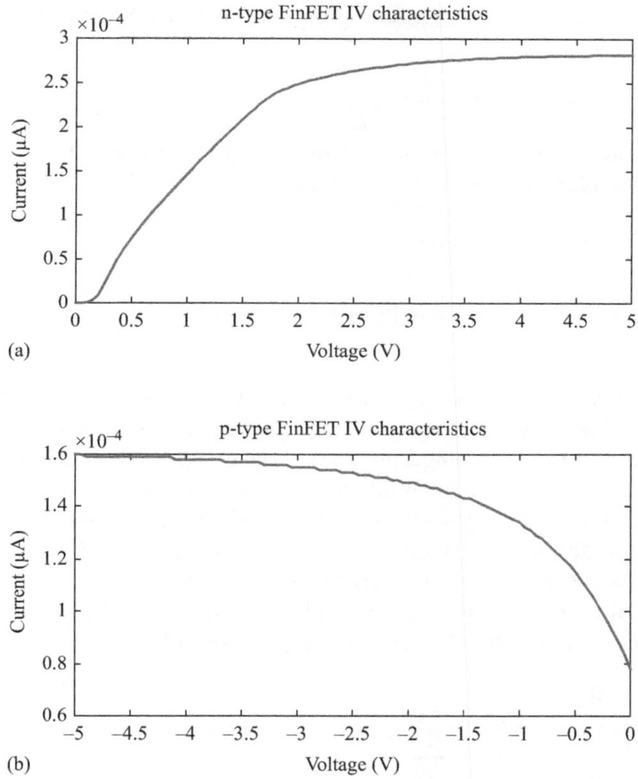

Figure 6.5 I–V characteristics of FinFET. (a) For n-type FinFET and (b) for p-type FinFET

$W = 2H_{fin}N_{fin} = 500\,\text{nm}$. For brevity, results for a n-type DG-FinFET device are presented while dual trends are observed for the p-type.

Different models for FinFETs are presented in the existing literature [2]. Here the ηth power law is used to compute the current of the FinFET. First the I–V characteristics of the devices are generated, and the current model parameters are extracted from them (Figure 6.5).

$$I_d = \left(\frac{W_{eff}}{L_{eff}}\right) B\left(V_{gs} - V_{th}\right)^n, \tag{6.7}$$

$$I_d = I_{dsat}\left(2 - \frac{V_{ds}}{V_{dsat}}\right)\frac{V_{ds}}{V_{dsat}}. \tag{6.8}$$

In the above expression, $V_{ds} < V_{dsat}$ is assumed for the linear region operation. In the above expression, the following are also assumed:

$$\begin{cases} V_{dsat} = K\left(V_{gs} - V_{th}\right) \\ V_{ds} > V_{dsat} : \text{for the saturation region.} \end{cases} \tag{6.9}$$

In an inverter, considering a falling output transition, the charge at the output node is [2]:

$$Q_{out} = Q_M + Q_L = (C_M + C_L)\,V_{dd}. \tag{6.10}$$

At $t = dt$, some charge is removed from the output node. Then the following expressions are obtained for the output charge:

$$Q_{out} = Q_M + Q_L + Q_{current}, \tag{6.11}$$

$$Q_{out} = (C_M + C_L)\,V_o - C_M V_i + \int_0^{dt} (I_n - I_p)\,dt. \tag{6.12}$$

To extend this model from an inverter to other logic gates like NAND and NOR, the position of the switching transistor is considered. In a NAND circuit, one of the NMOS devices will be switching from low to high or high to low, and the other device will be off. Thus the effective capability of the switching device is half that of the other on device. The effective width of the switching NMOS is half that of the other device. This mapping is applied for the bulk CMOS. In the case of a FinFET, the width is an integer multiple of the height. In order to consider the effect of unbalanced pull-up and pull-down currents in FinFET, a width correction term λ_w is introduced. The equivalent width is given by the following expressions [2]:

$$W_{peq} = W_{pk}\lambda_{wp}, \tag{6.13}$$

$$\frac{1}{W_{neq}} = \frac{1}{\lambda_{wn}}\left(\frac{1}{W_{n2}} + \frac{1}{2W_{n1}}\right). \tag{6.14}$$

The equivalent capacitance (C_{eq}) is given by the following:

$$R_{eq}C_{eq} = \left(\frac{R1}{2} + R_2\right)C_0 + R_2C_1. \tag{6.15}$$

Considering the inverse relation of resistance and the device width, the following expression is obtained:

$$\frac{C_{eq}}{W_{eq}} = \left(\frac{1}{2W_1} + \frac{1}{W_2}\right)C_0 + \frac{1}{W_2}C_1. \tag{6.16}$$

In the above expression, the following are assumed:

$$C_0 = C_L + C_{on1} + C_{op1}, \tag{6.17}$$

$$C_1 = C_{on1} + C_{on2}. \tag{6.18}$$

In the above expressions, C_{op} and C_{on} are gate-to-drain capacitance, respectively.

A compact model of a trapezoidal FinFET with complex fin cross-sections is presented in Reference 15. Primarily four parameters are needed to model a FinFET device accurately: A_{ch}, area of the channel, P_{ins}, C_{ins}, insulator capacitance per unit length, and N_{ch}, channel doping. The parameters for trapezoidal model are given by the following expressions [15]:

$$A_{ch} = \left(H_{fin}\frac{T_{fin,top} + T_{fin,base}}{2}\right), \tag{6.19}$$

$$P_{ins} = 2\sqrt{\frac{\left(T_{fin,top} + T_{fin,base}\right)^2}{2} + H_{fin}^2 + T_{fin,top}}, \tag{6.20}$$

$$C_{ins} = \frac{P_{ins}\varepsilon_{ins}}{EOT}. \tag{6.21}$$

6.5 The proposed methodology for logic library creation

The objective of a standard cell library is to design logic cells and characterize their figures of merit, while accounting for parameter variability that arises from process and temperature variations. This section discusses the variability and the proposed methodology to account for it during the cell library creation. Several logic cell or standard cell libraries are researched well, and several results are available in the current literature. In Reference 4, a process variation tolerant cell library containing four standard gates, such as INVX2, NAND2X1, NAND3X1, and XOR2X1, is presented for use with 65-nm technology. A 45-nm standard cell library for classical CMOS is presented in Reference 50. In References 53 and 54, an approach is presented that considers intra-cell process mismatch variations in standard cells (NAND, NOR, buffer, inverter, AND, and AIO), for 65-nm bulk CMOS. For extension of flexible electronics to complex digital circuitry standard cells containing inverter, NOR, NAND, MUX, Flip-Flops, and Latches are presented in Reference 59. In Reference 30, a double-via-driven standard cell library is presented which can be used for designing chips with maximum manufacturing yield. In Reference 28, an ultra-LP combinational standard cell library is presented which is designed using a new leakage reduction methodology. In Reference 46, a 65-nm standard cell library is presented which accounts for parasitics in the physical design. As evident from the above discussion, existing cell libraries are primarily for classical CMOS technology. In the future, characterized standard cells for non-classical technologies will be required [26, 37]. The high-κ transistors came into existence to replace the classical CMOS transistors with their low-leakage capabilities. The logic technology using 45-nm transistors is presented in Reference 33. In order to compensate for the SCEs introduced by high-κ, the FinFET transistors are introduced and used commercially by many applications. A FinFET-based logic gate design is presented in Reference 42. This section proposes a new methodology for HKMG transistors and DG-FinFET-based logic gate library.

6.5.1 Sources of variation and nature of variability

Process variation in nanoscale circuits originates from the uncertainties in the highly sophisticated lithographic processes in which feature size has reached the wavelength of light. Nanoscale manufacturing process steps, such as a CVD, ion implantation,

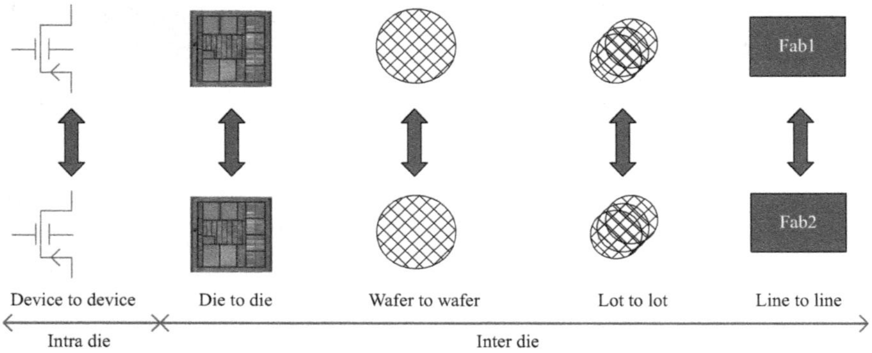

Figure 6.6 Process variation in nanoscale CMOS circuits

spin coating, and chemical mechanical polishing, contribute to process variation. These are manifested in various forms in the circuit and are classified in various ways by designers to facilitate modeling, as shown in Figure 6.6. Process variation can be broadly classified as intra-die and inter-die. The intra-die process variation is device-to-device variation and is also called mismatch. Inter-die process variation is die-to-die, wafer-to-wafer, lot-to-lot, and line-to-line variation.

For a specific process technology, different parameters are considered for process variation in the standard cells. It may be noted that these parameters may not be always available from the foundries. However, these are considered in order to obtain a very accurate logic library. The needed information is obtained from various published works. In the case of FinFET, a different set of parameters needs to be considered. For a DG-FinFET, the geometries of both gates are varied. All of device parameter variations are not independent. Hence, their correlations need to be considered for accurate modeling. For example, oxide thicknesses of n-type and p-type devices in MHMG technology are correlated. Similarly, the heights of the fins in the case of a DG-FinFET are assumed since both oxides are grown together. The statistical variability in most of these parameters can be modeled by normal or Gaussian distributions while the doping concentrations are Poisson distributed [34]. In a nano-CMOS device, the average distance between dopants is of the order $\sim$10 nm [14]. Hence, for most practical cases, the Poisson distribution can be approximated by a Gaussian distribution. Temperature variation can arise from on-chip thermal variation and ambient temperature.

6.5.2 Statistical logic library characterization flow

It is necessary to express each of the device responses as a function of process, voltage, and temperature (PVT) so that designers can use them for safe design. This can be expressed in the following form:

$$\hat{Y} = f(P, V, T), \tag{6.22}$$

where $\hat{Y}$ is the required response: power, leakage, or delay. Also, for nano-CMOS, a shift from deterministic to probabilistic design is required to accommodate the effects of device variability, which involves extensive use of statistical techniques.

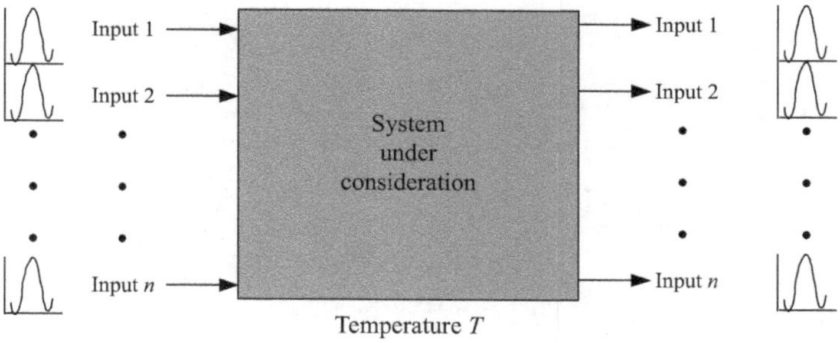

Figure 6.7 Logic level modeling for PVT-aware statistical library

There are some challenges for such modeling. There is a need (i) for realistic evaluation of circuit delay and power variability, considering process variations and correlations between them, (ii) to directly relate variability in circuit parameters to variability in process parameters, and (iii) to migrate from corner-based timing to statistical timing for accuracy.

To address these challenges, a Monte Carlo-based technique is proposed to create a PVT-aware library. The proposed flow has the following advantages: accurate estimation of power consumption, leakage, and delay in non-classical CMOS structures is possible. A closed form function relating the output to input is not required, which otherwise would have been cumbersome for the large number of parameters considered in this section and parameters take on more realistic or practical extreme values resulting in densely designed, reliable, manufacturable circuits.

Figure 6.7 shows the logic level modeling for a system under consideration for which a PVT-aware library is to be created. The inputs are considered as probability density functions (PDFs) of the sources of variability. Once the statistical distribution for all variability sources is determined, the probability distribution functions of these variability sources form the input to the system under consideration. In this section, the system under study is a set of standard logic gates, as the goal is a logic standard cell library creation. The input PDFs are denoted as $\hat{X}_j$, where $j = 1, \ldots, n$. Each logic gate is subjected to Monte Carlo simulations. The outputs of the simulation are the PDFs for outputs as functions of the inputs, which are process parameters (P) and voltage (V), at a specific temperature (T). The output PDFs are denoted by $\hat{Y}_i$, where $i = 1, \ldots, m$. These simulations run for different temperatures, and a PVT-aware library is obtained.

The proposed methodology for creating a statistical logic library is presented in Figure 6.8. The input is the HKMG or DG-FinFET model files. Initially, the currents, I_{GIDL}, I_{OFF}, and I_{ON}, are characterized at the device level. For this, the device is properly biased and the biasing conditions are discussed in Section 6.7. After biasing the devices properly, Monte Carlo simulations are performed on the setup, and the respective mean (μ) and standard deviation (σ) are calculated. Then the logic gates are designed using the characterized devices. The temperature is varied as mentioned above and the sub-threshold current (I_{sub}) and GIDL current (I_{GIDL}) are estimated by

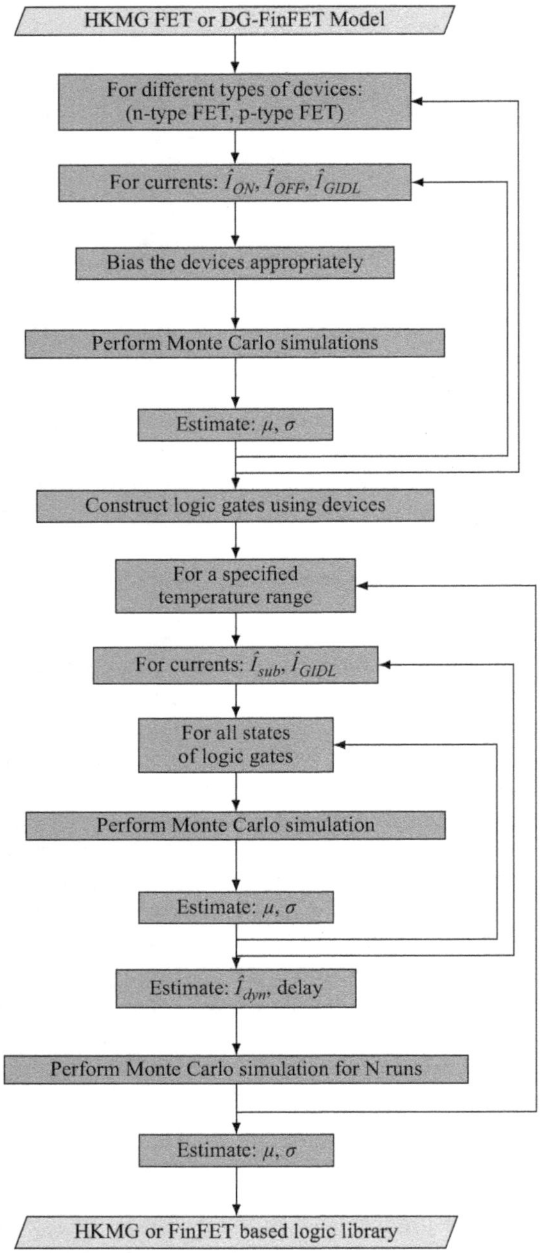

Figure 6.8 The proposed methodology for PVT-aware statistical logic library using HKMG or DG-FinFET-based technology

performing Monte Carlo simulations and calculating the mean and variance. After the calculation of the sub-threshold and the GIDL currents, the dynamic current and the delay are simulated and mean and variance are calculated. Then we will be getting a characterized logic gate library at the end of the PVT characterization.

6.6 Power, leakage, and delay models for HKMG and DG-FinFET technology

For HKMG technology-based library, dynamic power, sub-threshold leakage, GIDL, and propagation delay models are presented. At the same time for a DG-FinFET-based technology library, sub-threshold leakage, GIDL, and propagation delay models are presented.

6.6.1 For HKMG-based technology

6.6.1.1 Dynamic power

The dynamic power consumption of a CMOS circuit is given by the well-established model which predicts [35, 8]:

$$P_{dyn} = sC_L V_{dd}^2 f, \tag{6.23}$$

where the activity factor s depends on how many devices are active on any particular clock cycle, C_L is the total switched capacitive load at the circuit output, V_{dd} is the supply voltage, and f is the frequency of the clock. This power dissipation depends on loading conditions and not the device features. Thus, this model can also be used for DG-FinFET technology-based logic library. The current associated with P_{dyn} is I_{dyn} and is given by the average of the current through the load capacitance C_L [58]:

$$I_{dyn} = \lim_{T\to\infty} \frac{1}{T} \int_0^T I_c(t)dt, \tag{6.24}$$

where $I_c(t)$ is the current through the load capacitance C_L.

6.6.1.2 Sub-threshold leakage power

The sub-threshold leakage current of a CMOS device is modeled by the following expression [36, 47, 23]:

$$I_{sub} = I_0 \left(1 - \exp\left(\frac{-V_{ds}}{v_{therm}}\right)\right) \exp\left(\frac{V_{gs} - V_{Th} - V_{off}}{Sv_{therm}}\right), \tag{6.25}$$

where I_0 is a constant dependent upon device parameters for a given technology, V_{Th} is the threshold voltage, v_{therm} is the thermal voltage, V_{off} is the offset voltage which determines the channel current at $V_{gs} = 0$, and S is the sub-threshold swing factor.

The threshold voltage V_{Th} is typically scaled along with the supply voltage in order to maintain performance. However, the systematic reduction in V_{Th} causes the sub-threshold current to increase exponentially, as can be seen from (6.25).

6.6.1.3 GIDL power

The use of HKMG causes a significant GIDL current (I_{GIDL}) in addition to sub-threshold leakage (I_{sub}) [29]. I_{GIDL} is high mainly due to two physical effects:

1. The metal-gate introduces a high gate effective work function which leads to high electric field and a high GIDL current [43].
2. The high-κ gate dielectric and SiO_2 spacers meet at the surface of the drain region, thus causing a high electric field leading to a high GIDL current [9].

GIDL is caused by the high electric field at the drain junction [9]. In NMOS transistors, GIDL takes place when the gate is at a lower potential than the drain, which causes significant band bending in the drain, allowing electron–hole pair generation through avalanche multiplication and band to band tunneling. GIDL has a strong impact in HKMG transistors. Equation (6.26) shows the BSIM4 expression used for calculating GIDL [23]:

$$I_{GIDL} = AW_{effCJ}N_f\left(\frac{V_{ds}-V_{gs}-E}{3T_{gate}}\right)\exp\left(\frac{-3T_{gate}B}{-V_{ds}-V_{gs}-E}\right)\left(\frac{V_{db}^3}{C+V_{db}^3}\right), \tag{6.26}$$

where V_{db} is the drain to body voltage, V_{ds} is the drain to source voltage, V_{gs} is the effective gate voltage, W_{effCJ} is the effective width of the drain diffusions, N_f is the number of fingers of the transistor and A, B, C, and E are BSIM4 GIDL leakage-based parameters which have been fitted to existing data [43, 9, 29].

6.6.1.4 Propagation delay

The propagation delay (T_{PD}) is approximately given by the following expression [36, 51]:

$$T_{PD} = \beta\left(\frac{C_L V_{dd}}{\mu\left(\frac{\kappa_{gate}}{T_{gate}}\right)\left(\frac{W_{eff}}{L_{eff}}\right)(V_{dd}-V_{Th})^{\alpha}}\right), \tag{6.27}$$

where β is a technology-dependent constant, μ is the electron surface mobility, and α is the velocity saturation index. The propagation delay of logic cells is calculated using the following expression:

$$Delay = \left(\frac{T_{pdLH}+T_{pdHL}}{2}\right), \tag{6.28}$$

where T_{pdLH} refers to the low to high transition, and T_{pdHL} refers to the high to low transition of the output.

There is a sharp increase in the value of T_{pd} with an increase in the gate dielectric constant which continues until a value of around $\kappa = 6$ after which the slope is much lower. This increase can be attributed to the increase in capacitance per unit area C'_{gate}, in F/m^2 of gate oxide with dielectric constant which is expressed by the following relation [26, 39]:

$$C'_{gate} = \left(\frac{\kappa_{gate}}{T_{gate}}\right), \tag{6.29}$$

while the presence of a "knee" region around $\kappa \simeq 6$ is due to similar behavior of T_{pd} as in (6.27).

The propagation delay of a logic gate increases as the gate material thickness increases due to the increase in gate capacitance (C_{gate}, in F) with oxide thickness T_{ox} (in m or nm) for a particular dielectric, say SiO_2 with κ, as evident from the following discussion. The gate capacitance (C_{gate}) is expressed by the following [26, 39, 61]:

$$C_{gate} = \varepsilon_{ox}\left(\frac{L}{T_{ox}}\right)W, \tag{6.30}$$

$$= \varepsilon_{ox}\left(\frac{W}{L}\right)\left(\frac{L}{T_{ox}}\right)L. \tag{6.31}$$

In the above equations, ε_{ox} is the permittivity of the gate material (in F/m), and L and W are the length and width of the transistor, respectively (both in m or nm). Thus, with increase in T_{ox} a constant $(\frac{L}{T_{ox}})$ and $(\frac{W}{L})$ is maintained by increasing L, C_{gate} increases, and hence the propagation delay. This result is consistent with the experimental results presented in the existing literature [41, 61, 57].

The propagation delay shows a decreasing trend with an increase in the value of the supply voltage when T_{gate} and κ_{gate} are kept fixed due to the increase in the drive current resulting from the increase in supply voltage. A better insight of the situation can be obtained from the following discussion. For a technology parameter α (in s/F), the propagation delay is presented by the following [26, 39, 45]:

$$T_{pd} = \left(\frac{\alpha C_L V_{dd}}{V_{dd} - V_{Th}^*}\right) \text{(where } V_{Th}^* = V_{Th} + 0.5V_{dsat}\text{)}, \tag{6.32}$$

$$= \alpha C_L\left(\frac{1}{1 - V_{Th}^*/V_{dd}}\right) \text{(dividing by } V_{DD}\text{)}, \tag{6.33}$$

$$= \alpha C_L\left(\frac{1}{1 - V_{Th}^*/V_{dd}}\right). \tag{6.34}$$

In the above expressions, $V_{Th}^* = V_{Th} + 0.5V_{dsat}$, and V_{dsat} is the drain saturation voltage. All voltages are in volts (V). As $-1 < \frac{V_{Th}^*}{V_{dd}} < 1$, using a McLaurin series expansion, the following is obtained [26, 39]:

$$T_{pd} = \alpha C_L\left(1 + \frac{V_{Th}^*}{V_{dd}} + \left(\frac{V_{Th}^*}{V_{dd}}\right)^2 + \ldots\right), \tag{6.35}$$

$$\approx \alpha C_L\left(1 + \frac{V_{Th}^*}{V_{dd}}\right). \tag{6.36}$$

The above expressions clearly suggest that for fixed load and threshold voltage, as V_{dd} increases, T_{pd} decreases. As scaling continues, the trend is to scale down the supply along with other device features. This is compatible with the objective of decreasing the gate leakage current in the nanometer regime.

6.6.2 For DG-FinFET-based technology

6.6.2.1 Sub-threshold leakage power

The sub-threshold leakage current in the DG-FinFET is given by the following expression [20]:

$$I_{sub} = \alpha\left(\frac{H}{L}\right)\exp^{\left(\frac{V_{gs}-V_{Thnf}}{\beta}\right)}\left(1-\exp^{\frac{-qV_{ds}}{kT}}\right). \tag{6.37}$$

In the above expression, T is the temperature in kelvin, k is the Boltzmann constant, and α and β are fitting parameters. V_{Thnf} is the threshold voltage of the front gate. V_{Thnf} increases with the increase of back-gate reverse biasing voltage.

6.6.2.2 GIDL of FinFET

The introduction of high-κ in the transistors introduced the GIDL and other problems mentioned above. The GIDL is an SCE. This is reduced using the FinFET with the increase in the channel length. The fin width and length will affect the I_{off}, and it is given by the following expression [25]:

$$I_{off} = \beta\exp^{\left(-V_{th(variation)}\left(\frac{\alpha}{T_{si}}\right)\right)}. \tag{6.38}$$

The GIDL for FinFET used in the BSIM model is given the following expression [24]:

$$\begin{aligned} T_0 &= AGIDL_i W_{eff0}\left(\frac{V_{ds}-V_{gs}-EGIDL_i+V_{fbsd}}{\varepsilon_{ratio}EOT}\right)^{PGIDL_i} \\ &\quad\times\exp\left(-\frac{\varepsilon_{ratio}EOTBGIDL(T)}{V_{ds}-V_{gs}-EGIDL_i+V_{fbsd}}\right)NFIN_{total}, \end{aligned} \tag{6.39}$$

$$I_{GIDL}\begin{cases} T_0\frac{V_{de}^3}{CGIDL_i+V_{de}^3} & \text{for } BULKMOD=1 \\ T_0 V_{ds} & \text{for } BULKMOD=0. \end{cases} \tag{6.40}$$

AGIDL is the pre-exponential coefficient for GIDL, BGIDL is the exponential coefficient for GIDL, CGIDL is the parameter for body bias effect of GIDL, and EGIDL is the band bending parameter for GIDL.

6.6.2.3 Propagation delay in FinFET

The propagation delay in a FinFET device-based logic gate can be calculated by the following expression [55]:

$$d_i = T(gh+p), \tag{6.41}$$

where g is logical effort of gate, h is $\frac{C_{out}}{C_{in}} = n_j C_g = \frac{n_j}{n_i}$, p is αp_{im} in which α depends on gate type, and $n_i \subset 1, 2, 3, \ldots\ldots, n_{i_{max}}$. In the above equations, n_i represents the number of fins in the transistor, d_i is the gate delay, C_g is the fin gate capacitance, P_{inv} is the parasitic delay, and T is the intrinsic delay. The propagation delay caused by the thermal effects in the FinFET is presented in Reference 55. The geometry of

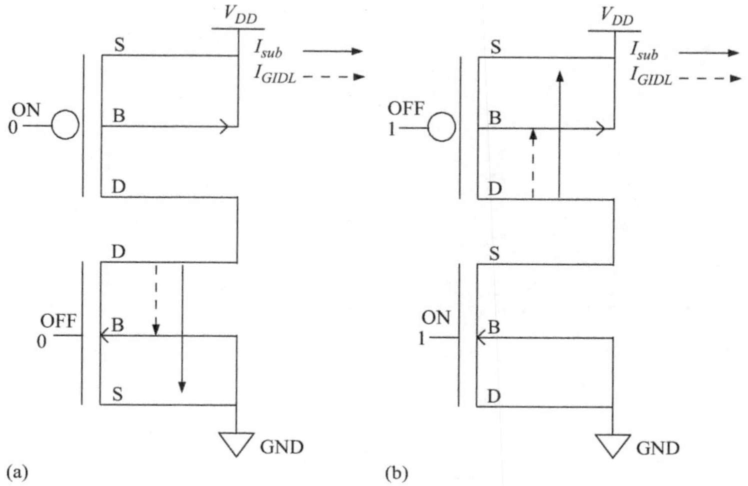

Figure 6.9 Directions of I_{sub} and I_{GIDL} for a high-κ inverter for different states. The dotted line shows I_{GIDL}, and the solid line represents I_{sub} flow. I_{GIDL} flows from drain to bulk, whereas I_{sub} flows from drain to source. (a) State "0" and (b) State "1"

the fin will create thermal problems. Besides the geometry self-heating problems, the wafer will have number of fins parallel to each other where it will produce a lot of heating effect and cause the delay.

6.7 Device level characterization of high-κ and FinFET

The previous section presented the characterization of the HKMG transistors and the DG-FinFETs at the device level. In this section, different current components present in the logic gates, inverter, and NAND are presented. Here the state-dependent logic level characterization is performed on each of the logic gates.

6.7.1 For HKMG CMOS

In this section, we present the results for HKMG logic cells at room temperature (27°C). The results are presented for inverter and NAND gates for brevity. The inverter has been chosen, because it represents the basic static CMOS logic. The NAND is an example of a universal gate. The state-dependent data are presented for $\hat{I}_{GIDL}$ and $\hat{I}_{sub}$ as they lead to accurate leakage estimation. Since I_{dyn} depends primarily on the switching of the logic gates, average data for $\hat{I}_{dyn}$ are presented, as per (6.24). C_L is assumed as 10 times C_{gg}, which is the gate capacitance of PMOS.

The directions of various currents in the inverter gate for each state are drawn in Figure 6.9(a) and 6.9(b). The distribution plots for these currents are shown in Figure 6.10(a)–6.10(f). The dotted line in the figure represents the GIDL current, and the solid line represents the sub-threshold leakage current of the transistors. Both

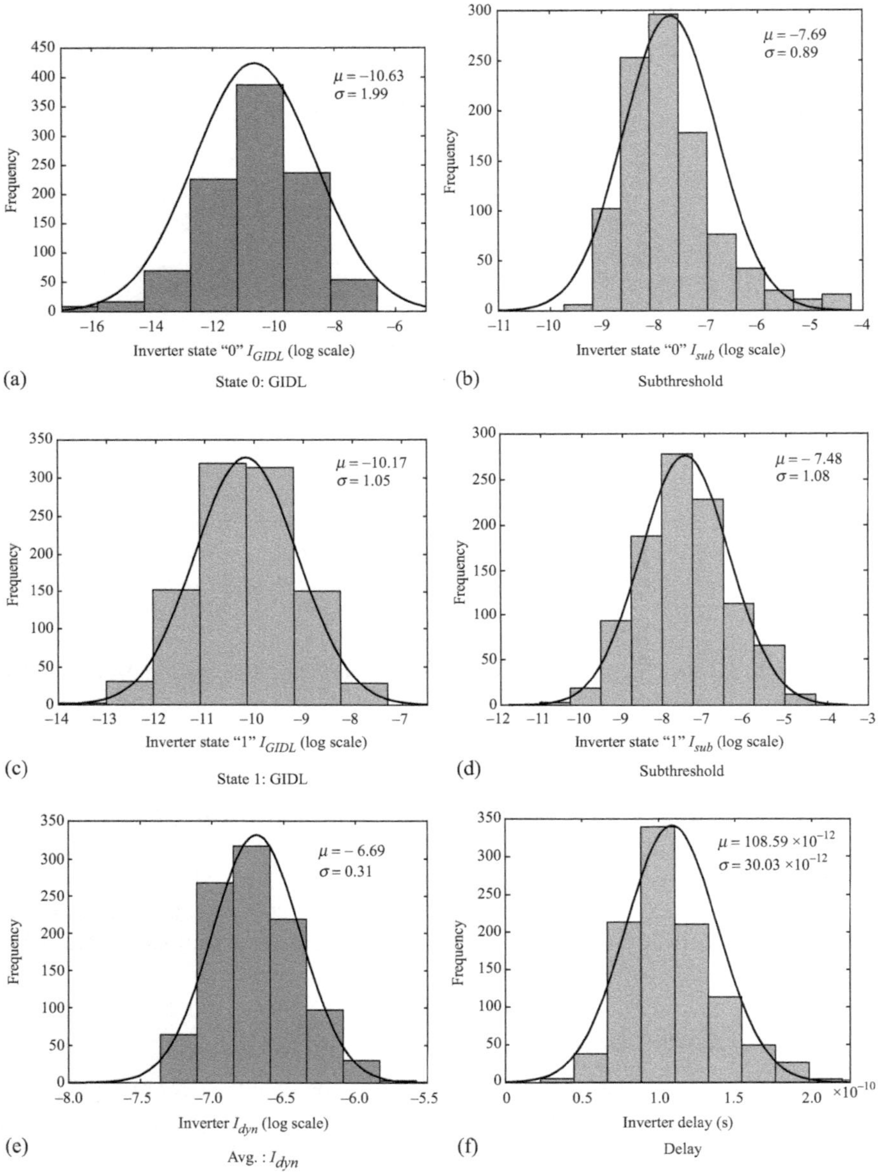

Figure 6.10 Distributions of GIDL ($\hat{I}_{GIDL}$), sub-threshold ($\hat{I}_{sub}$), and dynamic ($\hat{I}_{dyn}$) current, and delay for HKMG an inverter. (a) State "0": GIDL, (b) State "0": sub-threshold, (c) State "1": GIDL, (d) State "1": sub-threshold, (e) Inverter $\hat{I}_{dyn}$, and (f) Inverter delay

Table 6.3 Statistical state-dependent data for inverter [19]

Statistical distribution of current components or propagation delay		**"0"**	**"1"**
Dynamic current $\hat{I}_{dyn}$	μ	−6.69	
	σ	0.30	
Sub-threshold leakage current $\hat{I}_{sub}$	μ	−7.69	−7.48
	σ	0.89	1.08
GIDL current $\hat{I}_{GIDL}$	μ	−10.63	−10.17
	σ	1.99	1.05
Propagation delay	μ	108.59 ps	
	σ	30.03 ps	

Table 6.4 Statistical state-dependent data for NAND [19]

Statistical distribution of current components or propagation delay		**"00"**	**"01"**	**"10"**	**"11"**
Dynamic current $\hat{I}_{dyn}$	μ	−6.55			
	σ	0.22			
Sub-threshold leakage current $\hat{I}_{sub}$	μ	−8.48	−7.69	−7.92	−7.88
	σ	0.62	0.88	0.79	1.10
GIDL current $\hat{I}_{GIDL}$	μ	−10.54	−10.58	−13.36	−10.02
	σ	1.63	1.6377	1.67	1.63
Propagation delay	σ	29.72 ps			

these currents are represented for each of the states, State "0" and State "1" for the inverter in the figures. An input voltage of 0 V was used for State "0", and an input voltage of 0.7 V was used for State "1". V_{DD} is 0.7 V in both states. In the transistors, the GIDL current flows from drain to bulk, and the sub-threshold leakage current flows from drain to source. The statistical values were calculated by performing the process variation analysis using the Monte Carlo analysis as described in Section 6.7. Table 6.3 summarizes the statistical data for the various currents measured in an inverter. Results for only two gates are shown for brevity but the entire library is characterized following the same procedure.

The directions of various currents in the NAND gate for each state are shown in Figure 6.11(a)–6.11(d). Table 6.4 summarizes the statistical data for the various currents measured in an inverter and a NAND gate. The distribution plots for these currents are shown in Figures 6.12(a)–6.13(d).

The directions of various currents in the NOR gate for each state are shown in Figure 6.14(a)–6.14(d). Table 6.5 summarizes the statistical data for the various currents measured in a NOR gate. Results for only two gates are shown for brevity but the entire library is characterized following the same procedure. For State "1", 1 V is applied as the input to the transistors and a V_{DD} of 1 V is used.

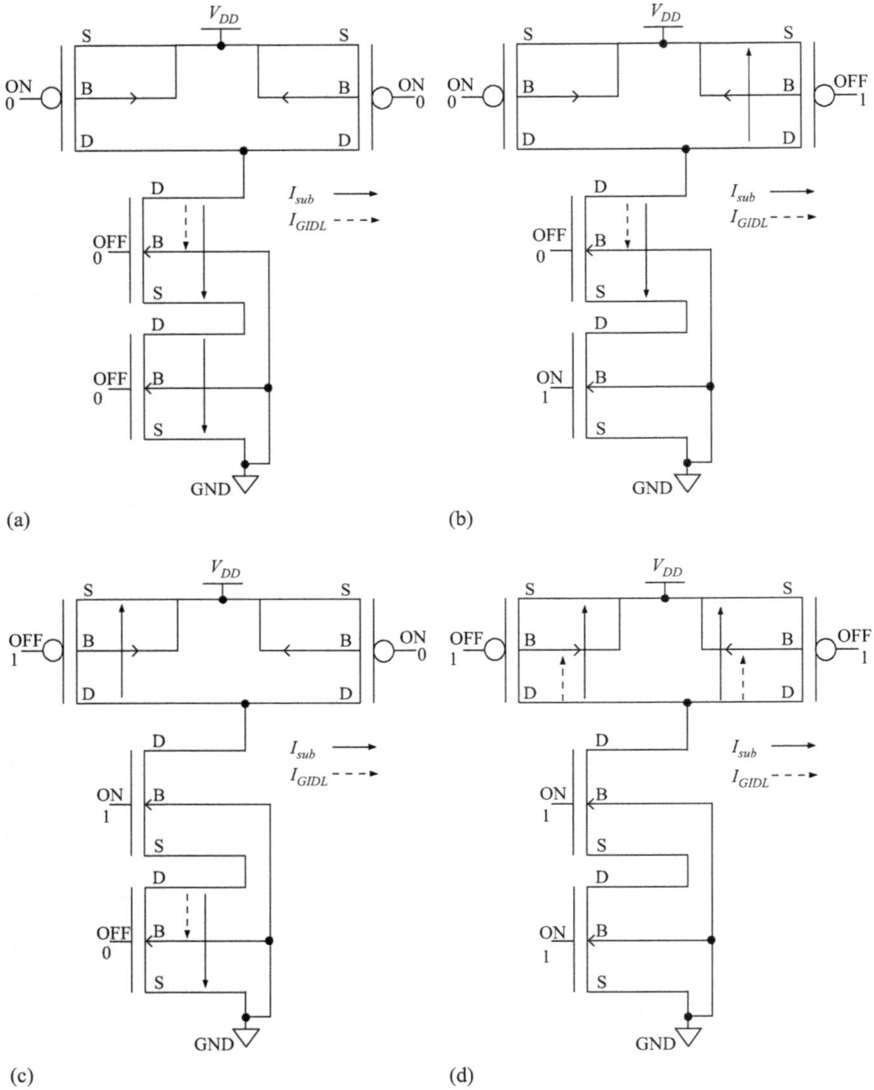

Figure 6.11 Directions of I_{sub} and I_{GIDL} for a high-κ NAND gate for different states. The dotted line shows I_{GIDL}, and the solid line represents I_{sub} flow. I_{GIDL} flows from drain to bulk, whereas I_{sub} flows from drain to source. (a) State "00", (b) State "01", (c) State "10", and (d) State "11"

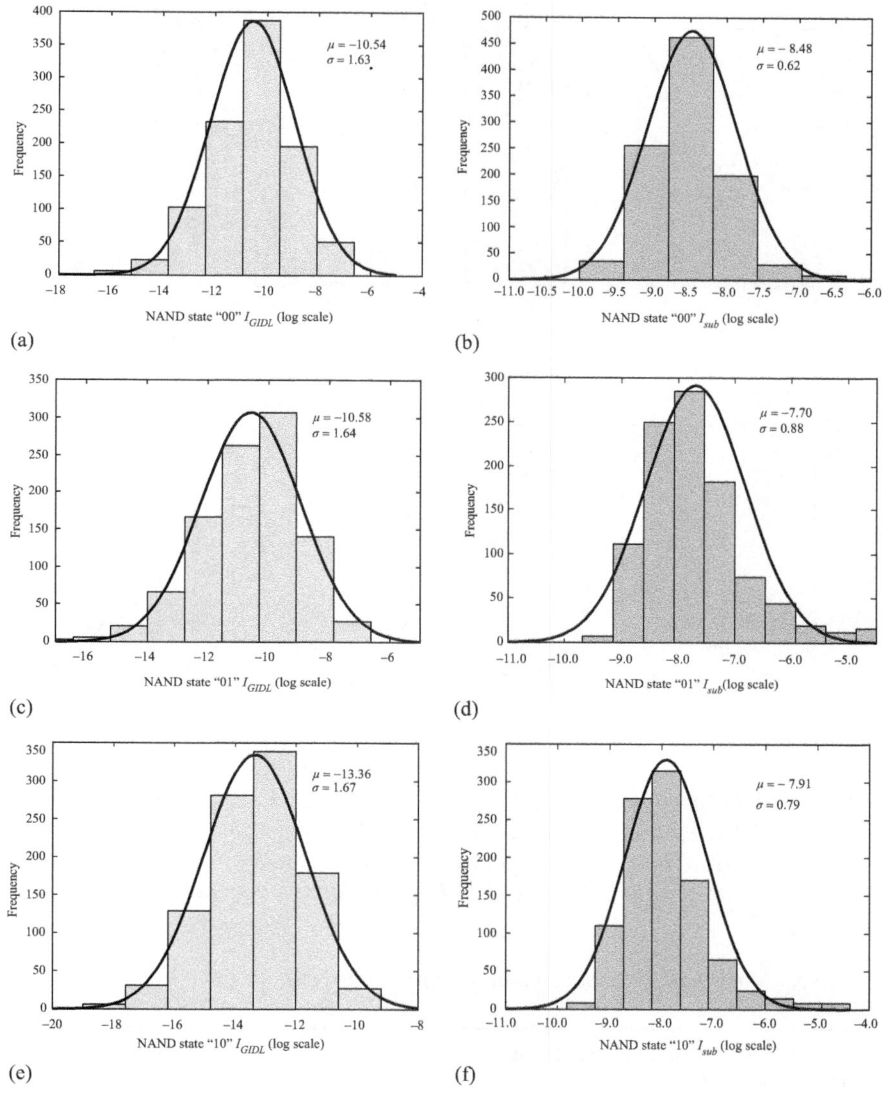

Figure 6.12 Distributions of GIDL current ($\hat{I}_{GIDL}$) and sub-threshold current ($\hat{I}_{sub}$) for HKMG-based two-input NAND gate for input 00, 01, and 10. (a) State 00: GIDL, (b) State 00: sub-threshold, (c) State 01: GIDL, (d) State 01: sub-threshold, (e) State 10: GIDL, and (f) State 10: sub-threshold

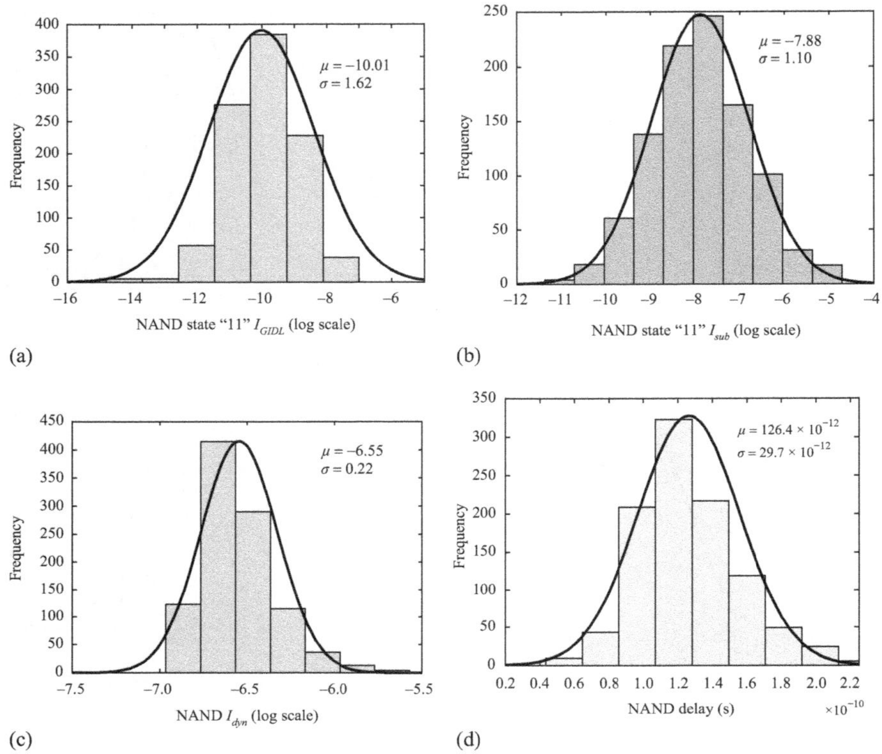

Figure 6.13 Distributions of GIDL current ($\hat{I}_{GIDL}$) and sub-threshold current ($\hat{I}_{sub}$) for input 11, and dynamic current ($\hat{I}_{dyn}$) and propagation delay for HKMG-based two-input NAND gate. (a) State 11: GIDL, (b) State 11: sub-threshold, (c) average: I_{dyn}, and (d) average: delay

Table 6.5 Statistical state-dependent data for NOR [19]

Statistical distribution of current components or propagation delay		"00"	"01"	"10"	"11"
Dynamic current $\hat{I}_{dyn}$	μ	−6.71			
	σ	0.29			
Sub-threshold leakage current $\hat{I}_{sub}$	μ	−7.40	−7.48	−7.63	−8.14
	σ	0.87	1.12	1.05	0.98
GIDL current $\hat{I}_{GIDL}$	μ	−10.41	−10.17	−12.72	−10.14
	σ	2.08	1.04	2.1842	1.01
Propagation delay	μ	115.14 ps			
	σ	31.40 ps			

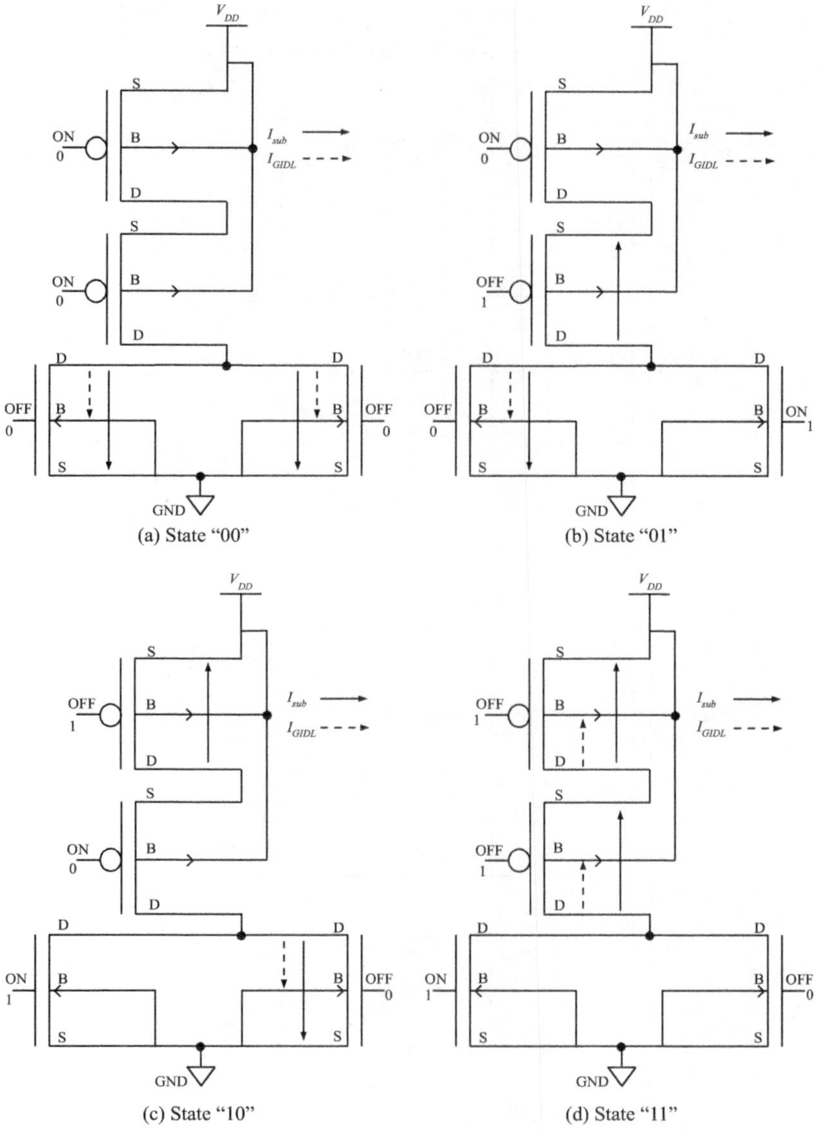

Figure 6.14 Directions of I_{sub} and I_{GIDL} for a high-κ NOR gate for different states. The dotted line shows I_{GIDL}, and the solid line represents I_{sub} flow. I_{GIDL} flows from drain to bulk, whereas I_{sub} flows from drain to source. (a) State "00", (b) State "01", (c) State "10", and (d) State "11"

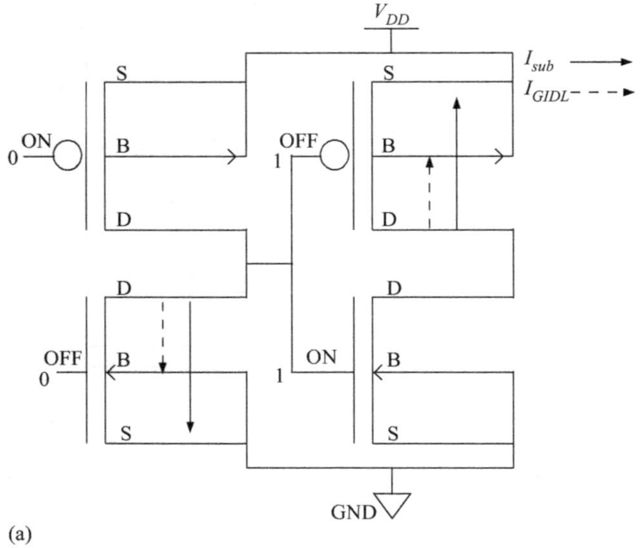

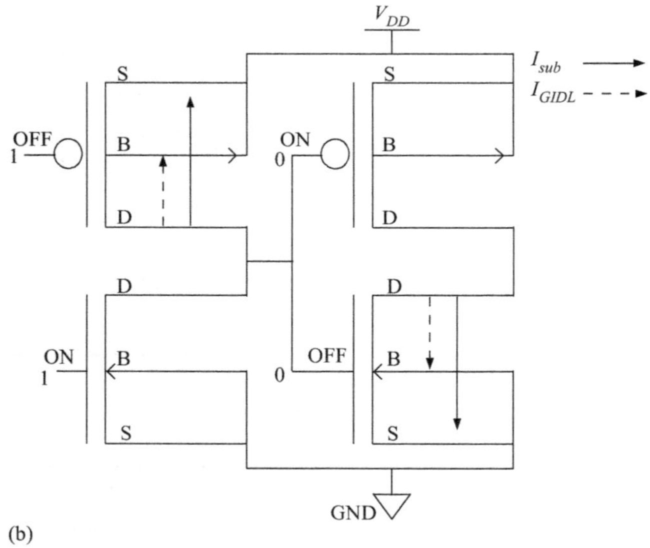

Figure 6.15 Directions of I_{sub} and I_{GIDL} for a high-κ buffer for different states. The dotted line shows I_{GIDL} and the solid line represents I_{sub} flow. I_{GIDL} flows from drain to bulk, whereas I_{sub} flows from drain to source. (a) State "0" and (b) State "1"

Table 6.6 Statistical state-dependent data for two-stage buffer [19]

Statistical distribution of current components or propagation delay		**"0"**	**"1"**
Dynamic current $\hat{I}_{dyn}$	μ	−6.79	
	σ	0.39	
Sub-threshold leakage current $\hat{I}_{sub}$	μ	−6.98	−6.94
	σ	0.79	0.93
GIDL current $\hat{I}_{GIDL}$	μ	−10.15	−9.78
	σ	1.99	1.12
Propagation delay	μ	50.14 ps	
	σ	12.54 ps	

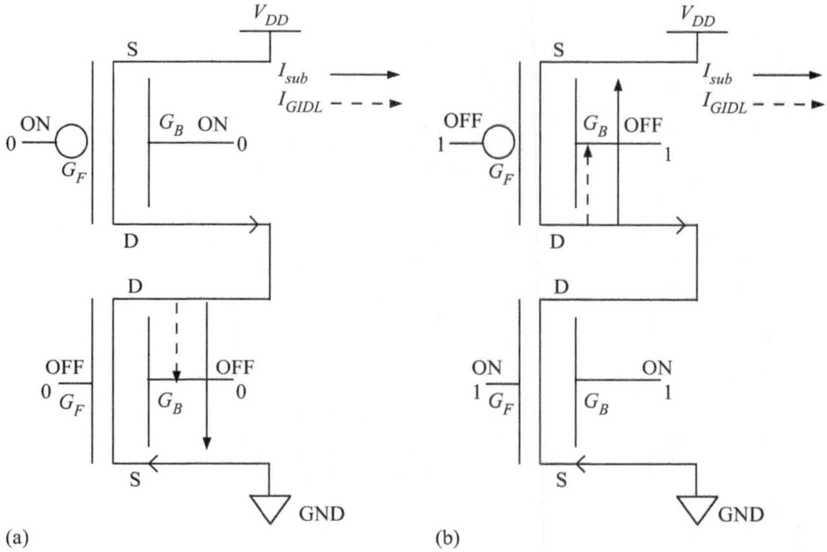

Figure 6.16 Directions of I_{sub} and I_{GIDL} for a DG-FinFET Inverter for different states. The dotted line shows I_{GIDL}, and the solid line represents I_{sub} flow. I_{GIDL} flows from drain to bulk, whereas I_{sub} flows from drain to source. (a) State "0" and (b) State "1"

The directions of various currents in a two-stage buffer for each state are shown in Figure 6.15(a) and 6.15(b). Table 6.6 summarizes the statistical data for the various currents measured in a two-stage buffer.

6.7.2 For DG-FinFET

This section presents the different currents flowing through the DG-FinFET transistors used in logic gates. The direction of currents flowing in a FinFET Inverter is shown in Figure 6.16(a) and 6.16(b). The figures show the schematic of the inverter with

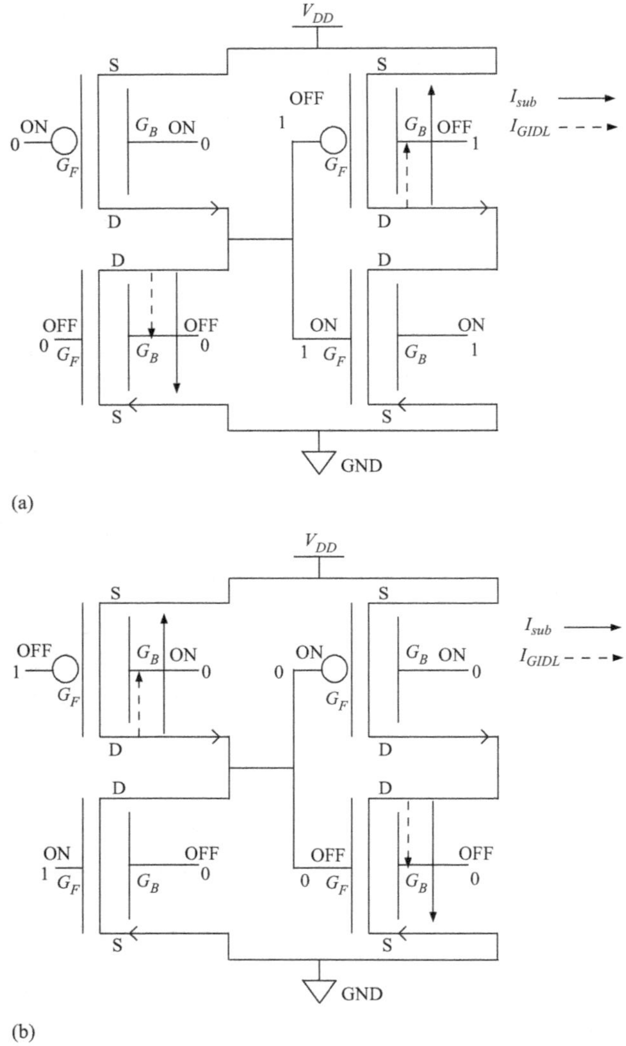

Figure 6.17 Flow of I_{sub} and I_{GIDL} for a DG-FinFET buffer for different states. The dotted line shows I_{GIDL}, and the solid line represents I_{sub} flow. I_{GIDL} flows from drain to bulk, whereas I_{sub} flows from drain to source. (a) State "0" and (b) State "1"

the DG-FinFET transistors operated with the same voltage given to both front and back gates G_F and G_B. For State 1, 1 V was applied to both the gates, and for State 0, 0 V was applied to both the gates. V_{DD} was 1 V for both the states. The dotted line represents the GIDL, and the solid line represents the sub-threshold leakage current. Table 6.7 shows the leakage current for the inverter in both states.

Table 6.7 Statistical state-dependent data for DG-FinFET inverter

Nominal distribution of current components	**State "0"**	**State "1"**
Sub-threshold leakage current (I_{sub})	0.9 pA	1.35 pA
Dynamic current (I_{dyn})	3.55 μA	
Propagation delay	53.42 ps	

Table 6.8 Statistical state-dependent data for DG-FinFET buffer

Nominal distribution of current components	**State "0"**	**State "1"**
Sub-threshold leakage current (I_{sub})	0.9 pA	1.35 pA
Dynamic current (I_{dyn})	2.28 μA	
Propagation delay	69.94 ps	

Table 6.9 Statistical state-dependent data for DG-FinFET NAND gate

Nominal distribution of current components	**State "00"**	**State "01"**	**State "10"**	**State "11"**
Sub-threshold leakage current (I_{sub})	10 fA	5.14 pA	4.498 pA	1.35 pA
Dynamic current (I_{dyn})		3.10 μA		
Propagation delay		127.73 ps		

The directions of currents flowing in a FinFET buffer are shown in Figure 6.17(a) and 6.17(b). The figures show the schematic of the inverter with the DG-FinFET transistors operated with the same voltage given to both the front and the back gates G_F and G_B. The sub-threshold leakage current, the dynamic current and the propagation delay are calculated for both State "1" and State "0". The respective experimental results are presented in Table 6.8.

The directions of different currents in a DG-FinFET-based NAND gate are shown in Figure 6.18(a)–6.18(d). The nominal values of leakage currents in each of the states for two logic gates are presented in this subsection. For the NAND gate using the DG-FinFET, both the gates, the front and the back gates, are given the same voltage in each of the states. One volt was applied for State 1, and 0 V was applied for State 0. V_{DD} was 1 V in all states. The sub-threshold leakage current for each of the states, dynamic current, and the propagation delay are calculated. The respective results are presented in Table 6.9.

The directions of different currents in a DG-FinFET-based NAND gate are shown in Figure 6.19(a)–6.19(d). The nominal values of leakage currents in each of the states for two logic gates are presented in this subsection. For the NAND gate using the DG-FinFET, both the gates, the front and the back gates, are given the same voltage in

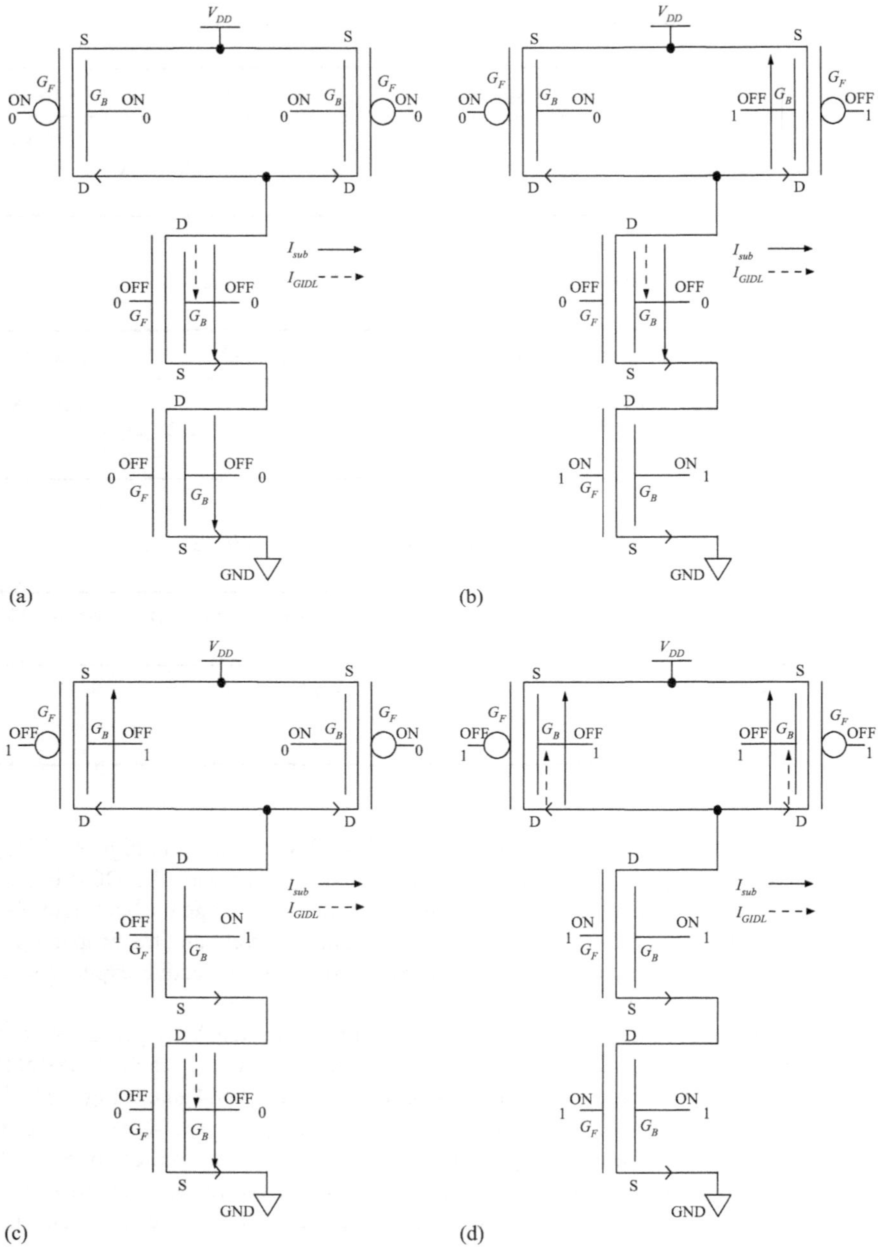

Figure 6.18 Flow of I_{sub} and I_{GIDL} for a FinFET NAND gate for different states. The dotted line shows I_{GIDL}, and the solid line represents I_{sub} flow. I_{GIDL} flows from drain to bulk, whereas I_{sub} flows from drain to source. (a) State "00", (b) State "01", (c) State "10", and (d) State "11"

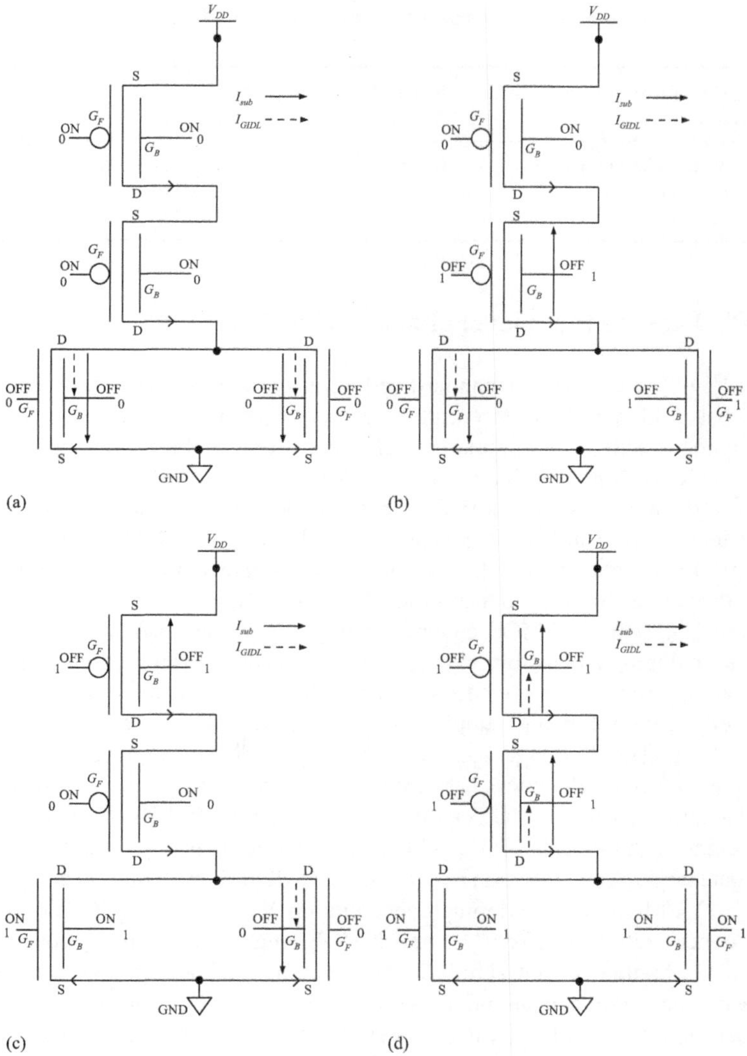

Figure 6.19 Flow of I_{sub} and I_{GIDL} for a FinFET NOR gate for different states. The dotted line shows I_{GIDL}, and the solid line represents I_{sub} flow. I_{GIDL} flows from drain to bulk, whereas I_{sub} flows from drain to source. (a) State "00", (b) State "01", (c) State "10", and (d) State "11"

each of the state. One volt was applied for State 1 and 0 V was applied for State 0. V_{DD} was 1 V in all states. The sub-threshold leakage current for each of the states, dynamic current, and the propagation delay are calculated. The respective results are presented in Table 6.10.

Table 6.10 Statistical state-dependent data for DG-FinFET NOR gate

Nominal distribution of current components	State "00"	State "01"	State "10"	State "11"
Sub-threshold leakage current (I_{sub})	0.8 pA	58.62 aA	58.62 aA	1.35 pA
Dynamic current (I_{dyn})		2.96 μA		
Propagation delay		197.77 ps		

6.8 PVT-aware logic level characterization

The results of the PVT-aware standard cell library are presented in this section. For HKMG-based library, 15 different parameters are considered for process variation in the standard cells. The parameters considered for variability are supply voltage V_{dd} (V), NMOS threshold voltage V_{Thn} (V), PMOS threshold voltage V_{Thp} (V), NMOS gate dielectric thickness t_{gaten} (nm), PMOS gate dielectric thickness t_{gatep} (nm), NMOS channel length L_{effn} (nm), PMOS channel length L_{effp} (nm), NMOS channel width W_{effn} (nm), PMOS channel width W_{effp} (nm), NMOS gate doping concentration N_{gaten} (cm^{-3}), PMOS gate doping concentration N_{gatep} (cm^{-3}), NMOS channel doping concentration N_{chn} (cm^{-3}), PMOS channel doping concentration N_{chp} (cm^{-3}), NMOS source/drain doping concentration N_{sdn} (cm^{-3}), and NMOS source/drain doping concentration N_{sdp} (cm^{-3}). In the case of FinFET-based library, the device parameters considered are the following: supply voltage V_{dd} (V), NMOS threshold voltage V_{Thn} (V), PMOS threshold voltage V_{Thp} (V), NMOS gate dielectric thickness t_{gaten} (nm), PMOS gate dielectric thickness t_{gatep} (nm), NMOS channel length L_{effn} (nm), PMOS channel length L_{effp} (nm), NMOS channel width W_{effn} (nm), PMOS channel width W_{effp} (nm), height of the Fin H_{fin} (nm), NMOS gate doping concentration N_{gaten} (cm^{-3}), PMOS gate doping concentration N_{gatep} (cm^{-3}), NMOS channel doping concentration N_{chn} (cm^{-3}), PMOS channel doping concentration N_{chp} (cm^{-3}), NMOS source/drain doping concentration N_{sdn} (cm^{-3}), and NMOS source/drain doping concentration N_{sdp} (cm^{-3}). As this is a DG-FinFET, the geometries of both gates are varied. All of these device parameters are not necessarily independent. A correlation coefficient of 0.9 between T_{gaten} and T_{gatep} is considered in this chapter. The heights of the fins in the case of a DG-FinFET are assumed since both oxides are grown together. In this section, it is assumed that all the process parameters follow Gaussian distributions. The ambient temperature is considered and modeled through the SPICE model card.

The effect of temperature on GIDL current, dynamic current, sub-threshold current, and the delay is presented as surface plots in Figure 6.20 [19]. Simulations are performed at 0°C, +50°C, +100°C, and +125°C. The mean and the variance for each of the currents at the above-mentioned temperatures are calculated. For brevity, the statistical results of NAND are presented. All the experimental results are presented for the input value of "00" for the two-input NAND for HKMG-based technology.

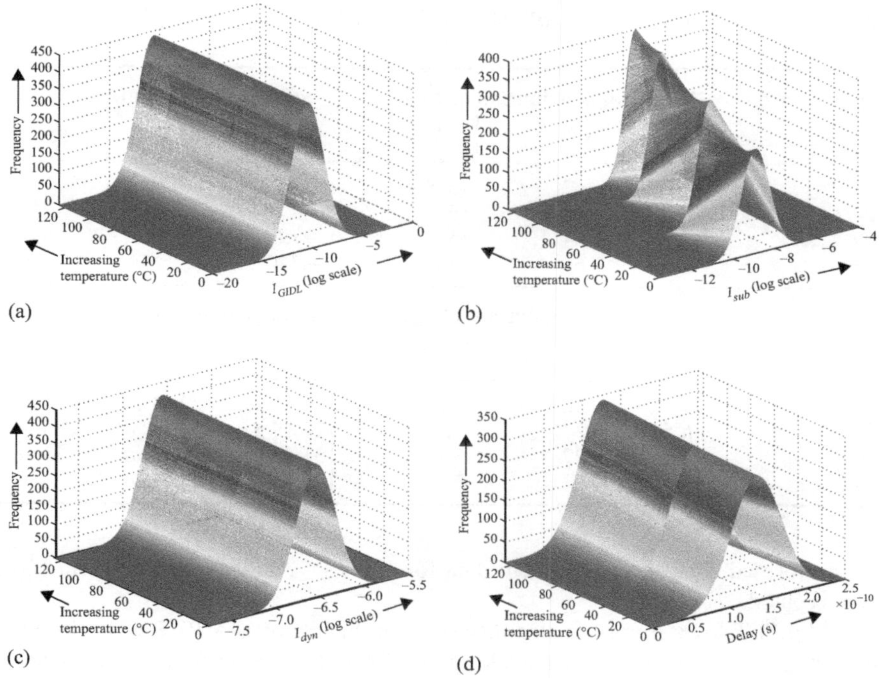

Figure 6.20 PVT plots for GIDL ($\hat{I}_{GIDL}$), sub-threshold ($\hat{I}_{sub}$), and dynamic (I_{dyn}) current, and delay for an HKMG NAND gate [19]. (a) I_{GIDL}, (b) I_{sub}, (c) I_{dyn}, and (d) delay

It can be seen that $\hat{I}_{GIDL}$ does not show a very strong dependence on temperature (Figure 6.20(a)), while $\hat{I}_{sub}$ shows an increase in the mean (μ) value with increasing temperature (Figure 6.20(b)) due to the strong temperature dependence of $\hat{I}_{sub}$ on V_{Th}. Delay also shows an increasing trend with temperature (Figure 6.20(d). $\hat{I}_{dyn}$ is measured over one cycle of operation, as per (6.24). $\hat{I}_{dyn}$ remains fairly constant with temperature, because for one cycle theoretically, $\hat{I}_{dyn}$ does not depend on frequency [21]. The statistical results are summarized in Table 6.11. The PVT-aware results for DG-FinFET can be presented in a similar manner, but have been skipped for brevity.

6.9 Conclusion and directions for future research

This chapter presented a statistical methodology for a PVT-aware HKMG logic cell library creation while taking the effect of process variations into consideration. Device level characterization for HKMG NMOS and PMOS transistors for drive current ($\hat{I}_{ON}$), off-current ($\hat{I}_{OFF}$), and GIDL current ($\hat{I}_{GIDL}$) was performed, modeled using 32 nm PTM models. In addition, PVT-aware statistical characterization of standard cells

Table 6.11 PVT-aware statistical data for HKMG NAND logic cell [19]

Temperature (°C)	Statistical distribution of $\hat{I}_{GIDL}$			Statistical distribution of $\hat{I}_{sub}$		
	μ	σ	$\left\lvert\frac{\sigma}{\mu}\right\rvert$ %	μ	σ	$\left\lvert\frac{\sigma}{\mu}\right\rvert$ %
0	−10.54	1.62	15.4	−8.90	0.68	7.6
50	−10.54	1.63	15.5	−8.17	0.58	7.1
100	−10.54	1.64	15.6	−7.64	0.51	6.6
125	−10.55	1.65	15.6	−7.42	0.48	6.4
	Statistical distribution of $\hat{I}_{dyn}$			Statistical distribution of delay		
	μ	σ	$\left\lvert\frac{\sigma}{\mu}\right\rvert$ %	μ	σ	$\left\lvert\frac{\sigma}{\mu}\right\rvert$ %
0	−6.55	0.22	3.3	126.41 ps	29.61 ps	23.42
50	−6.54	0.22	3.4	127.19 ps	30.12 ps	23.68
100	−6.54	0.22	3.4	131.55 ps	31.52 ps	23.96
125	−6.54	0.22	3.4	134.8 ps	32.32 ps	23.98

was performed. The state-dependent data for $\hat{I}_{sub}$, $\hat{I}_{GIDL}$, and dynamic current ($\hat{I}_{dyn}$) are presented.

A FinFET logic library is also discussed in this chapter. A PVT-aware FinFET standard cell library creation while taking the effect of process variation into account is also presented. For FinFETs, characterization for drive current ($\hat{I}_{ON}$), off-current ($\hat{I}_{OFF}$), and GIDL current ($\hat{I}_{GIDL}$) has been done, modeled using 32-nm PTM models. Further, PVT-aware statistical characterization of standard cells was completed. The state-dependent data for $\hat{I}_{sub}$, $\hat{I}_{GIDL}$, and dynamic current ($\hat{I}_{dyn}$) are presented.

As part of future research, we plan to implement similar logic libraries for other non-classical CMOS technologies such as multi-gate transistors, tunnel FET, and carbon nanotubes and analyze their performance. Memristor-based logic level and RTL libraries for digital design will also be investigated. The future research will also consider on-chip thermal variations.

Acknowledgment

A preliminary conference version of this research was presented at [19, 39].

References

[1] Akgul, B.E.S., Chakrapani, L.N., Korkmaz, P., Palem, K.V.: "Probabilistic CMOS technology: a survey and future directions". In: *Proceedings of the IFIP International Conference on Very Large Scale Integration*, pp. 1–6 (2006)

[2] Animesh, D., Goel, A., Cakici, R.T., Mahmoodi, H., Lekshmanan, D., Roy, K.: "Modeling and circuit synthesis for independently controlled double gate FinFET devices". In: *IEEE Transactions on Computer-Aided*

Design of Integrated Circuits and Systems, vol. 26, pp. 1957–1966 (2007). DOI 10.1109/TCAD.2007.896320

[3] Association, S.I.: International Technology Roadmap for Semiconductors (2002). http://public.itrs.net

[4] Basu, S., Thakore, P., Vemuri, R.: "Process variation tolerant standard cell library development using reduced dimension statistical modeling and optimization techniques". In: *Proceedings of the International Symposium on Quality Electronic Design*, pp. 814–820 (2008)

[5] Baumgartner, O., Karner, M., Kosina, H.: "Modeling of high-k-metal-gate-stacks using the non-equilibrium Green's function formalism". In: *Simulation of Semiconductor Processes and Devices*, pp. 353–356 (2008). DOI 10.1109/SISPAD.2008.4648310

[6] Bohr, M.T., Chau, R.S., Ghani, T., Mistry, K.: "The high-κ solution". *IEEE Spectrum* **10**(10), 29–35 (2007)

[7] Borkar, S., Karnik, T., De, V.: "Design and reliability challenges in nanometer technologies". In: *Proceedings of the Design Automation Conference*, pp. 75–75 (2004)

[8] A. Chandrakasan, A., Sheng, S., Brodersen, R.W.: "Low power CMOS digital design". *IEEE Journal of Solid-State Circuits* **27**(4), 473–484 (1992)

[9] Chang, S., Shin, H.: "Off-state leakage currents of MOSFETs with high-κ dielectrics". *Journal of the Korean Physical Society* **41**(6), 932–936 (2002)

[10] Chantem, T., Dick, R.P., Hu, X.S.: "Temperature-aware scheduling and assignment for hard real-time applications on MPSoCs". In: *Proceedings of the Design, Automation and Test in Europe (DATE)*, pp. 288–293 (2008)

[11] Charan, K., Panda, A.K., Noor, A., Sabharwal, S.: "Design of a humidity sensor with PVT variations using AMI C5 CMOS technology". In: *Proceedings of the International Conference on Recent Advances in Microwave Theory and Applications*, pp. 839–842 (2008)

[12] Chau, R., Datta, S., Doczy, M., Kavalieros, J., Metz, M.: "Gate dielectric scaling for high-performance CMOS: from SiO_2 to high-κ". In: *Proceedings of the International Workshop on Gate Insulator*, pp. 124–126 (2003)

[13] Chaudhuri, S., Jha, N.K.: "3D vs. 2D analysis of FinFET logic gates under process variations". In: *IEEE 29th International Conference on Computer Design (ICCD)*, pp. 435–436 (2011). DOI 10.1109/ICCD.2011.6081437

[14] Croon, J.A., Sansen, W., Maes, H.E.: *Matching Properties of Deep Sub-Micron Transistors*. Springer, New York, NY (2005)

[15] Duarte, J.P., Paydavosi, N., Venugopalan, S., Sachid, A., Hu, C.: "Unified FinFET compact model: modelling trapezoidal triple-gate FinFETs". In: *International Conference on Simulation of Semiconductor Processes and Devices (SISPAD)*, pp. 135–138 (2013). DOI 10.1109/SISPAD.2013.6650593

[16] Garitselov, O., Mohanty, S.P., Kougianos, E.: "Accurate polynomial metamodeling-based ultra-fast bee colony optimization of a nano-CMOS PLL". *Journal of Low Power Electronics* **8**(3), 317–328 (2012)

[17] Ghai, D.: "Variability aware low-power techniques for nanoscale mixed-signal circuits". Ph.D. thesis, University of North Texas, Denton, TX, USA (2009)

[18] Ghai, D., Mohanty, S.P., Kougianos, E.: "Parasitic aware process variation tolerant voltage controlled oscillator (VCO) design". In: *Proceedings of the 9th International Symposium on Quality Electronic Design (ISQED)*, pp. 330–333 (2008)

[19] Ghai, D., Mohanty, S.P., Kougianos, E., Patra, P.: "A PVT aware accurate statistical logic library for high-κ metal-gate nano-CMOS". In: *Proceedings of the 10th International Symposium on Quality of Electronic Design (ISQED 2009)*, pp. 47–54 (2009)

[20] Ghai, D., Mohanty, S.P., Thakral, G.: "Comparative analysis of double gate FinFET configurations for analog circuit design". In: *Proceedings of the 56th IEEE International Midwest Symposium on Circuits & Systems (MWSCAS)*, pp. 809–812 (2013)

[21] Golda, A., Kos, A.: "Temperature influence on power consumption and time delay". In: *Proceedings of the Euromicro Symposium on Digital System Design*, pp. 378–382 (2003)

[22] Gossmann, H.J.L., Agarwal, A., Parrill, T., Rubin, L.M., Poate, J.M.: "On the FinFET extension implant energy". In: *IEEE Transactions on Nanotechnology*, vol. 2, pp. 285–290 (2003). DOI 10.1109/TNANO.2003.820783

[23] Hu, C., Niknejad, A., Xi, X., *et al.*: BSIM4 MOS Models, Release 4.5.0. (2005). http://www.device.eecs.berkeley.edu/~bsim3/bsim4.html

[24] Hu, C., Sriramkumar, V., Paydavosi, N., *et al.*: BSIM FFinFET Model, Release 108.0.0 (2014). http://www-device.eecs.berkeley.edu/bsim/index.php/2012/03/15/the-first-standard-finfet-model/

[25] Kim, D., Kang, Y., Kim, Y.: "Simple and accurate modeling of double-gate FinFET fin body variations". In: *International Conference on Synthesis, Modeling, Analysis and Simulation Methods and Applications to Circuit Design (SMACD)*, pp. 265–268 (2012). DOI 10.1109/SMACD.2012.6339390

[26] Kougianos, E., Mohanty, S.P.: "A comparative study on gate leakage and performance of high-κ nano-CMOS logic gates". *International Journal of Electronics* **97**(9), 985–1005 (2010)

[27] Kranti, A., Armstrong, G.A.: "Design and optimization of FinFETs for ultra-low-voltage analog applications". In: *IEEE Transactions on Electron Devices*, vol. 54, pp. 3308–3316 (2007). DOI 10.1109/TED.2007.908596

[28] Lakshmikanthan, P., Sahni, K., Nunez, A.: "Design of ultra-low power combinational standard library cells using a novel leakage reduction methodology". In: *Proceedings of the IEEE International SOC Conference*, pp. 93–94 (2006)

[29] Lim, T., Kim, Y.: "Effect of band-to-band tunnelling leakage on 28 nm MOSFET design". *IEE Electronic Letters* **44**(2), 157–158 (2008)

[30] Lin, T.Y., Lin, T.H., Tung, H.H., Lin, R.B.: "Double-via-driven standard cell library design". In: *Proceedings of the Design, Automation & Test in Europe Conference & Exhibition*, pp. 1–6 (2007)

[31] Manchanda, L., Busch, B., Green, M.L., *et al.*: "High K gate dielectrics for the silicon industry". In: *International Workshop on Gate Insulator*, pp. 56–60 (2001)

[32] Misra, D., Iwai, H., Wong, H.: "High-κ gate dielectrics". *The Electrochemical Society Interface* **14**(2), 30–34 (2005)

[33] Mistry, K., Allen, C., Auth, C., *et al.*: "A 45 nm logic technology with high-k+metal gate transistors, strained silicon, 9 Cu interconnect layers, 193 nm dry patterning, and 100% Pb-free packaging". *IEEE International Electron Devices Meeting* (2007). DOI 10.1109/IEDM.2007.4418914

[34] Mizuno, T., Okamura, J., Toriumi, A.: "Experimental study of threshold voltage fluctuation due to statistical variation of channel dopant number in MOSFET's". *IEEE Transactions on Electron Devices* **41**(11), 2216–2221 (1994)

[35] Mohanty, S.P.: "Energy and transient power minimization during behavioral synthesis". Ph.D. thesis, University of South Florida, Tampa, FL, USA (2003)

[36] Mohanty, S.P.: *Nanoelectronic Mixed-Signal System Design*. 9780071825719. McGraw-Hill Education (2015)

[37] Mohanty, S.P., Kougianos, E.: "Steady and transient state analysis of gate leakage current in nanoscale CMOS logic gates". In: *Proceedings of the IEEE International Conference on Computer Design*, pp. 210–215 (2006)

[38] Mohanty, S.P., Kougianos, E., Pradhan, D.K.: "Simultaneous scheduling and binding for low gate leakage nano-CMOS datapath circuit behavioral synthesis". *Computers and Digital Techniques* **2**(2), 118–131 (2008)

[39] Mohanty, S.P., Mukherjee, V., Mahapatra, R.N.: "A comparative analysis of gate leakage and performance of high-κ nanoscale CMOS logic gates". In: *Proceedings of the 16th ACM/IEEE International Workshop on Logic and Synthesis (IWLS)*, pp. 31–38 (2007)

[40] Mohanty, S.P., Ranganathan, N., Krishna, V.: "Datapath scheduling using dynamic frequency clocking". In: *Proceedings of the IEEE Computer Society Annual Symposium on VLSI*, pp. 65–70 (2002)

[41] Mukherjee, V., Mohanty, S.P., Kougianos, E.: "A dual dielectric approach for performance aware gate tunneling reduction in combinational circuits". In: *Proceedings of the 23rd IEEE International Conference on Computer Design (ICCD 2005)*, pp. 431–437 (2005)

[42] Narendar, V., Dattatray, R.W., Rai, S., Mishra, R.A.: "Design of high-performance digital logic circuits based on FinFET technology". *International Journal of Computer Applications* (2012)

[43] Narendra, S.G., Chandrakasan, A.: *Leakage in Nanometer CMOS Technologies*. Springer (2005)

[44] Pei, G., Abraham Kedzierski, J.A., Oldiges, P., Leong, M., Kan, E.C.C.: "FinFET design considerations based on 3-d simulation and analytical modeling". In: *IEEE Transactions on Electron Devices*, vol. 49, pp. 1411–1419 (2002). DOI 10.1109/TED.2002.801263

[45] Rabaey, J.M., Chandrakasan, A., Nikolić, B.: *Digital Integrated Circuits*, 2nd edn. Prentice-Hall Publishers (2003)

[46] Rachit, I.K., Bhat, M.S.: "AutoLibGen: An open source tool for standard cell library characterization at 65nm technology". In: *Proceedings of the International Conference on Electronic Design*, pp. 1–6 (2008)

[47] Rastogi, A., Ganeshpure, K., Kundu, S.: "A study on impact of leakage current on dynamic power". In: *Proceedings of the International Symposium on Circuits and Systems*, pp. 1069–1072 (2007)

[48] Rastogi, A., Ganeshpure, K., Sanyal, A., Kundu, S.: "On composite leakage current maximization". *Journal of Electronic Testing: Theory and Applications* **24**(4), 405–420 (2008)

[49] Sairam, T., Zhao, W., Cao, Y.: "Optimizing FinFET technology for high-speed and low-power design". In: *ACM Great Lakes Symposium of VLSI*, pp. 73–77 (2007)

[50] Si2: Si2 Releases Open 45nm Library (2008). https://www.si2.org/openeda.si2.org/projects/nangatelib

[51] Sill, F., You, J., Timmerman, D.: "Design of mixed gates for leakage reduction". In: *Proceedings of the 17th Great Lakes Symposium on VLSI*, pp. 263–268 (2007)

[52] Singh, J., Mathew, J., Mohanty, S.P., Pradhan, D.K.: "Statistical analysis of steady state leakage currents in nano-CMOS devices". In: *Proceedings of the IEEE Norchip Conference*, pp. 1–4 (2007)

[53] Sundareswaran, S., Abraham, J.A., Ardelea, A.: "Characterization of standard cells for intra-cell mismatch variations". In: *Proceedings of the International Symposium on Quality Electronic Design*, pp. 213–219 (2008)

[54] Sundareswaran, S., Abraham, J.A., Panda, R., Ardelea, A.: "Characterization of standard cells for intra-cell mismatch variations". *IEEE Transactions on Semiconductor Manufacturing* **22**(1), 40–49 (2009)

[55] Swahn, B., Hassoun, S., Alam, S., Botha, D., Vidyarthi, A.: "Thermal analysis of FinFETs and its application to gate sizing". In: *TIMA Editions*, Grenoble, France (2006)

[56] Swahn, B., Hassoun, S.: "Gate sizing: FinFETs vs 32nm bulk MOSFETs". In: *43rd ACM/IEEE Design Automation Conference*, pp. 528–531 (2006). DOI 10.1109/DAC.2006.229286

[57] Taur, Y.: "CMOS design near the limit of scaling". *IBM Journal on Research and Development* **46**(2/3), 235–244 (2002)

[58] Thakral, G.: "Process-voltage-temperature aware nanoscale circuit optimization". Ph.D. thesis, University of North Texas, Denton, TX, USA. (2010)

[59] Uppili, S.G., Allee, D.R., Venugopal, S.M., Clark, L.T., Shringarpure, R.: "Standard cell library and automated design flow for circuits on flexible substrates". In: *Proceedings of the Flexible Electronics & Displays Conference and Exhibition*, pp. 1–5 (2009)

[60] Vasdaz, L.L., Grove, A.S., Rowe, T.A., Moore, G.E.: "Silicon gate technology". *IEEE Spectrum* **6**(10), 28–35 (1969)

[61] Weste, N.H.E., Harris, D.: *CMOS VLSI Design: A Circuit and Systems Perspective*, 4th edn. Addison Wesley (2005)

[62] Yang, M., Gusev, E.P., Ieong, M., *et al.*: "Performance dependence of CMOS on silicon substrate orientation for ultrathin and HfO_2 gate dielectrics". *IEEE Electron Device Letters* **24**(5), 339–341 (2003)

[63] Yuan, X., Park, J.E., Wang, J., *et al.*: "Gate-induced-drain-leakage current in 45-nm CMOS technology". In: *IEEE Transactions on Device and Materials Reliability*, vol. 8, pp. 501–508 (2008). DOI 10.1109/TDMR.2008.2002350

[64] Zhao, W., Cao, Y.: "New generation of predictive technology model for sub-45 nm design exploration". In: *Proceedings of the International Symposium on Quality Electronic Design*, pp. 585–590 (2006)

Chapter 7
FinFET and reliability considerations of next-generation processors

Ying Zhang[1], *Sui Chen*[1], *Lu Peng*[1], *and Shaoming Chen*[1]

Recent experimental studies reveal that Fin field-effect-transistor (FinFET) devices commercialized in recent years tend to suffer from more severe negative bias temperature instability (NBTI) degradation compared to planar transistors, necessitating effective techniques on processors built with FinFET for endurable operations. We propose to address this problem by exploiting the device heterogeneity and leveraging the slower NBTI aging rate manifested on the planar devices. We focus on modern graphics processing units in this study due to their wide usage in the current community. We validate the effectiveness of the technique by applying it to the warp scheduler and L2 cache, and demonstrate that NBTI degradation is considerably alleviated with slight performance overhead.

7.1 Introduction

As we shift into the deep submicron era, innovative materials and device architectures are becoming ever demanding to continue the trend toward smaller and faster transistors. Among all candidates in investigation, the Fin field-effect-transistor (FinFET) stands as one of the most promising substitutes for traditional devices at the ensuing technology nodes, since it presents several key advantages over its planar counterpart [1–4]. By wrapping the conducting channel with a thin vertical "fin" which forms the body of the device, the gate is coupled tighter with the channel, increasing the surface area of the gate–channel interface and allowing much stronger control over the conducting channel [1]. This effectively relieve the so-called short channel effects that are observed on planar transistors manufactured with sub-32 nm technology, which in turn implies that FinFET device can provide superior scalability in the deep submicron regime [1].

Another cornerstone motivating the realization of FinFET is the potential performance gain. FinFET transistors can be designed with lower threshold voltage (V_t) and operate with higher drive current, leading to faster switching speed compared

[1]Division of Electrical and Computer Engineering, School of Electrical Engineering and Computer Science, Louisiana State University, Baton Rouge, LA

to conventional planar devices [1]. Released documents from industry demonstrate that the FinFET transistor persistently demonstrates shorter delay than the planar 1, while the support voltage is varying, enabling the design and manufacturing of faster processors. Public documents from leading manufacturers also show that the FinFET structure is capable of largely decreasing leakage when the transistor is off [1]. Recently, the Ivy Bridge [5] and Haswell central processing units [6] released by Intel have commercialized this structure (i.e., referred to as "tri-gate transistor" by Intel), which is also expected to be adopted by other semiconductor manufacturers on their upcoming products [7].

Nonetheless, FinFET is not an impeccable replacement of traditional devices as it raises many challenges to the current industry. One of the most daunting conundrums is the increasing aging rate caused by negative bias temperature instability (NBTI). Recent experimental studies demonstrate that FinFET transistors are more vulnerable to NBTI, leading to a shorter lifetime than a planar device [8, 9]. The NBTI aging rate is evaluated by the increase of delay on the critical path after a certain amount of service time. A chip is considered as failed when the delay increment exceeds a predefined value after which the timing logic of the processor cannot function correctly. Under the same operation condition, the FinFET device is observed to degrade much faster than the planar counterpart, implying a significantly reduced service life span of the target processor. This clearly spurs the development of new techniques to circumvent this problem and prolong the lifetime of FinFET-made processors.

Fortunately, a brief comparison between the main features of FinFET and planar devices sheds some light on alleviating the NBTI effect on future processors. By effectively exploiting the device heterogeneity and leveraging the higher NBTI immunity of planar transistors, the aging of the FinFET structures can be largely suppressed. In this chapter, we propose a technique built on top of this principle to improve the durability of FinFET processors. In general, our technique is implemented by replacing an existing structure with a planar device equivalent. Along with minor modifications at the architectural level, our proposed technique is essentially transferring the "aging stress" from the vulnerable FinFET components to the more NBTI-tolerable planar structures, which in turn lower down the temperature on the structure in study, and thus considerably mitigate the NBTI degradation. Note that the proposed scheme is practically feasible because of the good compatibility between the FinFET and planar process technology [10–12].

Considering that the general-purpose graphics processing unit is becoming an increasingly important component in a wide spectrum of computing platforms, we choose a modern GPU as the target architecture to evaluate the effectiveness of our proposed strategy. In this chapter, we mainly concentrate on optimizing the reliability of the warp scheduler because of its importance. However, the technique described in this chapter can be simply applied to CPU for NBTI mitigation as well. In general, the main contributions of this work are as follows:

- We propose a hybrid-device warp scheduler for reliable operation. By decoupling the warp scheduling into two steps of operations and conducting the prerequisites evaluation in a planar-device structure, we eliminate a large amount of read accesses to the FinFET scheduler hardware and considerably alleviate the NBTI effect.

- We develop a hybrid-device sequential-access cache architecture. All memory requests to this cache hierarchy are handled in a serialized fashion that the tag array made of planar transistors is probed first and the matching block in the FinFET data array is only accessed on a cache hit. This significantly reduce the activity on the cache data array and improve its reliability.

7.2 Background

7.2.1 NBTI degradation mechanism

NBTI is becoming one of the dominant reliability concerns for nanoscale P-MOSFETs. It is caused by the interaction of silicon–hydrogen (Si-H) and the inversion charge at the Si/oxide interface [13, 14]. When a negative voltage is applied at the gate of PMOS transistors, the Si-H bonds are progressively dissociated and H atoms diffuse into the gate oxide. This process eventually breaks the interface between the gate oxide and the conducting channel, leaving positive traps behind. As a consequence, the threshold voltage of the PMOS transistor is increased, which in turn elongates the switching delay of the device through the alpha power law [15]:

$$T_s \propto \frac{V_{dd}L_{eff}}{\mu(V_{dd} - V_t)^{\alpha}} \tag{7.1}$$

where, μ is the mobility of carriers, α is the velocity saturation index and approximates to 1.3. L_{eff} denotes the channel length.

The process described above is termed the "stress" phase where the threshold voltage is persistently increasing with the service time, modeled by the following equation [9].

$$\Delta V_{tstress} = \left(\frac{qT_{ox}}{E_{ox}}\right)^{1.5} \cdot K \cdot \sqrt{C_{ox}\left(V_{gs} - V_t\right)} \cdot e^{\frac{-E_a}{4kT} + \frac{2\left(V_{gs}-V_t\right)}{T_{ox}E_{01}}} \cdot T_0^{-0.25} \cdot T_{stress} \tag{7.2}$$

However, when the stress voltage is removed from the gate, H atoms in the traps can diffuse back to the interface and repair the broken bond. This results in a decrease in the threshold voltage, thus termed the "recovery" stage. This iterative stress–recovery processes lead to a saw-tooth variation of the threshold voltage throughout the device's life span. The final V_t increase taking both stress and recovery into account can be computed as:

$$\Delta V_t = \Delta V_{tstress} \cdot \left(1 - \frac{2\xi_1 T_{ox} + \sqrt{\xi_2 e^{\frac{-E_a}{kT}} T_0 T_{stress}}}{(1+\delta)T_{ox} + \sqrt{e^{\frac{-E_a}{kT}}(T_{stress} + T_{recovery})}}\right) \tag{7.3}$$

Note that in (7.2) and (7.3), T_{stress} and $T_{recovery}$ denote the time under stress and recovery, respectively. Other parameters are either constants or material-dependent variables and are listed in Section 7.4.

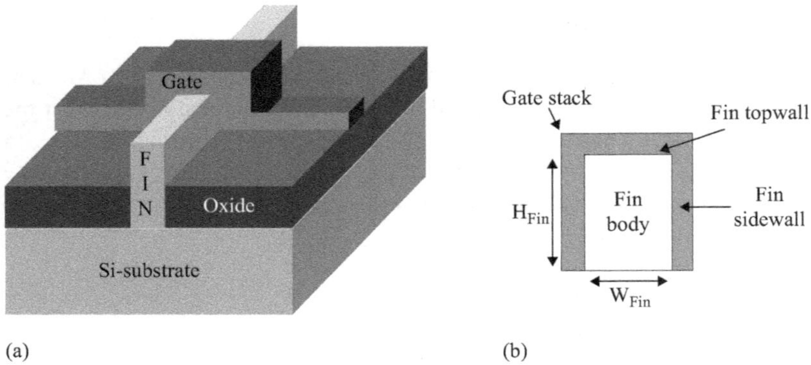

Figure 7.1 FinFET transistor structure: (a) overview (b) side view

FinFET devices are more vulnerable to NBTI that is generally attributed to its unique nonplanar architecture, which is visualized by Figure 7.1. As can be seen, compared to a traditional planar transistor, the FinFET structure is designed with additional fin sidewall surface with higher availability of Si-H bonds [8, 9], implying larger chances of forming interface trap and consequently expediting the device degradation.

The NBTI aging rate depends on multiple factors including both circuit parameters and workload execution patterns. In general, it is acknowledged that voltage, temperature, and the stress/recovery time have strong impact on the aging rate [16, 17]. In this work, our proposed techniques significantly reduce the accesses to the target structures, thus lowering down the localized activity and temperature, which is beneficial in enhancing the structure durability.

7.2.2 Target GPU architecture

The prevalence of unified programming language (e.g., CUDA, OpenCL) has made the general-purpose graphics processing unit a core component in a large variety of systems ranging from personal computers to high-performance computing clusters. Therefore, it is highly important to alleviate the NBTI degradation on this ever increasingly important platform.

Figure 7.2 visualizes the architectural organization of a representative GPU. Note that we follow the Nvidia terminology to depict the processor architecture. As can be seen, the major component of a modern GPU is an array of streaming multiprocessors (SMs), each of which contains an amount of CUDA cores (SPs), load/store units, and special function units (SFUs). A CUDA core is responsible for performing integer ALU and floating point operations, while the SFUs are devoted to conducting transcendental operations such as sine, cosine, and square root. Each stream multiprocessor also contains a register file, a shared memory, and a level 1 cache (usually including instruction/data/constant/texture caches) that are shared among all threads assigned to the SM. All stream multiprocessors connect to an interconnection network,

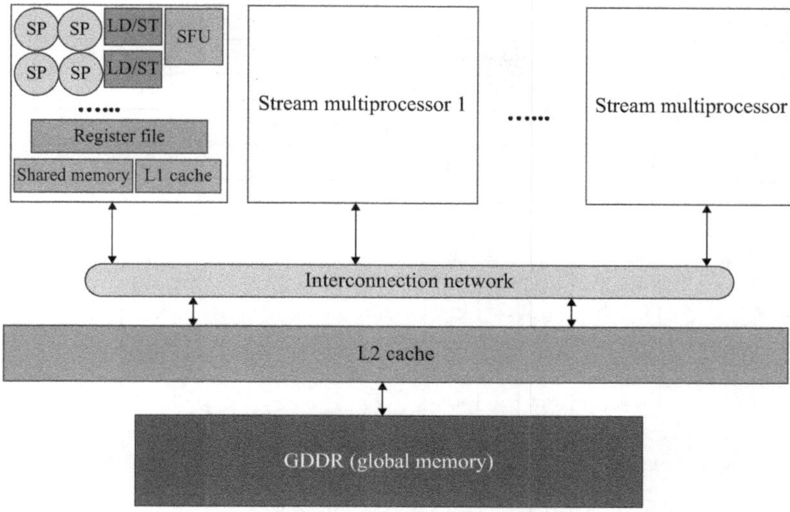

Figure 7.2 An illustration of typical GPGPU architecture

which transfers the memory requests/services between the SMs and the shared L2 cache.

An application developed in CUDA (or OpenCL) contains at least one kernel running on the GPU. A typical kernel includes several blocks composed of substantial threads. During a kernel execution, multiple blocks are assigned to an SM according to the resource requirement. A group of threads from the same block form a warp treated as the smallest scheduling unit to be run on the hardware function units (FUs) in an SIMT fashion.

7.3 Hybrid-device warp scheduler

As an emerging platform targeting for massively parallel computing domains, a modern GPU is designed with several unique characteristics different from a regular CPU. In this section, we concentrate on the warp scheduler because it is an important structure that is frequently accessed during program execution. By observing representative execution behaviors of a large collection of GPU applications, we propose a technique exploiting the device heterogeneity to alleviate the NBTI degradation. As we will demonstrate shortly, the proposed technique does not introduce any additional component to the existing GPU architecture, thus minimizing the hardware cost for the implementation.

7.3.1 Opportunity for improvement

To improve the thread-level parallelism and maximize the execution throughput, a modern GPU usually allows multiple warps to reside on the same streaming

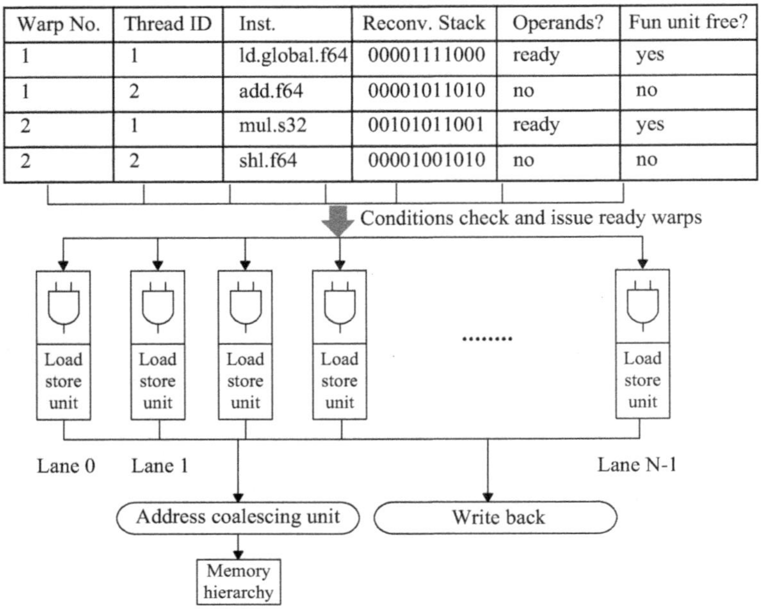

Figure 7.3 The architecture of the warp scheduler

multiprocessor and hide the execution latencies by switching among those resident warps. At any instant, a warp is considered as ready for execution only when several constraints are simultaneously satisfied.

A first-order prerequisite is the functional correctness, which is secured by ensuring data dependencies between warp instructions. When a warp cannot be dispatched because of unsatisfied data dependency, it should wait until all of its operands are ready. A scoreboard hardware structure is responsible for keeping track of data dependencies in a modern GPU. In addition, warps on a streaming multiprocessor contend for limited functional units. When the dispatch port of the functional unit that a warp needs to use is not vacant, the warp cannot be issued even when its data dependencies have been satisfied.

The warp scheduler is an SRAM hardware structure in charge of selecting candidates from all resident warps to dispatch. For the purpose of high performance, a warp scheduler is capable of dispatching one warp per clock cycle, requiring that scanning through all the scoreboard entries and querying the dispatch ports of all functional units should be performed at each cycle [18, 19]. Figure 7.3 illustrates the high-level organization of a warp scheduler equipped in an SM to elaborate the scheduling process. As shown in the figure, all entries, each of which stores complete information of a warp instruction, are going through the conditions checking in parallel in order to identify the candidates ready for execution. Note that to minimize the delay, the scheduler must read the detailed information of a warp (warp ID, opcode, etc.) while evaluating the constraints so that it can dispatch warps as soon as they are ready.

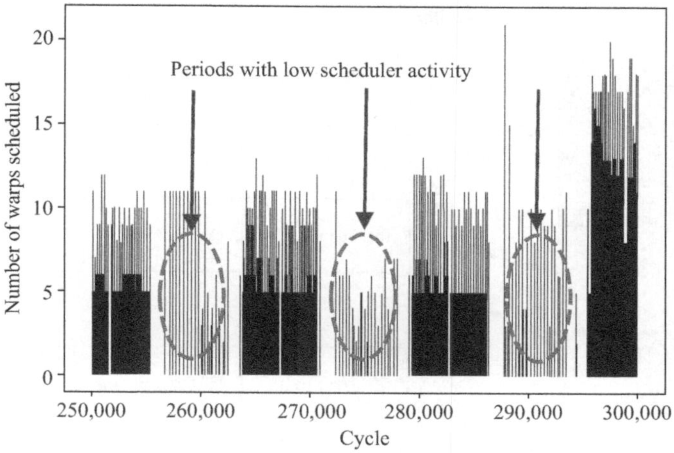

Figure 7.4 A snapshot of the scheduler activity while running WP

Selected warps are sent to the appropriate FUs according to the instruction opcode afterwards.

This particular design naturally inspires a technique to mitigate the NBTI degradation on the scheduler. If the readiness of all warp instructions are known ahead via a certain "predicate," then only the entries with all constraints met are accessed, which in turn decreases the localized activity and temperature, and improves the structure durability.

To justify the potential effectiveness of this strategy, we run a wide spectrum of GPU applications, aiming to observe typical behaviors on the warp scheduler. Figure 7.4 plots a snapshot of the warp scheduler's behavior when *WP* is running on a GPU in order to exemplify the activity on the scheduler. The horizontal axis corresponds to the elapsed time, and the vertical axis represents the accumulative number of ready warps at each time interval. The number is collected every 50 cycles. With this setting, the maximum number of ready warps cannot exceed 100 on each sampling point considering that two warp instructions can be issued at each cycle. As can be seen from the figure, there are a large amount of execution periods with a number of ready warps far less than the theoretical peak, implying a significant reduction in accesses to the scheduler entries in potential. We generally observe that at any given instant, less than 35% of all the warps have the two prerequisites satisfied for all the tested benchmarks. This observation confirms that there is a large headroom for us to optimize the reliability on the warp scheduler.

7.3.2 Two-stage scheduling

Our proposed technique to enhance the durability of the warp scheduler stems from the aforementioned fact at the first place. In order to identify the ready warps, the baseline scheduler is decoupled into two components as visualized in Figure 7.5.

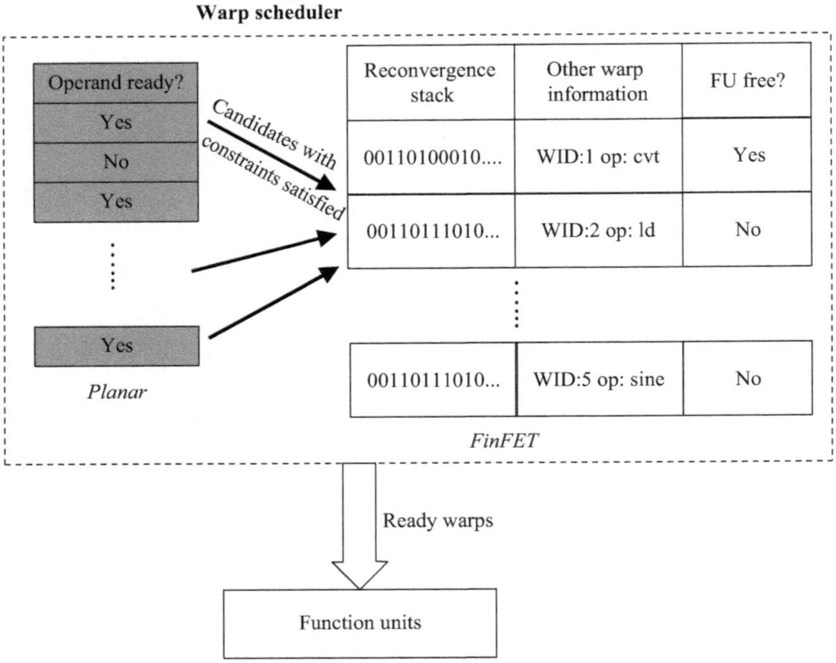

Figure 7.5 The architecture of hybrid-device 2-stage scheduler

By doing so, the prerequisites checking is extracted from the original parallel accesses and is performed prior to obtaining the detailed information of warp instructions. This checking operation outputs the ID of all available candidates resided on the SM, triggering the consequent accesses to the hardware structure which stores all necessary information to dispatch ready warps based on the specific scheduling policy. If a large amount of resident warps are eliminated from the candidate list due to the violation of scheduling constraints, substantial accesses to the scheduler hardware (i.e., the structure at the right side in Figure 7.5) can be avoided.

A nontrivial issue requiring careful consideration in this particular scheduler design is what information should be checked in the first stage. Theoretically, evaluating more scheduling prerequisites would filter larger number of accesses since only the common set of candidates that satisfy each individual constraints are allowed to continue the second stage. However, for certain conditions, checking them in the first stage would lead to undesirable execution behavior, because their evaluation results might be changed in the following cycle. The checking on FUs' availability falls into this category. This is because that the FU status is updated every cycle, and an FU that appears to be free in the current cycle is not necessarily available in the following cycle, if it is assigned to another warp instruction. Therefore in this work, we only check the data dependency in the first stage. As we will demonstrate in Section 7.6, this still results in sufficiently high filter rate for most benchmarks and largely alleviate the NBTI degradation.

On the other hand, considering that the failure of any structure located on the critical path will prevent the entire chip from working correctly, the component where the condition evaluations are conducted tends to become the bottleneck from the perspective of reliability, since all of its entries still needs to be scanned every cycle. To overcome this problem, we propose to manufacture this component with the more NBTI-tolerable planar devices. This hybrid device design effectively leverages the benefits of both devices, aiming to enhance the processor durability. Note that the planar/transistor-made component recording the data dependency and FU availability is unlikely to suffer from early failure because it only requires one bit for each entry and thus consume negligible power. Also recall that this design is technically feasible due to the good compatibility between FinFET and planar processes as demonstrated in patents [10, 12].

Another naturally arising concern with this design is the performance degradation resulted from the sequential scheduler access. Nevertheless, as we will demonstrate in Section 7.6, the performance overhead for most applications are fairly small because only actual accesses to the FinFET part of the scheduler introduces an extra cycle delay. In scenarios where none of the resident warps pass the constraints checking, the execution latency is not impacted.

7.4 Hybrid-device sequential-access L2 cache

It is widely acknowledged by the high performance computation (HPC) community that memory bandwidth is the main bottleneck in a large number of GPU applications. Due to this reason, the shared L2 cache is becoming an increasingly important component on a modern GPU to reduce the contention on the global memory bandwidth [4], implying that improving the reliability of the L2 cache is of great significance to ensure endurable operation of the GPU.

Typically, the L2 cache installed on a contemporary GPU is designed as a set-associative cache with a reasonable size, serving memory requests sent from the stream multiprocessors. To shorten the execution delay, all ways in the tag array and data array of the selected cache set are searched in parallel and if a stored tag equals to the tag in request, the matching cache block from the data array are returned. However, this access procedure is intrinsically unfriendly to reliable operation since it may introduce substantial unnecessary cache accesses in case the requested data block is not present. For example, the application *BlackScholes* demonstrates a close-to-100% miss rate on the L2 cache, meaning that approximately all the memory requests that are missed in the L1 cache need to be transferred to the global memory eventually. In other words, accesses to the L2 cache is completely unnecessary.

Based on this observation, it is straightforward to realize that filtering out the accesses resulting in cache misses is a simple yet effective approach to slow down the NBTI aging on the L2 cache. Since the data array is orders of magnitude larger than the tag array in both area and power consumption, we first concentrate on the

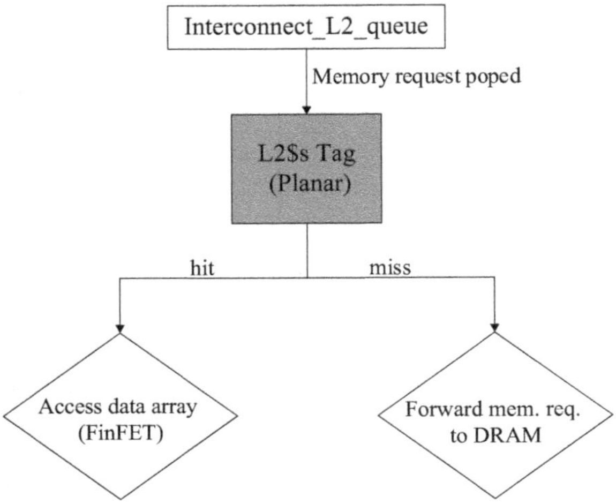

Figure 7.6 Workflow of hybrid-device sequential-access L2

optimization of the data array, which is achieved by applying a technique similar to that developed for the warp scheduler. In specific, we serialize the parallel tag/data access into a sequential procedure [20] with which the tag array in the selected cache set is probed first, and only in case an matching tag is found, the corresponding block in the data array is accessed. This particular design, as visualized by Figure 7.6, reduces the accesses to the data array in twofolds: (1) memory requests that results in cache misses (i.e., no matching tag is found) do not generate consequent accesses to the data array, and (2) only the cache block corresponding to the matching tag, instead of all ways in the set, is read to respond the memory request. With this technique, we expect that the accesses to the data array should be considerably reduced, thus the NBTI aging is largely suppressed due to the decreasing activity and temperature.

On the other hand, to prevent the tag array from becoming the reliability bottleneck, we exploit the device heterogeneity and propose to build the tag array with planar transistors. As we will show in later sections, this can effectively leverage the planar device's advantage in NBTI tolerance and guarantee reliable operations on the L2 tag array throughout the expected life span. Also note that in the remainder of this chapter, we may interchangeably use the terms planar-tag L2, hybrid-device L2, and sequential-access L2 to refer to this design.

7.5 Experimental setup

We validate the proposed techniques using a modified GPGPU-Sim 3.1 [21], a cycle-accurate GPGPU simulator. GPUWattch [22] and HotSpot 5.0 [23] are integrated in the simulator for power and temperature calculation, respectively. The chip floor plan required by HotSpot is calibrated against the one used in a recent paper focusing on

Table 7.1 Architectural parameters for the GPU in study

Parameter	**Values**
Number of SM	15
Number of SP	32/SM
LDST units	16/SM
Shared memory	32 kB/SM
L1 data cache	16 kB/SM
Scheduler	Greedy than oldest (GTO)
Core frequency	1400 MHz
Interconnection	1 crossbar/direction
L2 cache	768 kB: 128 cache line size, 16-way associativity. Access latency 5 cycles
L2 frequency	700 MHz
Memory	FR-FCFS scheduling, 64 max. requests/MC
SIMD lane width	16
Threads/warp	32
Technology	22 nm

Table 7.2 Benchmarks used in this work

Number	**Application**	**Domains**
1	*B+tree*	Search
2	*Backprop*	Pattern Recognition
3	*Blackscholes*	Financial Engineering
4	*Gaussian*	Linear Algebra
5	*Heartwall*	Medical Imaging
6	*LPS*	3D Laplace Solver
7	*Myocyte*	Biological Simulation
8	*NN*	Neural Network
9	*NW*	Bioinformatics
10	*WP*	Weather Prediction

GPU thermal management [24]. The target architecture is configured based on a Fermi GTX 480 [25] that is widely used in many high-performance computers. Table 7.1 lists the detailed architectural parameters for our simulation.

To evaluate the effectiveness of our techniques in practice, we choose a set of programs from several benchmark suites [21, 26, 27], representing typical HPC applications derived from different domains. A full list of applications used in this work is given in Table 7.2. For each program, we run them till completion and use the execution statistics to mimic distinct workload patterns. In specific, to model the NBTI degradation after a 7-year lifespan, we extrapolate the collected activity to represent the load in 7 years under the steady temperature. We report the final increase in the critical path delay as a measurement of the NBTI aging on the hardware.

Table 7.3 Parameter values for computing NBTI

Parameters	FinFET value	Planar value	Description
T_{ox}	1.2 nm	1 nm	Effective oxide thickness
V_t	0.179 V	0.3 V	Threshold voltage
E_o	0.335 V/nm	0.12 V/nm	Electrical field
Fixed parameters			
q	1.602×10^{-19}		Electron charge
V_{dd}	0.9 V		Operating voltage
ε_{ox}	1.26×10^{-19} F/m		Permittivity of gate oxide
ξ_1	0.9		Other constants
ξ_2	0.5		
k	8.6174×10^{-5} eV/K		
δ	0.5		
T_0	10^{-8} s/nm^2		

Equations (7.2) and (7.3) described in Section 7.2.1 are used to compute the variation in the threshold voltage, which in turn translates to the delay increase via (7.1). We set the parameters referred by the equations according to recent studies on device features [9, 23, 28]. Table 7.3 lists the specific parameter values used in this chapter.

7.6 Result analysis

7.6.1 Warp scheduler

7.6.1.1 Improvement on reliability

Figure 7.7 demonstrates the NBTI degradation in terms of the increase in scheduler delay on both the baseline GPU and the one with hybrid device 2-stage warp scheduler. Note that in the figure, the bars marked by "2-stage" refer to the proposed design. A higher delay increase indicates more severe NBTI degradation. As can be observed, the aging due to NBTI on the scheduler hardware is largely suppressed for all benchmarks under investigation when the proposed technique is applied. On average, the hybrid-device 2-stage scheduler presents merely 2.36% longer delay after the designed service life, reduced from 7.7% on the baseline GPU.

While the general improvement on the durability is significant, however, it is notable that the benefits corresponding to different workloads are obviously distinct. For example, the load represented by *NN* causes the scheduler delay to be prolonged by around 8.4% after 7 years services on the baseline GPU. With the adoption of the proposed technique, this degradation can be reduced to 1.96%. On the other hand, an execution pattern similar to *Backprop* prevents the scheduler obtaining the same amount of benefit from the technique. Specifically, the scheduler still suffers

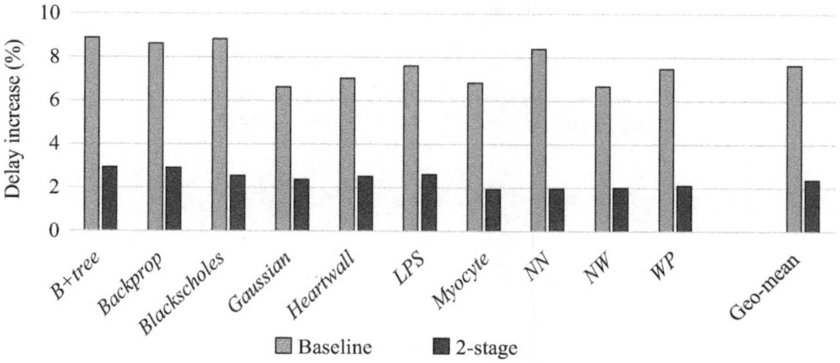

Figure 7.7 The NBTI degradation on the warp scheduler

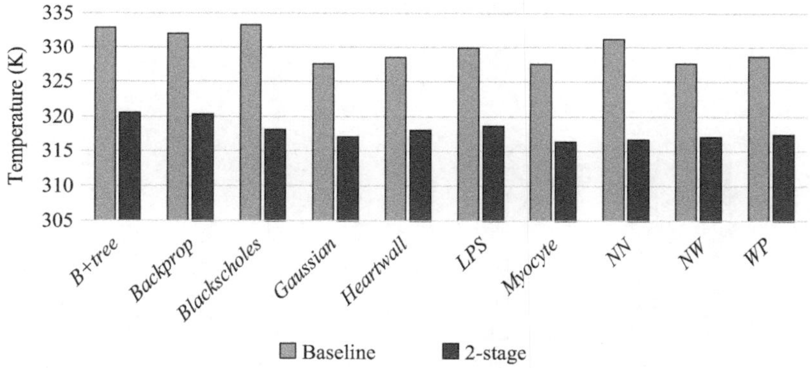

Figure 7.8 The steady temperature on the warp scheduler

from 2.9% longer delay after employing the hybrid-device design, while the baseline platform shows 8.6% longer delay that is similar to the degradation corresponding to *NN*.

Considering the exponential relationship between temperature and NBTI degradation, we collect the localized temperature on the scheduler hardware and demonstrate it in Figure 7.8 for further analysis. Not surprisingly, although the proposed technique can significantly cool down the scheduler in most cases, we note that the temperature reductions are apparently different among the evaluated programs, which is similar to the observation made from Figure 7.7. When executing *NN*, the temperature on the scheduler is reduced by up to 15°C, whereas the temperature reduction for *Backprop* is about 11°C. To gain more insights into the reason behind this phenomenon, let us recall the rationale of the 2-stage scheduler that is described in section 7.3.2. The essential reason for the reduced scheduler accesses is that a large amount of prerequisite evaluations turn out to be false, thus the unnecessary operations on the "unready warps" are avoided. In other words, how much benefit can

Table 7.4 Filter rate on the first stage of warp scheduler

Application	Filter rate (%)
B+tree	75.82
Backprop	76.93
Blackscholes	88.74
Gaussian	98.82
Heartwall	88.46
LPS	90.59
Myocyte	99.85
NN	97.41
NW	97.70
WP	99.49
Geo-mean	90.96

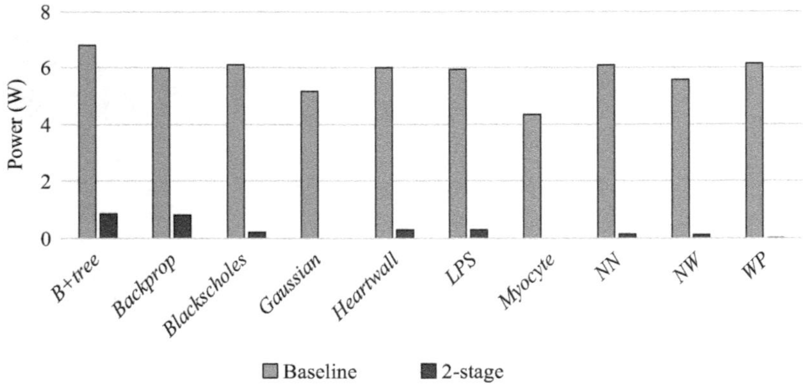

Figure 7.9 The power consumed by the warp scheduler

be obtained from the proposed technique largely depends on the amount of accesses that can be filtered. Table 7.4 lists the percentage of accesses saved by the constraint checking stage. As can be seen, the data dependency checking stage can generally filter out more than 92% of accesses to the scheduler, thus considerably enhancing the durability of the hardware. In particular, we note that 76.9% of scheduler accesses when executing *Backprop* are dispensable, while for *NN* this ratio rises up to 97.4%, implying higher possibilities to lower the power and temperature on the scheduler.

We also plot the power consumption of the scheduler in Figure 7.9 to visualize the changes on the scheduler activity. Clearly, the hybrid device 2-stage scheduler significantly reduces the scheduler power for all evaluated benchmarks, which in turn lowers the local temperature and improves the hardware durability.

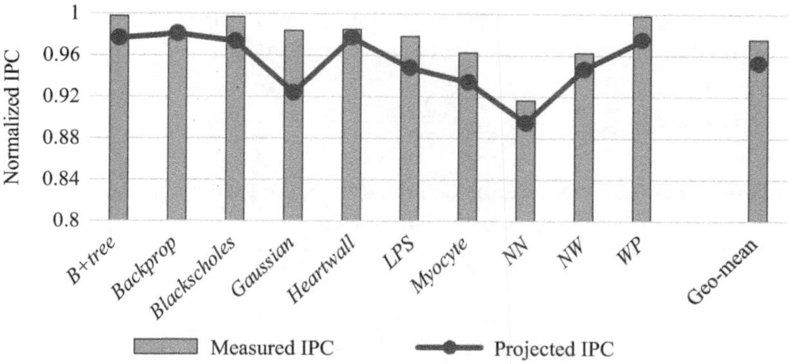

Figure 7.10 Normalized IPC on the GPU with 2-stage scheduler

7.6.1.2 Performance overhead

The extra cycle introduced by the 2-stage scheduler is likely to result in undesirable performance overhead for the program execution. Figure 7.10 shows the performance in terms of normalized IPC (normalized to the baseline GPU) of all benchmarks running on a GPU with the 2-stage scheduler. It is straightforward to note that the performance degradation is distinct among the program collection. In this subsection, we briefly analyze the possible impact on the performance due to the extra cycle and explain the different performance degradation.

The GPU's massive parallelism may be able to hide part of the extra latency during the execution depending on the features of applications. We use the terms "longest warp" and "longest-warp chain" to help explain the latency manifested in the results. We define "longest warp" as the warp with the longest running time during a kernel launch and "longest-warp-chain" as the set of longest warps in each of the sequence of kernel launches in the lifetime of an application. In a typical GPU application, the running time of a longest-warp chain is the sum of execution latencies of all warps in the chain because a) when a kernel is launched, all its warps are started simultaneously and b) a kernel is not launched until all warps of the previous kernel launch complete. In other words, latency on the longest warp could not be hidden as easily as that on other warps. Longest warps also do not overlap temporally. For each longest warp we can compute its average latency as:

$$AvgLatency = \frac{\sum_{n=1}^{N} C_n}{\sum_{n=1}^{N} I_n} \tag{7.4}$$

where, N is the number of kernel launches and C_n and I_n are respectively the number of cycles and warp instructions of the longest warp in each kernel launch.

The I_n instructions in a kernel launch are the instructions issued to and executed by a warp. The extra cycle introduced to the scheduler will be added before each of the instructions is executed. Since the instructions are executed in-order, this is equivalent to adding $\sum_{n=1}^{N} I_n$ extra cycles to the entire longest-warp chain. The average latency of the warp after adding the extra cycles should become:

$$AvgLatency_{delayed} = \frac{\sum_{n=1}^{N} C_n + \sum_{n=1}^{N} I_n}{\sum_{n=1}^{N} I_n} \tag{7.5}$$

The overhead indicators can be deducted from the two latencies shown above:

$$\begin{aligned} OverheadInd &= \frac{\Delta latency}{AvgLatency} = \frac{AvgLatency_{delayed} - AvgLatency}{AvgLatency} \\ &= \frac{\frac{\sum_{n=1}^{N} C_n + \sum_{n=1}^{N} I_n}{\sum_{n=1}^{N} I_n} - \frac{\sum_{n=1}^{N} C_n}{\sum_{n=1}^{N} I_n}}{\frac{\sum_{n=1}^{N} C_n}{\sum_{n=1}^{N} I_n}} = \frac{\sum_{i=1}^{N} C_n + \sum_{i=1}^{N} I_n - \sum_{i=1}^{N} C_n}{\sum_{i=1}^{N} C_n} \\ &= \frac{\sum_{i=1}^{N} I_n}{\sum_{i=1}^{N} C_n} = \frac{1}{AvgLatency} \end{aligned} \tag{7.6}$$

The normalized IPC (measured) and the one derived from the overhead indicator (projected) are both plotted in Figure 7.10. As the figure shows, they are closely correlated. The average latencies and the overheads are determined by the behaviors of the longest warps which are in turn closely related to the characteristics of individual applications. For example, *B+tree* involves a kernel launch with 48 warps on each SM and initiates many global memory transactions (159.26 per cycle). Its longest warp has an average delay of more than 100 cycles. *NN*, on the other hand, has a much smaller average delay (smaller than 10), because it generates much fewer global memory transactions (only 0.06 per cycle) and each SM executes only eight warps. With such few memory transactions and fewer warps, each of the warps, including the longest warp, does not have to wait for long-delay memory operations while sharing more computational resources. This different memory request intensities result in average latencies of the longest warp chains as 41.7 and 8.53 cycles for *B+tree* and *NN*, respectively. Consequently, we observe apparently different performance losses for these two benchmarks.

7.6.2 L2 cache

We now shift our concentration to the L2 cache. For this structure, we first focus on its data array. Figure 7.11 shows the NBTI degradation on the L2 cache data array on both the baseline GPU and the GPU with a planar-tag sequential-access L2. Note that the latter one is labeled as "with_Ptag" in the figure, where the capital letter P stands for planar device. As shown in the figure, the general trend is similar to what is observed in previous section that the proposed technique is capable of largely slowing down the aging due to NBTI on the target component throughout the service life. On average, the hybrid device design reduces the delay increase from 14.1% in the baseline situation to 2.8%.

We also note that the improvement on the durability is different among the programs in study. For example, the applications *Gaussian* and *LPS* causes approximately

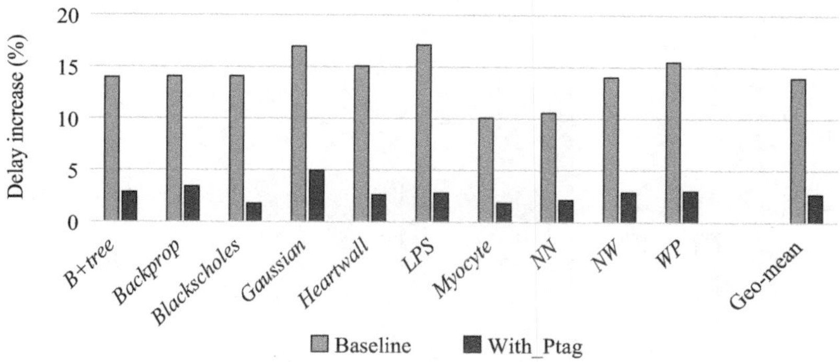

Figure 7.11 NBTI degradation on the L2 data array

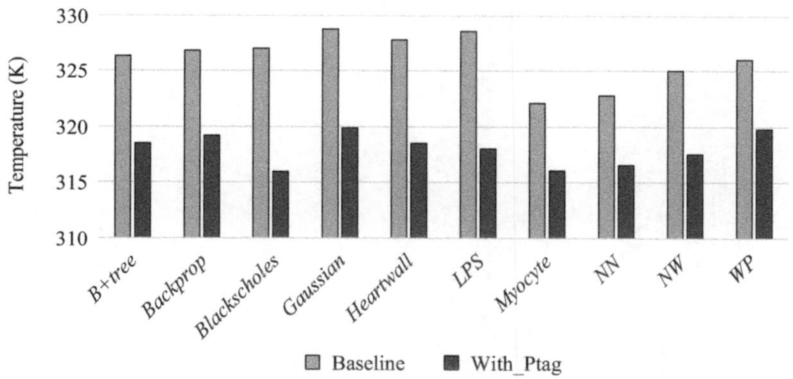

Figure 7.12 Steady temperature on the L2 data array

the same level of NBTI aging on the baseline platform. However, with the hybrid-device L2 cache, running *LPS* apparently leads to less significant NBTI degradation (2.7%) compared to the execution of Gaussian (4.93%). This is resulted from the distinct temperature variations on the L2 while running these programs. Figure 7.12 shows the steady L2 temperature for both the baseline and our proposed design. From the figure, we note that on the GPU with the hybrid device design, running *LPS* makes the L2 cache much cooler compared to the execution of Gaussian. The reason is as follows. Similar to accessing the 2-stage warp scheduler, memory requests sent to the L2 cache are served in a sequential tag-data access pattern, while the tag probing can eliminate the unnecessary accesses to the data array (i.e., cache misses). In other words, the different amount of cache accesses that are avoided are the essential reason for the distinct temperature and reliability changes. Figure 7.13(a,b), respectively, plot the L2 cache miss rates and comparison of L2 power for different applications. As can be seen, *LPS* demonstrates an L2 miss rate of 36%, thus resulting in impressive reduction in L2 power/temperature and great reliability enhancement as a consequence.

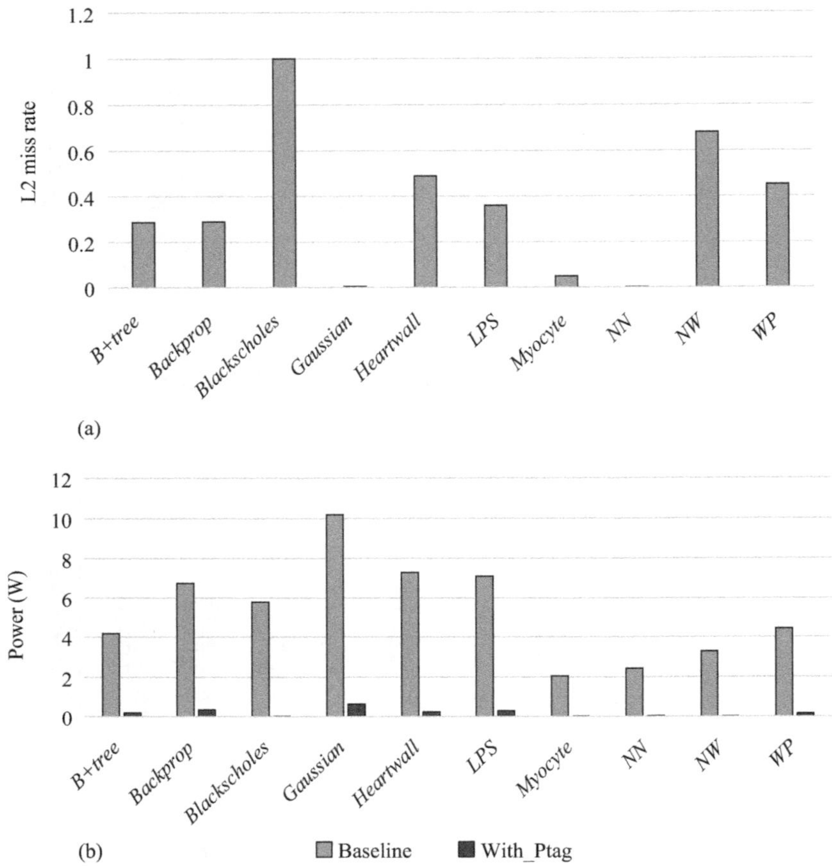

Figure 7.13 (a) L2 miss rate (b) power consumption of the L2 data array

For *Gaussian*, most of the accesses to the data array cannot be avoided because of the low L2 miss rate (4.7%). This eventually leads to the relatively smaller improvement on the NBTI degradation. Other benchmarks with high L2 miss rates including *Blackscholes* also present relatively larger improvement on device durability compared to those with low L2 miss rates such as *NN*. On the other hand, it is important to keep in mind that even for a cache hit, only the matching block is accessed afterwards. For caches with high associativity, which is the typical design in many modern processors, this provides another fold of reduction in the localized power and temperature. Due to this reason, the power consumption of L2 for all bench-marks is considerably reduced while running with sequential-access cache as shown in Figure 7.13(b).

The reliability of the tag array is becoming a major concern in the proposed cache design since the accesses to this structure have not been reduced. Fortunately,

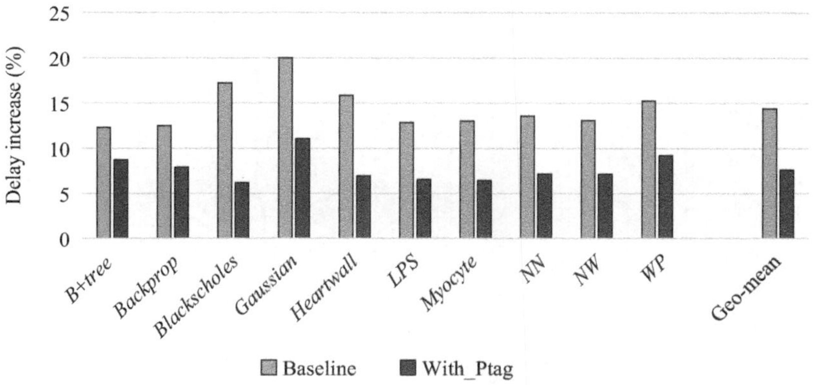

Figure 7.14 NBTI degradation on the L2 tag array

due to higher NBTI-immunity manifested by planar transistors and the small power consumed by the tag array, the L2 tag is not likely to suffer from significant NBTI degradation. Figure 7.14 compares the NBTI degradation in the tag array with both designs, which is essentially determined by the different NBTI tolerance of FinFET and planar transistors. As can be seen, the tag array made of planar device leads to much less degradation compared to the baseline plat-form, implying more endurable operation in the service life.

Another concern that deserves evaluation is the possible performance loss resulted from the extra delay spent on the cache tag probing. We demonstrate the normalized IPC of all programs with the sequential-access L2 cache in Figure 7.15(a) and find that the performance degradation for all benchmarks in investigation is within 1.5%. This does not go beyond our expectation due to the following reasons. First, only a cache hit introduces an extra cycle delay since misses will be promptly forwarded to the lower memory hierarchy after the tag probing, thus not wasting any cycles. Second, even an L2 cache hit takes multiple cycles to complete. This includes the 5 cycles to access the data array and the time spent on the interconnection network. Therefore, the extra one cycle does not weigh heavily and will not evidently impair the overall performance. Figure 7.15(b) plots the average L2 hits per cycle (i.e., actual accesses to the data array) for the program collection in order to briefly explain the different impacts on the performance caused by the extra cycle. As can be observed, applications such as *Blackscholes*, *Myocyte*, and *NW* have extremely low L2 hits intensity, so their performance is not notably degraded (close to zero loss) with the sequential-access L2 cache. On the contrary, Gaussian result in more frequent L2 hits, thus their execution speed is lowered by a relatively higher percentage (1.5%). Nonetheless, based on the evaluations made on the L2 cache, it is still reasonable for us to conclude that the proposed hybrid device sequential-access design can significantly slow down the NBTI aging on the L2 cache with slight performance overhead.

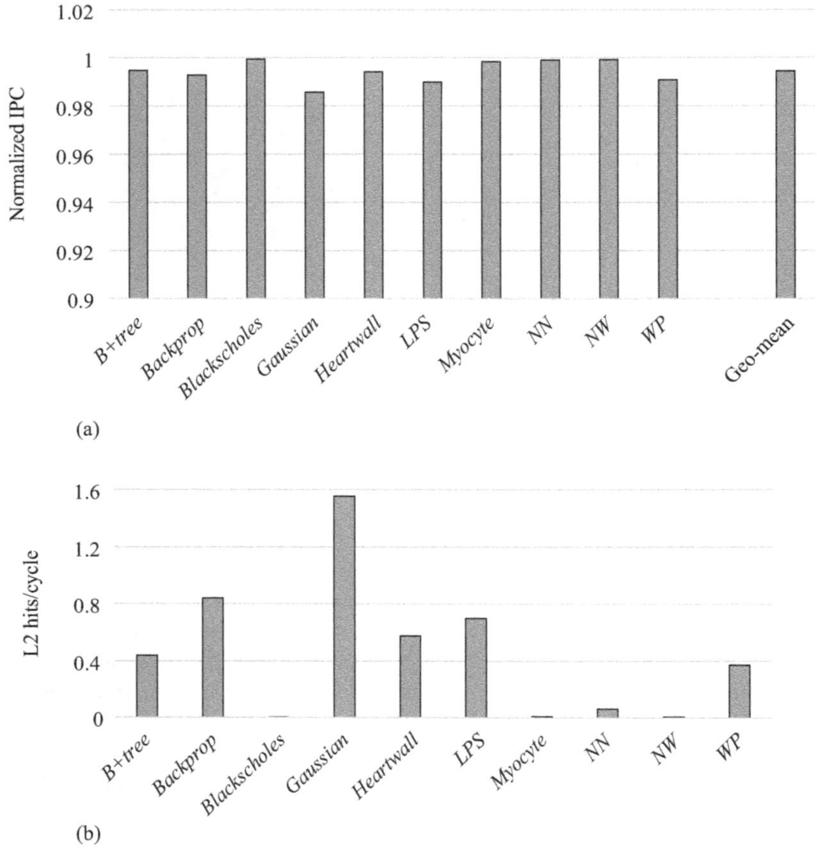

Figure 7.15 *(a) Normalized IPC with the sequential-access L2 cache (b) L2 hits/cycle*

7.7 Related work

7.7.1 NBTI mitigation

NBTI has been recognized as a major reliability concern as the semiconductor industry shifts into the deep submicron era. To mitigate the NBTI degradation and enhance the device's durability, researchers have conducted substantial works in the past years. Ramakrishnan *et al.* [29] introduce an approach to reduce the NBTI wearout in FPGAs by loading the reversing bit patterns in idle periods. Gunadi *et al.* [13] introduce a scheme called Colt to balance the utilization of devices in a processor for reliability improvement. Specifically focusing on the storage components, Shin *et al.* [30] propose to proactively set the PMOS transistors to recovery mode, and moving data around free cache arrays during operation.

Converse to these works which attempt to manipulate the time under stress and recovery, Tiwari *et al.* [14] propose a framework named facelift to combat NBTI degradation by adjusting higher level parameters including operating voltage, threshold voltage and the application scheduling policy. Fu *et al.* [31] concentrate on the NBTI mitigation in presence of process variation. They effectively utilize the interplay between NBTI aging and process variation to prevent early failure of specific structures.

There are few works aiming to alleviate the NBTI aging on GPUs in literature. Rahimi *et al.* [32] focus on the GPUs designed in VLIW fashion and present a technique to slow down the NBTI aging for this particular architecture. By exploring the unbalanced usage among FUs within a VLIW slot, their proposed strategy can uniformly as-sign the stress among all computation units and achieve an even aging rate.

7.7.2 Characterization of FinFET reliability

As FinFET is widely considered as an attractive replacement of planar transistors for the next few technology nodes, studies focusing on the reliability of this new structure are becoming fairly important. Lee *et al.* [33] investigate the NBTI characteristics on SOI and body-tied FinFETs and observe that a narrow fin width leads to more severe degradation than a wider fin width. Crupi *et al.* [34] compare the reliability of triple-gate and planar FETs. The author show that the behavior of time-dependent dielectric breakdown is not changed on the triple-gate architecture under different gate voltages and temperatures. This is also corroborated in the work conducted by Groeseneken *et al.* [8], which further demonstrate that FinFET devices tend to suffer from more severe NBTI degradation. In Reference 9, Wang *et al.* analyze the soft-error resilience of FinFET devices and conclude that FinFET circuit is more reliable than bulk CMOS circuit in terms of soft-error immunity.

7.7.3 Hybrid-device design

Exploiting device-level heterogeneity has been widely used for performance and energy efficiency optimization in computer architecture study. Saripalli *et al.* [35, 36] discuss the feasibility of technology-heterogeneous cores and demonstrate the design of mix-device memory. Wu *et al.* [37] presents the advantage of hybrid-device cache. Kultursay [38] and Swaminathan [39], respectively, introduce a few runtime schemes to improve performance and energy efficiency on CMOS-TFET hybrid CMPs. For the optimization on GPUs, Goswami *et al.* [40] propose to integrate resistive memory into the compute core for reducing the power consumption on GPU register file.

Our work deviates from the aforementioned studies in that we aim to alleviating the NBTI degradation on GPUs made of FinFET from the architectural level. In addition, compared to our prior work [41], this study extends the application of the proposed technique to more structures and thus provides more general guidance to the processor design.

7.8 Conclusion

FinFET technology is recognized as a promising substitute of conventional planar devices for building processors in the next decade due to its better scalability. However, recent experimental studies demonstrate that FinFET tends to suffer from more severe

NBTI degradation compared to the planar counterpart. In this work, we focus on the NBTI reliability issue of a modern GPU made of FinFET and propose to address this problem by exploiting the device heterogeneity. We introduce a set of techniques that merely involve minor modifications to the existing GPU architectures. The proposed techniques leverage planar devices' higher immunity to NBTI and are effective in slowing down the aging rate of the device. Our evaluation results demonstrate that the minor changes to the warp scheduler and the L2 cache can considerably alleviate the degradation due to NBTI with slight performance overhead.

References

[1] M. Bohr and K. Mistry. *Intel's Revolutionary 22 nm Transistor Technology*. Intel Corporation, Santa Clara, CA, May 2011.

[2] A. Asenov, C. Alexander, C. Riddet, and E. Towie. "Predicting future technology performance". In: *Proceedings of the 50th ACM/EDAC/IEEE Design Automation Conference (DAC)*, Austin, TX, June 2013. ACM, New York, NY.

[3] A. B. Kahng. "The ITRS design technology and system drivers roadmap: process and status". In: *Proceedings of the 50th ACM/EDAC/IEEE Design Automation Conference (DAC)*, Austin, TX, June 2013. ACM, New York, NY.

[4] V. B. Kleeberger, H. Graeb, and U. Schlichtmann. "Predicting future product performance: modeling and evaluation of standard cells in FinFET technologies". In: *Proceedings of the 50th ACM/EDAC/IEEE Design Automation Conference (DAC)*, Austin, TX, June 2013. ACM, New York, NY.

[5] Intel Corporation. *3rd Generation of Intel Core i7 Processor*. http://ark.intel.com/products/family/65505.

[6] Intel Corporation. *4th Generation of Intel Core i7 Processor*. http://ark.intel.com/products/family/75023.

[7] http://www.eetimes.com/document.asp?doc_id=1264668.

[8] G. Groeseneken, F. Crupi, A. Shickoya, *et al.* "Reliability issues in MUGFET nanodevices". In: *Proceedings of the 46th IEEE International Reliability Physics Symposium (IRPS)*, Phoenix, AZ, April 2008. IEEE, Piscataway, NJ.

[9] Y. Wang, S. D. Cotofana, and L. Fang. "Statistical reliability analysis of NBTI impact on FinFET SRAMs and mitigation technique using independent-gate devices". In: *Proceedings of the IEEE/ACM International Symposium on Nanoscale Architectures (NANOARCH)*, Amsterdam, The Netherlands, July 2012. IEEE, Piscataway, NJ.

[10] B. A. Anderson, A. J. Joseph, and E. J. Nowak. "Integrated circuit including FinFET RF switch angled relative to planar MOSFET and related design structure". U.S. Patent 8125007 B2, 2012.

[11] J. P. Colinge, "Multiple-gate SOI MOSFETs". *Solid-State Electronics*, vol. 48, no. 6, 2004.

[12] B. B. Doris, D. C. Boyd, M. Leong, T. S. Kanarsky, J. T. Kedzierski, and M. Yang. "Hybrid planar and FinFET CMOS devices". U.S. Patent 7250658 B2, 2007.

[13] E. Gunadi, A. A. Sinkar, N. S. Kim, and M. H. Lipasti. "Combating aging with the colt duty cycle equalizer". In: *Proceedings of 43rd Annual IEEE/ACM International Symposium on Microarchitecture (MICRO)*, Atlanta, GA, Dec. 2010. IEEE, Piscataway, NJ.

[14] A. Tiwari and J. Torrellas. "Facelift: Hiding and slowing down aging in multicores". In: *Proceedings of 41st IEEE/ACM International Symposium on Microarchitecture (MICRO)*, Lake Como, Italy, Nov. 2008. IEEE, Piscataway, NJ.

[15] T. Sakurai and R. Newton. "Alpha-power law MOSFET model and its applications to CMOS inverter delay and other formulas". In: *IEEE Journal of Solid-State Circuits*, vol. 25, no. 2, pp. 584–594, Apr. 1990.

[16] Predictive Technology Model. http://ptm.asu.edu

[17] F. Wang, Y. Xie, K. Bernstein, and Y. Luo. "Dependability analysis of nano-scale FinFET circuits". In: *Proceedings of the 2006 IEEE Computer Society Annual Symposium on Emerging VLSI Technologies and Architectures*. Karlsruhe, Germany, Mar. 2006. IEEE, Piscataway, NJ.

[18] J. Hennessy and D. A. Patterson. *Computer Architecture: A Quantitative Approach*. 5th edn. Morgan Kaufmann, Burlington, MA.

[19] H. Kim, R. Vuduc, S. Baghsorkhi, J. Choi, and W. Hu. "Performance analysis and tuning for general purpose graphics processing units (GPGPU)". DOI: 10.2200/S00451ED1V01Y201209CAC020

[20] Z. Chishti, M. D. Powell, and T. N. Vijaykumar. "Distance associativity for high performance energy efficient non-uniform cache architectures". In: *Proceedings of International Symposium on Microarchitecture (MICRO)*, 2003.

[21] A. Bakhoda, G. Yuan, W. Fung, H. Wong, and T. Aamodt. "Analyzing CUDA workloads using a detailed GPU simulator". In: *Proceedings of IEEE International Symposium on Performance Analysis of Systems and Software (ISPASS)*, Boston, MA, Apr. 2009. IEEE, Piscataway, NJ.

[22] J. Leng, T. Hetherington, A. Eltantawy, *et al.* "GPUWattch: enabling energy optimizations in GPGPUs". In: *Proceedings of the 40th ACM/IEEE International Symposium on Computer Architecture (ISCA)*, Tel Aviv, Israel, June 2013. IEEE, Piscataway, NJ.

[23] Hotspot 5.0 Temperature modeling tool.

[24] R. Nath, R. Ayoub, and T. S. Rosing. "Temperature aware thread block scheduling in GPGPUs". In: *Proceedings of the 50th ACM/EDAC/IEEE Design Automation Conference (DAC)*, Austin, TX, June 2013. ACM, New York, NY.

[25] GTX 480 Specifications. http://www.geforce.com/hardware/desktop-gpus/geforce-gtx-480/specifications

[26] Nvidia Corporation. CUDA Computing SDK 4.2.

[27] S. Che, M. Boyer, J. Meng, *et al.* "Prdinia: a benchmark suite for heterogeneous computing". In: *Proceedings of the IEEE International Symposium on Workload Characterization (IISWC)*, Austin, TX, Oct. 2009. IEEE, Piscataway, NJ.

[28] S. Chaudhuri and N. K. Jha. "3D vs. 2D analysis of FinFET logic gates under process variation". In: *Proceedings of the 29th IEEE International Conference on Computer Design (ICCD)*, Amherst, MA, Oct. 2011. IEEE, Piscataway, NJ.

[29] K. Ramakrishnan, S. Suresh, N. Vijaykrishnan, M. J. Irwin, and D. Degalahal. "Impact of NBTI on FPGAs". In: *Proceedings of 20th International Conference on VLSI Design*, Bangalore, India, Jan. 2007. IEEE, Piscataway, NJ.

[30] J. Shin, V. Zyuban, P. Bose, and T. M. Pinkston. "A proactive wearout recovery approach for exploiting microarchitectural redundancy to extend cache SRAM lifetime". In: *Proceedings of the 35th ACM/IEEE International*

Symposium on Computer Architecture (ISCA), Beijing, China. Jun. 2008. IEEE, Piscataway, NJ.

[31] X. Fu, T. Li, and J. Fortes. "NBTI tolerant microarchitecture design in the presence of process variations". In: *Proceedings of 41st IEEE/ACM International Symposium on Microarchitecture (MICRO)*, Lake Como, Italy, Nov. 2008. IEEE, Piscataway, NJ.

[32] A. Rahimi, L. Benini, and R. K. Gupta. "Aging-aware compiler-directed VLIW assignment for GPGPU architectures". In: *Proceedings of the 50th ACM/EDAC/IEEE Design Automation Conference (DAC)*, Austin, TX, Jun. 2013. ACM, New York, NY.

[33] H. Lee, C.-H. Lee, D. Park, and Y.-K. Choi. "A study of negative-bias temperature instability of SOI and body-tied FinFETs". *IEEE Electron Device Letters*, vol. 26, no. 5, pp. 326–328, May 2005.

[34] F. Crupi, B. Kaczer, R. Degraeve, *et al.* "Reliability comparison of triple-gate versus planar SOI FETs". *IEEE Transactions on Electron Devices*, vol. 53, no. 9, 2006.

[35] V. Saripalli, G. Sun, A. Mishra, Y. Xie, S. Datta, and V. Narayanan. "Exploiting heterogeneity for energy efficiency in chip multiprocessors". In: *IEEE Transactions on Emerging and Selected topics in Circuits and Systems*, 2011.

[36] V. Saripalli, A. K. Mishra, Y. Xie, S. Datta, and V. Narayanan. "An energy-efficient heterogeneous CMP based on hybrid TFET-CMOS cores". In: *Proceedings of the 48th ACM/EDAC/IEEE Design Automation Conference (DAC)*, San Diego, CA, Jun. 2011. ACM, New York, NY, USA.

[37] X. Wu, J. Li, L. Zhang, E. Speight, R. Rajamony, and Y. Xie. "Hybrid cache architecture with disparate memory technologies". In: *Proceedings of the 36th ACM/IEEE International Symposium on Computer Architecture (ISCA)*, Austin, TX, Jun. 2009. IEEE, Piscataway, NJ.

[38] E. Kultursay, K. Swaminathan, V. Saripalli, V. Narayanan, M. Kandemir, and S. Datta. "Performance enhancement under power constraints using heterogeneous CMOS-TFET multi-cores". In: *Proceedings of the 8th IEEE/ACM/IFIP International Conference on Hardware/software Codesign and System Synthesis (CODES+ISSS)*, Tampere, Finland, Oct. 2012. ACM, New York, NY, USA.

[39] K. Swaminathan, E. Kultursay, V. Saripalli, V. Narayanan, M. Kandemir, and S. Datta. "Improving energy efficiency of multi-threaded applications using heterogeneous CMOS-TFET multi-cores". In: *Proceedings of the 17th IEEE/ACM International Symposium on Low-power Electronics and Design (ISLPED)*, Fukuoka, Japan, Aug. 2011. ACM, New York, NY, USA.

[40] N. Goswami, B. Cao, and T. Li. "Power-performance co-optimization of throughput core architecture using resistive memory". In: *Proceedings of 19th IEEE International Symposium on High Performance Computer Architecture (HPCA)*, 2013.

[41] Y. Zhang, S. Chen, L. Peng, and S.-M. Chen. "Mitigating NBTI degradation on FinFET GPUs through exploiting device heterogeneity". In: *Proceedings of IEEE Computer Society Annual Symposium on VLSI (ISVLSI)*, Tampa, FL, 2014.

Chapter 8

Multiple-independent-gate nanowire transistors: from technology to advanced SoC design

Pierre-Emmanuel Gaillardon[1], *Jian Zhang*[2], *Luca Amaru*[2], *and Giovanni De Micheli*[2]

The focus of this chapter is a novel class of devices such as *Multiple-Independent-Gate Field-Effect Transistor* (MIGFET) which can provide better functionality and flexibility as compared to classic *Metal-Oxide-Semiconductor Field-Effect Transistors* (MOSFETs). Specific emphasis is given to *Three-Independent-Gate Field-Effect Transistor* (TIGFET).

8.1 Introduction

Since the introduction of the MOSFET, the semiconductor industry has been able to reduce the dimensions of the transistors at a regular pace, as captured Moore's Law [1]. Devices with feature size below 20 nm have been fabricated using new materials and processing steps. A recent important innovation is the introduction of non-planar geometries with fins and tri-gate structures [2, 3]. Following this trend, *Silicon NanoWires* (SiNW) coupled to *Gate-All-Around* (GAA) structures appear as a promising solution to further reduce the dimensions of the transistors [4] and to improve electrostatic control.

Device downscaling has correlated with increased performances of *Integrated Circuits* (ICs) and systems [1], as well as increased functionality per unit area. As downscaling becomes increasingly more expensive in terms of fabrication facility costs, we propose an alternative path to Moore's Law where, instead of scaling the dimensions of the transistors further, we focus on increasing their functionality for a given area.

In this chapter, we introduce MIGFETs, a novel class of devices that demonstrates enhanced functionality and flexibility as compared to standard MOSFETs. MIGFETs are GAA *Schottky-Barrier* (SB) NWFETs with multiple gate regions that control independently the properties of the semiconductor channel [5, 6, 7, 8, 9, 10, 12, 13]. For practical reasons, we focus on a device with three independent gate

[1]The University of Utah, Salt Lake City, UT, USA
[2]EPFL, Lausanne, Switzerland

regions called TIGFET [27]. In addition to a conventional gate, the device has two gated regions over the SB. Independently controlling the SB at the source and drain of the device allows us to modulate the barrier heights and therefore to select the carrier type to inject into the device. The modulation of the SB enables a dynamic control of both the polarity and the threshold of the device at runtime. Having a device with different modes of conduction that can be changed at runtime leads to many circuit-level opportunities. In particular, a dynamic control of the device polarity enables the realization of compact binate operators, such as a four-transistor *Exclusive-OR* (XOR) operator [28, 11]. Similarly, it is possible to create compact unate combinational logic gates, such as NAND [26] or *MAJority* (MAJ) [29] gates, as well as compact memory primitives, such as a compact *True-Single Phase Clock* (TSPC) flip-flop [36] or four-transistor *Static Random Access Memory* (SRAM) [25]. Moreover, this device can be used fruitfully within low-power circuit design. Its multiple-V_T control allows designers to realize high-V_T (HVT) or low-V_T (LVT) logic gates with a unique type of transistor [26], while power-gating architectures can be realized with no additional sleep transistors [35]. In order to demonstrate the capabilities of the presented technology, we consider an application case study, where different design approaches are used to implement a telecommunication circuit. The realization of a Polar code decoder with a 22-nm TIGFET technology leads to 20% faster and 32% more energy-efficient system compared to its FinFET counterpart, at a moderate area overhead of 15%.

The remainder of the paper is organized as follows. In Section 8.2, we present our TIG-SiNWFET technology, and we discuss the physics behind it. In Section 8.3, we present the circuit design opportunities with compact arithmetic logic gate implementations, low-power design methodologies, and compact memory circuits. We also showcase the interest of all these techniques on the design of a System-on-Chip (SoC), targeting a contemporary telecommunication application. Finally, we conclude the chapter in Section 8.4.

8.2 Multiple-independent-gate field-effect transistors

MIG devices are transistors whose electrostatic properties are dynamically controlled via additional gate terminals. MIG devices have been successfully fabricated using carbon nanotube [6], graphene [7], and SiNW [10, 12] technologies. As the natural evolution of the FinFET structure, vertically stacked SiNWs are a promising platform for MIG controllable-polarity devices thanks to their high I_{on}/I_{off} ratio and CMOS compatible fabrication process [10]. Among this family, we emphasize on *three-independent-gate* (TIG) [27] SiNWFETs.

8.2.1 TIG device overview and operation

The conceptual sketch of the TIG SiNWFET is shown in Figure 8.1. Four vertically stacked nanowires are confined within the source and drain pillars. They are surrounded by TIG structures, named *Polarity Gate* at *S*ource (PG_S), *Control Gate*

Figure 8.1 Conceptual sketch of a TIG SiNWFET

(CG), and *P*olarity *G*ate at *D*rain (PG_D). Metallic source and drain form Schottky junctions with the silicon channel.

Note that DIGFETs [10] are fabricated by connecting both PG_S and PG_D.

With the TIG structure, the operation of the device is as follows. PG_S and PG_D independently modulate the height of the corresponding SB. The desired type of carriers is selected to tunnel into the channel through the thin SB, and the other type of carriers is blocked by the thick SBs. The electrostatic polarization is thereby achieved. CG induces a potential barrier in the inner region of the channel to control the selected carriers flow through the channel.

Eight operation states of this device are obtained by independently biasing the three gates to either GND ("0") or V_{DD} ("1"). We can identify two ON states, two LVT OFF states, two HVT OFF states, and two uncertain states which will not be used. Figure 8.2 illustrates the six most important operation modes and their corresponding band diagrams when $V_{DS} = V_{DD}$ (i.e., S = "0" and D = "1").

(1) ***ON states:*** When $PG_S = PG_D = CG$ (States (1)(4) in Figure 8.2), one of the SBs is thin enough to allow hole tunneling from drain (p-type) or electron tunneling from source (n-type), and there is no barrier in the channel. Thus, majority carriers flow through the device easily.

(2) ***LVT OFF states:*** The opposite biasing of control gate and polarity gates blocks the current flow in the channel (States (2)(5) in Figure 8.2). Nevertheless, small number of carriers can still tunnel through the thin barrier into the channel. This mode is similar to DIG SiNWFET [10].

(3) ***HVT OFF states:*** When $PG_S = S$ and $PG_D = D$ indicated in States (3)(6) in Figure 8.2, thick barriers prevent carriers from tunneling at both source and drain and ensure minimum leakage in the device. This mode corresponds to the two-gate SiNWFET [12].

(a)

	PG_S	CG	PG_D	State
(1)	"0"	"0"	"0"	ON (p-type)
(2)	"0"	"1"	"0"	OFF (LVT)
(3)	"0"	"0"	"1"	OFF (HVT)
(4)	"1"	"1"	"1"	ON (n-type)
(5)	"1"	"0"	"1"	OFF (LVT)
(6)	"0"	"1"	"1"	OFF (HVT)

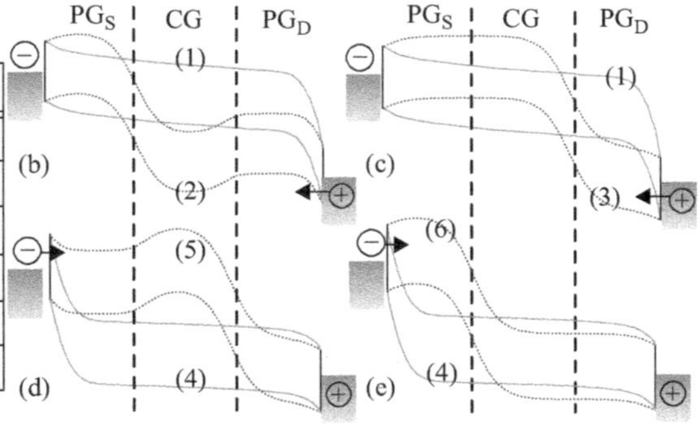

Figure 8.2 (a) Operation states of TIG SiNWFET under different bias conditions. Note that, two other possible configurations do not yield useful operation states. (b)–(e) Band diagrams of corresponding configurations: (b) LVT p-FET, (c) HVT p-FET, (d) LVT n-FET, and (e) HVT n-FET. Numbers represent the corresponding states in (a)

(4) ***Uncertain states:*** When PG_S = "1" and PG_D = "0", barriers are thin enough for tunneling. However, this condition may also create an unexpected barrier in the inner region that will block the current flow and cause signal degradation. Hence, the uncertain states should be prohibited by always fixing PG_D = "1" (PG_S = "0") for n-FET (p-FET), or using $PG_D = PG_S$.

By combining ON and OFF states, the TIG SiNWFET can be configured as LVT p-FET, HVT p-FET, LVT n-FET, or HVT n-FET under different bias conditions (Figure 8.2b–e). As observed in the configurations, the ON states in LVT and HVT configurations are exactly the same. Therefore, the *on*-state currents of both HVT and LVT modes are exactly the same value, regardless of the supply voltage. Although a lower V_T, i.e., earlier turn-*on*, is helpful for improving the circuit speed, the HVT mode with the same *on*-state current will not significantly degrade the circuit performance. This property provides a finer tradeoff between performance and standby power consumption, which is not achievable in conventional multi-V_T techniques, and represents one of the key advantages of our approach.

8.2.2 Device fabrication and electrical characterization

In order to experimentally demonstrate the dual-V_T operation, the TIG SiNWFET is fabricated on a lightly p-type doped ($\sim 10^{15}$ cm^{-3}) *Silicon-On-Insulator* (SOI) substrate with a 340-nm thick silicon device layer. First, the nanowires are defined using electron-beam lithography. The length and the diameter of the defined nanowires are 350 nm and 50 nm, respectively. Then, four vertically stacked nanowires are formed in a top-down fashion, using a single *Deep Reactive Ion Etching* (DRIE) process step [10, 14] (Figure 8.3a). The typical vertical spacing between the nanowires is 40 nm.

Fifteen nanometer SiO_2 is formed on the vertically stacked nanowires as gate dielectric. Following a conformal deposition of polycrystalline silicon, two GAA

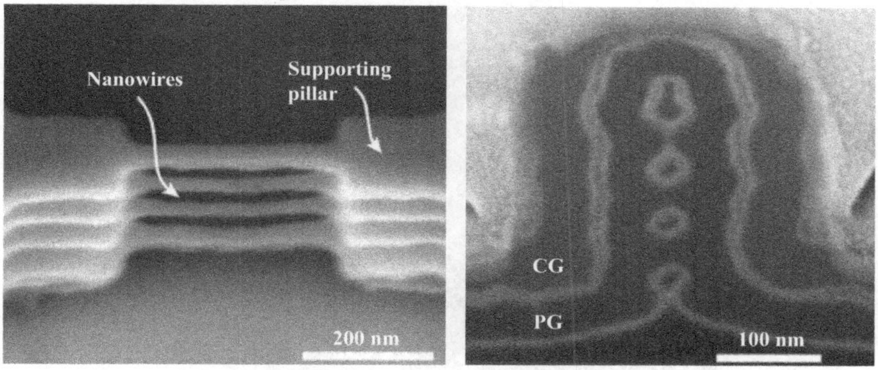

(a) (b)

Figure 8.3 (a) SEM image of vertically stacked nanowires fabricated using a single DRIE process step. (b) Cross-sectional SEM image of the SiNWFET at the region where CG and PG are overlapped

structures with a length of 120 nm are patterned as PG_S and PG_D. Then, a second 15-nm gate oxidation is performed, and a polycrystalline silicon CG is patterned in a self-aligned way. The diameter of the resulting nanowires is around 30 nm considering the silicon consumed during oxidation. Figure 8.3b shows the cross-section of the SiNW stack and the gates. Within an academic clean room facility, the thick gate oxide used in the fabrication reduces the risk of gate leakage, thereby maximizing the fabrication yield. Nevertheless, no physical constraints limit gate oxide scaling in this device with state-of-the-art high-κ dielectric stacks directly implementable in the fabrication process.

Silicon nitride spacers are used to isolate the structures. After necessary cleaning steps, a 20-nm nickel layer is subsequently deposited with sputtering to perform an in situ cleaning and keep the layer uniformity. Then the deposited nickel is annealed to form nickel silicide on source/drain pillars and polycrystalline silicon gates. The silicide creates Schottky junctions with the silicon channel and also reduces the resistance of the gate contacts. By controlling the annealing temperature and duration, the preferred Ni_1Si_1 phase is selected to utilize its near mid-gap workfunction ($\sim$4.8 eV) and low resistivity [15, 16]. Figure 8.4 shows the Scanning Electron Microscopy (SEM) image of the final device. Note that the device may be more aggressively scaled with the GAA channel geometry, which is best suited for strong suppression of short channel effects. In addition, the absence of abrupt doping profiles in the channel relaxes constraints on doping levels down to the current nanoscale technology nodes (22 nm and beyond).

According to the working principle introduced in Section 8.2.1, the transfer characteristics of the fabricated device are measured and shown in Figure 8.5. Both n-type and p-type behaviors with different threshold voltages (LVT and HVT) are observed in the same device.

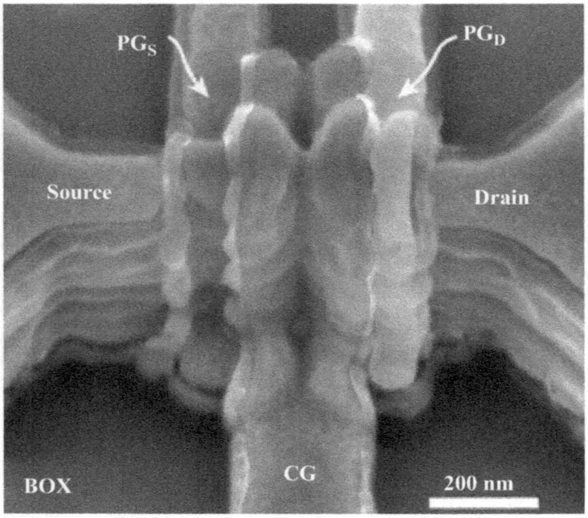

Figure 8.4 SEM image of the fabricated TIG SiNWFET

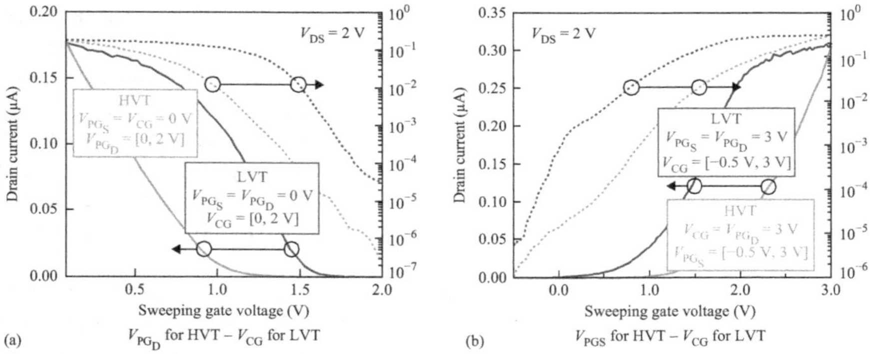

Figure 8.5 Measured characteristics of a TIG SiNWFET. (a) LVT and HVT p-type transfer characteristics and (b) LVT and HVT n-type transfer characteristic in the same device

In all demonstrated characteristics in Figure 8.5, the applied voltages at source and drain are set to 0 V and 2 V, respectively. When V_{PG_S} and V_{PG_D} are set to 0 V, the TIG SiNWFET is configured as LVT p-FET (LVT curves in Figure 8.5a). It is observed in Figure 8.2b that electrons are blocked at source, while holes can tunnel from drain into the channel through the thin SB induced by band bending. The CG modulates the barrier in the channel, thereby turning the device *on* or *off* as in conventional MOSFETs [17, 18, 19]. When V_{CG} is below 0.5 V, the current gradually saturates.

In contrast, HVT p-FET configuration (HVT curves in Figure 8.5a) is obtained when V_{PG_S} and V_{CG} are set to 0 V to block the electrons tunneling from source.

PG_D controls the current flow by modulating the hole tunneling through the Schottky junction at drain. The *on*-state currents of both HVT and LVT modes are exactly the same. The reason is already discussed in Section 8.2.1 as that the bias conditions at *on* states are identical in LVT and HVT configurations. Moreover, the *off*-state current suppression in HVT configuration is more effective than in the LVT configuration, since the opposite band bending at the Schottky contacts prevents both electron and hole injection into the channel (curve (3) in Figure 8.2c) and also ensures the whole channel to be unpopulated [12]. Therefore, the *off*-state current is reduced by two orders of magnitude as compared to LVT configuration and reaches a leakage floor of 10.5 pA/μm (315 fA) normalized to the nanowire diameter.

Similarly, LVT n-FET configuration (LVT curves in Figure 8.5b) is reached by applying 3 V to V_{PG_S} and V_{PG_D}. HVT n-FET configuration (HVT curves in Figure 8.5b) is reached for V_{PG_D} and V_{CG} fixed to 3 V. In the same principle as p-FET configurations, the *on*-state currents of LVT and HVT n-FET configurations are the same since they share the same *on* state (State (4) in Figure 8.2a), and the reduction of leakage current is also observed in HVT configurations.

To summarize the performance of the fabricated device, the *on*-state currents of p-FET and n-FET configurations are 177 nA (5.9 μA/μm) and 310 nA (10.3 μA/μm), respectively, which are comparable to recent works on polarity-controllable devices [10, 13]. Extracted at 1 nA drain current [20], the threshold differences in p-FET configurations and in n-FET configurations are 0.48 V and 0.86 V, respectively. The *off*-state currents of HVT p-FET and n-FET configurations reach 315 fA (10.5 pA/μm) and 1 pA (33.3 pA/μm) compared to 30 pA (1 nA/μm) and 4.6 pA (153.3 pA/μm), respectively, in LVT configurations. Thus, the total I_{ON}/I_{OFF} ratio for the HVT p-FET and n-FET configurations are 6×10^5 and 3×10^5, respectively. On the other hand, LVT configurations demonstrate better subthreshold slopes of 155 mV/dec (p-FET) and 217 mV/dec (n-FET).

To further improve the performance of the device, the fabrication process needs optimization toward better electrostatic control. High-κ gate dielectric and metal gates, together with channel strain techniques, can be directly applied to the presented structure. Moreover, the proposed device concept may be applied to other materials (e.g., carbon nanotube, graphene, and MoS_2 [6, 7, 8]), giving the opportunities for continuous scaling down.

8.2.3 Physical understanding

To better understand the physics involved in the proposed device, we simulate a single TIG SiNWFET with Sentaurus Device [24]. The simulation employs drift-diffusion transport in the silicon channel, while thermionic emission and quantum mechanical tunneling with Wentzel–Kramers–Brillouin approximation are used at the junctions. The dimensions of the simulated SiNWFET are the same as the fabricated one except an optimized 2-nm gate oxide and a fine-adjusted SB height (0.35 eV for electrons and 0.75 eV for holes).

The simulated characteristics are illustrated in Figure 8.6. The device with an optimized gate oxide demonstrates the performance at levels of regular advanced

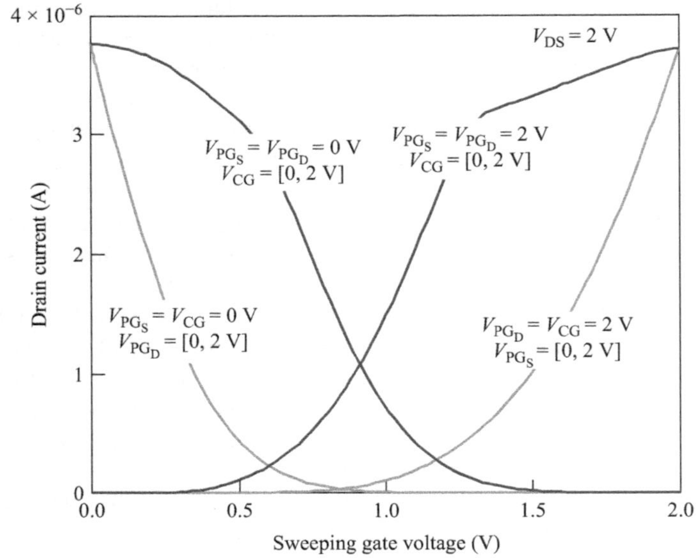

Figure 8.6 Simulation dual-V_T characteristics of an optimized TIG SiNWFET

MOSFETs. Fine-adjusted SB height also gives symmetric n-type and p-type characteristics, which is important to achieve energy-efficient circuits with balanced delay [13]. The simulated device reproduces all the key properties in the measured characteristics, including the shared *on*-state current in dual-V_T configurations, the suppressed leakage current in HVT configurations, as well as the current saturation in LVT configurations.

In the following part, we will discuss the reason of this dual-V_T characteristic and also show the effect of the device geometry and physical parameters on the V_T difference with the help of Technology Computer-Aided-Design (TCAD) simulation. We choose the n-FET configurations for example, but the analysis is also applicable for p-FET configurations.

In n-FET configurations, band bending induced by a positive voltage on PG_S reduces the thickness of the source SB and enhances the tunneling of electrons through the source barrier. This leads to a reduction of the effective barrier height [17]. In LVT configuration (Figure 8.2d), the effective SBs at the source are fully suppressed by sufficiently positive voltage configured on PG_S. V_{CG} is swept to tune the conduction of the device. Therefore, the current transport is dominated by thermionic emission of electrons over a potential barrier induced by CG [21], i.e.,

$$I_D = AA^*T^2 \exp\left(-\frac{q\phi_B}{k_B T}\right) \tag{8.1}$$

where A is the junction area, A^* is the effective Richardson constant, T is the temperature, q is the elementary charge, k_B is the Boltzmann constant, and ϕ_B is the effective barrier height. This barrier height is determined by the electrostatic potential in the CG-controlled region. If we assume that there is no induced charge in the channel under subthreshold operation, the applied voltage on CG directly translates into a reduction of ϕ_B. Therefore,

$$\Delta\phi_B = -\Delta V_{CG} \tag{8.2}$$

In contrast, in HVT configuration (Figure 8.2e), a sufficiently positive voltage is applied to CG and PG_D to make sure there is no barrier inside the silicon channel. The current is therefore determined by an effective SB height at source. A positive voltage on PG_S reduces the thickness of the SB at source, and the consequent enhancement of tunneling by V_{PG_S} leads to a reduction of ϕ_B. Thus,

$$\Delta\phi_B = -\lambda\Delta V_{PG_S} \tag{8.3}$$

where the coefficient λ represents the dependence of the effective barrier height on V_{PG_S}. Since the tunneling probability is smaller than 1, λ is also smaller than 1 [17, 18, 19].

Therefore, a higher V_T is required in this configuration to turn on the device due to the lower efficiency of tuning the effective barrier height by PG_S than CG.

According to the analysis, we simulated a series of devices with different parameters related to this efficiency, including the oxide thickness (T_{ox}), the radius of nanowire (R_{nw}), the tunneling effective mass (m_h^*) and the SB height (SBH_h). The V_T difference of p-FET configurations of various devices is plotted in Figure 8.7. The reduction of T_{ox} and R_{nw} enhances the electrostatic control of the gate, thus resulting in a thinner SB at a given gate voltage in HVT configurations. Tunneling current is thereby improved and reduces the effective barrier height further [17], implying a larger λ and a decreased V_T difference. A smaller effective mass also results in a larger tunneling probability and the V_T difference is consequently reduced. The V_T difference is also proportional to the SB height. Other than the previous three parameters, a reduction of SB height for holes leads to an increase of SB height for electrons. Thus, tuning of SB height can achieve a trade-off of V_T differences between n-type configurations and p-type configurations.

Regarding the current saturation in LVT n-FET configuration (e.g., V_{CG} from 2.0 V to 3.0 V in Figure 8.5b), first, electrons are induced in the channel with V_{CG} above V_T, and the electrostatic potential in the channel gradually saturates [22, 23]. More importantly, when the bent conduction band edge is lower than the Fermi level in the source, the current starts to be dominated by the source injection and cannot be further modulated by CG. Therefore, the current saturates at a large positive V_{CG} due to the "source exhaustion" [40].

8.2.4 Performance predictions

Device evolution allows the semiconductor industry to keep the pace towards more performances, less short-channel effects, and ultimately reduced leakage floor. In

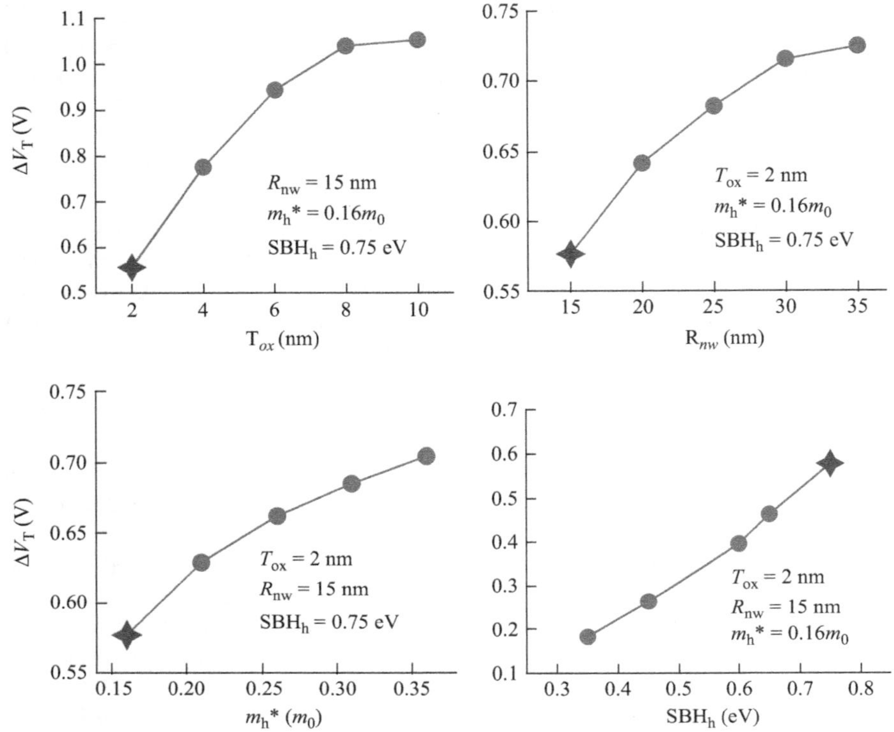

Figure 8.7 The V_T difference of p-FET configurations in simulated devices. m_0 is the free electron mass. Stars represent the simulated device in Figure 8.6

order to assess the performances/power capabilities of the presented MIGFETs with regards to current mainstream technologies, Figure 8.8 shows the frequency of operation along with the power consumption of an 11-stage ring oscillator realized with 28-nm bulk, 28-nm *Fully-Depleted Silicon-On-Insulator* (FDSOI) [39], 22-nm FinFETs [3], and 22-nm TIG nanowire FETs [27] in both HVT and LVT options. Exploiting an improved electrostatic control over the channel structure, these devices are capable of either increasing the performance level as compared to bulk technologies, while keeping under control the power consumption, or drastically reducing the power budget. Device-level optimizations, i.e., HVT or LVT options, can be done for the different technologies discussed above, in order to trade-off between speed and power consumption. In addition to technology boosters, we recently observed a growing interest for devices, whose properties can be fine tuned through an external electrostatic control, such as UTBB FDSOI or MIGFETs. In the evaluation of Figure 8.8, this extra control flexibility is not leveraged. In the next section, we will see how these tuning knobs can be exploited at the circuit level.

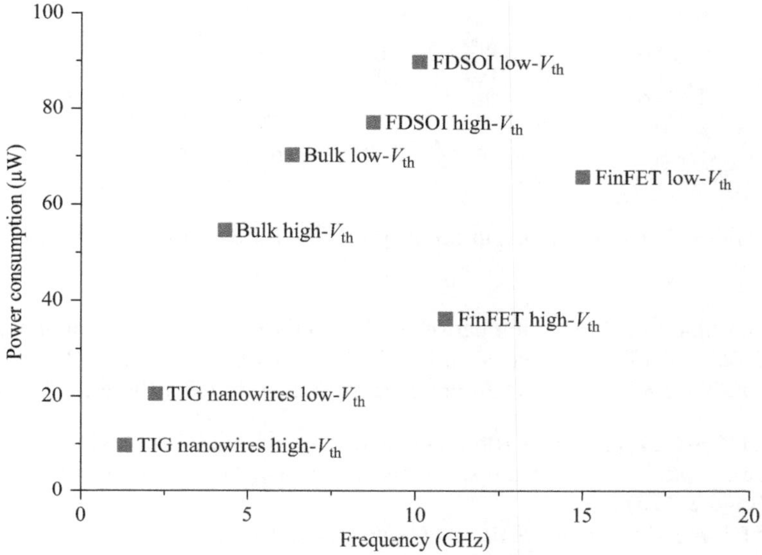

Figure 8.8 Simulated power consumption/frequency of an 11-stage ring oscillator implemented with 28-nm bulk, 28-nm FDSOI, 22-nm FinFET, and 22-nm TIG NWFETs in HVT and LVT options

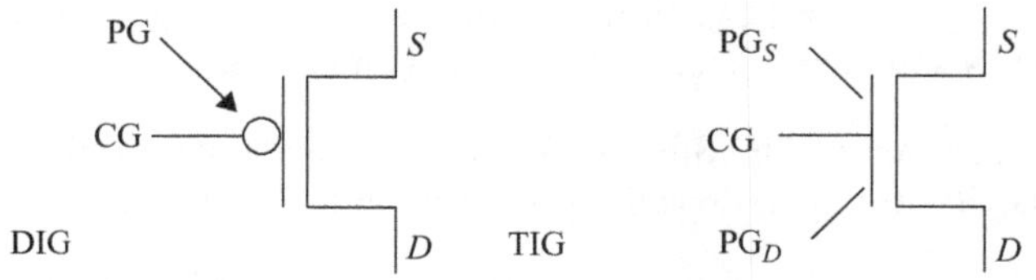

Figure 8.9 Circuit-level symbols of DIG and TIG transistors

8.3 Circuit design opportunities

Recently, design with MIG devices has been widely investigated. In this section, we review their design opportunities compared to CMOS.

8.3.1 Generalities

In this section, the devices are represented by their circuit-level symbols, shown in Figure 8.9. The TIG symbol consists of five terminals, source/drain contacts, and three-gate regions, while the DIG symbol has only four terminals. PG_S and PG_D are tied together in DIG devices and result in the PG terminal.

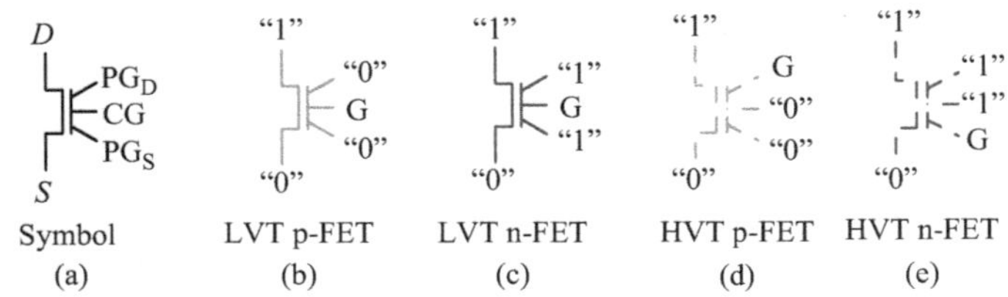

Figure 8.10 Bias configurations of TIG SiNWFET with a single input

According to the transition between ON and OFF states, four configurations of TIG SiNWFET are also depicted in Figure 8.10, including LVT n-FET/p-FET and HVT n-FET/p-FET. The uncertain states are naturally avoided in these configurations.

(1) ***LVT p-FET*** (Figure 8.10b): PG_S and PG_D are biased to GND. The voltage sweep on CG makes a transition between p-type ON and standard OFF states (Figure 8.2b).
(2) ***LVT n-FET*** (Figure 8.10c): PG_S and PG_D are biased to V_{DD}. The voltage sweep on CG makes a transition between n-type ON and standard OFF states (Figure 8.2d).
(3) ***HVT p-FET*** (Figure 8.10d): GND is applied to CG and PG_S, and a voltage sweep is applied on PG_D. In this configuration, the device switches between p-type ON and low-leakage OFF states (Figure 8.2c).
(4) ***HVT n-FET*** (Figure 8.10e): V_{DD} is applied to CG and PG_D, and a voltage sweep is applied on PG_S. In this configuration, the device switches between n-type ON and low-leakage OFF states (Figure 8.2e).

The configuration of TIG SiNWFET can be further extended for two inputs by combining the bias configurations for a single input in Figure 8.4.

(1) ***2-series n-FETs*** (Figure 8.11a): By combining the LVT and HVT n-FET configurations, two inputs on CG and PG_S implement the function of 2-series n-FETs.
(2) ***2-series p-FETs*** (Figure 8.11b): Similarly, the configuration of 2-series p-FETs is obtained by combining the LVT and HVT p-FET configurations.
(3) ***DG configuration*** (Figure 8.11c): The TIG SiNWFET work in DIG mode. In this configuration, the device is ON when G1 = G2. Thus, this configuration is efficient for implementation of XOR function [28].

Even though a specified gate is used for polarization in TIG SiNWFETs, these two-input configurations efficiently utilize the extra gates without source/drain region between two inputs, thereby mitigating the area overhead compared to conventional CMOS devices. In addition, the internal node capacitance between two inputs does not exist in TIG SiNWFETs. This helps to reduce the delay of circuits.

Finally, note that TIG devices have symmetric performances for both n-type and p-type polarities. As a result, the transistor sizing in circuit design is simplified, as

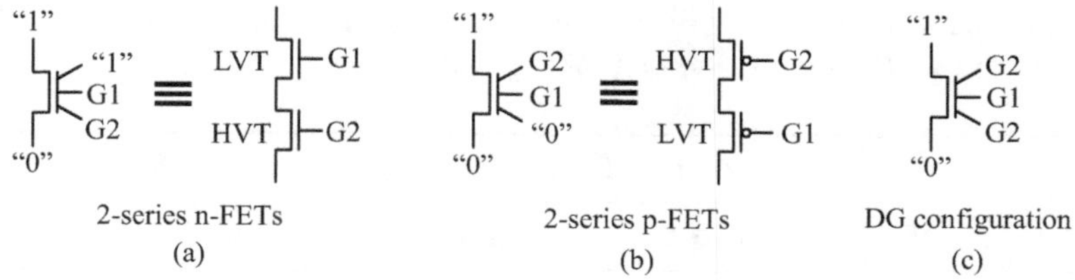

Figure 8.11 Bias configurations of TIG SiNWFET for two inputs

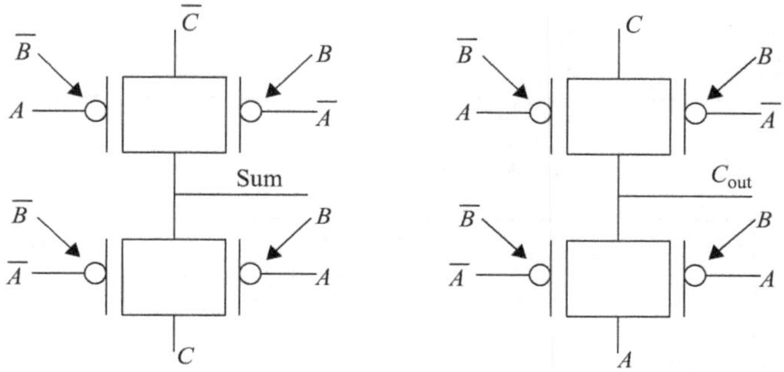

Figure 8.12 Full-adder with eight controllable-polarity devices [34]

minimum size transistors can be used in both pull-up and pull-down branches. In the following, dimensions of the transistors are assumed to be minimal.

8.3.2 Compact data path design

Arithmetic logic is critical in most of today's ICs. Indeed, arithmetic operations are the basis of data paths that form the reasoning core of logic applications in silicon. XOR and *MAJority* (MAJ) logic functions are extensively used in arithmetic circuits; consequently their physical realization is of paramount importance. In this context, MIG transistors open up new opportunities to implement XOR- and MAJ-based logic gates with few resources [28, 30, 31]. Based on transmission-gates, the implementation of three-input XOR and three-input MAJ gates, depicted in Figure 8.12, enables a full-adder realization with only eight devices (input inverters apart) and only one transistor per stack. The full-adder forms a fundamental building block for many arithmetic circuits.

Such advantageous full-adder design extends to a whole range of arithmetic primitives, as reported in Table 8.1. The area is normalized to the number of transistors employed to realize the function, whereas the delay is normalized to the delay of a two-input XOR gate.

Compared to an equivalent transmission-gate FinFET implementation, the proposed full-adder shows 22% improvement in area and 40% improvement in normalized delay. When used in arithmetic compressors, the proposed full-adders also

Table 8.1 Arithmetic cell library – (Area normalized to the transistor count – Delay normalized to two-input XOR)

Gate	MIGFET logic		CMOS logic	
	Area	Delay	Area	Delay
Two-input XOR	8	1	12	1
Three-input XOR	10	1	24	2
Half-adder	12	1	16	1
Full-adder	14	1	18	2
4-2 Compressor	26	2	42	4
5-3 Compressor	38	3	64	4

Table 8.2 Arithmetic benchmark circuits

Benchmarks	MIGFET logic		CMOS logic	
	Area	Delay	Area	Delay
5-bit ripple-carry adder	68	5.76	88	9
16-bit carry-select adder	392	6.76	504	10
(29, 3)-compressor	408	12	680	16
(11, 2)-reduction tree (16 × 16 MAC)	124	6	226	10
Wallace tree (54 × 54 multiplier)	338	8	546	16
(31, 5) Parallel counter	364	8.33	468	14

enable area and delay savings. By employing the arithmetic elements in Table 8.1, we report on various study various industry standard benchmark circuits comprising of adders, multipliers, compressors, and counters as listed in Table 8.2. Note that, for the transmission-gate style, we included additional buffering every four-stacked devices. From Table 8.2, we observe that the use of MIG devices consistently fares well when compared to the conventional CMOS logic in both area (32% on average) and delay (38% on average).

The compact implementations of XOR and MAJ functions with controllable-polarity transistors bear a promise for superior automated design of data paths. However, conventional logic synthesis tools are not adequate to fully harness the advantages led by the controllable-polarity feature in arithmetic logic, missing some optimization opportunities. To overcome these limitations, it is required to develop new approaches that better integrate the efficient primitives of controllable-polarity FETs. On the one hand, it is possible to propose innovations in the data representation form. For instance, *Biconditional Binary Decision Diagrams* (BBDDs) [29, 32] are a canonical logic representation form based on the biconditional (XOR) expansion. They provide a one-to-one correspondence between the functionality of a controllable-polarity transistors and its core expansion, thereby enabling an efficient mapping of the devices onto BBDD structures. On the other hand, it is also possible to identify the logic primitives efficiently realized by controllable-polarity FETs in existing data structures. In particular, BDD Decomposition System based on MAJority decomposition (BDS-MAJ) [30] is a logic optimization system driven by binary decision diagrams that supports integrated MUX, XOR, AND, OR, and MAJ logic decompositions. Since it provides both XOR and MAJ decompositions, BDS-MAJ is an effective alternative to standard tools to synthesize data path circuits. In the controllable-polarity transistor context, BDS-MAJ natively and automatically highlights the efficient implementation of arithmetic gates. Finally, very efficient logic optimization can be directly performed on data structures supporting MAJ operator. In Reference 33, a novel data structure, called *Majority-Inverter-Graph* (MIG), exploiting only MAJ and INV operators has been introduced. Such data structure is supported by an expressive Boolean algebra allowing for powerful logic optimization of both standard general logic and arithmetic oriented logic.

8.3.3 Advanced low-power techniques

Thanks to their good electrostatic control (coming from the GAA architecture), MIG devices are promising candidates for low-power applications. However, the gain brought by the technology does not reduce to intrinsic device performances. Indeed, the enhanced set of functionalities enables simple implementation of advanced low-power techniques. In this section, we review a dual-V_T mode of operation and a power-gating technique.

8.3.3.1 Dual-threshold voltage operations

As briefly introduced in Section 8.2, TIGFETs can be configured in terms of polarity but also in terms of threshold voltage. The dual-V_T characteristics of TIG SiNWFET are depicted in Figure 8.2. For LVT configuration (solid lines), PG_S and PG_D are biased with the same voltage. In this configuration, the device is switching between *on* and standard *off* states [26]. For HVT configuration (dash lines), the device is wired unconventionally, as compared to a DIG device. Indeed, fixed bias voltages are now applied to CG and PG_S for p-type (CG and PG_D for n-type), while a voltage sweep is applied on PG_D (PG_S). Here, the device is switching between on and low-leakage off states. Such properties are used to create multi-V_T circuits in a simplified way. Indeed, traditional multi-V_T circuits require extra technological steps to build devices with different threshold voltages, which affect the layout regularity and increase the process costs as compared to single-V_T design [37]. Here, the same transistors support the two configurations and lead to a drastic cost reduction. Figure 8.13 illustrates two different NAND gate realizations for HP and LL applications, implemented with only three transistors. In Figure 8.13a, the HP gate is obtained by connecting inputs to the CGs of p-FETs. Thus, the performance for pulling the logic gate up is improved by applying the LVT configuration of the devices (low line in Figure 8.5). In contrast, the LL gate (Figure 8.13b) is obtained by controlling the p-FETs from the PG_D. Leakage power is thereby reduced by forcing the devices into HVT operation (Figure 8.5). In both HP and LL gates, PG_S and CGs of n-FETs are connected to input signals. Hence, delay and leakage in pull-down paths cannot be further tuned. Extensively studied

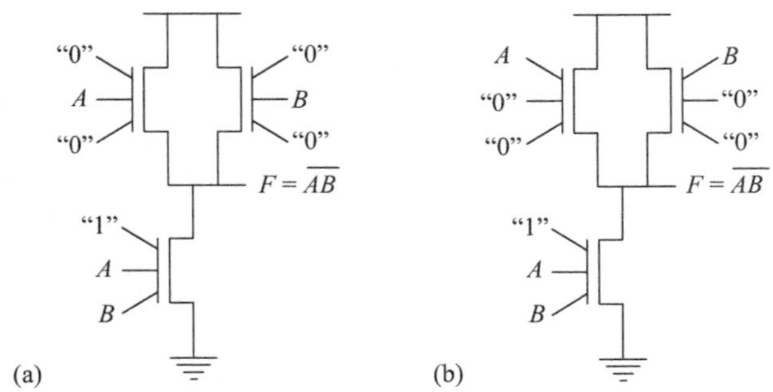

Figure 8.13 NAND gates realization using TIGFETs (a) HP and (b) LL configurations

in Reference 26 by TCAD simulations, such an approach demonstrates the ability to reach the same level of performances than FinFET Low-STandby Power (LSTP) transistors at 22-nm technology node for a slight area overhead of 8%. Therefore, the circuit-level opportunities compensate the initial limitations, such as the increase of parasitic capacitances, found at the device level.

8.3.3.2 Embedded power-gating

An efficient power-gating implementation is also unlocked by the enhanced functionality offered by MIGFETs. From a system perspective, power-gating is a common and effective technique to reduce leakage power in conjunction with multi-V_T design. Power-gating uses sleep transistors to disconnect the power supply from the rest of the circuit during idle time. The main drawbacks of power-gating are due to the series sleep transistor that (i) reduces the speed during normal operation and (ii) increases the circuit area. An efficient power-gating implementation is also achieved by the enhanced functionality offered by MIGFETs. From a system perspective, power-gating is a common and effective technique to reduce leakage power in conjunction with multi-V_T design. Power-gating uses sleep transistors to disconnect the power supply from the rest of the circuit during idle time. The main drawbacks of power-gating are due to the series sleep transistor that (i) reduces the speed during normal operation and (ii) increases the circuit area. By exploiting the on-line control of the MIGFET devices polarity, it is possible to create logic gates with power-gating capabilities with no series sleep transistors [35]. Based on *Differential Cascade Voltage Switch Logic* (DCVSL), pull-up MIGFET devices are not fixed to behave as p-type but their polarity is on-line modulated by a sleep signal, connected to the polarity gates.

The global concept is depicted in Figure 8.14. In standby mode, i.e., when *Sleep* = 1, the pull-up devices are switched to n-type through the PGs. The CGs are tied to ground by the two additional n-type devices. Therefore, both pull-up devices are in the *off* state. This provides the desired disconnection from the power supply. In the active operation mode, i.e., when *Sleep* = 0, the pull-up devices act as p-type. The CGs (connected to the gate outputs) are not anymore tied to ground since the

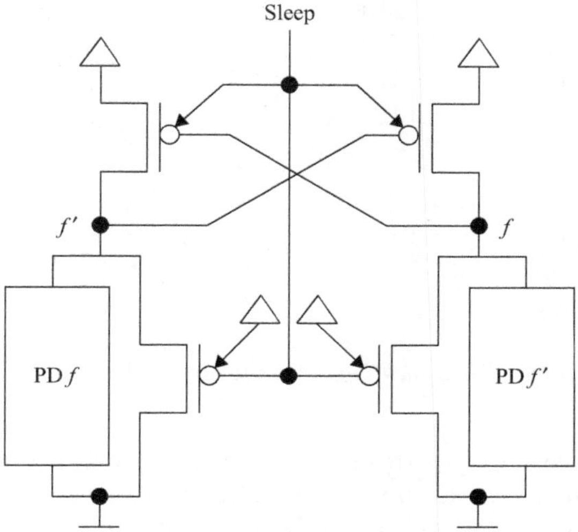

Figure 8.14 DIGFETs-based DCVSL style with advanced power-gating scheme

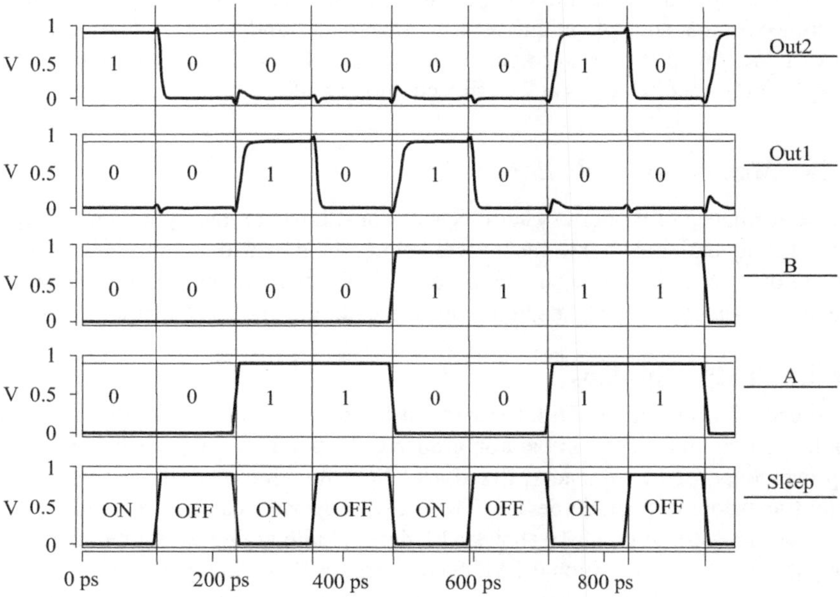

Figure 8.15 Simulation waveforms for a DCVSL XOR/XNOR-2 gate exploiting the proposed power-gating

two additional n-type devices are in the *off* state. The pull-down networks are now enabled to drive the outputs and close the standard feedback in DCVSL gates.

Simulation waveforms for the proposed power-gating scheme applied to a two-input DCVSL XOR/XNOR gate are shown in Figure 8.15. A proper XOR/XNOR-2

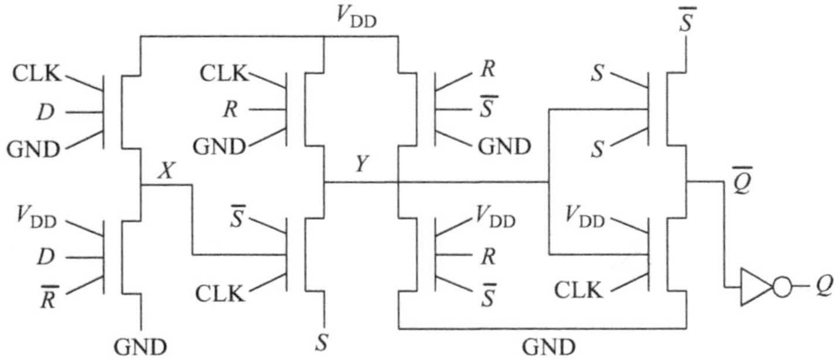

Figure 8.16 TSPC flip-flop design using TIGFETs compactness

function is noted when *Sleep* = 0 (active mode) while when *Sleep* = 1 (standby mode) both outputs assume the logic 0 value regardless of the other inputs. Note that the wake-up time is comparable with the regular logic gate delay permitting a fast standby to active mode transition.

Applied to arithmetic and computation intensive circuits, it has been shown in Reference 35 that such technique leads to area, delay, and leakage power savings of 1.4×, 1.3×, and 1.9× on average, respectively, compared to power-gated circuits using FinFET LSTP devices at 22-nm technology node.

8.3.4 Memory opportunities

The performance of modern systems System-on-Chips are mainly influenced by the different sequential elements encountered along the data path. Controllable-polarity devices open again promising new approaches in this field. In this section, we describe two novel memory designs adapted to local registering and large memory planes.

8.3.4.1 TSPC flip-flops

As already introduced, MIGFETs enable a large compactness of different circuits thanks to their intrinsic comparator property. In practice, they enable two major improvements: (i) the compact realization of XOR functions and (ii) the merge of two serial transistors in a single device. These two properties can be efficiently used in TSPC design [36]. Figure 8.16 shows a FF design build with only eight transistors as compared to 15 in its traditional CMOS counterpart. By reducing the number of transistor stacked in pull-up and pull-down networks and by using the larger functionality set offered by the controllable polarity, it has been shown in Reference 36 that the proposed design leads to data path storage elements with on average area and delay savings of 20% and 43%, respectively, compared again to FinFET LSTP transistors at 22-nm technological node.

To illustrate the correct behavior of the cell under asynchronous reset, we run electrical simulations and provide transient waveforms in Figure 8.17. In Figure 8.6, the output Q is observed to be correctly pulled down, when reset operation is

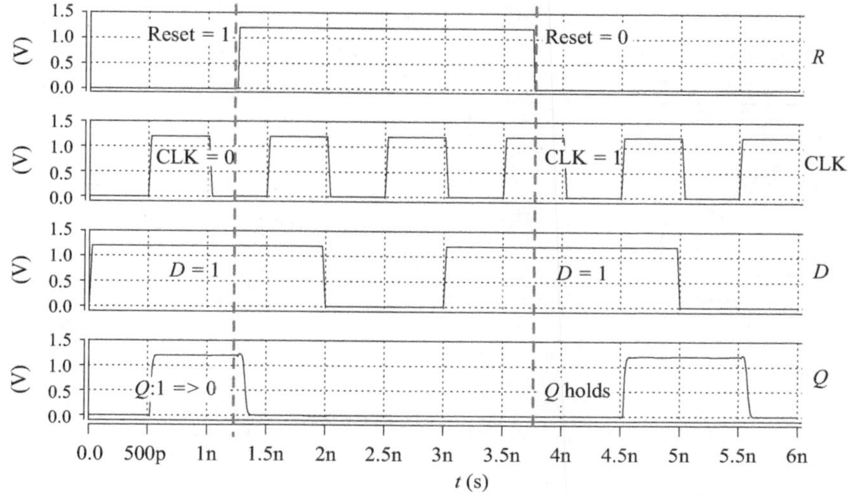

Figure 8.17 TIGFET TSPC flip-flop transient simulation with asynchronous reset

triggered, even in the most challenging case CLK $= 0$, $D = 1$, and $Q = 1$. Once the reset signal is de-asserted, the output Q switches again accordingly to the next clock rising edge.

8.3.4.2 Four-transistor pseudo-SRAM cells

MIGFET also introduces novel opportunities to design versatile memory arrays. In particular, we introduce a memory cell that can play an active role in future systems-on-chip by enabling a dual operation mode between dynamic RAM and traditional SRAMs. As the memory cell has both static and dynamic latching modes, we call it a pseudo-SRAM [25].

The memory cell, depicted in Figure 8.18a, consists of four transistors, realizing two cross-coupled inverters with special properties. First, the bottom transistors are not standard FETs but four independent gate FETs, where one gate is still connected as in usual inverters while the others provide enhanced controllability. Second, the bottom terminals of the cross-coupled inverters are not grounded, but are connected to *BitLines* (BLs). By exploiting the controllability of the bottom multi-gate FETs, it is possible to let the BLs write/read the cell by directly forcing/sensing the logic value at the output nodes of the cross-coupled inverter.

The proposed cells have three operation modes as highlighted in Figure 8.18b–d. The signals W (write) and EN (enable) control the bottom multi-gate FETs and thus impose the operation mode.

- When $W =$ EN $= 1$, the memory cell is in writing or static latch mode. Indeed, if both BLs are grounded, then the cell behaves as a static latch, as depicted in Figure 8.18b. When instead the BLs assume non-identical value in this specific

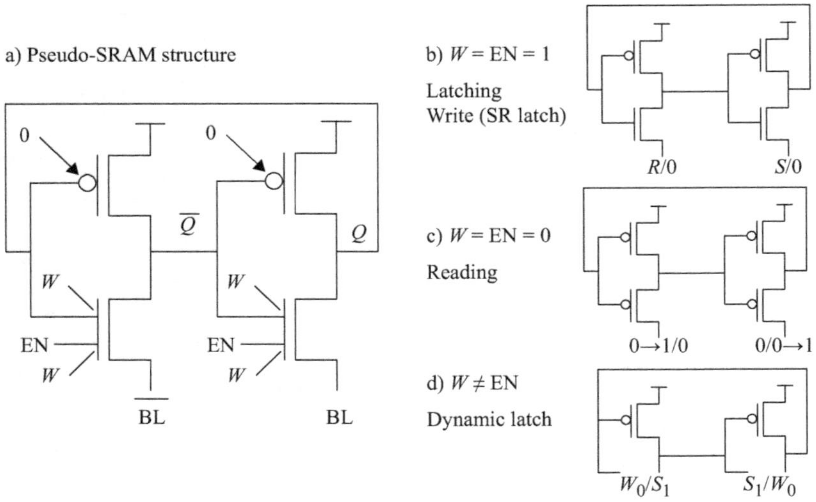

Figure 8.18 Pseudo-SRAM circuit structure (a) and operation modes (b–d)

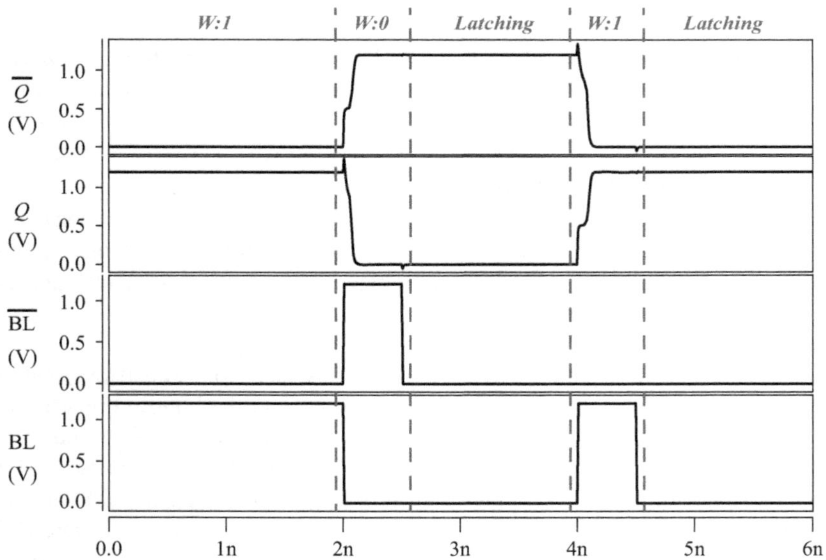

Figure 8.19 Pseudo-SRAM write/latching transient simulations

operation mode, the internal nodes of the cross-coupled inverter are forced to assume such values, thereby operating in a similar way than an SR latch. Note that after a short period of time (typically few tens of picoseconds), the memory cells naturally stabilize to the written values. Simulation waveforms, in a 22-nm SiNW technology, for write/latching operations are reported in Figure 8.19.

- When $W = \text{EN} = 0$, the cell is in reading mode (Figure 8.18c). The BLs are initially discharged to ground. Subsequently, the bottom FETs charge the BLs to the values stored in the internal nodes. Similarly as in standard SRAMs, the reading process can be speeded-up by using sense amplifiers and related circuitry.
- When W and EN are different, the memory operates as a dynamic latch (Figure 8.18d). The bottom transistors disconnect the cell internal nodes from the BLs. Therefore, the value stored inside the cell is stored as in a dynamic latch. This mode enables an intrinsic power-gating configuration. In this mode, the cell needs periodically refresh operation (typically every ms) through a configuration in its static latch mode.

Regarding reading and writing times, simulation results in a 22-nm SiNW technology show that these operations can be accomplished in 14.67 ps and 14.25 ps, respectively. As compared to a standard SRAM cell in 22-nm FinFET technology, the proposed cell is 14% smaller and 16% faster. The enhanced configurability of the proposed pseudo-SRAM cell is especially interesting in modular systems, where a memory array either can be used as regular dynamic memory or can be employed as static registers array. This selection can be dynamic and therefore give more flexibility to increase the configurability of some portions of the chip. From a system-level perspective, we expect the pseudo-SRAM cell to provide a valuable alternative to standard SRAM cells in circuits where low-power and high-performance operations are of paramount importance.

8.3.5 Case study: implementation of a Polar code decoder with MIGFETs

In this section, we showcase the proposed MIGFET technology by exploring the performance benefits that they can bring to the design of a contemporary telecommunication circuit.

8.3.5.1 Methodology

We consider a SoC system inspired from a promising class of linear block error-correcting code: Polar codes. In particular, we use the 1024-bit Polar code decoder design presented in Reference 38, with a parallelism degree of 64. The architecture for the Polar code decoder is depicted in Figure 8.20 and divides into four major units: an array of arithmetic processing elements (64 in parallel), a regular SRAM-based memory, a partial-sum logic, and an FSM in charge to schedule the decoding algorithm. The initial performance of such SoC platform is coarse-grain estimated for a CMOS 22-nm technology node, scaling original data from Reference 38, using tri-gate FinFETs with LSTP option [3].

We employ the techniques presented so far to design an optimized Polar code decoder using TIG controllable-polarity SiNWFET technology. We apply diverse optimization techniques for each major unit in Figure 8.20, and we sketch their effect in comparison to an un-optimized design but also to the traditional CMOS technology. Note that all evaluations are done at the block level, i.e., considering the expected

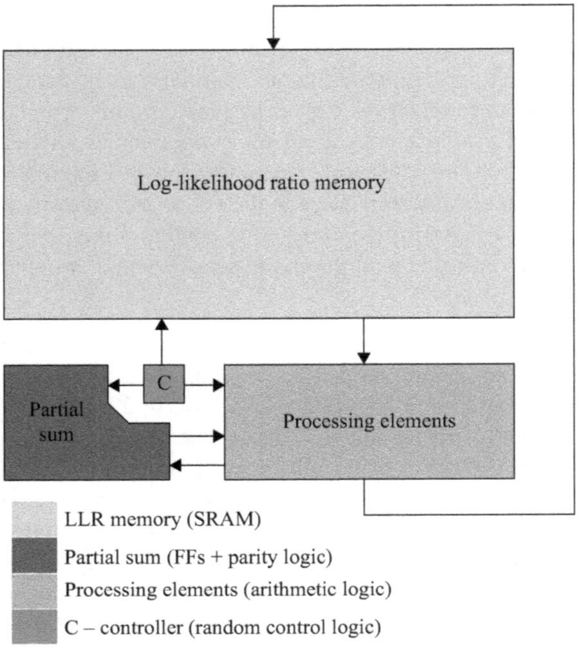

Figure 8.20 Polar code decoder architecture [38]

impact on major units, without taking into account physical design. For this reason, the data presented hereafter is not pretending to be fully accurate, but more an indicator of a new technology potential.

8.3.5.2 Improvement evaluation and discussions

Figure 8.21 anticipates the results for CMOS technology and MIG-SiNWFETs, with different levels of optimization. In a standard CMOS design, the Polar code decoder has an area of 20.17 μm^2, a critical path delay of 0.35 ns (corresponding to a throughput of 933 Mbps) and a power consumption of 8.58 mW (corresponding to an energy efficiency of 9.19 pJ/bit) 20.17 μm^2, 0.35 ns, and 8.58 mW. By simply substituting each FinFET transistor with a TIG-SiNWFET, the design characteristics are 28.04 μm^2, 0.37 ns, 9.61 mW (Figure 8.21 – TIG). This is worse than the traditional CMOS design. However, this is not surprising because the MIG-SiNWFETs are bigger (three-gate regions) than FinFET and introduces significant parasitic capacitances. To get smaller and faster designs with MIG-SiNWFETs, we need to fully exploit their enhanced functionality. For this purpose, the design techniques presented so far come to help. First, one can exploit the polarity-control to have more compact arithmetic gates (Figure 8.12), which are of paramount importance in the processing elements and partial-sum logic of the Polar code decoder. Also traditional negative unate gates are more compact thanks to the polarity-control of TIG-SiNWFETs (Figure 8.13). The combined effect in the decoder design corresponds to 26.66 μm^2, 0.29 ns, and

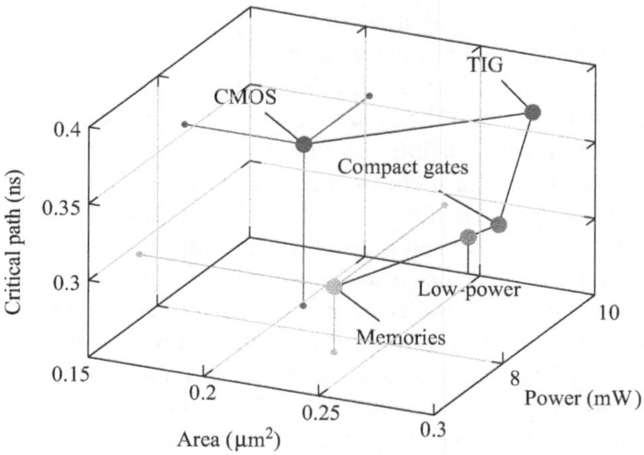

Figure 8.21 Design space for the Polar code decoder in CMOS FinFET or TIG-SiNWFET technologies (22 nm). Different optimizations are applied to the TIG-SiNWFET design showing their impact as compared to standard CMOS

9.57 mW (Figure 8.21 – compact gates). Already at this point, the predicted critical path in the SiNWFET design is much shorter than in CMOS, allowing higher throughput. Focusing instead on power numbers, we can exploit the power-gating opportunity to disconnect portion of the circuit in idle time during the decoding process. Following to the chosen parallelism degree, we can shutdown, on average, half of the processing elements, partially active during the decoding time. This corresponds in a reduction of the power consumption to 8.72 mW (Figure 8.21 – low-power) at minor area and delay penalties. Considering the memories, the pseudo-SRAM cells of Figure 8.18 find use in the Log-likelihood Ratio (LLR) memory region of the Polar code decoder, while the TSPC FFs of Figure 8.16 enable an efficient implementation of the partial-sum memories. These techniques reduce both area and power consumption (Figure 8.21 – memories).

Considering all the optimization techniques for TIG-SiNWFETs, we obtain a design having 23.47 μm^2, 0.29 ns, 6.98 mW thus a throughput of 1126 Mbps and an energy efficiency of 6.19pJ/bit. With our design techniques, a TIG-SiNWFET-based Polar code decoder can be notably faster (20%) and more energy efficient (32%) than in CMOS, at a moderate area overhead cost (15%).

8.4 Summary and conclusions

In this chapter, we surveyed the main results associated with MIG-SiNWFETs from technology to circuit design. We have focused on a recently demonstrated SiNW transistor technology with TIGs. The uniqueness of the proposed device lays in the high-degree of configurability reachable by a unique device. By biasing separately the TIGs, this device is configured as n-type or p-type transistor in either HVT or LVT mode. The threshold voltage tuning of this device is achieved by independently modulating the carrier transport at source/drain interface and in the channel. By fully suppressing the SBs and controlling the potential barrier in the channel, LVT configuration with earlier turn-*on* is helpful for improving the circuit speed. In contrast, by efficiently controlling the SBs at both source and drain, HVT configuration achieves a suppression of leakage current by two orders of magnitude without sacrificing the *on*-state current, showing advantages over the conventional multi-V_T techniques. The large configurability of the device leads to several opportunities at the circuit design level. Therefore, we also reviewed a complete design framework exploiting MIG transistors to support the next generations of System-on-Chips. Thanks to their enhanced functionality, MIGFETs enable a superior design of critical components in a SoC, such as the processing units and memories, while also providing native solutions to control the power consumption. Many techniques introduced recently were reviewed, and their opportunities were evaluated on a complex arithmetic intensive SoC. We showed that, with our proposals, the designed SoC can be 20% faster and is 32% more energy efficient than its FinFET counterpart, at 22-nm node, at a moderate area overhead of 15%.

Acknowledgment

This work has been supported by the *European Research Council* (ERC) advanced grant ERC-2009-AdG-246810.

References

[1] International Technology Roadmap for Semiconductors: Executive Summary – 2013 Edition.

[2] C.-H. Jan *et al.*, "A 22 nm SoC platform technology featuring 3-D tri-gate and high-k/metal gate, optimized for ultra low power, high performance and high density SoC applications," *IEDM Tech. Dig.*, pp. 3.1.1–3.1.4, 2012.

[3] C. Auth *et al.*, "A 22 nm high performance and low-power CMOS technology featuring fully-depleted tri-gate transistors, self-aligned contacts and high density MIM capacitors," *VLSI Tech. Dig.*, pp. 131–132, 2012.

[4] S. Bangsaruntip *et al.*, "High performance and highly uniform gate-all-around silicon nanowire MOSFETs with wire size dependent scaling," *IEDM Tech. Dig.*, 2009.

[5] S.-M. Koo, Q. Li, M. D. Edelstein, C. A. Richter, and E. M. Vogel, "Enhanced channel modulation in dual-gated silicon nanowire transistors," *Nano Lett.*, vol. 5, pp. 2519–2523, 2005.

[6] Y.-M. Lin, J. Appenzeller, J. Knoch, and P. Avouris, "High-performance carbon nanotube field-effect transistor with tunable polarities," *IEEE Trans. Nanotechnol.*, vol. 4, pp. 481–489, 2005.

[7] N. Harada, K. Yagi, S. Sato, and N. Yokoyama, "A polarity-controllable graphene inverter," *Appl. Phys. Lett.*, vol. 96, 012102, 2010.

[8] S. Sutar, P. Agnihotri, E. Comfort, T. Taniguchi, K. Watanabe, and J. U. Lee, "Reconfigurable p–n junction diodes and the photovoltaic effect in exfoliated MoS2 films," *Appl. Phys. Lett.*, vol. 104, 122104, 2014.

[9] F. Wessely, T. Krauss, and U. Schwalke, "CMOS without doping: multi-gate silicon-nanowire field-effect-transistors," *Solid-State Electron.*, vol. 70, pp. 33–38, 2012.

[10] M. De Marchi *et al.*, "Polarity control in double-gate, gate-all-around vertically stacked silicon nanowire FETs," *IEDM Tech. Dig.*, pp. 8.4.1–8.4.4, 2012.

[11] M. De Marchi *et al.*, "Configurable logic gates using polarity controlled silicon nanowire gate-all-around FETs," *IEEE Electron Device Lett.*, vol. 35, no. 8, pp. 880–882, 2014.

[12] A. Heinzig, S. Slesazeck, F. Kreupl, T. Mikolajick, and W. M. Weber, "Reconfigurable silicon nanowire transistors," *Nano Lett.*, vol. 12, pp. 119–124, 2012.

[13] A. Heinzig, T. Mikolajick, J. Trommer, D. Grimm, and W. M. Weber, "Dually active silicon nanowire transistors and circuits with equal electron and hole transport," *Nano Lett.*, vol. 13, pp. 4176–4181, 2013.

[14] R. Ng *et al.*, "Vertically stacked silicon nanowire transistors fabricated by inductive plasma etching and stress-limited oxidation," *IEEE Electron Device Lett.*, vol. 30, pp. 520–522, 2009.

[15] Y.-J. Chang and J. L. Erskine, "Diffusion layers and the Schottky barrier height in nickel silicide-silicon interface," *Phys. Rev. B*, vol. 28, pp. 5766–5773, 1983.

[16] Q. T. Zhao, U. Breuer, E. Rije, S. Lenk, and S. Smantl, "Tuning of NiSi/Si Schottky barrier heights by sulfur segregation during Ni silicidation," *Appl. Phys. Lett.*, vol. 86, 062108, 2005.

[17] J. Appenzeller, M. Radosavljevic, J. Knoch, and P. Avouris, "Tunneling versus thermionic emission in one-dimensional semiconductors," *Phys. Rev. Lett.*, vol. 92, 048301, 2004.

[18] M. Mongillo, P. Spathis, G. Katsaros, P. Gentile, and S. De Franceschi, "Multifunctional devices and logic gates with undoped silicon nanowires," *Nano Lett.*, vol. 12, pp. 3074–3079, 2012.

[19] J. Appenzeller, Y.-M. Lin, J. Knoch, and P. Avouris, "Band-to-band tunneling in carbon nanotube field-effect transistors," *Phys. Rev. Lett.* vol. 93, 196805, 2004.

[20] L. Chang, S. Tang, T.-J. King, J. Bokor, and C. Hu, "Gate length scaling and threshold voltage control of double-gate MOSFETs," *IEDM Tech. Dig.*, pp. 31.2.1–31.2.4, 2000.

[21] M. S. Lundstrom and D. A. Antoniadis, "Compact models and the physics of nanoscale FETs," *IEEE Trans. Electron Devices*, vol. 61, no. 2, pp. 225–233, 2014.

[22] W. Shangguan *et al.*, "Surface-potential solution for generic undoped MOSFETs with two gates," *IEEE Trans. Electron Devices*, vol. 54, pp. 169–172, 2007.

[23] J. Zhang, L. Zhang, J. He, and M. Chan, "A noncharge-sheet channel potential and drain current model for dynamic-depletion silicon-on-insulator metal-oxide-semiconductor field-effect transistors," *J. Appl. Phys.*, vol. 107, 054507, 2010.

[24] Synopsys; http://www.synopsys.com, 2009.

[25] P.-E. Gaillardon, L. Amaru, J. Zhang, and G. De Micheli, "Advanced systems on a chip design based on controllable-polarity FETs," *DATE Tech. Dig.* (Invited), 2014.

[26] J. Zhang, X. Tang, P.-E. Gaillardon, and G. De Micheli, "Configurable circuits featuring dual-threshold-voltage design with three-independent-gate silicon nanowire FETs," *IEEE Trans. Circuits Syst. I: Regular Paper*, vol. 61, no. 10, pp. 2851–2861, 2014.

[27] J. Zhang *et al.*, "Polarity-controllable silicon nanowire transistors with dual threshold voltages," *IEEE Trans. Electron Devices*, vol. 61, no. 11, pp. 3654–3660, 2014.

[28] M. H. Ben Jamaa, K. Mohanram, and G. De Micheli, "Novel library of logic gates with ambipolar CNTFETs: opportunities for multi-level logic synthesis," *DATE Tech. Dig.*, 2009.

[29] L. Amaru, P.-E. Gaillardon, and G. De Micheli, "Biconditional BDD: a new canonical BDD for logic synthesis targeting amipolar transistors," *DATE Tech. Dig.*, 2013.

[30] L. Amar, P.-E. Gaillardon, and G. De Micheli, "BDS-MAJ: a BDD-based logic synthesis tool exploiting majority decomposition," *DAC Tech. Dig.*, 2013.

[31] A. Zukovski, Y. Xuebei, and K. Mohanram, "Universal logic modules based on double-gate carbon nanotube transistors," *DAC Tech. Dig.*, 2011.

[32] L. Amaru, P.-E. Gaillardon, and G. De Micheli, "An efficient manipulation package for biconditional binary decision diagrams," *DATE Tech. Dig.*, 2014.

[33] L. Amaru, P.-E. Gaillardon, and G. De Micheli, "Majority-inverter graph: a novel data-structure and algorithms for efficient logic optimization," *DAC Tech. Dig.*, 2014.

[34] O. Turkyilmaz, L. Amar, F. Clermidy, P.-E. Gaillardon, and G. De Micheli, "Self-checking ripple-carry adder with ambipolar silicon nanowire FET," *ISCAS Tech. Dig.*, 2013.

[35] L. Amaru, P.-E. Gaillardon, J. Zhang, and G. De Micheli, "Power-gate differential logic style based on double-gate controllable polarity transistors," *IEEE Trans. CAS-II*, vol. 60, no. 10, pp. 672–676, 2013.

[36] X. Tang, J. Zhang, P.-E. Gaillardon, and G. De Micheli, "TSPC flip-flop circuit design with three-independent-gate silicon nanowire FETs," *ISCAS Tech. Dig.*, 2013.

[37] T. Matsukawa *et al.*, "Dual metal gate FinFET integration by Ta/Mo diffusion technology for Vt reduction and multi-Vt CMOS application," *ESSDERC Tech. Dig.*, pp. 282–285, 2008.
[38] A. Mishra *et al.*, "A successive cancellation decode ASIC for a 1024-bit polar code in 180 nm CMOS," *A-SSCC Tech. Dig.*, 2012.
[39] Q. Liu *et al.*, "Ultra-thin-body and BOX (UTBB) fully depleted (FD) device integration for 22 nm and beyond," *VLSI Tech. Dig.*, 2010.
[40] S. Fregonese, C. Maneux, and T. Zimmer, "A compact model for dual-gate one-dimensional FET: application to carbon-nanotube FETs," *IEEE Trans. Electron Devices*, vol. 58, no. 1, pp. 206–215, 2011.

Chapter 9

Exploration of carbon nanotubes for efficient power delivery

Aida Todri-Sanial[1]

Carbon nanotubes (CNTs) due to their unique mechanical, thermal, and electrical properties are being investigated as promising candidate material for on-chip and off-chip interconnects. The attractive mechanical properties of CNTs, including high Young's modulus, resiliency, and low thermal expansion coefficient offer great advantage for reliable and strong interconnects, and even more so for three-dimensional (3D) integration. Through-silicon-vias (TSVs) enable 3D integration and implementation of denser, faster, and heterogeneous circuits, which also lead to excessive power densities and elevated temperatures. Due to their unique properties, CNTs present an opportunity to address these challenges and provide solutions for reliable power delivery networks in two-dimensional (2D) and 3D integration. In this chapter, we perform detailed analyses of horizontally aligned CNTs and report on their efficiency to be exploited for both 2D and 3D power delivery networks.

9.1 Introduction

CNTs are a class of nanomaterials with unique mechanical, thermal, and electrical properties [7]. CNTs can be classified into two types: single-wall (SWCNTs) and multi-wall (MWCNTs). SWCNTs are rolled graphitic sheets with diameters on the order of 1 nm. MWCNTs consist of several rolled graphitic sheets nested inside each other and can have diameters as large as 100 nm. Depending on their chirality, the CNTs can be metallic or semiconductors. Metallic CNTs (m-CNTs) are ballistic conductors, which show promise for use as interconnects in nanoelectronics. On the other hand, semiconducting CNTs have a diameter-dependent band-gap and do not have surface states that need passivation, thus can be used to make devices such as diodes and transistors [10, 9, 7, 6].

CNTs are cylindrical carbon molecules formed by one-atom-thick sheets of carbon, or graphene [6–12]. CNTs, both SWCNT and MWCNT, are being investigated for a variety of nanoelectronics applications because of their unique properties [6–8]. Their extraordinary large electron mean free paths and resistance to electromigration

[1]CNRS-LIRMM, Montpellier, France

make them potential candidates for interconnects in large-scale systems. During the past decade, most of research is focused on CNT growth, synthesis, modeling and simulation, and characterizing contact interfaces [2]. Detailed simulation for signal interconnects has been performed by the authors of References 6–12, and it has been shown that CNTs have lower parasitics than Cu metal lines, however, the contact resistance between CNT-to-CNT and CNT-to-metal is large and can be detrimental for timing issues. Additionally, researchers are looking into different CNT growth techniques that are compatible with CMOS process, and lab measurements indicate the potential of integrating CNTs on-chip [13].

One application of the nanotubes in microelectronics is as interconnects using the ballistic (without scattering) transport of electrons and the extremely high thermal conductivity along the tube axis [1]. Electronic transport in SWCNTS and MWCNTS can go over long nanotube lengths, 1 μm, enabling CNTs to carry very high currents (i.e. $>10^9$ A/cm^2) with essentially no heating due to nearly one-dimensional (1D) electronic structure.

In the literature, the comparison of copper and CNTs has been limited to signal interconnects. Investigation of CNTs for power and clock delivery would also have a significant importance. It would reveal whether or not CNTs can potentially replace both signal and power/ground copper wires. Additionally, clock and power networks are most vulnerable to electromigration, it is therefore critical to know whether or not CNTs improve their reliability.

There are many works in the literature that investigate CNT interconnects. The first group of works focuses on modeling aspects of CNT interconnects [1, 7, 6, 11]. The second group of works focuses on performance comparison of CNT interconnects versus copper (Cu) interconnects [10, 15, 12, 6]. Almost all these works have considered the application of CNT interconnects for signaling, and few works focus on power delivery [3, 9]. Complementary to these efforts, in this chapter we investigate the application of horizontally aligned CNTs for power delivery network (i.e. both 2D and 3D Integrated Circuits (ICs)) while exploiting their unique electrical and thermal properties.

The rest of this chapter is organized as follows. Section 9.2 describes the modeling techniques that we utilize in this work. In Section 9.3, we explore 2D power delivery networks with CNTs. In Section 9.4, we present the analysis of CNT TSVs for 3D power delivery networks. Section 9.5 concludes this chapter.

9.2 Modeling of CNTs

There are many papers in the literature that focus on CNT modeling and understanding its transport properties [1, 10, 9, 7, 6]. In this section, we provide a brief description of CNT modeling that we utilize in this work. A generalized model for CNT interconnects is depicted as in Figure 9.1. In Figure 9.1a, the model of an individual MWCNT is shown with parasitics that represent both dc conductance and high-frequency impedance, that is, inductance and capacitance effects. Multiple shells of an MWCNT are presented by the individual parasitics of each shell. Such model can also be applicable to SWCNTs where only a single shell is represented.

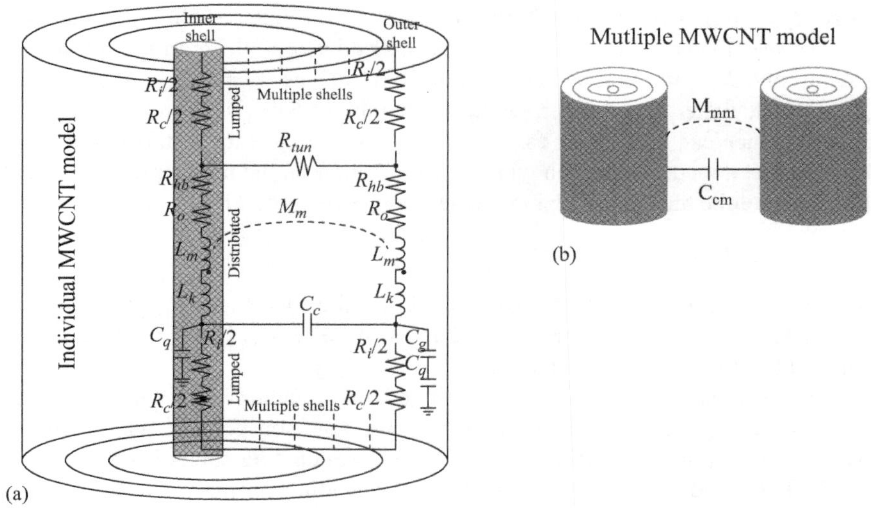

Figure 9.1 (a) Circuit model of an individual MWCNT and (b) multiple MWCNTs. This is general enough model to be applicable to MWCNTs of different diameters and shell numbers. It can also be applicable to SWCNTs where the model of a single shell can be utilized

Each shell has a lumped ballistic resistance (R_i) and lumped contact resistance (R_c) due to imperfect metal–nanotube contacts. These contacts are typically constructed of gold, palladium, or rhodium [7]. The nanotubes have also distributed ohmic resistance (R_o), which is dependent on length (l_b) and mean free path of acoustic phonon scattering (λ_{ap}). Overall CNT resistance depends also on the applied bias voltage, $R_{hb} = V_{bias}/I_o$, where I_o is the maximum saturation current (I_o values 15–30 μA [11]). Between shells in MWCNTs, there is also an intershell tunneling resistance (R_{tun}). As the applied bias voltage to each shell is the same, the impact of R_{tun} is relatively small. All the aforementioned ballistic, ohmic, and contact resistances depend on the number of 1D conducting channels, N_c. For metallic SWCNTs, the number of conducting channels is always $N_c = 2$ due to lattice degeneracy [1]. Whereas, for semiconducting SWCNTs and small-diameter semiconducting MWCNTs, $N_c = 0$. For any conducting shells in an MWCNT, the intrinsic resistance, $R_i = R_q/N_c$, where R_q is the quanta conductance for a 1D conduction channel ($R_q = 12.9\,\Omega$) [1]. Also, contact resistance is $R_c = 2R_{co}/N_c$ where R_{co} is the nominal contact resistance [1]. Ohmic resistance is derived as $R_o = R_qL/N_cd_sC_\lambda$, where L is the length of the MWCNT, d_s is the diameter of the shell, and C_λ is the acoustic phonon scattering mean free path (λ_{ap}). Thus, the total resistance of an individual nanotube (R_t) can be obtained by computing resistance of each shell, $R_{shell} = R_i + R_c + R_{hb} + R_o$:

$$G_t = \sum_{i=0}^{N} \frac{1}{R_{shell_i}} = \sum_{i=0}^{N} \frac{1}{\dfrac{R_f}{N_c} + \dfrac{R_qL}{C_\lambda(d_{in} + iS_a)N_c}} \tag{9.1}$$

where $R_f = R_i + R_c + R_{hb}$, d_{in} is the diameter of the inner shell, S_a is the space between shells where typically 0.34 nm is shell thickness, 0.34 nm is shell-to-shell spacing, and N is the number of shells in MWCNT as $N = (d_{out} - d_{in})/S_a$ where d_{out} is the diameter of the outer shell. In a bundle of SWCNTS or MWCNTS, the total resistance can be derived as, $R_b = R_t/n_b$, where n_b is the number of bundles. For example, a metal track with width, w, and height, h, the number of horizontally aligned nanotube bundles can be expressed as in Reference 11:

$$n_b = P_m(n_h n_w - \lfloor n_h/2 \rfloor) \tag{9.2}$$

where P_m is the probability that a nanotube is metallic and usually $P_m = 0.3$ [1], n_h is the number of nanotubes in vertical direction as $n_h = \lfloor h/d_{out} \rfloor$, and n_w is the number of nanotubes in horizontal direction as $n_w = \lfloor w/d_{out} \rfloor$.

The capacitance of nanotubes consists of both quantum, C_q, and electrostatic capacitances, C_e, that can impact power supply noise on power tracks. Additionally, there is coupling capacitance between: (1) conducting shells in an individual MWCNT, C_c and (2) individual MWCNTs depending on the proximity between them, C_{cm}. Using Luttinger liquid theory [1], quantum capacitance can be derived as $4e^2/h_p v_F \approx 193$ aF/μm per conducting channel where h_p is Planck's constant, e is charge of single electron, and v_F is Fermi velocity in graphene. Therefore, for each shell, quantum capacitance is as $C_q = \frac{4e^2}{h_p v_F} N_c L$ and the total quantum capacitance of a CNT bundle is as:

$$C_{q_t} = n_b \sum_{i=1}^{N} C_{q_i} \tag{9.3}$$

Electrostatic coupling depends on the geometry of the CNT and also the bundle density (i.e. number of bundles, n_b). It is shown in Reference 12 that CNT bundles have slightly smaller electrostatic capacitance compared to Cu interconnects with same dimensions. Capacitance of CNT bundles would decrease slowly with increase of bundle density [1]. However, for an MWCNT, these capacitances cannot be assumed equal due to the fringing coupling effects between shells. The electrostatic capacitance of an MWCNT which is equivalent to ground capacitance from the outer shell to the ground plane, distance y, can be obtained as:

$$C_{e_t} = \frac{2\pi\varepsilon}{\ln(y/d_{out})} \tag{9.4}$$

The shell-to-shell coupling capacitance is as in References 10 and 12:

$$C_c = \frac{2\pi\varepsilon}{\ln(d_{out}/d_{in})} \tag{9.5}$$

and coupling capacitance C_{cm} between two CNT bundles with space, s, can expressed as:

$$C_{cm} = \frac{2\pi\varepsilon}{s/d_{out}} \tag{9.6}$$

As for inductance, CNTs have both kinetic and magnetic inductances that impact power supply noise and high-frequency effects on power tracks. Again, based on the Luttinger liquid theory, the kinetic inductance per conducting shell can be theoretically expressed as $L_k = h_pL/4e^2v_FN_c$ or ≈ 8 nH/μm per conducting shell. Thus, the total kinetic inductance for all shells in a CNT bundle is derived as:

$$L_{k_t} = \frac{1}{n_b \sum_{i=1}^{N} \frac{1}{L_{k_i}}} \tag{9.7}$$

where L_{k_i} is the kinetic inductance of each shell i. Magnetic inductance, L_m, and mutual inductance M_m, are also of importance as they can have an impact on dynamic voltage drop behavior. For each shell $L_m = \frac{\mu}{2\pi}\ln(y/d)$ and for a CNT bundle is derived as:

$$L_{m_t} = \frac{1}{n_b \sum_{i=1}^{N} \frac{1}{L_{m_i}}} \tag{9.8}$$

Scalable mutual inductance model between any two shells i and $i+1$ with space distance, S_a, was presented in References 7, 6, and 12 and can be estimated as:

$$M_{m_i} = \frac{\mu_o l}{\pi}\ln(S_a/(d_{i+1} - d_i)) \tag{9.9}$$

and mutual inductance, M_{mm}, between two CNT bundles with space, s, can be similarly expressed as:

$$M_{mm} = \frac{\mu_o l}{\pi}\ln(s/d_{out}) \tag{9.10}$$

Resistance, capacitance, and inductance models for MWCNTs are further utilized to study the dynamic voltage drop behavior on power delivery networks and TSVs.

9.3 CNTs for 2D power delivery network

In this section, we explore CNTs as global tracks for on-chip power distribution.

Typically, power distribution networks are hierarchical in nature and can be classified as local, intermediate, and global power delivery networks. In this work, we focus solely on global power delivery networks as already pointed out from References 9 and 12 that compared to Cu interconnect CNTs would be beneficial at the global interconnects.

Additionally, power delivery networks are structured as meshes with power tracks running in parallel with each other, and vias are inserted on their perpendicular intersections also as shown in Figure 9.2a. Also depending on the circuit current demand, these meshes can be designed as uniform (i.e. all branches are equal lengths) and non-uniform grids (i.e. branches of different lengths) while still being regular meshes as shown in Figure 9.2b.

Power branches as shown in Figure 9.2a and 9.2b are the simplest composing element of the power delivery network, which is the segment between two intersecting metal tracks that can be composed of horizontally aligned CNTs. To check the power

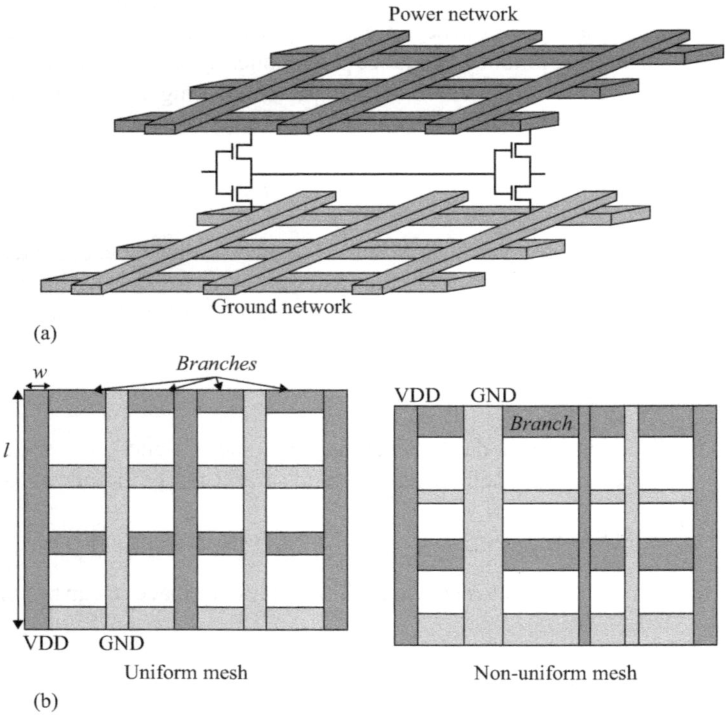

Figure 9.2 (a) Illustration of global power delivery network for a single tier and (b) description of uniform and non-uniform power delivery networks. The meshes have the same area and regular structure but some tracks have different widths, thus varying the branch lengths

integrity of the network, we check the voltage drop along the branch and derive it as:

$$V_{branch} = I_{branch}R_{branch} + L_{branch}\frac{dI_{branch}}{dt} + R_{branch}C_{branch}\frac{dV_{branch}}{dt} \tag{9.11}$$

where R_{branch}, C_{branch}, and L_{branch} represent the parasitics of the branch segment that can be either a copper metal layer or CNTs (i.e. SWCNTs, MWCNTs), and I_{branch} is the current flow on the branch. Regardless of the mesh (i.e. uniform or non-uniform), V_{branch} can be computed for each branch and serves as a quality metric for the power delivery network.

9.3.1 Branch analysis with CNTs

To predict the voltage drop on a power delivery network, we analyze a single branch implemented with MWCNT bundles. The branch width 100 nm and height 100 nm

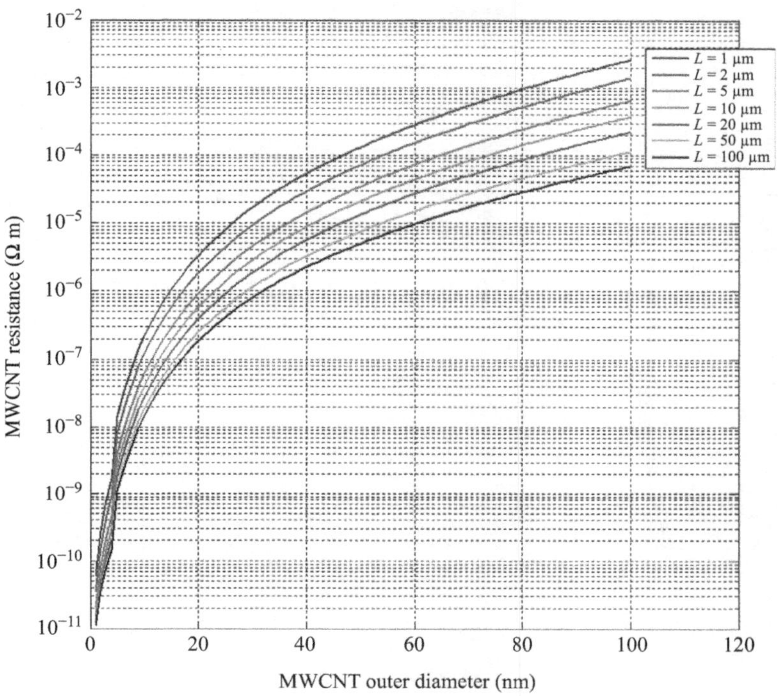

Figure 9.3 Branch resistivity of MWCNT bundle for various diameters and lengths

are fixed which are typical values for global power grid branches on advanced scaled technologies, that is, 28 nm or 32 nm. Whereas, the branch length and diameter of MWCNT bundles are varied. Figure 9.3a shows the resistivity of MWCNT bundles computed with (9.1) and (9.2). Bundle lengths are varied from 1 μm to 100 μm and MWCNT outer diameter varying from 1 nm to 100 nm. Figure 9.3 shows that for a fixed length, the smallest possible diameter, d_{out}, provides the lowest resistivity. Whereas, for a fixed diameter, the largest possible length, L, provides the smallest resistivity. Therefore, selecting optimal diameter in terms of resistance depends on MWCNT bundle length.

Similarly, we analyze the capacitance (both quantum and electrostatic) of MWCNT bundle as a function of bundle length and diameter as shown in Figure 9.4. We utilize (9.3)–(9.6) to derive branch capacitance, and we observe that for branches of fixed bundle length, small-diameter bundles provide smaller capacitance. Whereas, for branches with fixed diameter, large bundle lengths provide the smallest capacitance. Both quantum and electrostatic capacitances are equally important for deriving branch capacitance. In Figure 9.5, the kinetic and magnetic inductance values are plotted for MWCNT bundles of various diameters and lengths. Utilizing (9.7)–(9.10), kinetic inductance was derived while assuming 1D structure of MWCNT bundle. As length of MWCNT bundles is larger than the mean free path (λ), kinetic inductance has

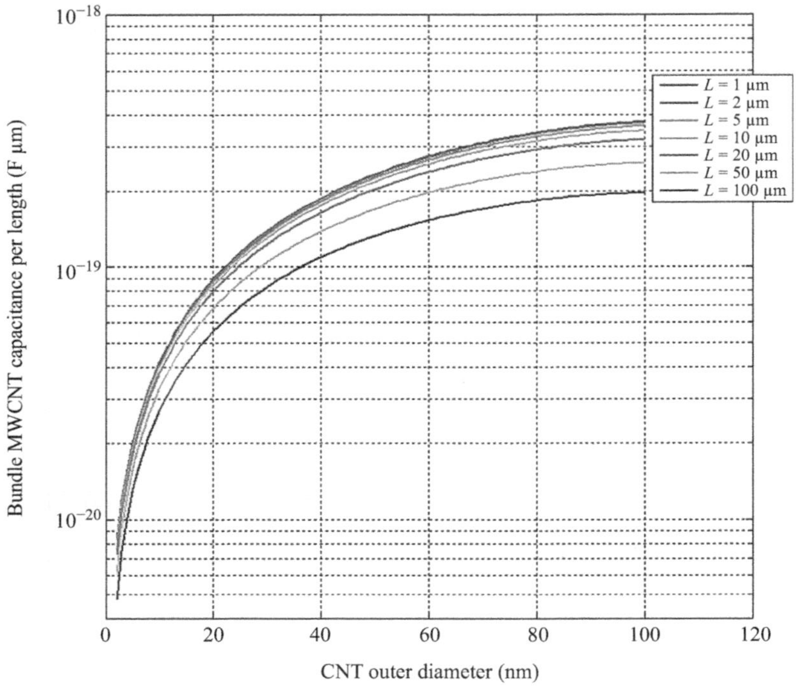

Figure 9.4 Branch capacitance (quantum and electrostatic capacitances) of MWCNT bundle for various diameters and lengths

small impact [11]. Whereas, magnetic inductances have larger values which decrease with the diameter of MWCNT bundles. For short-length MWCNT bundles, magnetic inductance is somewhat similar for bundles of different diameters. However, for large-length bundles, small magnetic inductance values are obtained at larger diameters. The amount of voltage drop contributed independently from resistance, capacitance, and inductance is shown in Figure 9.6. To compute voltage drop on the branch, we make assumptions that current flowing on the branch, $I_{branch} = 1\ \mu A$, $dI_{branch} = 1\ \mu A$, $dt = 1$ ns (or 1 GHz switching frequency), and $dV_{branch} = 0.1$ V. Note that these values are simply chosen to quantify branch voltage drop when 1 µA current is flowing on the branch for varying MWCNT bundle lengths and diameters. We notice that resistance impact (IR) on branch voltage drop increases with the MWCNT bundle diameter and decreases with bundle length. A similar but less dominant effect is obtained from capacitance ($RCdV_{branch}/dt$) impact on branch voltage drop. Whereas, inductive effects (LdI_{branch}/dt) are very minimal at this frequency. In Figure 9.7, the branch voltage drop contour plots are shown with respect to MWCNT bundle length and outer diameter. Note that the branch length is represented by the MWCNT bundle length. Voltage drop is computed as in (9.11). Each contour represents the amount of branch voltage drop which can vary from 2 mV to 960 mV for large diameters and lengths. It is important to note that contours can indicate a region for which MWCNT

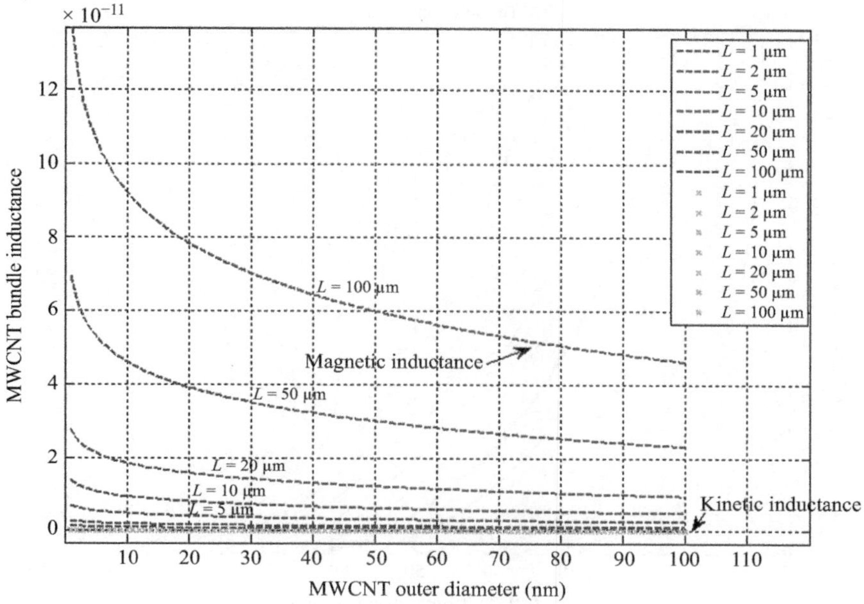

Figure 9.5 Branch inductance (kinetic and magnetic inductance) of MWCNT bundle for various diameters and lengths

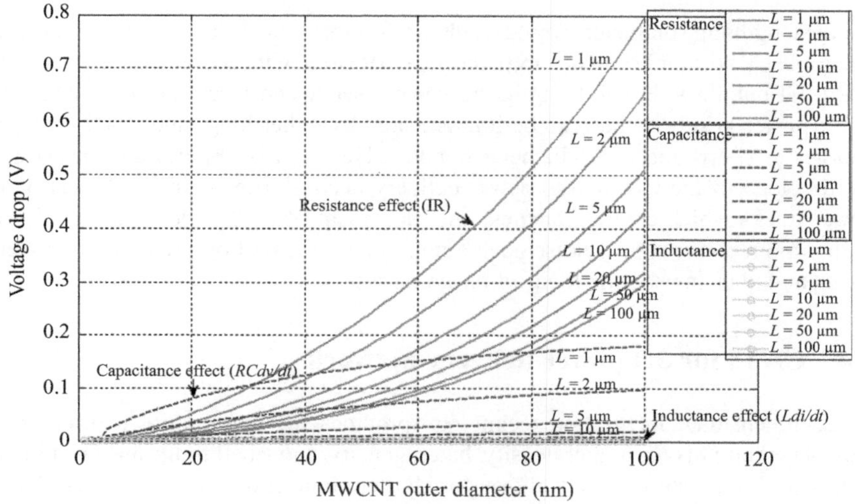

Figure 9.6 Individual impact of parasitic resistance, inductance, and capacitance to voltage drop on a power grid branch

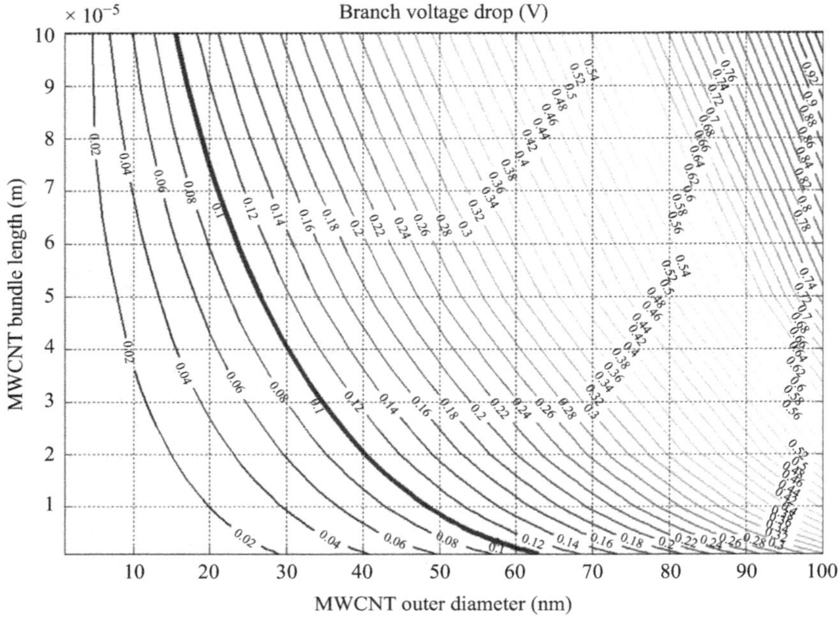

Figure 9.7 Contour plot of voltage drop on a power grid branch *as a function of MWCNT bundle length (1 μm to 100 μm) and diameter (1 nm to 100 nm)*

bundle lengths and diameters would result in allowable branch voltage drop. For example, for 0.1 V allowable voltage drop, the area to the left of thick contour represents the MWCNT bundle widths and lengths that can be used to construct a power grid branch. Thus, power grid branches can be implemented by either long- and small-diameter bundles or short- and large-diameter bundles. Hence, it is important to investigate the topology of the CNT-based power delivery network (i.e. sizing and location of power tracks which create branches) and placement of CNT tubes that would lead to minimum power supply noise generation. The problem of optimal power delivery sizing and placement is the focus of on-going research.

9.4 CNTs for 3D power delivery network

Three-dimensional integration technology provides the opportunity to implement multi-layer circuits for higher density, heterogeneity, and small footprint. The utilization of TSVs as interconnects allows for shorter connections with improved delays and increased bandwidths. As wire width continues to shrink, copper interconnects in high-performance systems will suffer from significant increase in resistivity due to surface roughness and grain boundary scattering and from electromigration problems due to the low current densities supported by copper conductors. Hence, despite the advantages of 3D integration, copper (Cu)-based interconnects, that is, Cu TSVs

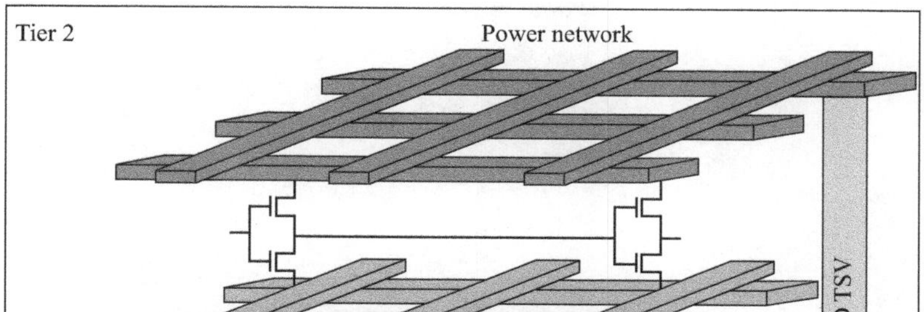

Figure 9.8 *(a) Illustration of global power delivery network for a single tier and (b) description of uniform and non-uniform power delivery networks. The meshes have the same area and regular structure but some tracks have different widths, thus varying the branch lengths. (c) Illustration of 3D power delivery network for two tiers connected via TSVs*

will hinder performance and reliability of interconnects, thus motivating the need for alternative interconnect materials for future process technologies.

Typically, 3D power distribution networks are hierarchical in nature and can be classified as local, intermediate and global 3D power delivery networks. In this work, we focus solely on global power delivery networks as already pointed out from References 9 and 12 that compared to Cu interconnect CNTs would be beneficial at the global interconnects.

In Figure 9.8, 3D power delivery network for two tiers is shown connected using TSVs for power and ground. For this work, we assume that TSVs connect global to global interconnects. Depending on the stacking configuration such as face-to-face or face-to-back and processing approach such as via-first, -middle, or -last, TSVs can vary in pitch. Here, we study high-density TSVs that can vary from

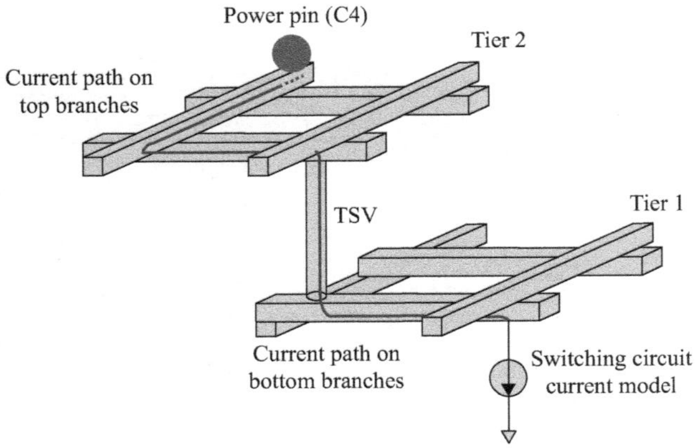

Figure 9.9 Current path from power pin to switching circuit passing through the shortest path on branches of Tier 2, Tier 1, and TSVs

1 nm to 5 nm in diameter and 5 μm to 100 μm in thickness, thus reaching density up to 10,000 TSVs/mm^2. Regardless of the structure, the power networks should deliver voltage with minimum voltage drop. Power branches as shown in Figure 9.2a and 9.2b are the simplest composing element of the power delivery network, which is the segment between two intersecting metal tracks that can be composed of horizontally aligned CNTs. Whereas, TSVs can be implemented using vertically grown CNTs, which make them suitable for TSV process. In comparison with copper (Cu)- or tungsten (W)-filled TSV, CNT-based TSVs provide a long mean free path, a large thermal conductivity, and high current-carrying capacity. CNTs can provide competitive solution for high-density vertical interconnects.

To check the power integrity of the network, we check the voltage drop from the power supply pin to the voltage node of the switching circuit. For example, for a two-tier stack, such current path would include the voltage drop on the branches of the top tier (i.e. Tier 2), the voltage drop on TSVs between top and bottom tiers, and the voltage drop on the branches of the bottom tier (i.e. Tier 1). This is also illustrated in Figure 9.9.

To compute the worst-case voltage drop that the switching circuits will experience when CNT interconnects and TSV are utilized, we first compute the voltage on top branches (i.e. Tier 2) with respect to different dimensions of CNT interconnect lengths and diameters. Voltage drop on a single branch can be computed based on the *RLC* parasitics of the CNT bundle interconnect as:

$$V_{top\_branch} = I_{top\_branch}R_{top\_branch} + L_{top\_branch}\frac{dI_{top\_branch}}{dt} + R_{top\_branch}C_{top\_branch}\frac{dV_{top\_branch}}{dt} \quad (9.12)$$

where R_{top_branch}, C_{top_branch}, and L_{top_branch} represent the parasitics of the top branch segment that are in Tier 2 and are CNT bundles (i.e. either SWCNTs or MWCNTs),

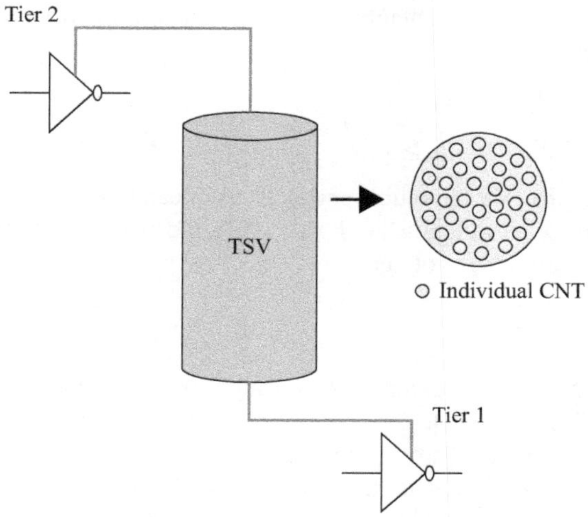

Figure 9.10 Cross-sectional view of a CNT TSVs made of several individual CNTs

and I_{top_branch} is the current flow on the branch representing the current demand of the switching circuit. Note that the voltage drop can also be computed as a summation of individual branch voltage drop to obtain the total voltage drop on top branches.

Similarly, the voltage drop on bottom branches (i.e. Tier 1) can be computed as:

$$V_{bot\_branch} = I_{bot\_branch}R_{bot\_branch} + L_{bot\_branch}\frac{dI_{bot\_branch}}{dt} + R_{bot\_branch}C_{bot\_branch}\frac{dV_{bot\_branch}}{dt} \tag{9.13}$$

where R_{bot_branch}, C_{bot_branch}, and L_{bot_branch} represent the parasitics of the bottom branch segment that are in Tier 1 and are CNT bundles (i.e. either SWCNTs or MWCNTs), and I_{bot_branch} is the current flow on the branch representing the current demand of the switching circuit. Similarly, if the current path flows through several branches, then the voltage drop can also be represented as the summation of voltage drops from each branch.

Cross-section of a power CNT TSV is shown in Figure 9.10. The m-CNT bundles can be densely packed to create many parallel connections and reduce the overall resistance of the TSV. Metallic nanotubes are distributed in the sparcely bundle with probability 1/3 since approximately one-third of tubes are metallic [2]. It has been shown that nanotubes with diameter less than 0.5 nm tend to be metallic regardless of the chirality due to the steep angle of curvature in the graphene sheet [4, 8]. The resistance of an individual nanotube depends on the applied bias voltage. If low bias voltage (i.e. $V_b \leq 0.1$) is applied, then the resistance is the summation of lumped intrinsic ($R_i \approx 6.5\,\mathrm{k}\Omega$) and contact ($R_c$) resistances and distributed per unit length ohmic resistance (R_o).

$$R_{low} = R_i + R_c \quad \text{if} \quad l_b \le \lambda_{ap} \tag{9.14}$$

$$R_{low} = R_o + R_c \quad \text{if} \quad l_b > \lambda_{ap} \tag{9.15}$$

where l_b is the length of the nanotube, and λ_{ap} is the mean free path for acoustic phonon scattering. For high bias voltages (i.e. $V_{bias} \ge 0.1$), the resistance of an individual tube depends on the applied bias voltage as:

$$R_{high} = R_{low} + \frac{V_b}{I_o} \tag{9.16}$$

where I_o is the maximum current flow through an individual nanotube, which is approximately 20–25 μA [14].

Thus, the resistance of an individual CNT TSV for power and ground lines will have resistance as R_{high}. Contact resistance of CNT TSV models the increased resistance due to imperfect metal contacts. Recent studies have shown that R_c of an SWCNT greatly increases when the diameter of the nanotube (d_t) is less than 1 nm [13, 5]. Ohmic resistivity of an SWCNT is defined as [11]:

$$\rho_t = \frac{h}{4e^2 C_\lambda d_t} \tag{9.17}$$

where C_λ is a mean-free-path-to-nanotube-diameter proportionality constant defined as $C_\lambda = \lambda_{ap}/d_t$. The ohmic resistance is proportional to its diameter $1/d_t$, while for standard copper conductors, resistance has $1/d^2$ dependence. Thus, ohmic resistance of CNTs is proportional to surface area of the tube, whereas resistance of metallic conductor is proportional to its cross-sectional area of the conductor.

Thus, the overall TSV resistance made of SWCNT bundle is defined as the parallel combination of the individual SWCNT resistances as

$$R_{tsv} = \frac{R_o + R_c + \frac{V_b}{I_o}}{n_b} \tag{9.18}$$

where n_b is the number of bundles and it was defined in (9.2). Here, we compute the voltage drop on a TSV based on its resistance as:

$$V_{tsv} = R_{tsv} I_o \tag{9.19}$$

Hence, the total voltage drop on a current path passing through the CNT-based branches on Tier 1, TSVs, and Tier 2 can be computed as:

$$V_{drop} = V_{top\_branch} + V_{tsv} + V_{bot\_branch} \tag{9.20}$$

9.4.1 CNT TSV analysis

Here, we analyze the CNT TSV resistance when SWCNT bundles are used. Assuming there is no current redistribution due to magnetic inductance, the resistance of CNT TSVs (or SWCNT bundle) is defined as the parallel combination of individual SWCNT resistances. This is also described in (1.18). We compute TSV resistance

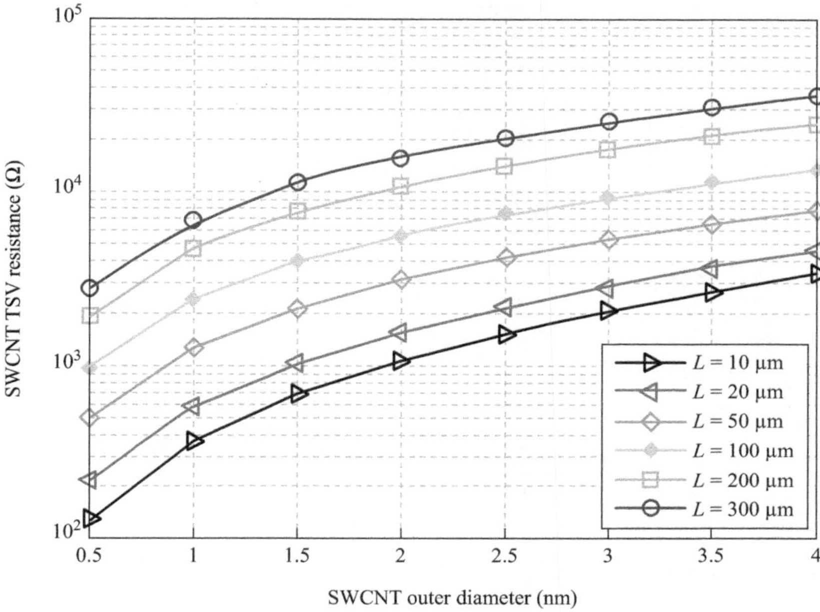

Figure 9.11 Resistance of TSV made of SWCNTs with varying diameters and lengths

while varying the diameter and length of individual SWCNTs. The experiments are performed on $d_t = 0.5$–4 nm diameter range for lengths (or TSV heights) varying from 10 μm to 300 μm where $R_c = 20\,\text{k}\Omega$ and $I_o = 25\,\mu\text{A}$. We assume supply voltage is $V_{bias} = 0.8\,\text{V}$ and driver resistance is 2.5 kΩ which corresponds to predictions of International Technology Roadmap for Semiconductors (ITRS) for 28 nm process technology. The CNT-based TSV resistance is displayed in Figure 9.11.

As shown in Figure 9.11, the TSV resistance can be categorized based on the diameter and length of the SWCNT bundles. For long SWCNT bundles, the TSV resistance increases with the increase in diameter size. SWCNT bundles are also at a disadvantage for short TSV lengths and large diameters. Whereas for short bundle lengths with small diameters, SWCNT TSV becomes more favorable. As SWCNT TSV resistance varies significantly due to both diameter and length, selecting the optimal TSV dimensions is not trivial as it can lead large voltage drop. Figure 9.12 displays the voltage drop on SWCNT TSVs for various diameter and length sizes.

We observe that the TSV voltage drop varies significantly due to SWCNT length. For the same size diameter SWCNT (i.e. $d_t = 2$ nm), voltage drop varies from 26.6 mV to 400 mV when length varies from 10 μm to 300 μm, respectively. We also note that for short SWCNT lengths, the impact of diameter on voltage drop is minimal such as a variation of 3 mV to 85 mV for SWCNT length of 10 μm. However, for longer SWCNT lengths, the impact of SWCNT diameter becomes more dominant on voltage

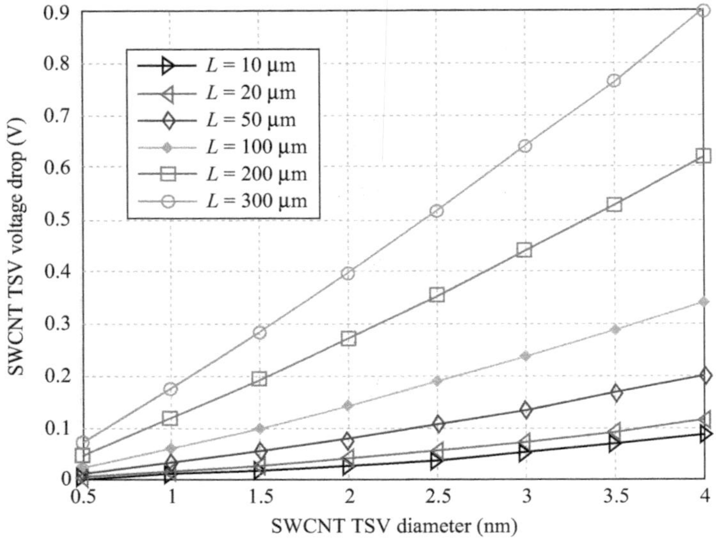

Figure 9.12 Voltage drop on TSV made of SWCNTs with varying diameters and lengths

drop such as 5 mV to 620 mV for SWCNT lengths of 200 μm. Hence, for a given allowed voltage drop threshold, one can easily determine the length and diameter ranges of SWCNT TSVs to satisfy such voltage drop threshold. For example, for a TSV voltage drop threshold of 0.1 V, we can determine that TSV lengths of 10 μm and 20 μm can be utilized for any diameter size SWCNTs. However, for longer TSV lengths, the range of SWCNT diameter sizes becomes more limited in order to satisfy the 0.1 V voltage drop threshold.

9.4.2 Voltage drop analysis on a 3D PDN

Here, we analyze the voltage drop on a 3D power delivery network composed of two stacked tiers as depicted in Figures 9.8 and 9.9. Assuming that there is a single switching circuit, we can identify the current path from the power pin to the circuit traversing branches on top tier, TSVs, and bottom tier. The same principle would also apply if several switching circuits were present. This is due to the superposition principle on linear systems.

As heterogeneous technologies and functionalities can be implemented in 3D, each tier can have its individual topology which might differ from the rest of the tiers. In this work, we assume that both Tier 1 and Tier 2 have uniform topologies. However, the applied analysis method and analytical formulas are general enough to be applied to any type of topology.

We consider different geometries for SWCNT branches for both Tier 1 and Tier 2. The experiments are performed on CNT-based branches (i.e. either SWCNTs

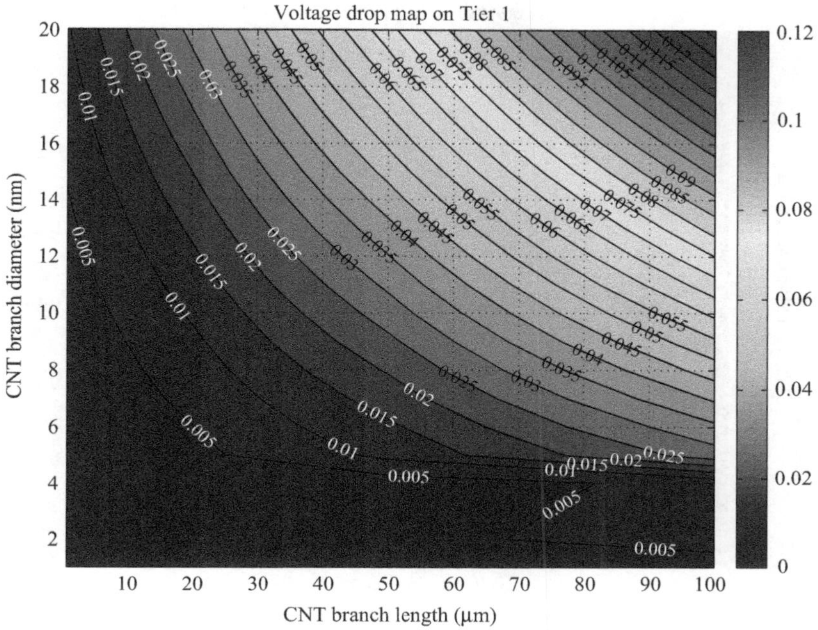

Figure 9.13 Voltage drop on CNT branches located in Tier 1

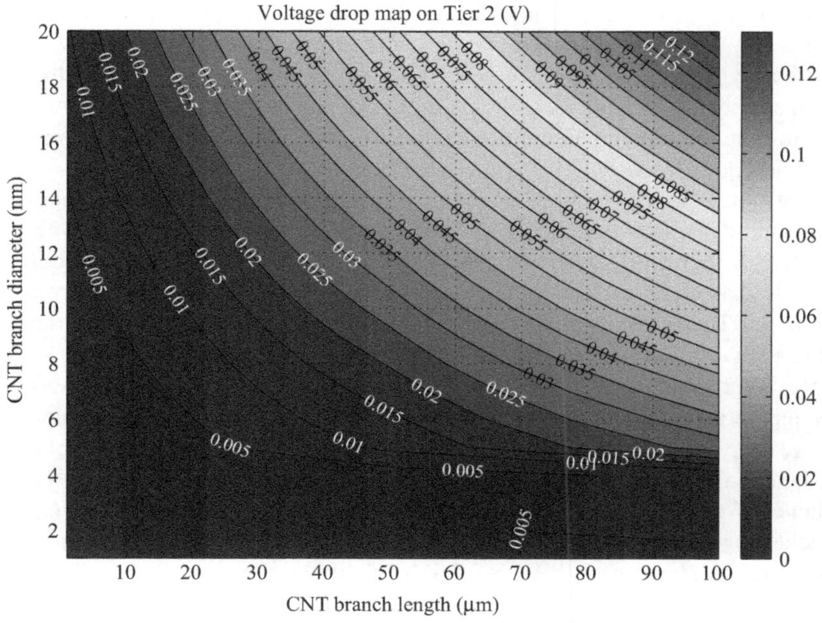

Figure 9.14 Voltage drop on CNT branches located in Tier 2

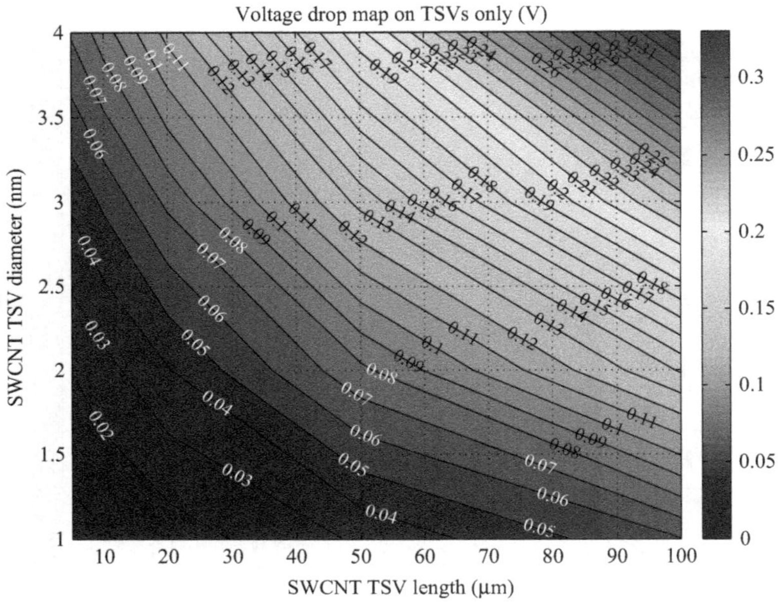

Figure 9.15 Voltage drop on CNT TSVs for varying diameter and height dimensions

or MWCNTs) with lengths of 1 μm to 100 μm with inner wall diameter of 1 nm and outer diameter ranges from 1 nm to 20 nm. Whereas, SWCNT parameters with lengths of 5–100 μm and diameter range 1–4 nm are applied to SWCNT TSVs. Figure 9.13 displays the voltage drop on Tier 1 CNT-based branches with respect to branch length and diameter. Note that voltage drop is computed based on (1.12). As we are applying the same parameters for branches in Tier 2, the voltage drop is similar as shown in Figure 9.14.

We note a large variation in branch voltage drop due to both its length and CNT bundle diameter size. As power grid branches can be implemented with either SWCNTs or MWCNTs, diameter-dependent voltage drop becomes more dominant for large diameters and longer length branches. Similarly, one can determine the optimal CNT-based branch lengths and diameter for a given voltage drop threshold that is allowed on the power grid branches.

In Figure 9.15, the TSV only voltage drop is plotted with respect to SWCNT TSV length and diameter. In comparison to CNT-based branches, larger amount of voltage drop can occur on SWCNT TSVs. This could be due to the impact of contact resistances and geometry of the SWCNT TSV which are different from CNT-based branches. We also plot the total voltage drop that occurs on Tier 1, TSV, and Tier 2 branches for different geometries of branches and TSVs. Figure 9.16 shows different voltage drops maps when SWCNT TSVs of different sizes are used. The x- and y-axis of each sub-figure represent the CNT-based branch diameter and lengths. Each sub-figure represents different voltage drops with respect to SWCNT TSV

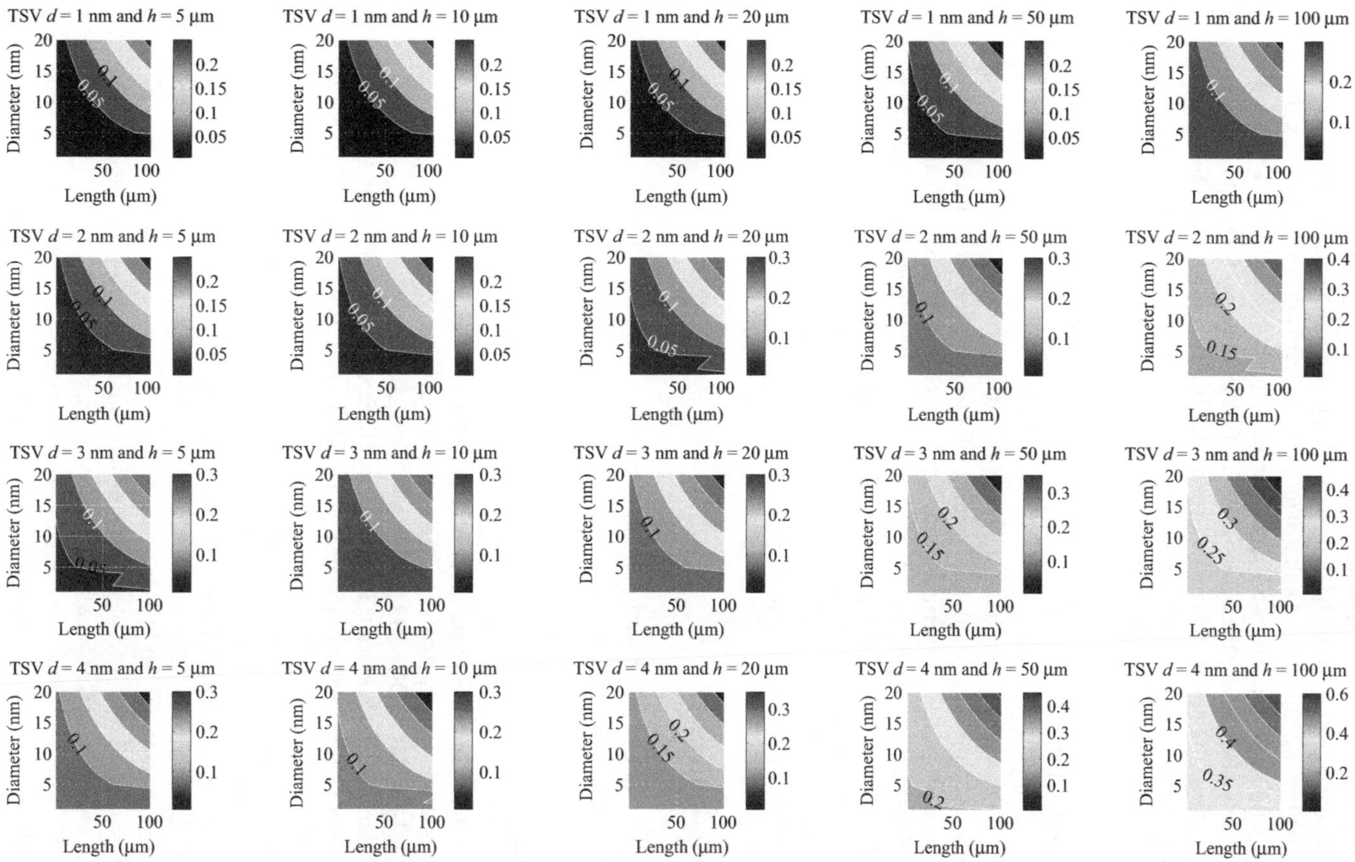

Figure 9.16 Total voltage drop on 3D power delivery network considering the voltage drop on Tier 1, TSVs and Tier 2. Each figure shows the total voltage drop as a function of branch CNT length and diameters (both Tier 1 and Tier 1 branches are uniform) and also as dimensions of TSV diameter and height

diameter and length noted on top of each sub-figure. Such analysis helps to determine the TSV dimensions that would be suitable for a given topology on each tier while providing minimal voltage drop. As shown in each sub-figure, the overall voltage drop gets worse with the increase of TSV diameter and length. From these analyses, we are able to understand better the potential of CNT-based interconnects and TSVs for designing optimal 3D power delivery networks. Additionally, such early-on analyses are valuable for performing design space exploration of 3D Power Delivery Networks (PDNs) with CNT interconnects.

9.5 Conclusion

CNTs due to their unique mechanical, thermal, and electrical properties are being investigated as promising candidate material for power delivery. The attractive mechanical, electrical, and thermal properties of CNTs offer great advantage for reliable and strong interconnects, and even more so for 3D integration. In this chapter, we performed a detailed design space exploration of horizontally and vertically aligned CNTs for implementing power delivery and TSVs. The analyses demonstrate that CNTs can be efficiently exploited for both 2D and 3D power delivery networks while resulting in minimal voltage drop.

Acknowledgment

A part of the work has been performed in the Project H2020 CONNECT, which is funded by European Commission.

References

[1] Burke, P.: "Luttinger liquid theory as a model of the gigahertz electrical properties of carbon nanotubes". *IEEE Transactions on Nanotechnology*, **1**(3), 129–144 (2002). DOI 10.1109/TNANO.2002.806823

[2] Wilder, J.W.G., Venema, L.C., Rinzler, A.G., Smalley, R.E., Dekker, C.: "Electronic structure of atomically resolved carbon nanotubes". *Nature*, **391**(6662), 59–61 (1998).

[3] Khan, N., Hassoun, S.: "The feasibility of carbon nanotubes for power delivery in 3-D integrated circuits". In: *Design Automation Conference (ASP-DAC)*, 53–58 (2012). DOI 10.1109/ASPDAC.2012.6165010

[4] Qin, L.-C., Zhao, X., Jirahara, K., *et al.*: "The smallest carbon nanotube". *Nature*, **408**(6808), 50 (2000).

[5] Leonard, F., Talin, A.: "Electrical contacts to nanotubes and nanowires: Why size matters". *ArXiv Condensed Matter e-prints* (2006).

[6] Li, H., Liu, W., Cassell, A., Kreupl, F., Banerjee, K.: "Low-resistivity long-length horizontal carbon nanotube bundles for interconnect applications 2014 – Part II – characterization". *IEEE Transactions on Electron Devices*, **60**(9), 2870–2876 (2013). DOI 10.1109/TED.2013.2275258

[7] Li, H., Xu, C., Srivastava, N., Banerjee, K.: "Carbon nanomaterials for next-generation interconnects and passives: Physics, status, and prospects". *IEEE Transactions on Electron Devices*, **56**(9), 1799–1821 (2009). DOI 10.1109/TED.2009.2026524

[8] Wang, N., Tang, Z.K., Li, G.D., Chen, J.S.: "Single-walled carbon 4 Å carbon nanotube arrays". *Nature*, **408**(6808), 50–51 (2000).

[9] Naeemi, A., Huang, G., Meindl, J.: "Performance modeling for carbon nanotube interconnects in on-chip power distribution". In: *Electronic Components and Technology Conference*, 420–428 (2007). DOI 10.1109/ECTC.2007.373831

[10] Naeemi, A., Sarvari, R., Meindl, J.: "Performance comparison between carbon nanotube and copper interconnects for gigascale integration (GSI)". *IEEE Electron Device Letters*, **26**(2), 84–86 (2005). DOI 10.1109/LED.2004.841440

[11] Nieuwoudt, A., Massoud, Y.: "Evaluating the impact of resistance in carbon nanotube bundles for VLSI interconnect using diameter-dependent modeling techniques". *IEEE Transactions on Electron Devices*, **53**(10), 2460–2466 (2006). DOI 10.1109/TED.2006.882035

[12] Srivastava, N., Banerjee, K.: "Performance analysis of carbon nanotube interconnects for VLSI applications". In: *IEEE/ACM International Conference Computer-Aided Design*, 383–390 (2005). DOI 10.1109/ICCAD.2005.1560098

[13] Kim, W., Javey, A., Tu, R., Cao, J., Wang, Q., Daia, H.: "Electrical contacts to carbon nanotubes down to 1 nm in diameter". *Applied Physics Letters*, **87**(17), 173101 (2005).

[14] Yao, Z., Kane, C.L., Dekker, C.: "High-field electrical transport in single-wall carbon nanotubes". *Physical Review Letters*, **84**(13), 2941–2944 (2000).

[15] Zhu, L., Sun, Y., Xu, J., Zhang, Z., Hess, D., Wong, C.: "Aligned carbon nanotubes for electrical interconnect and thermal management". In: *Electronic Components and Technology Conference*, 1, 44–50 (2005). DOI 10.1109/ECTC.2005.1441243

Chapter 10
Timing driven buffer insertion for carbon nanotube interconnects

Lin Liu[1], Yuchen Zhou[1], and Shiyan Hu[1]

In the nanoscale technology, both the device and interconnect performances affect the overall performance of the integrated circuits and systems in which they are used. So, it is quite natural to explore various solutions for devices as well as interconnects to mitigate the challenges of technology scaling and meet high-speed demand. This chapter discusses the use of carbon nanotubes (CNTs) as a potential high-speed high-performance interconnect as compared to the metal interconnects.

10.1 Introduction

Buffer insertion is one of the most effective timing optimization techniques on interconnects. In the existing techniques, copper buffering is indispensable in physical design [1–5]. However, copper interconnects based technology is reaching the bottleneck due to the fundamental physical limit of copper material. The interconnect delay due to ever increasing wire resistivity has greatly limited the circuit miniaturization. In addition, due to the inherently low tolerable current density, the electromigration induced interconnect reliability issue aggravates the problem. Therefore, the novel materials to replace copper interconnects are highly desirable in nanoscale high-frequency circuit design. As one of the promising replacement materials, CNTs alleviate the above severe timing and reliability issues in copper interconnects based design due to their superior conductivity and current-carrying capabilities. According to existing experiments, CNTs have significantly larger carrier mean free paths and can conduct larger currents without deterioration comparing to copper interconnects [6]. As a result, the issues such as increasing interconnect delay and electromigration that plaque the copper interconnects are mitigated. In addition to the above advantages,

[1]Michigan Technological University, Houghton, MI, USA

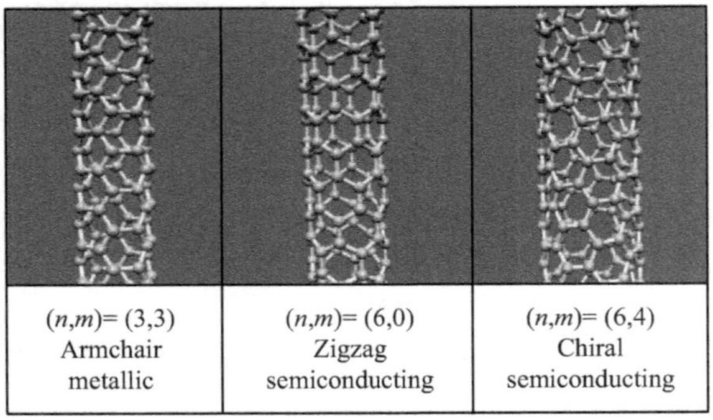

Figure 10.1 Three types of CNTs with different (n,m) which are generated using software in Reference 7

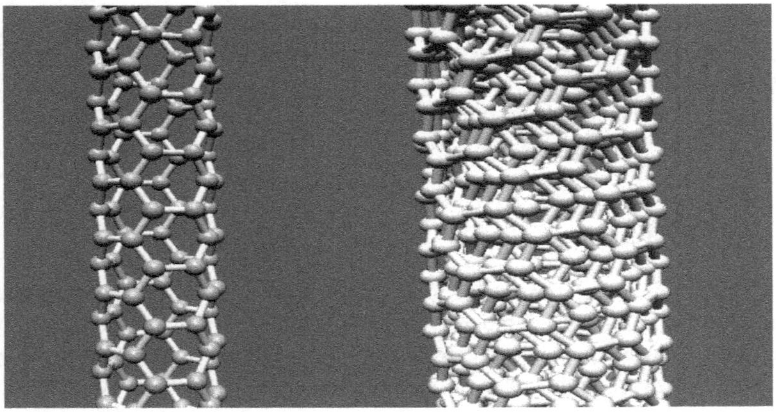

Figure 10.2 Structures of SWCNT and MWCNT which are generated using software in Reference 7

CNTs have high thermal conductivity and mechanical stability which makes CNTs more desired in the nanoscale design.

CNTs are miniaturized tubes which can be viewed as rolled up sheets of carbon hexagons. A pair of indices (n, m) is used to represent different CNTs with different electrical properties. The integers n and m denote the numbers of unit vectors along two directions which are horizontal and 60° directions in the honeycomb crystal lattice of graphene, respectively. Refer to Figure 10.1. If $n = m \neq 0$, the CNTs are called armchair nanotubes which are metallic. If $n \neq m$ and $m = 0$, the CNTs are called zigzag nanotubes which are semiconducting. If $n \neq m$ and $m \neq 0$, they are called chiral which are semiconducting as well. Since CNTs are used as interconnects in this work, metallic CNTs are more desirable. On the other hand, there are two main structures of CNTs, which are single-walled CNTs (SWCNTs) and multi-walled CNTs (MWCNTs). Refer to Figure 10.2. SWCNT is composed of a single graphite sheet wrapped into a cylindrical tube while MWCNTs are composed

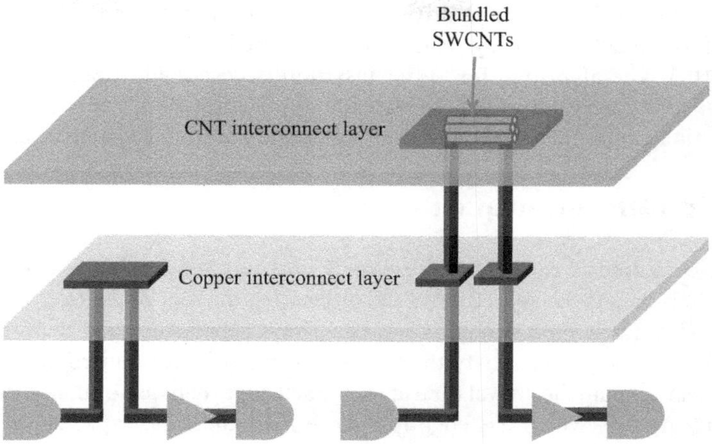

Figure 10.3 Copper buffering and CNT buffering

of an array of concentrically nested CNTs. According to Reference 8, using MWCNTs for long-length ballistic transport is not desirable which is why they are not used in this work. Since a single SWCNT has much larger resistance than copper for global interconnects [8], it is desired to bundle SWCNTs in parallel, resulting in *bundled SWCNTs* for better performance. Various research efforts have been spent in CNT fabrication. Most existing works, such as References 9–14, explore chemical vapor deposition technologies and achieve successful fabrication experience on CNTs.

According to some existing works [8, 15–18], the bundled SWCNTs can outperform copper interconnects in signal wave transportation along long global interconnects. For example, Srivastava *et al.* [8] show that the resistance of dense bundled SWCNTs can be 50% smaller than that of copper at the same size of long interconnects at 22 nm technology node. Despite this, buffer insertion is still necessary to improve the timing of the bundled SWCNTs based design, which is the main topic of this work. Thus, it is desirable to use SWCNTs to replace copper in global interconnects [19]. Refer to Figure 10.3. Although the authors of the References 8 and 15 consider CNTs interconnects, they always use a two pin model since they are from the perspective of the device and interconnect modeling. The main contribution of this work is summarized as follows.

- The timing driven buffer insertion technique for the bundled SWCNTs interconnect is proposed through adapting the timing driven buffering algorithm for copper interconnect.
- Our experiments are conducted with 500 scaled industrial nets and 10 different types of scaled buffers and inverters at 22 nm technology. With the same timing constraint, CNT buffering can save over 50% buffer area compared to copper buffering. In addition, it is demonstrated that CNT buffering can effectively reduce the delay by up to 32%.

In Section 10.2, the timing driven buffer insertion problem for CNT interconnects is formulated. The delay model for CNT-based interconnects is presented in Section 10.3. The algorithm for buffer insertion is described in Section 10.4. One example is designed to illustrate the algorithm in Section 10.5. Experimental results and analysis are presented in Section 10.6. A summary of work is given in Section 10.7.

10.2 Problem formulation

Consider a routing tree $T = (V, E)$ where $V = s_0 \cup V_s \cup V_n$, and $E \in V \times V$. Let $|V| = n$. Vertex s_0 is the source node and also called the root of the tree. V_s is the set of sink nodes. Each sink, denoted by s, has a sink capacitance and required arrival time $RAT(s)$. T is said to satisfy the timing constraint if its required arrival time at root is no earlier than the arrival time at root. Each edge, denoted by e, in E represents a segment of wire, which has edge resistance $R(e)$ and edge capacitance $C(e)$. V_n refers to the candidate buffer locations where the buffers can be inserted. In practice, they are discrete locations and are specified before buffer insertion algorithm by, e.g., wire segmenting technique [20].

A buffer library B which consists of a set of different types of buffers is given to the buffering problem. Let $|B| = m$. Each buffer, denoted by b, has cost $W(b)$, input capacitance $C(b)$, driving resistance $R(b)$, and intrinsic delay $t(b)$. Following most existing buffering works [1–5], the underlying routing tree can be assumed to be binary since trees in other topologies can be converted to a binary one using the technique in Reference 3. Given a tree in CNT interconnect layer, a buffer assignment is to determine the locations and the types of buffers which will be inserted to the routing tree. Our buffer insertion problem is formulated as follows.

Timing constrained minimum cost buffering for CNT interconnects: Given a binary routing tree with n candidate buffer locations in the CNT interconnect layers and a buffer library, to compute a buffer assignment solution such that the timing constraint is satisfied, and the total buffer cost is minimized.

10.3 CNT interconnects

To tackle the fundamental physical limits on copper interconnects, CNTs have emerged as promising replacements for copper interconnects due to their better conductivity and current-carrying capabilities. Table 10.1 from References 21 and 22 summarizes some major advantages of CNTs over copper materials. In fact, similar

Table 10.1 Comparison between CNT and Cu interconnects [21, 22]

Properties	**CNT**	**Cu**
Max. current density (A/cm^2)	10^{10}	10^6
Mean free path (nm)	1000	40
Thermal conductivity (W/mK)	6000	400

observations have been made from many other works [23–27]. Figure 10.4 shows an equivalent circuit model for the bundled SWCNTs, which is originally proposed in Reference 8. It will be described in detail as follows.

10.3.1 Resistance for CNT

10.3.1.1 Resistance for an isolated SWCNT

The resistance of an isolated SWCNT, denoted by $R_{isolated}$, is divided into two parts, the quantum resistance R_Q and scattering resistance R_S as shown in Figure 10.4.

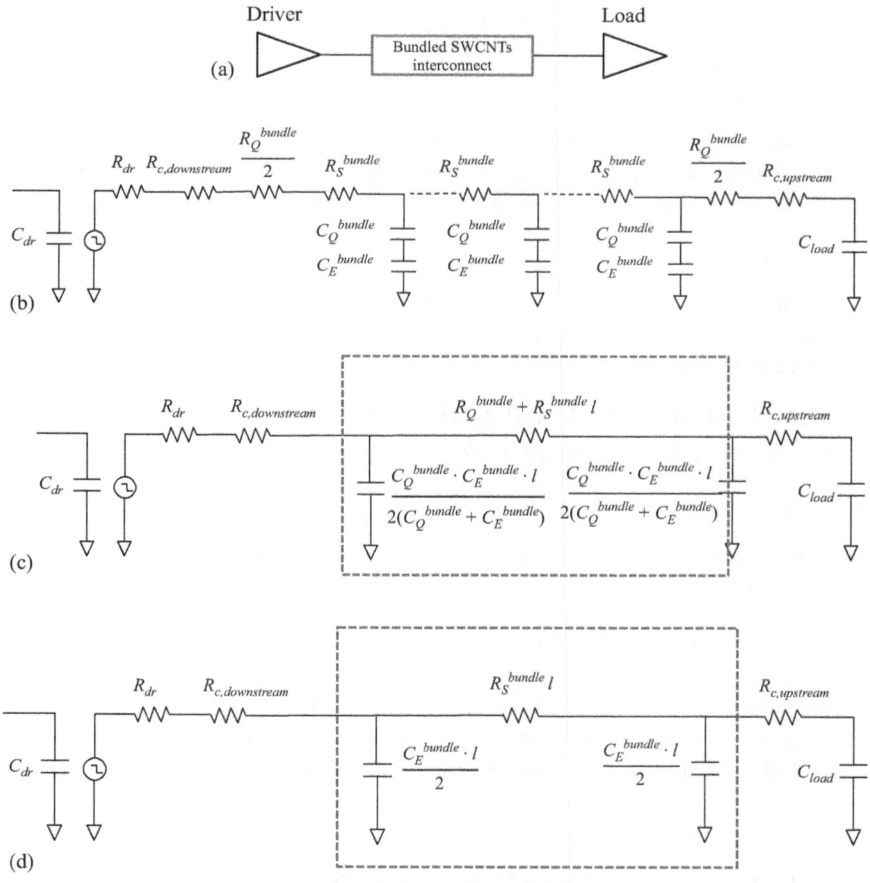

Figure 10.4 Equivalent circuit model for a bundled SWCNT interconnects [8]. (a) Schematic of bundled SWCNTs interconnects, (b) distributed equivalent circuit model for bundled SWCNTs interconnects, (c) equivalent π circuit model for bundled SWCNTs interconnects, and (d) simplified equivalent π circuit model for bundled SWCNTs interconnects

Recall that the mean free path, denoted by λ, refers to the average distance between two subsequent collisions of electrons. The mean free path of electrons for a CNT is about 1 μm as shown in Table 10.1, i.e., $\lambda = 1\ \mu\text{m}$. When $l \leq \lambda$ where l is the length of a CNT, we have [28]

$$R_Q = \frac{h}{4e^2} = 6.45\,\text{k}\Omega, \tag{10.1}$$

where ε is the electronic charge and h is Plank's constant. Thus, if the length l of a CNT is less than $\lambda = 1\ \mu\text{m}$, the resistance of CNT is independent of length.

For the length greater than the mean free path, the distributed scattering resistance for an interconnect with length l is [28, 29]

$$R_S l = \frac{hl}{4e^2\lambda}. \tag{10.2}$$

For simplicity, one defines $R_S = 0$ when $l \leq \lambda$. In practice, the total resistance of a single CNT, denoted by $R_{isolated}$, is expressed as the sum of quantum resistance and scattering resistance as shown in the following equation [8]:

$$R_{isolated} = R_Q + R_S l. \tag{10.3}$$

Comparing to copper global interconnect, a single SWCNT global interconnect has resistance of $6.45\,\text{k}\Omega/\mu\text{m}$, which is too large for timing minimization. However, if bundled SWCNTs are used, the resistance can be significantly reduced.

10.3.1.2 Resistance for bundled SWCNTs interconnects

The resistance of a bundle, denoted by R_{bundle}, is given by the following equation [29]:

$$R_{bundle} = R_{isolated}/N_{cnt}, \tag{10.4}$$

where N_{cnt} is the number of CNTs contained in the bundle. It is clear that the resistance decreases with increasing N_{cnt}.

10.3.1.3 Contact resistance

Due to the presence of imperfect metal and CNT contacts, contact resistance needs to be considered. According to Reference 18, some research groups have accomplished to fabricate the contact resistances ranging from a few hundred ohms to a few kilo-ohms which have similar magnitude with quantum resistance and scattering resistance.

10.3.2 Capacitance for CNT

10.3.2.1 Capacitance for an isolated SWCNT

The capacitance of the CNT comes from two aspects. One is the electrostatic capacitance denoted by C_E, and the other is quantum capacitance denoted by C_Q.

The quantum capacitance $C_Q l$ is obtained by [30]:

$$C_Q l = \frac{2e^2}{hv_f} l. \tag{10.5}$$

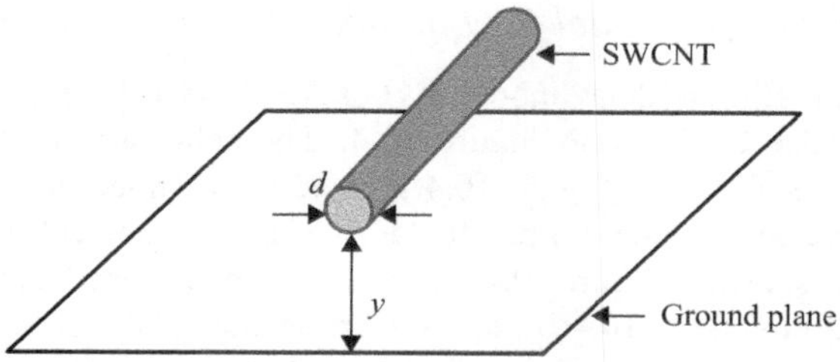

Figure 10.5 Isolated SWCNT with diameter d and over ground plan by y

Since an SWCNT has four conducting channels, the net quantum capacitance of an isolated SWCNT is

$$C_Q^{CNT} l = 4C_Q l. \tag{10.6}$$

The quantum capacitance for bundled SWCNTs interconnects can be computed as

$$C_Q^{bundle} l = N_{cnt} C_Q^{CNT} l. \tag{10.7}$$

The electrostatic capacitance C_E is calculated by treating the CNT as a thin wire, with diameter d and the distance to the ground plane y. $C_E l$ can be calculated as follows:

$$C_E l = \frac{2\pi\varepsilon}{\cosh^{-1}(y/d)} l, \tag{10.8}$$

where ε is the permittivity of free space. The electrostatic capacitance for bundled SWCNTs interconnects C_E^{bundle} is given by a parallel combination of all SWCNTs in the bundle. The electrostatic capacitance can be calculated using FastCap (Figure 10.5) [31].

According to Reference 8, the effect of the quantum capacitance is small, the effective capacitance of bundled SWCNTs interconnects is nearly equal to its electrostatic capacitance.

$$C_{bundle} l = C_E^{bundle} l. \tag{10.9}$$

10.3.3 Inductive impact is not important

According to Reference 8, the inductive impact is not important. It shows that an RC model for interconnect delay is accurate when the following inequality does not hold.

$$R_{dr} Cl < \frac{1}{2} RlCl < \sqrt{LC} l, \tag{10.10}$$

where R_{dr} is the driver impedance, and R, C, and L are the per unit length interconnect resistance, capacitance, and inductance, respectively. According to the simulation conducted in Reference 8 for different sizes of driver and SWCNTs, Inequality 10.10 is never satisfied. Therefore, RC model is sufficient to handle the bundled SWCNTs interconnects delay.

10.3.4 Elmore delay model for bundled SWCNTs interconnects

This work uses the Elmore delay model for bundled SWCNTs interconnects as presented in Reference 8. Refer to Figure 10.4. The schematic of the driver, load and interconnect is shown in Figure 10.4(a). The interconnect is made of bundled SWCNTs. Elmore delay model for bundled SWCNTs interconnects with a driver and a load capacitor is shown in Figure 10.4(c) which is derived from the distributed equivalent circuit model in Figure 10.4(b). R_{dr} is the resistance of the driver and C_{load} is the load capacitance connecting to the interconnect. $R_{c,downstream}$ is the contact resistance between the driver and the bundled SWCNTs interconnect, and $R_{c,upstream}$ is the contact resistance between the bundled SWCNTs interconnect and load capacitor. R_Q^{bundle} and R_S^{bundle} are the quantum and scattering resistances of bundled SWCNTs interconnects, respectively. C_Q^{bundle} and C_E^{bundle} are the quantum and electrostatic capacitances of bundled SWCNTs interconnects, respectively. Since the capacitance of bundled SWCNTs interconnects is approximately equal to the quantum capacitance of bundled SWCNTs interconnects and quantum resistance is not important for long global interconnect, the π model can be simplified to Figure 10.4(d).

10.4 Timing buffering for CNT interconnects

Our algorithm for CNT interconnects timing driven buffer insertion problem is based on the dynamic programming algorithm in Reference 2. In the algorithm, a three-tuple (Q, C, W) is used to characterize each buffering solution. Q represents the required arrival time for each buffering solution, C represents the downstream capacitance for each buffering solution, and W is the cumulative buffer cost of the buffering solution. Working under the dynamic programming framework [2], the tree is processed in a bottom-up fashion and a set of candidate buffering solutions and the corresponding three-tuples are propagated from sinks to driver. Let γ denote a buffer solution and Γ denote a solution set. Precisely, a routing tree is traversed in the post order. The algorithm will compute Q, C, and W from sinks up to the driver. The algorithmic flow is shown in Figure 10.6. During the dynamic programming, there are four operations, namely, add wire, add buffer, branch merge, and add driver. They are described as follows.

10.4.1 Add wire

Since one considers long global interconnect, it can be simply assumed that the distance between two consecutive buffers is larger than 1 μm. Under this assumption, the resistance of bundled SWCNTs interconnects can be simplified to $\frac{6.45\,\text{k}\Omega/\mu\text{m}}{N_{cnt}}$. In this operation, one is to add a wire from location v to its upstream location u for a candidate buffering solution as shown in Figure 10.7. Recall that the capacitance for the wire (u, v) is computed as $C(u, v) = C_E^{bundle} \cdot l(u, v)$ and the resistance for

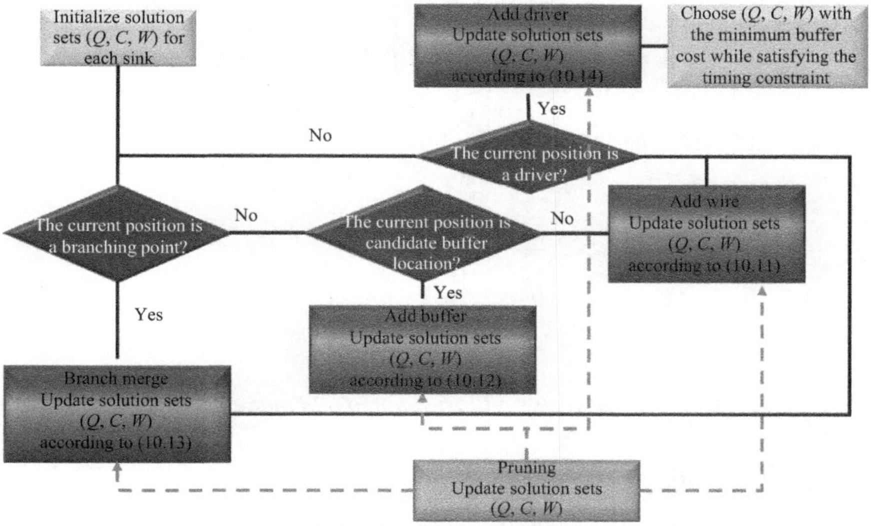

Figure 10.6 The algorithmic flow for the CNT buffering algorithm

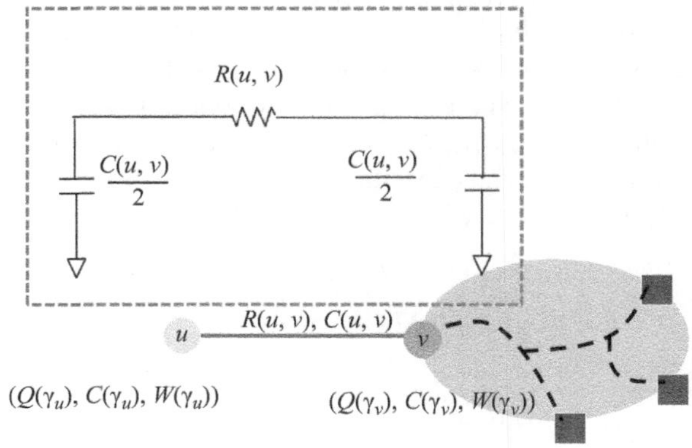

Figure 10.7 Illustration of add wire

the wire (u, v) is computed as $R(u, v) = R_{bundle} = R_S l(u, v)/N_{cnt}$, where $l(u, v)$ is the length of wire (u, v). One has

$$\begin{aligned} Q(\gamma_u) &= Q(\gamma_v) - R(u, v) \cdot \left[\frac{C(u, v)}{2} + C(\gamma_v) \right] \\ C(\gamma_u) &= C(\gamma_v) + C(u, v) \\ W(\gamma_u) &= W(\gamma_v). \end{aligned} \tag{10.11}$$

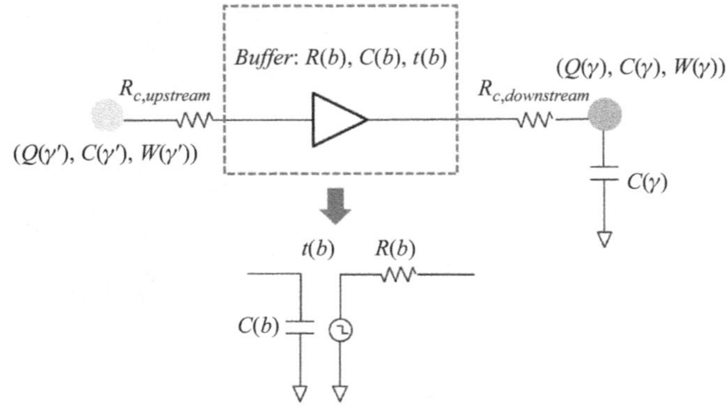

Figure 10.8 Illustration of add buffer

10.4.2 Add buffer

This operation is invoked when a buffer is to be inserted at a candidate buffer location v. In any buffering solution γ, after a buffer insertion, a new solution γ' will be generated. The cost $W(\gamma')$ will be computed as $W(\gamma') = W(\gamma) + W(b)$ if the buffer b is inserted. Refer to Figure 10.8. Recall that the buffer resistance is $R(b)$, buffer capacitance is $C(b)$, and buffer intrinsic delay is $t(b)$. To handle the contact resistance, recall that the contact resistance for the contact linking the buffer b with the downstream CNT wire is $R_{c,downstream}(b)$, and the contact resistance for the contact linking the upstream CNT wires with the buffer b is $R_{c,upstream}(b)$. The required arrival time needs to be updated considering the buffer delay, and capacitance needs to be set to the input capacitance of the buffer. Sinks can be similarly handled. We have

$$\begin{aligned} Q(\gamma') &= Q(\gamma) - R(b) \cdot C(\gamma) - R_{c,downstream}(b) \\ &\quad \cdot C(\gamma) - R_{c,upstream}(b) \cdot C(b) - t(b) \\ C(\gamma') &= C(b) \\ W(\gamma') &= W(\gamma) + W(b). \end{aligned} \tag{10.12}$$

10.4.3 Branch merge

Refer to Figure 10.9. This operation is to merge the solutions in two branches connected by a branching point. Since the solutions along each branch have been computed, one will compute the combinations among them. Suppose that there are a solution $(Q(\gamma_1), C(\gamma_1), W(\gamma_1))$ at left branch and a solution $(Q(\gamma_2), C(\gamma_2), W(\gamma_2))$ at right branch. After merging, we have

$$\begin{aligned} Q(\gamma) &= \min\{Q(\gamma_1), Q(\gamma_2)\} \\ C(\gamma) &= C(\gamma_1) + C(\gamma_2) \\ W(\gamma) &= W(\gamma_1) + W(\gamma_2). \end{aligned} \tag{10.13}$$

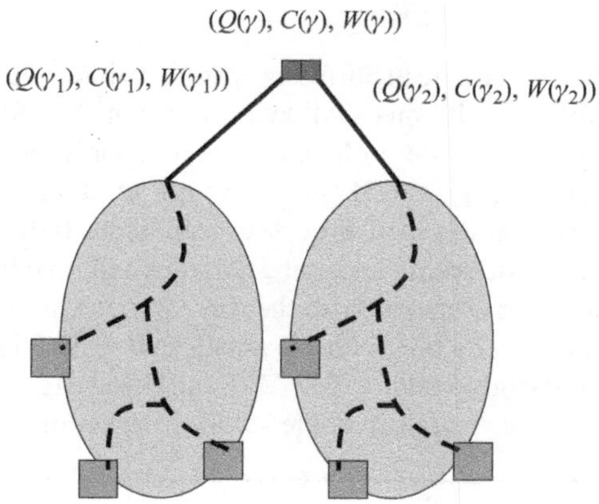

Figure 10.9 Illustration of branch merge

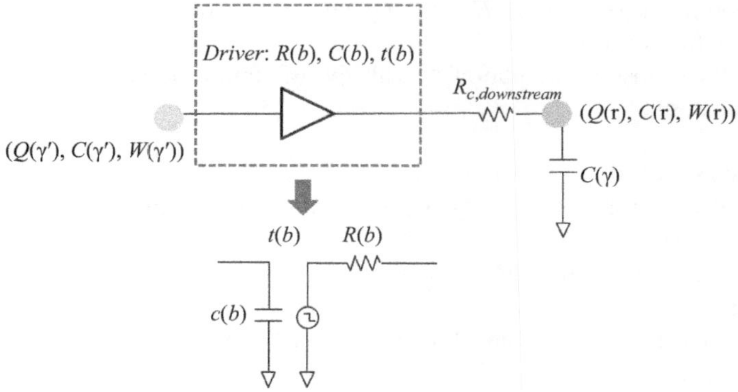

Figure 10.10 Illustration of add driver

That is, one needs to set the merged required arrival time to be smaller required arrival time of the two branches, the total downstream capacitance to be the sum of the downstream capacitance of the two branches, and the total buffer cost to be the sum of buffer costs of the two branches.

10.4.4 Add driver

Refer to Figure 10.10. This operation is to add the driver b to the candidate buffering solution. It is similar to the add buffer operation with the difference that one does not compute the delay due to the upstream contact resistance of the driver and one does not update the cumulative buffer cost.

$$\begin{aligned} Q(\gamma') &= Q(\gamma) - R(b) \cdot C(\gamma) - R_{c,downstream}(b) \cdot C(\gamma) - t(b) \\ C(\gamma') &= C(b) \\ W(\gamma') &= W(\gamma). \end{aligned} \tag{10.14}$$

10.4.5 Pruning

Pruning is an important technique in buffer insertion technique due to its effectiveness in reducing the number of solutions. Following Reference 2, for any two solutions denoted by γ_1, γ_2 at the same node, γ_2 is said to be inferior to γ_1 and is thus pruned if $Q(\gamma_1) \geq Q(\gamma_2)$, $C(\gamma_1) \leq C(\gamma_2)$, and $W(\gamma_1) \leq W(\gamma_2)$. In other words, one will compare γ_1 and γ_2 with the same set of processed candidate buffer locations by their required arrival time, downstream capacitance, and cumulative buffer cost.

When the solutions are propagated all the way up to the driver, one can obtain all the non-inferior solutions. The one with the smallest W satisfying timing constraint will be returned. The pseudo-code of the CNT buffering algorithm is summarized in Algorithm 10.1. Given a routing tree, the solutions are propagated in a bottom-up

---

Algorithm 10.1 CNT buffering algorithm

Input: Routing tree $T = (V, E)$, timing constraints of sinks, buffer library B, candidate buffer locations n
Output: Buffer insertion solution γ with the minimum buffer cost satisfying the timing constraint

for each sink s, **do**
    build a solution set $\Gamma_s = (Q_s, C_s, W_s)$, where Q_s is timing constraints of sink s, C_s is
    load capacitance of s and $W_s = 0$;
endfor
/* From each sink, prorogate along the ordered nodes given by a post-order traversal
  of routing tree T */
$k = 0$; /* k is the index of node */
while the node k is not the driver, **then do**
  if the node k is a branching point, **then do**
    $\Gamma_k = BranchMerge(\Gamma_{k_1}, \Gamma_{k_2}, k)$;
    $\Gamma_k = Pruning(\Gamma_k)$;
    else if current node k is a candidate buffer location, **then do**
      $\Gamma_k = AddBuffer(\Gamma_k, k)$;
      $\Gamma_k = Pruning(\Gamma_k)$;
      else do
        $k \leftarrow k + 1$;
        $\Gamma_k = AddWire(\Gamma_{k-1}, k - 1)$;
        $\Gamma_k = Pruning(\Gamma_k)$;
endwhile
$\Gamma_k = AddDriver(\Gamma_k, k)$;
$\Gamma_k = Pruning(\Gamma_k)$;
Pick the best solution γ from Γ_k at the driver

---

fashion from sinks to driver. The solution set of each sink is updated according to the buffer insertion approaches which are add wire, add buffer, branch merge, and add driver. To accelerate the process, the inferior solutions are deleted according to the pruning technique. At root, the best solution which has the least buffer cost and satisfies the timing constraint is returned as the buffer insertion solution.

10.5 An example

To explain how this algorithm works, an example is designed in this section. Refer to Figure 10.11. In this example, there are two sinks, one driver and one branching point. For each sink, there are corresponding required arrival time and load capacitance. For example, the required arrival time of sink 1 $RAT(s_1) = 60$ ps and load capacitance $C(s_1) = 1$ fF. There are three candidate buffer locations which are marked by small rectangular boxes with strips. There are several segments of wires, and the length of each wire is labeled in the tree. In addition, there are three types of buffers in the buffer library which are shown in Table 10.2. Each buffer has different resistance, capacitance, and area cost. For simplicity, the intrinsic delay of buffers is assumed to be zero. The CNT wire unit resistance is 6.45 kΩ/μm, and unit capacitance is 0.16 fF/μm.

According to the algorithm, one starts from sinks and propagates solutions to the driver. The procedure is shown in Figure 10.12–10.14. The first step is to initialize the tuple set at sink 1. Refer to Figure 10.12(a). The three-tuple is $(Q, C, W) = (60.000, 1.00, 0)$ in which $Q = RAT(s_1) = 60.000$ ps, and the downstream capacitance $C = C(s_1) = 1.00$ fF. Since there is no buffer inserted, buffer cost $W = 0\ \mathrm{nm}^2$. The next step is to add wire as shown in Figure 10.12(b). The operation AddWire() is described in Section 10.4.1. According to (10.11), the three-tuple set is updated to $(Q', C', W') = (53.034, 1.16, 0)$ where $Q' = Q - r \cdot l \cdot [cl/2 + C] = 60.000 - 6.45 \cdot 1.00 \cdot [0.16 \cdot 1/2 + 1] = 53.034$, $C' = C + cl = 1.00 + 0.16 \cdot 1 = 1.16$, and $W' = W = 0$.

When the current node is a candidate buffer location, operation AddBuffer() is performed which is described in Section 10.4.2. We have three types of buffers, together with no buffer insertion, there are four three-tuples associated with the buffer

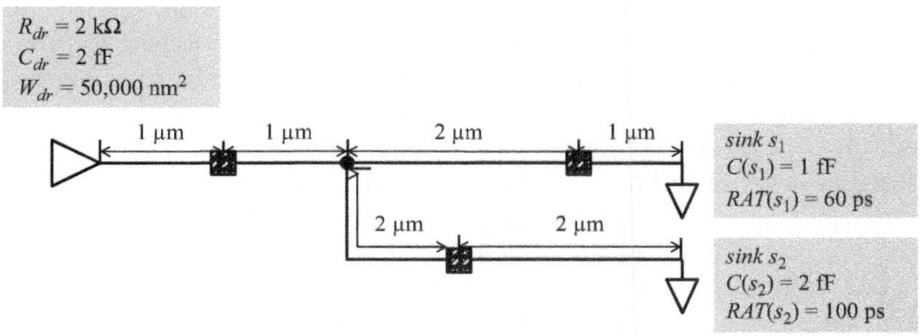

Figure 10.11 An example

Table 10.2 Buffer library

Buffer type	r_b(kΩ)	c_b(fF)	Area (nm^2)
1	1.0	0.5	30,000
2	0.5	1.0	60,000
3	0.2	3.0	80,000

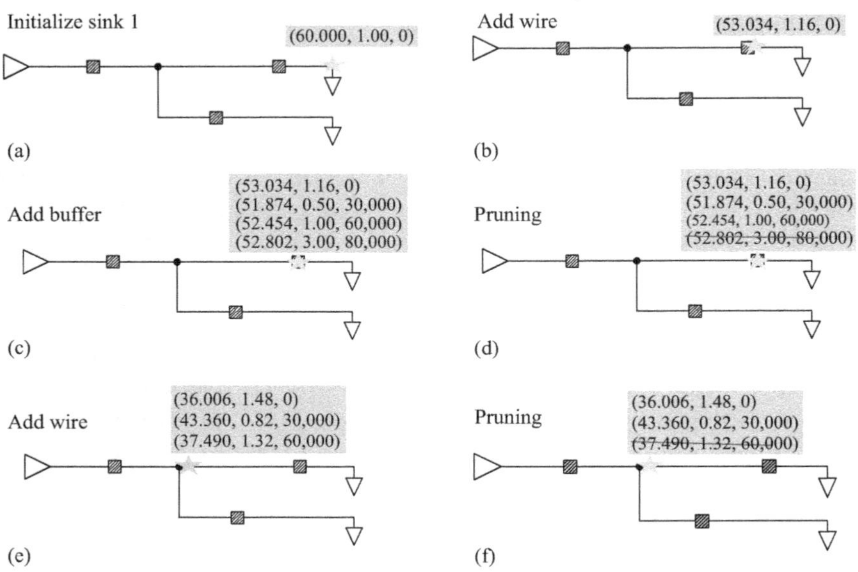

Figure 10.12 Illustration of add buffer and add wire in first branch. (a) Initialize sink 1, (b) add wire, (c) add buffer, (d) pruning after add buffer, (e) add wire, and (f) pruning after add wire

insertion. Refer to Figure 10.12(c). The first three-tuple $(Q', C', W') = (Q, C, W) = (53.034, 1.16, 0)$ is that there is no buffer inserted at current node, and it is exactly same as the solution of the node before add buffer. Take buffer type 1 in Table 10.2 as an example. The three-tuple set $(Q, C, W) = (53.034, 1.16, 0)$ is updated to $(Q', C', W') = (51.874, 0.50, 30,000)$ according to (10.12) where $Q' = Q - r_b \cdot C = 53.034 - 1.0 \cdot 1.16 = 51.874$, $C' = c_b = 0.50$, and $W' = W_b = 30,000$.

Pruning is performed after add buffer as shown in Figure 10.12(d). For example, according to the pruning technique presented in Section 10.4.5, the tuple $(52.802, 3.00, 80,000)$ is pruned since it is inferior to $(53.034, 1.16, 0)$. Comparing to $(53.034, 1.16, 0)$, $(52.802, 3.00, 80,000)$ has smaller required arrival time, larger downstream capacitance, and larger buffer cost. AddWire() is then performed in this branch until reaching the branching point. The results are shown in Figure 10.12(e). Subsequently, the nodes along the second branch are handled in a similar fashion to

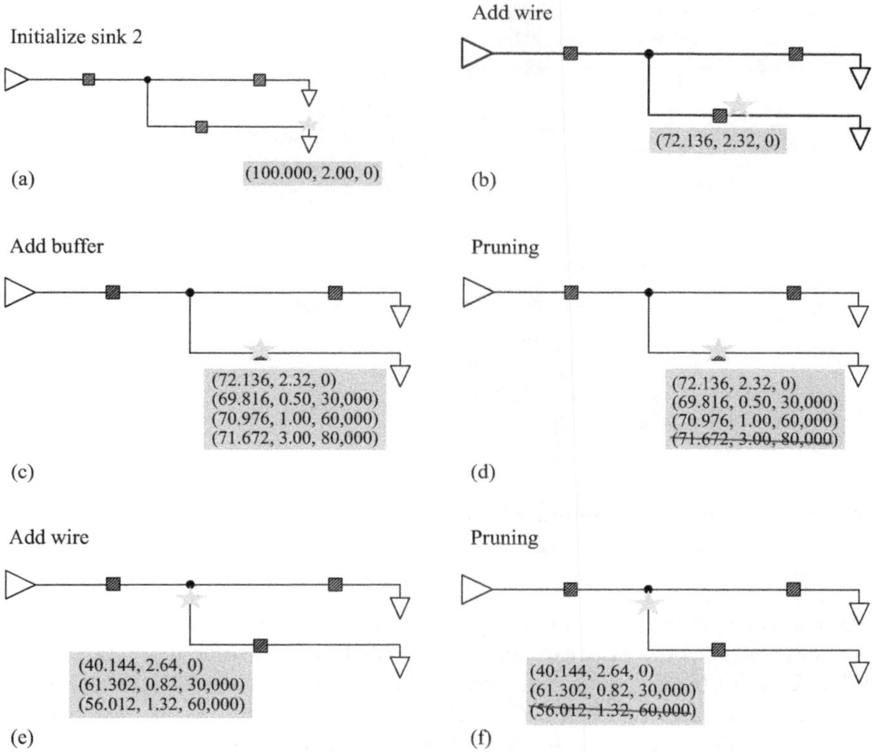

Figure 10.13 Illustration of add buffer and add wire in second branch. (a) Initialize sink 2, (b) add wire, (c) add buffer, (d) pruning after add buffer, (e) add wire, and (f) pruning after add wire

the first one. The steps are shown in Figure 10.13. One can obtain the solution set at the branching point in the second branch.

The current node is a branching point of the two branches, therefore operation BranchMerge() is performed. According to Sections 10.4.3 and 10.4.5, the solution set after branch merging and pruning can be obtained which is shown in Figure 10.14. Each three-tuple in the first branch should be merged with all the three-tuples in the second branch. Since there are two three-tuples in each branch, four combinations can be generated. Take the first three-tuple $(Q', C', W') = (36.006, 4.12, 0)$ after branch merge as an example. It is merged from two three-tuples (36.006, 1.48, 0) and (40.144, 2.64, 0) according to (10.13) where $Q' = \min\{36.006, 40.144\} = 36.006$, $C' = 1.48 + 2.64 = 4.12$, and $W' = 0 + 0 = 0$. After branch merging, there are two wires and one candidate buffer location which can be similarly handled. The steps are shown in Figure 10.15.

At root, the driver is inserted. The resistance and capacitance of the driver are shown in Figure 10.11. According to Section 10.4.4, the solution set can be

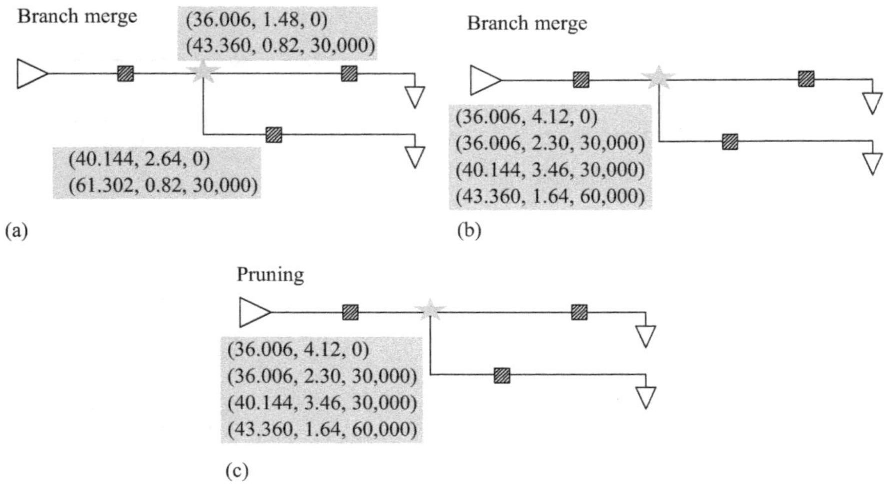

Figure 10.14 Illustration of branch merge. (a, b) Branch merge and (c) pruning after branch merge

obtained which are shown in Figure 10.16(a)(b). Take the three-tuple $(Q', C', W') = (-0.968, 2.00, 80,000)$ as an example. It is updated from the three-tuple $(Q, C, W) = (4.272, 2.62, 30,000)$ before add driver. According to (10.14), $Q' = Q - R_{dr} \cdot C = 4.272 - 2 \cdot 2.62 = -0.968$, $C' = C_{dr} = 2.00$, and $W' = W + W_{dr} = 30,000 + 50,000 = 80,000$. However, this three-tuple is pruned since it does not satisfy the timing constraint where $Q' = -0.968 < 0$. Finally, the solution with the minimum buffer cost satisfying the timing constraint is chosen at the driver. In this example, solution $(16.220, 2.00, 110,000)$ is returned as shown in Figure 10.16(c).

10.6 Experimental results

10.6.1 Experimental setup

Our CNT interconnects based timing driven minimum cost buffer insertion algorithm is implemented in C language and tested on a machine with 3.40 GHz Intel Pentium CPU and 3GB memory. The results of CNT buffering are compared with copper buffering. In this paper, the buffer cost is measured by buffer area.

Our buffer library consists of 10 buffer types including 5 buffers and 5 inverters. Due to the lack of industrial buffer library at 22 nm technology, a buffer library of 45 nm technology [32] is scaled to 22 nm technology. To calculate the resistance, capacitance, and intrinsic delay of different types of buffers and inverters at 22 nm node, the simulation is performed using ngspice [33]. The resistance, capacitance, intrinsic delay, and gate area are shown in Table 10.3. Linear fitting is applied to obtain resistance and intrinsic delay. The capacitance of buffer is simulated using method in Reference 34 (Figure 10.17).

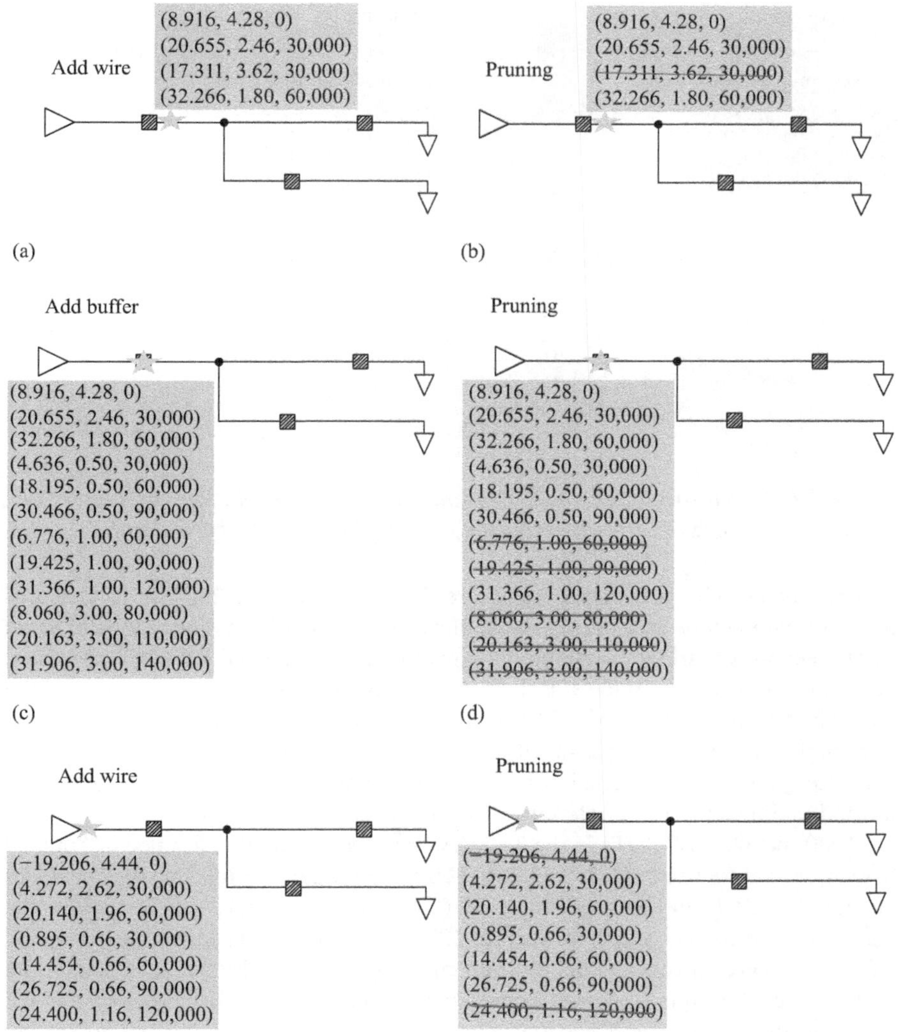

Figure 10.15 Illustration of add wire and add buffer after branch merge. (a) Add wire, (b) pruning after add wire, (c) add buffer, (d) pruning after add buffer, (e) add wire, and (f) pruning after add wire

Our experiments are performed to 500 global nets extracted from an industrial Application Specific Integrated Circuit (ASIC) chip in an old technology. Due to the lack of industrial nets in 22 nm technology, we scale wire lengths of these old technology nets to 22 nm technology.

The parameters of copper interconnects and bundled SWCNTs interconnects are presented in Table 10.4. The unit resistance and unit capacitance are for 1 μm.

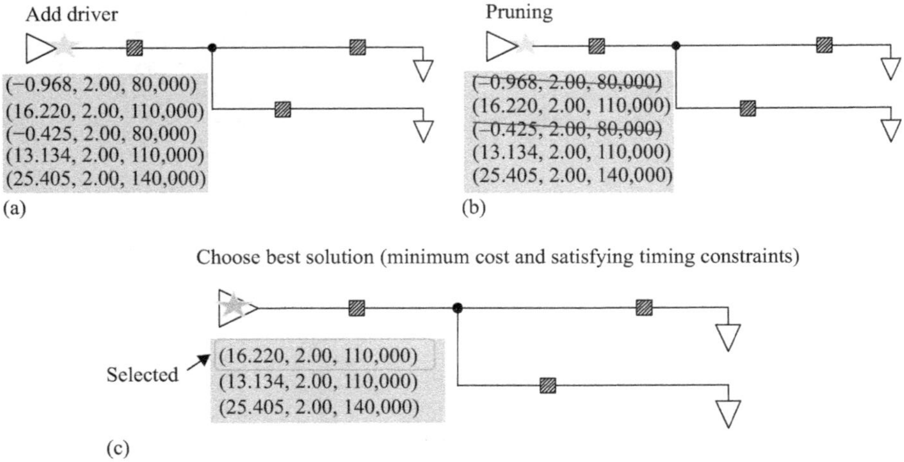

Figure 10.16 Illustration of add driver and choosing best solution. (a) Add driver, (b) pruning after add driver, and (c) choose the best solution

The parameters of copper interconnects are obtained from ITRS 2007 [35].[1] One uses the ITRS parameters since the resistance and capacitance information of the industrial 22 nm technology are not available. The parameters of bundled SWCNTs interconnects are calculated as follows. Refer to Figure 10.18. The cross-section area of the global interconnects is set to be $33 \times 88\,\text{nm}^2$. For global interconnects, the resistance of a single SWCNT is approximately $6.45k\Omega/\mu\text{m}$ since the effect of quantum resistance for global interconnects is small. The impact of different numbers of SWCNTs in the bundle to the CNT resistance can be observed from Figure 10.18. If there are 1000 metallic SWCNTs in the $33 \times 88\,\text{nm}^2$ area, the total resistance of bundled SWCNTs interconnects is $6.45\,\text{k}\Omega/\mu\text{m}/1000 = 6.45\Omega/\mu\text{m}$. Note that the density of bundled SWCNTs interconnects is $1000/(33 \cdot 88) = 0.34\,\text{nm}^{-2}$ which is below the maximum density $0.66\,\text{nm}^{-2}$ from ITRS 2011 [37]. The unit capacitances of bundled SWCNTs interconnects and copper interconnects are set to be the same according to Reference 8. In this work, one considers both the ideal contact resistance and the practical contact resistance. The ideal contact resistance means no contact resistance. In the following discussion, without considering contact resistance is identical to ideal contact resistance. The practical contact resistance is set to 100Ω which is achievable according to Reference 18.

10.6.2 Experimental results

Two sets of experiments are conducted which are timing constrained minimum cost buffering and timing minimization without cost minimization.

[1] Note that the feature sizes predicted by ITRS 2007 are smaller than those in the industrial 22 nm technology according to Reference 36.

Table 10.3 Parameters of different inverters and buffers types at 22 nm node. (Note that the inverters in BUF are different from those in INV.)

	BUF_X1	BUF_X2	BUF_X4	BUF_X8	BUF_X16	INV_X1	INV_X2	INV_X4	INV_X8	INV_X16
Resistance (Ω)	2310.0	1201.0	618.9	315.5	159.6	1846.0	976.5	514.8	270.2	139.7
Capacitance (fF)	0.21	0.44	0.88	1.76	3.51	0.44	0.87	1.74	3.49	6.97
Intrinsic delay (ps)	2.93	2.91	2.87	2.87	2.87	0.59	0.62	0.61	0.61	0.61
Area (nm^2)	15,197.6	30,395.2	60,790.4	121,580.8	243,161.6	10,115.6	20,231.2	40,462.4	80,924.8	161,849.6

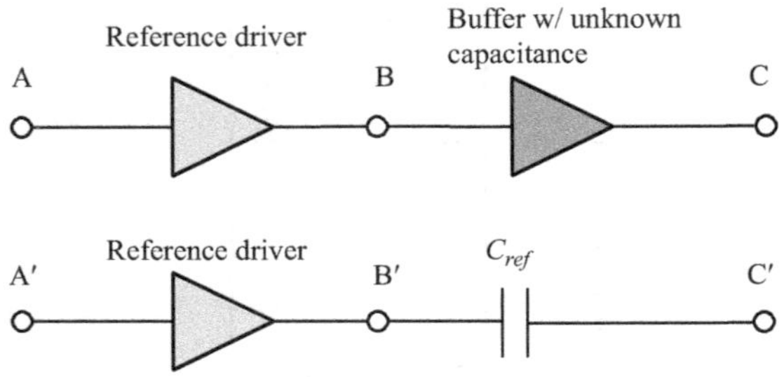

Figure 10.17 Spice method to calculate the capacitance of buffer

Table 10.4 Unit resistance and capacitance (for 1 μm) of global interconnects with Cu and bundled SWCNTs at 22 nm node

Properties	Cu	CNT
Unit resistance (Ω)	14.50	6.45
Unit capacitance (fF)	0.16	0.16

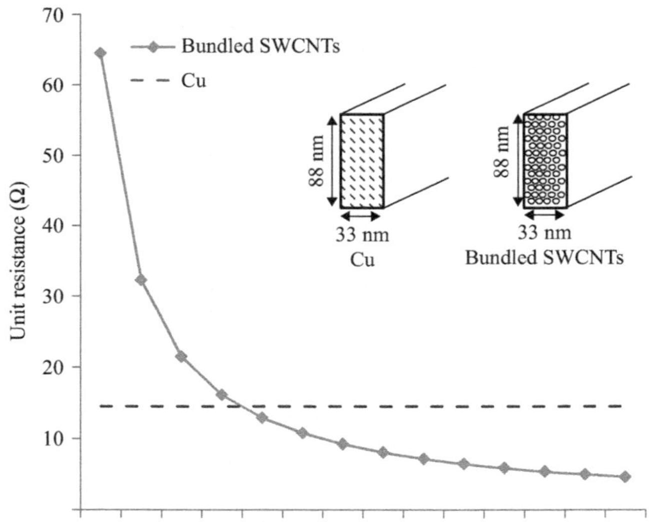

Figure 10.18 Resistance comparison and cross-section area of Cu and bundled SWCNTs global interconnects in 22 nm technology

Table 10.5 Timing constrained minimum cost buffering results on five representative nets

Test cases	CNT w/o contact resistance			CNT w/contact resistance (100Ω)			Cu		
	Area (nm^2)	# Buffers	Delay (ps)	Area (nm^2)	# Buffers	Delay (ps)	Area (nm^2)	# Buffers	Delay (ps)
1	318,666.0	7	754	379,359.0	7	762	955,997.0	18	766
2	364,162.0	5	611	424,855.0	6	599	819,412.0	17	611
3	222,543.0	5	676	222,543.0	5	691	475,433.0	12	702
4	50,578.0	3	1019	80,924.8	4	927	202,312.0	10	994
5	40,462.4	2	722	40,462.4	2	736	91,040.4	5	870

Table 10.6 Average result for timing constrained minimum cost buffering on 500 nets

Test cases	Area (nm^2)	Area ratio	# Buffers	Delay (ps)	# Solutions	CPU (s)
CNT w/o contact resistance	107,816.70	0.42	3.4	1125.8	2193.2	3.79
CNT w/ contact resistance (100Ω)	105,494.80	0.41	3.5	1127.9	1827.9	3.15
Cu	255,110.10	1.00	7.7	1248.9	2250.0	3.54

For timing constrained minimum cost buffering, the results on five representative nets are shown in Table 10.5, and the results on 500 nets are shown in Table 10.6. We make the following observations.

- One can see that in order to achieve the similar delay, the CNT buffering saves more than 50% buffer area over copper buffering. Averaging over 500 nets, CNT buffering without considering contact resistance saves 58% buffer area, and CNT buffering with 100Ω contact resistance saves 59% buffer area. Take net 1 in Table 10.5 as an example, CNT buffering without considering contact resistance saves 67% buffer area, and CNT buffering with 100Ω contact resistance saves 60% buffer area.
- The total number of buffers in CNT buffering is much (about 2×) smaller than that of copper buffering thanks to the fact that wire resistivity of bundled SWCNTs interconnects is much lower than that of copper for global interconnects as shown in Table 10.4.
- One can see that the contact resistance does not have significant impact on the performance for CNT interconnects timing constrained minimum cost buffering.
- It would be interesting in investigating the delay–area tradeoff between copper buffering and CNT buffering. For this, four nets in benchmark are chosen to run the buffering algorithm while keeping all non-dominated solutions. One can

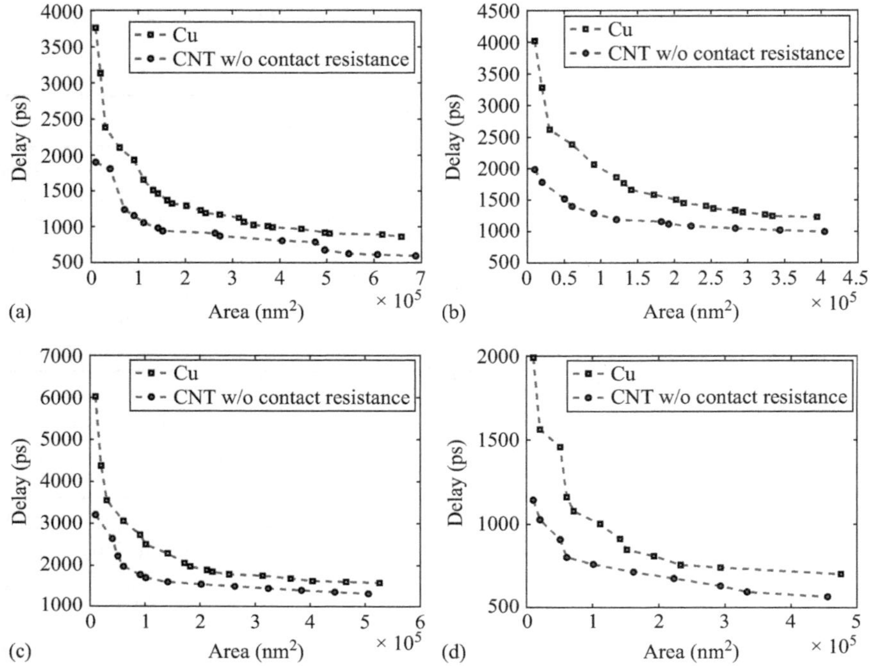

Figure 10.19 Area and delay comparison between Cu and CNT of some representative nets

generate delay–area tradeoff curves for copper buffering and CNT buffering. Refer to Figure 10.19. It is clear that CNT buffering always outperforms the copper buffering in terms of timing and buffer area.

The above results are obtained through setting certain timing constraint and compute the minimum area solutions. One may be interested in the best achievable timing in both of CNT buffering and copper buffering. The results of five representative nets for buffering timing minimization without considering cost are shown in Table 10.7. It demonstrates that CNT buffering can reduce timing by up to 32% which is obtained from net 5. In addition, the contact resistance has some impact on the performance of CNT buffering such as area and timing.

10.7 Conclusions

CNT interconnects have become a promising replacement material for copper interconnects thanks to their superior conductivity. This chapter discusses the development of the first timing driven buffer insertion technique for CNT interconnects. A timing driven buffer insertion algorithm is designed for bundled SWCNTs interconnects

Table 10.7 Timing minimization (without considering cost) on five nets

Test cases		1	2	3	4	5
CNT w/o contact resistance	Area (nm^2)	3,307,950.0	2,867,260.0	2,477,160.0	3,039,520.0	1,945,290.0
	# Buffers	50	51	44	44	32
	Delay (ps)	376	216	314	249	188
CNT w/ contact resistance (100Ω)	Area (nm^2)	1,463,910.0	1,468,890.0	1,408,250.0	1,458,970.0	1,094,230.0
	# Buffers	36	31	31	24	18
	Delay (ps)	423	263	347	302	229
Cu	Area (nm^2)	2,851,920.0	2,745,490.0	2,269,040.0	2,872,350.0	2,142,860.0
	# Buffers	65	55	56	48	36
	Delay (ps)	479	317	382	363	276

through adapting the traditional buffer insertion algorithm including add wire, add buffer, branch merge, add driver, and pruning. Our experimental results demonstrate that with the same timing constraint, CNT buffering can save over 50% buffer area compared to copper buffering. In addition, CNT buffering can effectively reduce the delay by up to 32% without considering cost.

Acknowledgment

This work was supported in part by NSF CAREER Award CCF-1349984.

References

[1] L. P. P. P. van Ginneken, "Buffer placement in distributed RC-tree networks for minimal Elmore delay," in *Proceedings of the IEEE International Symposium on Circuits and Systems*, pp. 865–868, 1990.

[2] J. Lillis, C. K. Cheng, and T. T. Y. Lin, "Optimal wire sizing and buffer insertion for low power and a generalized delay model," *IEEE Journal of Solid State Circuits*, vol. 31, no. 3, pp. 437–447, 1996.

[3] W. Shi and Z. Li, "A fast algorithm for optimal buffer insertion," *IEEE Transactions on Computer-Aided Design (TCAD)*, vol. 24, no. 6, pp. 879–891, 2005.

[4] S. Hu, C. J. Alpert, J. Hu, *et al.*, "Fast algorithms for slew constrained minimum cost buffering," in *Proceedings of ACM/IEEE Design Automation Conference (DAC)*, vol. 26, no. 11, pp. 2009–2022, 2006.

[5] S. Hu, Z. Li, and C. J. Alpert, "A fully polynomial time approximation scheme for timing driven minimum cost buffer insertion," in *Proceedings of ACM/IEEE Design Automation Conference (DAC)*, pp. 424–429, 2009.

[6] C. Xu, H. Li, and K. Banerjee, "Graphene nano-robbon (GNR) interconnects: A genuine contender or a delusive dream?" in *Proceedings of IEEE International Electron Devices Meeting (IEDM)*, pp. 15–17, 2008.
[7] [Online]. Available: http://www.jcrystal.com/products/wincnt/.
[8] N. Srivastava, H. Li, F. Kreupl, and K. Banerjee, "On the applicability of single-walled carbon nanotubes as VLSI interconnects," *IEEE Transactions on Nanotechnology*, vol. 8, no. 4, pp. 542–559, 2009.
[9] F. Kreup, A. Graham, M. Liebau, G. Duesberg, R. Seidel, and E. Unger, "Carbon nanotubes for interconnect applications," in *Proceedings of IEEE International Electron Devices Meeting (IEDM)*, pp. 683–686, 2004.
[10] S. W. Lee, D. S. Lee, R. E. Morjan, *et al.*, "A three-terminal carbon nanorelay," *Nano Letters*, pp. 2027–2030, 2004.
[11] S. Sukirno, Z. Bisri, L. Hasanah, M. Mursal, I. Usman, and A. B. Suryamas, "Low temperature carbon nanotube fabrication using very high frequency-plasma enhanced chemical vapour deposition method," in *Proceedings of IEEE International Conference on Semiconductor Electronics*, pp. 155–159, 2006.
[12] J. Wu, M. Eastman, T. Gutu, *et al.*, "Fabrication of carbon nanotube-based nanodevices using a combination technique of focused ion beam and plasma-enhanced chemical vapor deposition," *Applied Physics Letters*, vol. 91, no. 17, pp. 173122-1–173122-3, 2007.
[13] S. H. Lee, M. Bumki, S. Park, K. C. Lee, and S. S. Lee, "Fabrication of carbon nanomechanical resonators with embedded single walled carbon nanotube stiffening layers," in *Proceedings of 2010 IEEE International Conference on Micro Electro Mechanical Systems (MEMS)*, pp. 268–271, 2010.
[14] K. Chikkadi, M. Haluska, C. Hierold, and C. Roman, "Process control monitors for individual single-walled carbon nanotube transistor fabrication processes," in *Proceedings of 2013 IEEE International Conference on Microelectronic Test Structures (ICMTS)*, pp. 173–177, 2013.
[15] A. Nieuwoudt and Y. Massoud, "On the optimal design, performance, and reliability of future carbon nanotube-based interconnect solutions," *IEEE Transactions on Electron Devices*, vol. 55, no. 8, pp. 2097–2110, 2008.
[16] A. Srivastava, A. K. Sharma, and Y. Xu, "Carbon nanotubes for next generation very large scale integration interconnects," *Journal of Nanophotonics*, vol. 4, no. 1, pp. 1–26, 2010.
[17] G. Close and H.-S. Wong, "Assembly and electrical characterization of multiwall carbon nanotube interconnects," *IEEE Transactions on Nanotechnology*, vol. 7, no. 5, pp. 596–600, 2008.
[18] A. Naeemi and J. D. Meindl, "Design and performance modeling for single-wall carbon nanotubes as local, semi-global, and global interconnects in gigascale integrated systems," *IEEE Transactions on Electron Devices*, vol. 54, no. 1, pp. 26–37, 2008.
[19] L. Liu, Y. Zhou, and S. Hu, "Buffering single-walled carbon nanotubes bundle interconnects for timing optimization," in *Proceedings of IEEE Computer Society Annual Symposium on VLSI (ISVLSI)*, pp. 362–367, 2014.

[20] C. J. Alpert and A. Devgan, "Wire segmenting for improved buffer insertion," in *Proceedings of ACM/IEEE Design Automation Conference (DAC)*, pp. 588–593, 1997.

[21] B. Wei, R. Vajtai, and P. Ajayan, "Reliability and current carrying capacity of carbon nanotubes," *Applied Physics Letters*, vol. 79, no. 8, pp. 1172–1174, 2001.

[22] B. Radosavljevic, J. Lefebvre, and A. T. Johnson, "High-field electrical transport and breakdown in bundles of single-wall carbon nanotubes," *Physical Review B*, vol. 64, no. 24, pp. 241307.1–241307.4, 2001.

[23] H. S. Gokturk, "Electrical properties of ideal carbon nanotubes," in *Proceedings of 5th IEEE Conference on Nanotechnology*, pp. 800–803, 2005.

[24] L. K. Scheffer, "CAD implications of new interconnect technologies," in *Proceedings of ACM/IEEE Design Automation Conference (DAC)*, pp. 576–581, 2007.

[25] N. Srivastava and K. Banerjee, "Interconnect challenges for nanoscale electronic circuits," *TMS Journal of Materials (JOM), Special Issue on Nanoelectronics*, vol. 56, no. 10, pp. 30–31, 2004.

[26] A. Raychowdhury and K. Roy, "A circuit model for carbon nanotube interconnects: Comparative study with Cu interconnects for scaled technologies," in *Proceedings of International Conference on Computer Aided Design*, pp. 237–240, 2004.

[27] Y. Xu and A. Srivastava, "A model for carbon nanotube interconnects," *International Journal of Circuit Theory and Applications*, vol. 38, no. 6, pp. 559–575, 2010.

[28] S. Datta, "Electrical resistance: An atomistic view," *Nanotechnology*, vol. 15, pp. S433–S451, 2004.

[29] K. N. Srivastava, "Performance analysis of carbon nanotube interconnects for VLSI applications," in *Proceedings of IEEE/ACM International Conference on Computer-Aided Design (ICCAD)*, pp. 383–390, 2005.

[30] P. J. Burke, "Luttinger liquid theory as a model of the gigahertz electrical properties of carbon nanotubes," *IEEE Transactions on Nanotechnology*, vol. 1, no. 3, pp. 129–144, 2002.

[31] K. Nabors and J. White, "FastCap: A multipole accelerated 3-D capacitance extraction program," *IEEE Transactions on Computer-Aided Design of Integrated Circuits and Systems*, vol. 10, no. 11, pp. 1447–1459, 1991.

[32] [Online]. Available: http://www.nangate.com.

[33] [Online]. Available: http://ngspice.sourceforge.net/.

[34] [Online]. Available: https://www.pagiamtzis.com/articles/how-to-find-input-capacitance-using-spice.

[35] *International Technology Roadmap for Semiconductors,* 2007 [Online]. Available: http://www.itrs.net/Links/2007ITRS/Home2007.htm.

[36] R. Aitken, G. Yeric, B. Cline, *et al.*, "Physical design and FinFETs," in *Proceedings of International Symposium on Physical Design*, pp. 65–68, 2014.

[37] *International Technology Roadmap for Semiconductors,* 2011 [Online]. Available: http://www.itrs.net/Links/2011ITRS/Home2011.htm.

Chapter 11

Memristor modeling – static, statistical, and stochastic methodologies

Hai (Helen) Li[1], Miao Hu[1], and Beiye Liu[1]

Memristor, the fourth passive circuit element, has attracted increased attention since it was rediscovered by HP Lab in 2008. Its distinctive characteristic to record the historic profile of the voltage/current creates a great potential for future neuromorphic computing system design. However, at the nano-scale, process variation control in the manufacturing of memristor devices is very difficult. The impact of process variations on a memristive system that relies on the continuous (analog) states of the memristors could be significant. In addition, the stochastic switching behaviors have been widely observed. To facilitate the investigation on memristor-based hardware implementation, we compare and summarize different memristor modeling methodologies, from the simple static model to statistical analysis by taking the impact of process variations into consideration and the stochastic behavior model based on the real experimental measurements. In this work, we use the most popular TiO_2 thin-film device as an example to analyze the memristor's electrical properties. Our proposed modeling methodologies can be easily extended to the other structures/materials with necessary modifications.

11.1 Introduction

In the circuit theory before 1971, there were only three fundamental passive circuit elements – resistor, capacitor, and inductor. Professor Chua observed its incompleteness and predicted the existence of memristor, the fourth basic circuit element which can build the bridge between the magnetic flux ϕ and the electronic charge q [5]. In 2008 – 37 years after Professor Chua's prediction, the first experimental realization of memristor was demonstrated in a TiO_2 thin-film two-terminal device by HP Labs [25]. The memristive effect was achieved by moving the doping front along the device [25].

Besides the solid-state device, magnetic technology provides other possible solutions to build a memristive system [19, 27]. Three spintronic memristor structures have been proposed in Reference 27. They are spin valve with spin-torque-induced

[1] University of Pittsburgh, Pittsburgh, PA 15260, USA

domain-wall motion in the free layer, MTJ (magnetic tunneling junction) with spin-torque-induced magnetization switching, and thin-film element with spin-torque-induced domain-wall motion. Compared to the solid-state thin-film device [25], the behavior of a spintronic memristor, e.g., the relationship between the memristance and the current through the memristor, can be controlled more flexibly. Also, the technology to integrate magnetic device on the top of CMOS device has become mature in the development of magnetic memory [13].

Memristors have many unique properties, such as simple physical structure, high-density, non-volatility, historic behavior, low power consumption, and good scalability [7]. The non-volatile nature and the good scalability (down to 10 nm and below with an integration density of 100 Gbits/cm^2) of memristor make it an attractive candidate for the next-generation memory technology [7, 32]. Because it can record the historic behavior of the current through it, memristor is expected to have a great potential in electronic neutral network [4, 10, 20]. Applications in analog circuitries, such as Op-Amp and UWB receiver, have also been investigated recently [26, 28, 31].

As process technology shrinks down to decananometer (sub-50 nm) scale, device parameter fluctuations incurred by process variations have become a critical issue affecting the electrical characteristics of devices [1]. The situation in a memristive system could be even worse when utilizing the analog states of the memristors in design: variations of device parameters, e.g., the instantaneous memristance, can result in the shift of electrical responses, e.g., current. The deviation of the electrical excitations will affect memristance, because the total charge through a memristor indeed is the historic behavior of its current profile. In this work, we explore the implications of the device parameters of memristors to the circuit design by taking into account the impact of memristor geometry variations. The evaluations were conducted based on both theoretical analysis and Monte Carlo simulations.

The device geometry variations significantly influence the electrical properties of nano-devices [23]. For example, the random uncertainties in lithography and patterning processes lead to the random deviations of line edge print-images from its ideal pattern, which is called *line edge roughness* (LER) [18]. Thickness fluctuation is caused by deposition process in which mounds of atoms form and coarsen over time. As technology shrinks, the geometry variations do not decrease accordingly. In this work, we propose an algorithm to generate a large volume of *three-dimensional* (3D) memristor structures to mimic the geometry variations. The LER model is based on the latest LER characterization method for *electron beam lithography* (EBL) technology from top-down scanning electron microscope measurement [12]. Other process variations such as *random discrete doping* (RDD) could also result in the fluctuations of the electrical properties of devices. However, because the existing memristors are all based on the thin-film deposition technology, the local randomness of RDD is not as significant as geometry variations, and therefore, is not covered in this work.

Moreover, metal oxide–based memristor behaves stochastically, and hence even a single memristive device demonstrates large variations in performance. More specifically, the static states of a single memristor are not fixed, but have large variations with skewed distributions and heavy tails [30]. The switching mechanism of a memristor, that is, its dynamic behavior, performs as a stochastic process [29], which has been widely demonstrated in various materials [6, 33]. Thus, we built a stochastic behavior model of TiO_2 memristive devices based on the real measurement results [16, 30] to better facilitate the exploration of memristors in hardware implementation. The model bypasses material-related parameters while directly linking the device analog behavior to stochastic functions. Simulations show that the proposed stochastic device model fits well to the existing device measurement results.

Note that memristive function can be achieved by various materials and device structures. For its popularity, TiO_2-based memristor is analyzed and evaluated. However, our proposed modeling methodologies and design philosophies are not limited by the specific types of devices and can be easily extended to the other structures/materials with necessary modifications.

The organization of this paper is as follows: Section 11.2 briefly introduces the physical mechanisms of TiO_2 thin-film memristors and describes its simple static model; Section 11.3 analyzes the memristor model under geometry variations; Section 11.4 presents a stochastic modeling based on the real device measurement; Section 11.5 describes the neuromorphic system composed of bidirectional synapses and analyzes its performance for pattern recognition; at last, Section 11.6 concludes this work.

11.2 Static modeling

11.2.1 TiO_2 thin-film memristor

In 2008, HP Lab demonstrated the first intentional memristive device by using a $Pt/TiO_2/Pt$ thin-film structure [25]. The conceptual view is illustrated in Figure 11.1(a): two metal wires on Pt are used as the top and bottom electrodes, and a thick titanium dioxide film is sandwiched in between. The stoichiometric TiO_2 with an exact 2:1 ratio of oxygen to titanium has a natural state as an insulator. However, if the titanium dioxide is lacking a small amount of oxygen, its conductivity becomes relatively high like a semiconductor. We call it *oxygen-deficient titanium dioxide* (TiO_{2-x}) [17]. The memristive function can be achieved by moving the doping front: A positive voltage applied on the top electrode can drive the oxygen vacancies into the pure TiO_2 part and therefore lower the resistance continuously. On the other hand,

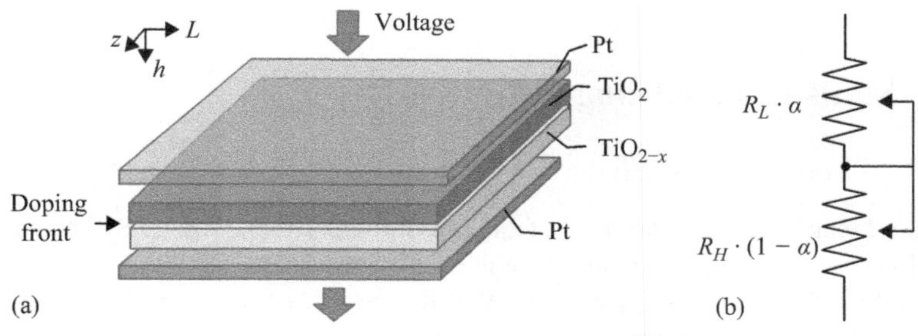

Figure 11.1 TiO_2 thin-film memristor. (a) Structure and (b) equivalent circuit

a negative voltage applied on the top electrode can push the dopants back to the TiO_{2-x} part and hence increase the overall resistance. For a TiO_2-based memristor, R_L (R_H) is used to denote the low (high) resistance when it is fully doped (undoped).

11.2.2 Memristor static (bulk) model

Figure 11.1(b) illustrates a coupled variable resistor model for a TiO_2-based memristor. It is equivalent to two series-connected resistors with an overall resistance:

$$M(\alpha) = R_L \cdot \alpha + R_H \cdot (1 - \alpha). \tag{11.1}$$

Here α ($0 \leq \alpha \leq 1$) is the relative doping front position, which is the ratio of doping front position over the total thickness of TiO_2 thin film. The velocity of doping front movement $v(t)$, which is driven by the voltage applied across the memristor $V(t)$, can be expressed as:

$$v(t) = \frac{d\alpha}{dt} = \mu_v \cdot \frac{R_L}{h^2} \cdot \frac{V(t)}{M(\alpha)}, \tag{11.2}$$

where μ_v is the equivalent mobility of dopants, h is the total thickness of the TiO_2 thin film, and $M(\alpha)$ is the total memristance when the relative doping front position is α.

Bulk model is a general model derived from the mathematical definition of memristor, which assumes a flat doping front moving up or down. However, in reality, filamentary conduction has been observed in nano-scale semiconductors: the current travels through some high conducting filaments rather than passes the device evenly [14, 15]. The doping front is formed so randomly that a few filaments dope much faster than others, observed as hot spots on the device. This is called as *filament conduction phenomenon*. The way we solved the conflict between the bulk and filament models in this work can be explained as follows: when cutting the device into many tiny filaments as we shall describe in Section 11.3, it is reasonable to assume a small flat doping front exists in each filament. Therefore, bulk model can be used for each small flat doping front movement.

Recent experiments showed that μ_v is not a constant but grows exponentially when the bias voltage goes beyond certain threshold voltage [24]. Nevertheless, the structure of TiO_2 memristor model, i.e., (11.2), still remains valid.

11.3 Statistical modeling

11.3.1 Theoretical analysis

The actual length (L) and width (z) of a memristor are affected by LER. The variation of thickness (h) of a thin-film structure is usually described by *thickness fluctuation*. As a matter of convenience, we define that the impact of process variations on any given variable can be expressed as a factor $\theta = \omega'/\omega$, where ω is its ideal value, and ω' is the actual value under process variations.

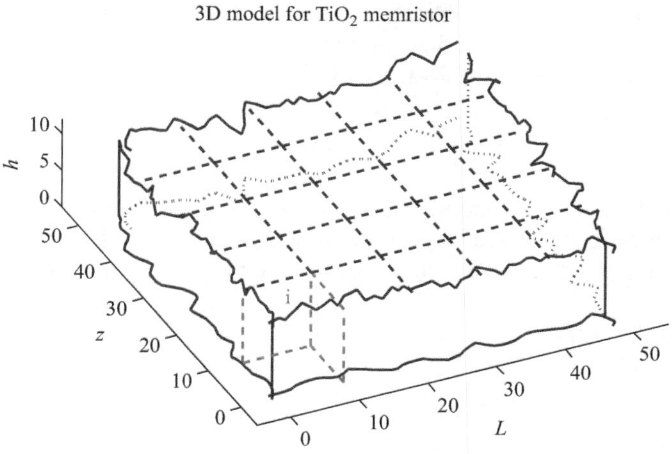

Figure 11.2 An example of 3D TiO_2 memristor structure, which is partitioned into many filaments in statistical analysis

In TiO_2 thin-film memristors, the current passes through the device along the thickness (h) direction. Ideally the doping front has an area $S = L \cdot z$. To simulate the impact of LER on the electrical properties, the memristor device is divided into many small filaments between the two electrodes. Each filament i has a cross-section area ds and a thickness h. Figure 11.2 demonstrates a non-ideal 3D structure of a TiO_2 memristor (i.e., with geometry variations in consideration), which is partitioned into many filaments in statistical analysis.

Ideally, the cross-section area of a filament is ds/s of the entire device area, and its thickness is h. For filament i, the ideal upper bound and lower bound of the memristance can be expressed as:

$$R_{i,H} = R_H \cdot S/ds, \text{ and} \tag{11.3a}$$

$$R_{i,L} = R_L \cdot S/ds. \tag{11.3b}$$

Here, $\theta_{i,s}$ represents the variation ratio on the cross-section area ds, which is caused by two-dimensional (2D) LER. Similarly, $\theta_{i,h}$ is the variation ratio on thickness h due to Thickness Fluctuations (TF). The resistance of a filament is determined by its section area and thickness, i.e., $R = \rho \cdot h/s$, where ρ is the resistance density. Therefore, the actual upper and the lower bounds under the process variations can be expressed as:

$$R'_{i,H} = R_{i,H} \cdot \theta_{i,h}/\theta_{i,s}, \text{ and} \tag{11.4a}$$

$$R'_{i,L} = R_{i,L} \cdot \theta_{i,h}/\theta_{i,s}. \tag{11.4b}$$

If a filament is small enough, we can assume that it has a flat doping front. The actual doping front velocity in filament i after considering process variations can be calculated by replacing the ideal values with actual values in (11.2), such as:

$$v_i'(t) = \mu_v \cdot \frac{R_{i,L}'}{h'^2} \cdot \frac{V(t)}{M_i'(\alpha_i')}. \tag{11.5a}$$

Here h' and M_i' are the actual thickness and memristance of filament i. As such, we can get a set of related equations for filament i, including the doping front position $\alpha_i'(t)$, the corresponding memristance $M_i'(\alpha_i')$, and the current through the filament i:

$$\alpha_i'(t) = \int_0^t v'(\tau) \cdot d\tau, \tag{11.5b}$$

$$M_i'(\alpha_i') = \alpha_i' \cdot R_{i,L}' + (1 - \alpha_i') \cdot R_{i,H}', \tag{11.5c}$$

$$I_i'(t) = V(t)/M_i'(\alpha_i'). \tag{11.5d}$$

By combining (11.5a)–(11.5d), the doping front position in every filament i under process variations $\alpha_i'(t)$ can be obtained by solving the differential equation:

$$\frac{d\alpha_i'(t)}{dt} = \mu_v \cdot \frac{R_{i,L}'}{h'^2} \cdot \frac{V(t)}{\alpha_i'(t) \cdot R_{i,L}' + (1 - \alpha_i'(t)) \cdot R_{i,H}'}. \tag{11.6}$$

Equation (11.6) indicates that the doping front movement is dependent on the specific electrical excitations, e.g., $V(t)$. For instance, applying a sinusoidal voltage source to the TiO_2 thin-film memristor such as:

$$V(t) = V_m \cdot \sin(2\pi f \cdot t), \tag{11.7}$$

the corresponding doping front position of filament i can be expressed as:

$$\alpha_i'(t) = \frac{R_{i,H} - \sqrt{R_{i,H}^2 - A \cdot B(t) \cdot \frac{2}{\theta_{i,h}^2} + 2C_0 \cdot A \cdot \frac{\theta_{i,s}}{\theta_{i,h}}}}{A}, \tag{11.8}$$

where $A = R_{i,H} - R_{i,L}$, $B(t) = \mu_v \cdot R_{i,L} \cdot V_m \cdot \cos(2\pi f \cdot t)$, and C_0 is an initial state constant. The term $B(t)$ accounts for the effect of electrical excitation on doping front position. The terms $\theta_{i,s}$ and $\theta_{i,h}$ represent the effect of both LER and TF on memristive behavior. Moreover, the impact of the geometry variations on the electrical properties of memristors could be affected by the electrical excitations. For example, we can set $\alpha(0) = 0$ to represent the case that the TiO_2 memristor starts from $M(0) = R_H$. In such a condition, $C[1]$ becomes 0, and hence, the doping front position $\alpha_i'(t)$ can be calculated as:

$$\alpha_i'(t) = \left\{ R_{i,H} - \sqrt{R_{i,H}^2 - 2A \cdot B(t)/\theta_{i,h}^2} \right\} /A, \tag{11.9}$$

which is affected only by TF and electrical excitations. LER will not disturb $\alpha_i'(t)$ if the TiO_2 thin-film memristor has an initial state $\alpha(0) = 0$.

The overall memristance of the memristor can be calculated as the total resistance of all n filaments connected in parallel. When n goes to ∞, we can have

$$R'_H = \frac{R_H}{\int_0^{\infty} \theta_{i,h}/\theta_{i,s} \cdot di}, \text{ and} \tag{11.10a}$$

$$R'_L = \frac{R_L}{\int_0^{\infty} \theta_{i,h}/\theta_{i,s} \cdot di}. \tag{11.10b}$$

The instantaneous memristance of the overall memristor can be defined as:

$$M'(t) = \frac{V(t)}{I'(t)} = \frac{1}{\int_0^{\infty} 1/M'_i \cdot di}. \tag{11.11}$$

Since the doping front position movement in each filament will not be the same because h'_i varies, we define the average doping front position of the whole memristor as:

$$\alpha'(t) = \frac{R'_H - M'(t)}{R'_H - R'_L}. \tag{11.12}$$

11.3.2 3D device sample generation flow

Analytic modeling is a fast way to estimate the impact of process variations on memristors. However, we noticed that in modeling some variations analytically, e.g., simulating the LER may be beyond the capability of analytic model [12]. The data on silicon variations, however, is usually very hard to obtain simply due to intellectual property protection. To improve the accuracy of our evaluations, we create a simulation flow to generate 3D memristor samples with the geometry variations including LER and thickness fluctuation. The correlation between the generated samples and the real silicon data are guaranteed by the sanity check of the LER characterization parameters. The flow is shown in Figure 11.3.

Many factors affecting the quality of the line edges show different random effects. Usually statistical parameters such as the *auto-correlation function* (ACF) and *power spectral density* (PSD) are used to describe the property of the line edges.

ACF is a basic statistical function of the wavelength of the line profile, representing the correlation of point fluctuations on the line edge at different position. PSD describes the waveform in the frequency domain, reflecting the ratio of signals with different frequencies to the whole signal.

Considering that LER issues are related to fabrication processes, we mainly target the nano-scale pattern fabricated by EBL. The measurements show that under such a condition, the line edge profile has two important properties: (1) the line edge profile in ACF figure demonstrates regular oscillations, which are caused by periodic composition in the EBL fabrication system; and (2) the LER mainly concentrates in a low frequency zone, which is reflected by PSD figure [12].

To generate line edge samples close to the real cases, we can equally divide the entire line edge into many segments, say, n segments. Without losing the LER

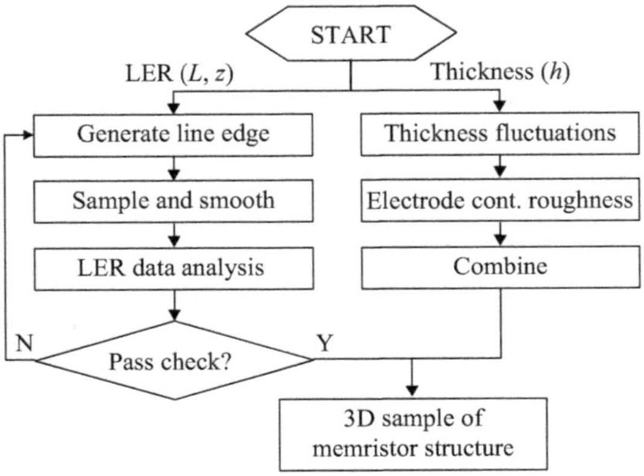

Figure 11.3 The flow of 3D memristor structure generation including geometry variations

properties in EBL process, we modified the random LER modeling proposed in [2] to a simpler form with less parameters. The LER of the ith segment can be modeled by

$$\mathrm{LER}_i = L_{LF} \cdot \sin(f_{\max} \cdot x_i) + L_{HF} \cdot p_i. \tag{11.13}$$

The first term on the right side of (11.13) represents the regular disturbance at the low frequency range, which is modeled as a sinusoid function with amplitude L_{LF}. $f_{\max}$ the mean of the low frequency range derived from PSD analysis. Without loss of generality, a uniform distribution $x_i \in U(-1, 1)$ is used to represent an equal distribution of all frequency components in the low frequency range. The high-frequency disturbances are also taken into account by the second term on the right side of (11.13) as a Gaussian white noise with amplitude L_{HF}. Here p_i follows the normal distribution $N(0, 1)$ [12]. The actual values of L_{LF}, L_{HF}, and $f_{\max}$ are determined by ACF and PSD.

To ensure the correlation between the generated line edge samples with the measurement results, we introduce four constraints to conduct a sanity check of the generated samples:

- σ_{LER}: the *root mean square* (RMS) of LER;
- σ_{LWR}: the RMS of *line width roughness* (LWR);
- *Sk*: skewness, used to specify the symmetry of the amplitude of the line edge; and
- *Ku*: kurtosis, used to describe the steepness of the amplitude distribution curve.

The above four parameters are widely used in LER characterization and can be obtained from measurement results directly [12]. Only the line edge samples that satisfy the constraints will be taken as valid device samples. Table 11.1 summarizes the parameters used in our algorithm, which are correlated with the characterization method and experimental results in [12]. And Figure 11.4 shows the LER characteristic parameters distribution among 1000 Monte Carlo simulations.

Table 11.1 The parameters/constraints in LER characterization

Parameters		Constraints	
L_{LF}	0.8 nm	σ_{LER}	2.5–3.5 nm
$f_{\max}$	1.8 MHz	σ_{LWR}	4.0–5.0 nm
L_{HF}	0.4 nm	Sk	0.1–0.2 nm
/	/	Ku	2.5–3.5 nm

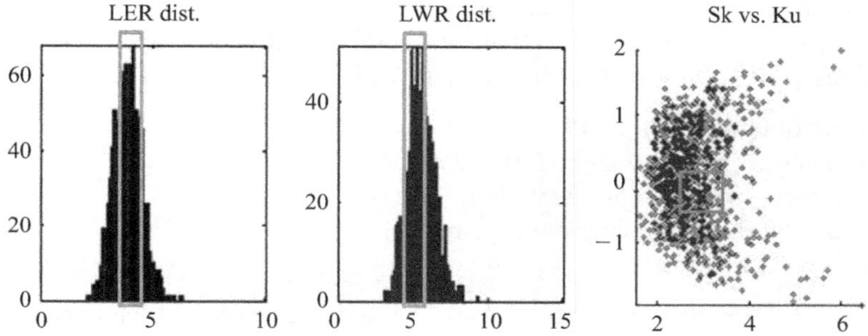

Figure 11.4 LER characteristic parameters distribution among 1000 Monte Carlo simulations. Constraints are shown in rectangles

Even the main function has captured the major features of LER, it is not enough to mimic all the LER characteristics. The difference between real LER distribution and our modeling function results in the fact that some generated samples are not qualified compared to the characteristic parameters, or the constraints of the real LER profile. Thus, sanity check which screens out the unsuccessful results is necessary. Only those samples in rectangles shown in Figure 11.4 satisfy the constraints and will be used for the device electrical property analysis. The criteria of the sanity check are defined based on the measurement results of real LER data.

The thickness fluctuation is caused by the random uncertainties in sputter deposition or atomic layer deposition. It has a relatively smaller impact than the LER and can be modeled as a Gaussian distribution. We also considered roughness of electrode contact in our simulation: The means of the thickness of each memristor is generated by assuming it follows the Gaussian distribution. Each memristor is then divided into many filaments between the two electrodes. The roughness of electrode contracts is modeled based on the variations of the thickness of each filament. Here, we assume that both thickness fluctuations and electrode contact roughness follow Gaussian distributions with a deviation $\sigma = 2\%$ of thin-film thickness.

Figure 11.2 indeed is an example of 3D structure of a TiO_2 thin-film memristor generated by the proposed flow. It illustrates the effects of all the geometry variations on a TiO_2 memristor device structure. According to Section 11.3.1, a 2D partition is required for the statistical analysis. In the given example, we partition the device into 25 small filaments with the ideal dimensions of length $L = 10$ nm, width $z = 10$ nm, and height $h = 10\,\text{nm}$. Each filament can be regarded as a small memristor, which is affected by either only TF or both LER and TF. The overall performance of device can be approximated by paralleled connecting all the filaments.

11.3.3 The impact of process variations

To evaluate the impact of process variations on the electrical properties of memristors, we conducted Monte Carlo simulations with 10,000 qualified 3D device samples generated by our proposed flow. A sinusoidal voltage source in (11.7) with $V_m = 1V$ and $f = 0.5\,\text{Hz}$ is applied as the external excitation. The initial state of the memristor is set as $M(\alpha = 0) = R_H$. Both separate and combined effects of geometry variations on various properties of memristors are analyzed, including:

- the distribution of R_H and R_L;
- the change of memristance $M(t)$ and $M(\alpha)$;
- the velocity of wall movement $v(\alpha)$;
- the current through memristor $i(t)$; and
- the I–V characteristics.

The $\pm 3\sigma$ (minimal/maximal) values of the device/electrical parameters as the percentage of the corresponding ideal values are summarized in Table 11.2. For those parameters that vary over time, we consider the variation at each time step of all the devices. The simulation results considering only either LER or TF are also listed. To visually demonstrate the overall impact of process variations on the memristive behavior of TiO_2 memristors, the dynamic responses of 100 Monte Carlo simulations are shown in Figure 11.5.

Table 11.2 shows that the static behavior parameters of memristors, i.e., R_H and R_L, are affected in a similar way by both LER and thickness fluctuations. This is consistent to our analytical results in (11.10), which show that θ_s and θ_h have the similar effects on the variation of R'_H and R'_L.

However, thickness fluctuation shows a much more significant impact on the memristive behaviors such as $v(t)$, $\alpha(t)$, and $M(\alpha)$ than LER does. It is because the doping front movement is along the thickness direction: $v(t)$ is inversely proportional

Table 11.2 3σ min./max. of TiO_2 memristor parameters

Sinusoidal voltage	Only LER		Only TF		Overall	
	−3σ (%)	+3σ (%)	−3σ (%)	+3σ (%)	−3σ (%)	+3σ (%)
R_H&R_L	−5.4	4.1	−5.5	4.8	−6.4	7.3
$M(\alpha)$	−5.4	4.1	−37.1	20.8	−36.5	24.1
$\alpha(t)$	0.0	0.0	−13.3	27.5	−14.7	27.4
$v(\alpha)$	0.0	0.0	−9.3	15.6	−10.4	16.9
$i(\alpha)$	−4.7	5.7	−9.3	15.7	−10.7	17.2
Power	−4.7	5.7	−8.8	14.1	−10.1	15.6

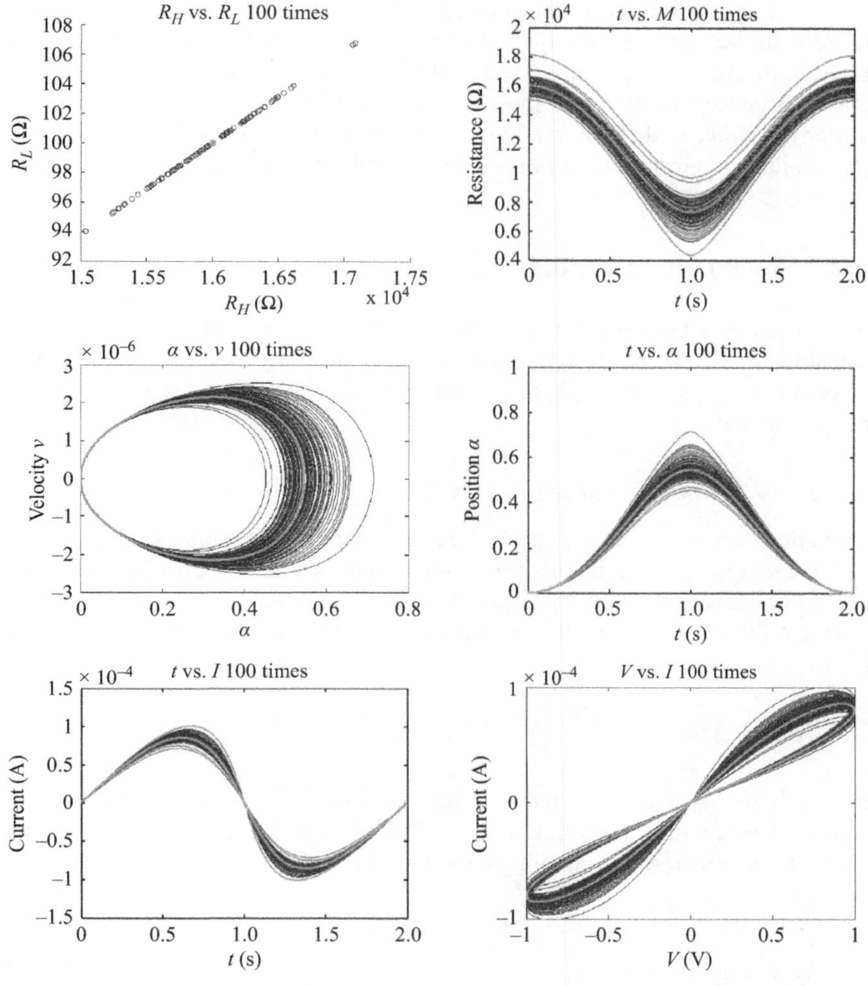

Figure 11.5 Simulation results for TiO_2 thin-film memristors. The dark thin curves are from 100 Monte Carlo simulations, and light thick lines are the ideal condition. From top left to right bottom, the figures are R_H vs. R_L; $M(t)$ vs. t; v vs. α; α vs. t; I vs. t; and I–V characteristics

to the square of the thickness, and $\alpha(t)$ is the integral of $v(t)$ over time as shown in (11.5a) and (11.5b). For the same reason, thickness fluctuations significantly affect the instantaneous memristance $M(\alpha)$ as well.

Because the thickness of the TiO_2 memristor is relatively small compared to other dimensions, we assume the doping front cross-section area is a constant along the thickness direction in our simulation. The impact of LER on $\alpha(t)$ or $v(t)$, which is relatively small compared to that of the thickness fluctuations, is ignored in Table 11.2.

An interesting observation in Figure 11.5 is that as the doping front α moves toward 1, the velocity v regularly grows larger and reaches its peak at the half period of the sinusoidal excitation, i.e., $t = 1$ s. This can be explained by (11.5c): the memristance is getting smaller as α moves toward 1. With the same input amplitude, a smaller resistance will result in a bigger current and therefore a bigger variation on $v(t)$. Similarly, memristance $M(\alpha)$ reaches its peak variance when α is close to 1.

11.4 Stochastic modeling

To better describe the stochastic memristive switching in both static states and dynamic switching process, we proposed a stochastic model for TiO_2 memristive switch based on both the inspection of the physical mechanisms [21, 29] and the statistical analysis of experimental data [6, 33].

11.4.1 ON and OFF static states

The static stochastic behavior can be described by the distributions of R_L and R_H. In TiO_2 memristor, the initial barrier width w follows a normal distribution, and the device resistance exponentially depends on w. Therefore, the distribution of state resistance follows the lognormal *probability density function* (pdf) function, which is [29]:

$$f_x(x; \mu, \sigma) = \exp\left(-\frac{(\ln x/\mu)^2}{2\sigma^2}\right) / \left(x\sigma\sqrt{2\pi}\right), x > 0. \tag{11.14}$$

Here, μ is the normal mean and σ is the standard deviation. Note that R_L or R_H does not change within a given static state. Therefore, we can use lognormal function (Lognorm) to generate the sampling data, such as:

$$R_L = \text{Lognorm}(\mu_{R_L}, \sigma_{R_L}), \text{ and} \tag{11.15a}$$

$$R_H = \text{Lognorm}(\mu_{R_H}, \sigma_{R_H}). \tag{11.15b}$$

11.4.2 Dynamic switching process

The dynamics in TiO_2 memristor is a complex oxide electroforming process. It can be explained as an electro-reduction and vacancy creation process caused by high electric fields and enhanced by electrical Joule heating. Usually, the barrier width w is used to model the vacancy channeling mechanism. Although the vacancy channeling mechanism has been evidenced by experiments [29], it is difficult to match it to a pure physical model. Instead, our model is based on the analysis of three major behaviors; we start with a mathematical analysis of the analog stochastic switching behavior from the statistical aspect and then bridge the parameters in mathematical expression with the physical excitation. At last, the impact of over tune is integrated into the stochastic model.

Analog stochastic switching behavior: The stochastic resistance changing has been observed in high-frequency measurement at low voltage [21]. The time dependency of switching probability can be approximated by the *cumulative density function* (CDF) of lognormal distribution, such as [16]:

$$P(\text{Success switch}) = \frac{1}{2}\text{erfc}\left[-\frac{(\ln t_{switch}/\mu_t)^2}{\sqrt{2}\sigma_t^2}\right]. \tag{11.16}$$

Here, t_{switch} represents the pulse width of activation time. And μ_t and σ_t are related to the external voltage V.

Instead of studying the complicated physical mechanism and its impact, we use mathematical method to analyze the ON–OFF switching probability. According to (11.16), the ON–OFF switching probability can be approximated by a CDF of lognormal distribution, differentiation of P(Success switch) at t_{switch}, then is a pdf of the lognormal distribution, such as:

$$\frac{dP(\text{Success switch})}{dt_{switch}} = f_{t_{switch}}(t_{switch}; \mu_t, \sigma_t). \tag{11.17}$$

Equation (11.17) describes the distribution of the increment of switching probability dP(Success switch) at time t_{switch} when applying a signal with a short pulse width dt_{switch}.

The switching mechanism of a memristive device is intrinsic. Hence, the characteristic of the stochastic behavior remains unchanged and follows the same probability function during its switching process. From its physical meaning perspective, (11.17) reflects the increment of switching probability at time t_{switch}, which can be associated with the resistance change ΔR. Physically, a successful switching event with a pulse of t_{switch} indicates that the device resistance changes from R_L to R_H, or vice versa, that is, $\Delta R = |R_H - R_L|$.

Considering that both ON and OFF switching are the cumulative results of the analog resistance changing and the increment of switching probability is directly reflected by the change of resistance, the change of analog resistance at time t_{switch} can be generated by mapping to the distribution of the increment of switching probability, leading to

$$\frac{dR}{dt} = (R_H - R_L) \cdot f_{t_{switch}}(t_{switch}; \mu_t, \sigma_t). \tag{11.18}$$

Time and voltage dependency of switching probability: This describes the switching probability of memristive switch under applied voltage V and activation time t_{switch}. The switching process resulted from the cumulative impact of input signals can be modeled with CDF function. The lognormal switching time distribution comes from the nonlinear switching dynamics of the devices. Considering that the median switch time (μ_t) is exponentially dependent on the applied voltage amplitude V, we approximate μ_t as an exponential function, such as:

$$\mu_t = \exp(aV + b), \tag{11.19}$$

where a and b are fitting parameters.

Since σ_t has only a weak dependence on V, we can approximate the relationship between σ_t and V by a hard threshold squashing function, such as:

$$\sigma_t = \begin{cases} \sigma_{thres\_H} & (\sigma_t \geq \sigma_{thres\_H}) \\ c \cdot V + d & (\sigma_{thres\_L} < \sigma_t < \sigma_{thres\_H}), \\ \sigma_{thres\_L} & (\sigma_t \leq \sigma_{thres\_L}) \end{cases} \tag{11.20}$$

where c and d are fitting parameters. σ_{thres_H} and σ_{thres_L} are the upper and lower boundaries, respectively. Our model applies two individual sets of fitting parameters to ON and OFF switching processes.

The resistance shifting due to over tune: Over tune stands for the behavior when one or more external voltage pulses continue being applied in the switching direction after the state switching of memristor already succeeds. For example, apply an ON switching signal to a device already in ON state. Based on the vacancy channeling mechanism, the over tune in OFF state continues eliminating the oxygen vacancy until all the oxygen vacancies disappear and the device becomes an insulator. In ON state, the over tune creates more oxygen vacancies to form more conducting channels. The device mechanism becomes less appropriate to be modeled with barrier width w since the channel frontier no longer exists. The resistance shifting in real devices is even more complex after including thermal, electron kinetic energy, and other physical issues. During over tune, a memristor device remains in the same static state and the resistance shifting follows the static resistance distribution. However, a systematic impact on μ_{R_L} and μ_{R_H} has been observed [30].

Here, we use a statistical method to analyze the impact of over tune on the resistance shifting. The charge q flowing through the device is used as the input variable, which has a direct impact on the number of oxygen vacancies and the device resistance. To exhibit the trend of resistance shifting, a linear approximation can be assumed between the passing charge q and the mean shifting μ_{shift} as [25]:

$$\mu_{shift} = e \cdot q = e \cdot \frac{V}{M} \cdot t. \tag{11.21}$$

Here, e is the fitting parameter that describes the shift speed of mean, M is the current memristor resistance. The new μ_{R_L} and μ_{R_H} can be calculated from (11.20):

$$\mu'_{R_L} = \mu_{R_L} - \mu_{on\text{-}shift} = \mu_{R_L} - e_{on} \cdot q, \ \mu'_{R_L} \geq 0, \tag{11.22a}$$

$$\mu'_{R_H} = \mu_{R_H} + \mu_{off\text{-}shift} = \mu_{R_H} + e_{off} \cdot q. \tag{11.22b}$$

Though more complicated fitting equations can be established, such an approach is impractical and unnecessary at current stage considering insufficient experimental data available. The resistance shifting caused by over tune is constrained within the target resistance state, demonstrating less impact on the overall memristor characteristic compared to the ON–OFF switching.

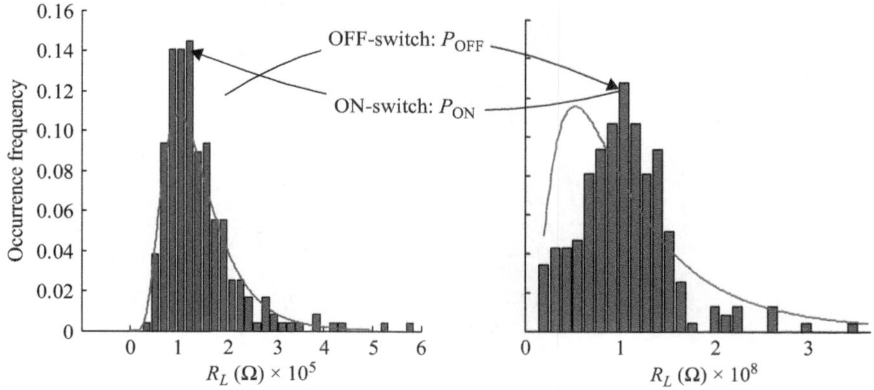

Figure 11.6 The static state distribution of a TiO_2 memristor device

11.4.3 Stochastic model verification

We verified the proposed stochastic model from perspectives of static states and dynamic switching process.

Static states: Figure 11.6 shows the resistance distributions of a memristive switch in ON and OFF states. The bars in the figure are real measurement data of a TiO_2 memristor device [30]. The results show that the lognormal distribution fits well to the real device data in ON state. However, in OFF state, the heavy tail is captured but the median value is slightly skewed. Though the distribution of R_H is not perfectly fitted, the error in distribution fitting of R_H has ignorable impact in the circuit simulation since R_H is more than two orders of magnitude higher than R_L.

Dynamic switching process: Figure 11.7 shows the time dependencies of ON and OFF switching probability at different applied voltages. The results have high approximation to the experimental results [16]. The error mainly comes from the approximation of the relationship between σ_t and V. As aforementioned, establishing a more reliable estimation of σ_t requires more experimental data.

Figure 11.8 shows the simulated analog resistance changing process of a TiO_2 memristor to better demonstrate the time and voltage dependency of switching probability and the resistance shifting due to over tune. The external voltage is set as 3.0 V to switch the memristor from R_H to R_L. The 100 curves in the figure represent the resistance changings by repeating 100 times of the ON switching procedure for the same device. The distribution of 100 tests agrees well with the switching probability curve at −3.0 V in Figure 11.7(a): about 40% of the curves reach R_L before 0.1 s.

Considering the obvious stochastic behavior of memristive device at nanometer regime, traditional device modeling based on curve fitting is not enough. In this work, we built a stochastic model for TiO_2 memristor by bridging the key physical mechanisms and the experimental data fitting. The model combines the stochastic

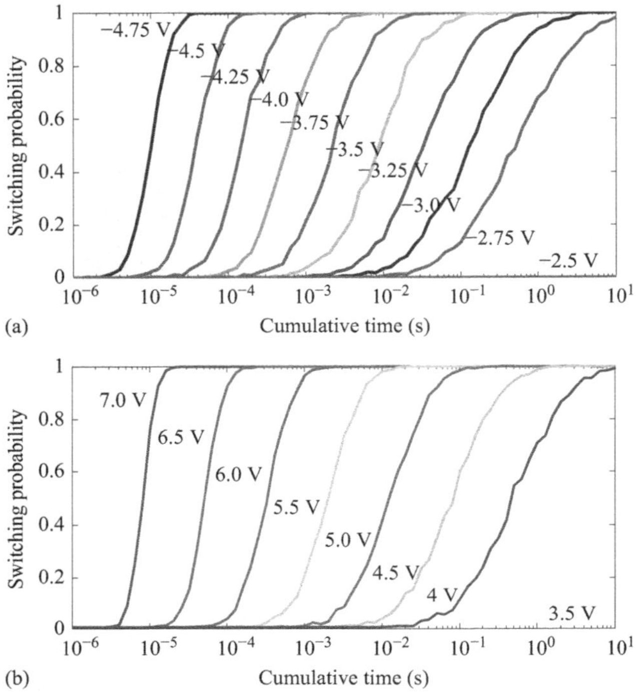

Figure 11.7 The time dependency of ON (a) and OFF (b) switching at different external voltage V

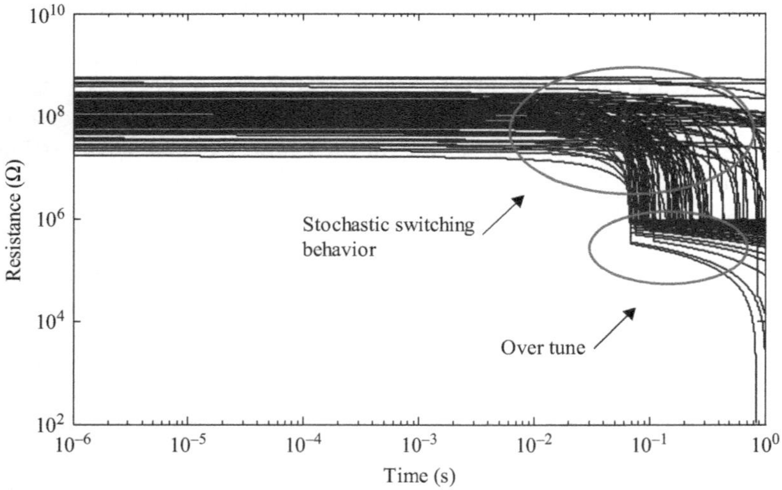

Figure 11.8 The analog switching process of a TiO_2 memristor

characteristics in static states and dynamic switching process together and extends the stochastic study to the analog state while still holding high approximation to the existing data. The accurate and fast estimation on the distribution of device's analog states makes the proposed model more meaningful for higher level circuit and system designs. This model can be generalized to other metal oxide memristors [6, 33] for the same stochastic nature, that is, the percolation property of the thin dielectric soft breakdown [3]. The proposed model can be further enhanced by integrating with reliable physical model that precisely describes the stochastic switching mechanism. The complex and slow physical model generates the required distribution data to develop the proposed fast stochastic model.

11.5 Robustness of a neuromorphic system

A memristor behaves similarly to a synapse in biological systems and hence can be easily used as the weighted connections in neural networks. Based on the memristor-based bidirectional synapse design, we implement a network serving as neuromorphic computing system with *units* (artificial neurons) and *weighted connections* (synapses). The neuron in this network is a binary threshold unit that produces only two different values to represent its state. A synapse works as a weighted connection to transmit a signal from one neuron to another. The activation function can be described as:

$$N_0 = \begin{cases} 1, & \text{if } \sum_{i=1}^{n} (N_i \times W_i) \geq \text{threshold} \\ 0, & \text{otherwise} \end{cases} \tag{11.23}$$

Here, the neuron N_0 collects signals from all the other neurons N_i through the weighted connections W_i. The state of N_0 could be *excitation* ($N_0 = 1$) or *inhibition* ($N_0 = 0$) that is determined by the relation between the summed weighted signals and the threshold. Here, we use bidirectional synapses in the design to build a fully connected neural network, in which any two connected neurons interact each other.

The proposed neural network can be used for pattern recognition: first, multiple standard input images are used to train the connection weights of the system till they reach convergence; after that, any input pattern will produce to a local minimum, which is a stable state corresponding to one of the stored standard patterns. Such a network system can even be used to recognize the input image with defects.

In our experiment, we build a network with 100 (10×10) neurons and store the character images "A", "B", and "C" shown in Figure 11.9(a) as the standard patterns. Each neuron in the network represents a pixel of the image. Then the defected images in Figure 11.9(b) are applied as inputs to initialize the network's state. Figure 11.9(c) shows that each input has 13 defects compared to its corresponding standard images (see black bars). The proposed system can completely eliminate the difference to zero and converge to one of the standard patterns, as demonstrated by the write bars in Figure 11.9(c).

The maximal allowed stored standard patterns (*capacity*) of this neural network design is determined by the amounts of neurons and connections. Moreover, the

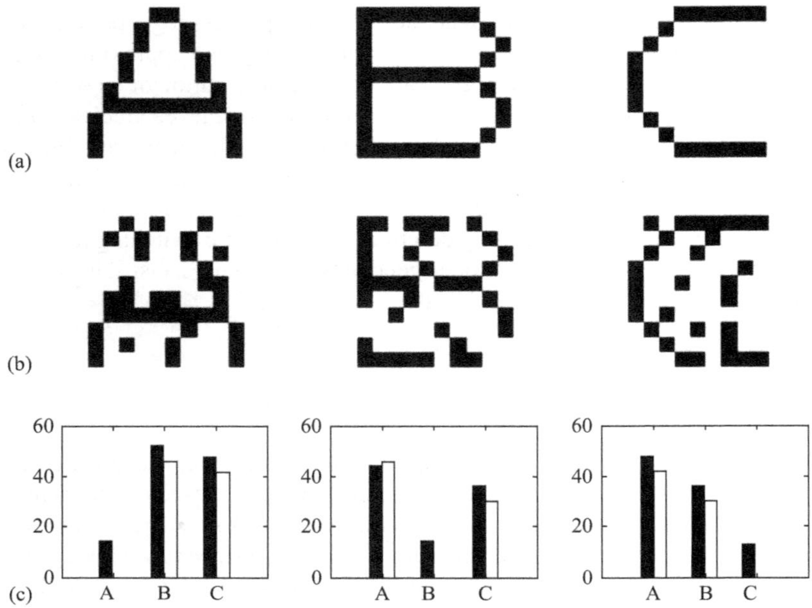

Figure 11.9 The neural network in pattern recognition: (a) the standard patterns, (b) the noised input patterns, and (c) the comparison of the input noised images (black bars) or the output converged images (white bars) from their corresponding standard patterns

more patterns stored in the system, the higher precision of the connection weights is needed. Therefore, a large number of stored patterns and the high process variation on memristances will result in a higher failure probability (P_f).

To quantitatively evaluate the impact of memristance variations and robustness of the proposed neural network design, we conducted Monte Carlo simulations for the network with 100 (10×10) neurons. Random variations following Gaussian distribution have been injected to the memristors. And σ is the standard deviation of the memristance. The system could fail to recognize the noised patterns or mismatch an input with other standard patterns due to the inaccurate connection weights. To test the failure probability under different conditions, we ran 10,000 Monte Carlo simulations by varying the memristance variation σ when 7, 8, 9, or 10 patterns are stored in the system. In this experiment, each input image contains 21 defects among 100 pixels.

The simulation results in Figure 11.10 demonstrate that the proposed memristor-based neuromorphic system has a high tolerance on memristance variations. When $\sigma < 0.4$, which already exceeds the upper bound of memristance variation in Table 11.2, P_f of all the four configuration are close to the ideal condition at $\sigma = 0$. This indicates that even a large process variation exists in memristor devices, the performance of the proposed neuromorphic system is not affected much. Further

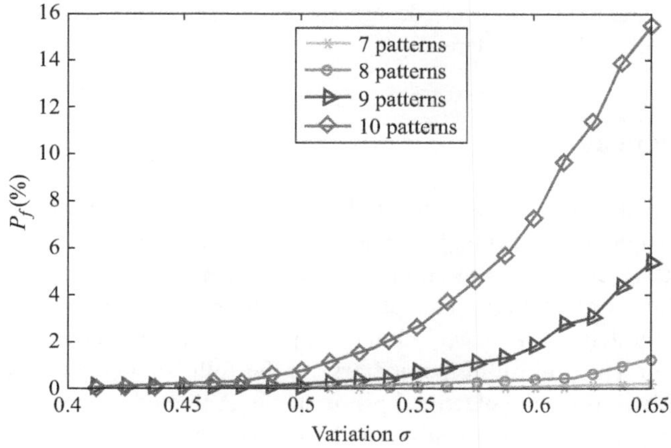

Figure 11.10 The impact of memristance variations on the probability of failure (P_f)

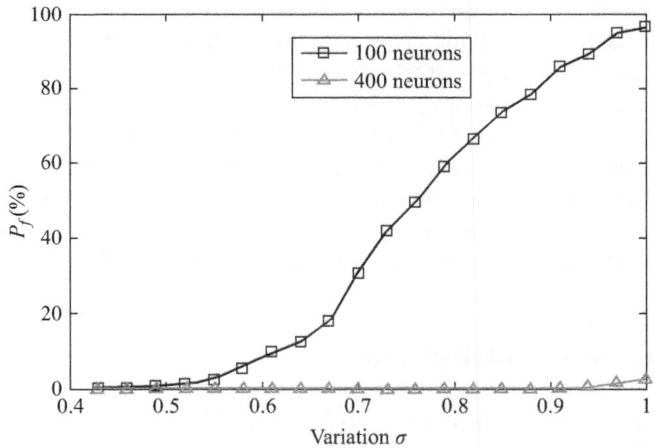

Figure 11.11 Increasing the network size can reduce P_f

increasing $\sigma > 0.5$, P_f grows significantly. As expected, under the same process variation condition, the system suffers a higher P_f when more patterns are stored.

For the same amount of stored patterns, a larger network with more neurons is more robust to process variations. Figure 11.11 compares the performance of the systems with 100 neurons (square marker) and with 400 neurons (triangle marker). Both systems have 10 standard patterns. And the input defect rate remains at 21% for the two designs. The simulations show that the impact of process variations is smaller, and therefore the required precision of connection weights is lower in a bigger network.

Hence, in a neural network system design, the tradeoff between network capacity and robustness needs to be considered.

11.6 Conclusion

Our research related to memristor technology can be concluded in two major aspects: (1) the device modeling including variations and stochastic features for efficient large-scale memristive circuit design and (2) the circuits and applications of memristors for "*brain-like*" computations. This paper summarizes our major attempts on memristor device modeling. We start with the simple static model based on the physical mechanisms of TiO_2 thin-film memristors. In the following, the impact of various geometry variations on the electrical properties is evaluated by conducting analytic modeling analysis and Monte Carlo simulations [8]. A statistical modeling method was proposed which successfully speeds up the simulations in circuit and system levels by 3–4 orders of magnitude [22]. Lately, we developed a stochastic model from the macro perspective of stochastic characteristics that were discovered in memristor devices recently [11]. The model bypasses material-related parameters while directly linking the device analog behavior to stochastic functions. Simulations show that the proposed stochastic device model fits well to the existing device measurement results. All these variations-aware and stochastic models are independent generic flows, so they can be integrated and adopted to other memristor materials.

At the applications layer, we investigated the usage of memristor-based crossbar array as synapse network in neuromorphic circuits. Our work demonstrated very high tolerance of hardware variations and signal noises in recall functions [10, 22]. And a complete neuromorphic circuit embedded with training circuits is also carried out to further reduce the impact of variations and stochastic issues [9].

Acknowledgment and disclaimer

This work was supported in part by DARPA D13AP00042, NSF EECS-1311747, and CNS-1342566. Any opinions, findings, and conclusions or recommendations expressed in this material are those of the authors and do not necessarily reflect the views of DARPA, NSF, or their contractors.

References

[1] Asenov, A., Kaya, S., Brown, A.R.: "Intrinsic parameter fluctuations in decananometer MOSFETs introduced by gate line edge roughness". *IEEE Transaction on Electron Devices* **50**, 1254–1260 (2003)

[2] Ban, Y., Sundareswaran, S., Panda, R., Pan, D.: "Electrical impact of line-edge roughness on sub-45nm node standard cell". In: *Proc. SPIE*, vol. 7275, pp. 727518-1–727518-10 (2009)

[3] Blonkowski, S.: "Filamentary model of dielectric breakdown". *Journal of Applied Physics* **107**(8), 084109 (2010)

[4] Choi, H., Jung, H., Lee, J., *et al.*: "An electrically modifiable synapse array of resistive switching memory". *Nanotechnology* **20**(34), 345201 (2009). URL http://stacks.iop.org/0957-4484/20/i=34/a=345201

[5] Chua, L.: "Memristor – the missing circuit element". *IEEE Transaction on Circuit Theory* **18**, 507–519 (1971)

[6] Gaba, S., Sheridan, P., Zhou, J., Choi, S., Lu, W.: "Stochastic memristive devices for computing and neuromorphic applications". *Nanoscale* **5**(13), 5872–5878 (2013)

[7] Ho, Y., Huang, G., Li, P.: "Nonvolatile memristor memory: device characteristics and design implications". In: *IEEE/ACM International Conference on Computer-Aided Design – Digest of Technical Papers, 2009. ICCAD 2009*, pp. 485–490 (2009)

[8] Ban, Y., Sundareswaran, S., Pan, D.: "Electrical impact of line-edge roughness on sub-45-nm node standard cells". *Journal of Micro/Nanolithography, MEMS, and MOEMS* 9(4), 041206-041206 (2010)

[9] Hu, M., Li, H., Chen, Y., Wu, Q., Rose, G.S., Linderman, R.W.: "Memristor crossbar-based neuromorphic computing system: a case study". *IEEE Transactions on Neural Networks and Learning Systems (TNNLS)* **25**(10), 1864–1878 (2014)

[10] Hu, M., Li, H., Wu, Q., Rose, G.S.: "Hardware realization of BSB recall function using memristor crossbar arrays". In: *49th Annual Design Automation Conference*, pp. 498–503 (2012)

[11] Hu, M., Wang, Y., Qiu, Q., Chen, Y., Li, H.: "The stochastic modeling of TiO_2 memristor and its usage in neuromorphic system design". In: *19th Asia and South Pacific Design Automation Conference (ASP-DAC)*, pp. 831–836 (2014)

[12] Jiang, Z., *et al.*: "Characterization of line edge roughness and line width roughness of nano-scale typical structures". In: *International Conference on Nano/Micro Engineered and Molecular Systems*, pp. 299–303 (2009)

[13] Kawahara, T., Takemura, R., Miura, K., *et al.*: "2 Mb SPRAM (spin-transfer torque RAM) with bit-by-bit bi-directional current write and parallelizing-direction current read". *IEEE Journal of Solid-State Circuits* **43**(1), 109–120 (2008). DOI 10.1109/JSSC.2007.909751

[14] Kim, D., Seo, S., Ahn, S., *et al.*: "Electrical observations of filamentary conductions for the resistive memory switching in NiO films". *Applied Physics Letters* **88**(20), 202102 (2006)

[15] Kim, K., Choi, B., Shin, Y., Choi, S., Hwang, C.: "Anode-interface localized filamentary mechanism in resistive switching of TiO thin films". *Applied Physics Letters* **91**, 012907 (2007)

[16] Medeiros-Ribeiro, G., Perner, F., Carter, R., Abdalla, H., Pickett, M.D., Williams, R.S.: "Lognormal switching times for titanium dioxide bipolar memristors: origin and resolution". *Nanotechnology* **22**(9), 095702 (2011)

[17] Niu, D., Chen, Y., Xu, C., Xie, Y.: "Impact of process variations on emerging memristor". In: *Design Automation Conference (DAC)*, pp. 877–882 (2010)

[18] Oldiges, P., Lin, Q., Petrillo, K., Sanchez, M., Ieong, M., Hargrove, M.: "Modeling line edge roughness effects in sub 100 nanometer gate length devices". In: *2000 International Conference on Simulation of Semiconductor Processes and Devices, 2000. SISPAD 2000*, pp. 131–134 (2000). DOI 10.1109/SISPAD.2000.871225

[19] Pershin, Y.V., Di Ventra, M.: "Spin memristive systems: spin memory effects in semiconductor spintronics". *Physical Review B* **78**, 113309 (2008). DOI 10.1103/PhysRevB.78.113309. URL http://link.aps.org/doi/10.1103/PhysRevB.78.113309

[20] Pershin, Y.V., Di Ventra, M.: "Experimental demonstration of associative memory with memristive neural networks". *Neural Networks* **23**(7), 881–886 (2010). DOI 10.1016/j.neunet.2010.05.001. URL http://dx.doi.org/10.1016/j.neunet.2010.05.001

[21] Pickett, M.D., Strukov, D.B., Borghetti, J.L., *et al.*: "Switching dynamics in titanium dioxide memristive devices". *Journal of Applied Physics* **106**(7), 074508 (2009)

[22] Pino, R.E., Li, H., Chen, Y., Hu, M., Liu, B.: "Statistical memristor modeling and case study in neuromorphic computing". In: *49th Design Automation Conference (DAC)*, pp. 585–590 (2012)

[23] Roy, G., Brown, A., Adamu-Lema, F., Roy, S., Asenov, A.: "Simulation study of individual and combined sources of intrinsic parameter fluctuations in conventional nano-MOSFETs". *IEEE Transactions on Electron Devices* **53**(12), 3063–3070 (2006). DOI 10.1109/TED.2006.885683

[24] Strukov, D., Williams, R.: "Exponential ionic drift: fast switching and low volatility of thin-film memristors". *Applied Physics A: Materials Science & Processing* **94**(3), 515–519 (2009)

[25] Strukov, D.B., Snider, G.S., Stewart, D.R., Williams, R.S.: "The missing memristor found". *Nature* **453**, 80–83 (2008)

[26] Wang, W., Yu, Q., Xu, C., Cui, Y.: "Study of filter characteristics based on PWL memristor". In: *International Conference on Communications, Circuits and Systems, 2009. ICCCAS 2009*, pp. 969–973 (2009). DOI 10.1109/ICCCAS.2009.5250355

[27] Wang, X., Chen, Y., Xi, H., Li, H., Dimitrov, D.: "Spintronic memristor through spin-torque-induced magnetization motion". *IEEE Electron Device Letters* **30**, 294–297 (2009)

[28] Witrisal, K.: "A memristor-based multicarrier UWB receiver". In: *IEEE International Conference on Ultra-Wideband, 2009. ICUWB 2009*, pp. 679–683 (2009). DOI 10.1109/ICUWB.2009.5288703

[29] Yang, J.J., Miao, F., Pickett, M.D., *et al.*: "The mechanism of electroforming of metal oxide memristive switches". *Nanotechnology* **20**(21), 215201 (2009)

[30] Yi, W., Perner, F., Qureshi, M.S., *et al.*: "Feedback write scheme for memristive switching devices". *Applied Physics A* **102**(4), 973–982 (2011)

[31] Yu, Q., Qin, Z., Yu, J., Mao, Y.: "Transmission characteristics study of memristors based op-amp circuits". In: *International Conference on*

Communications, Circuits and Systems, 2009. ICCCAS 2009, pp. 974–977 (2009). DOI 10.1109/ICCCAS.2009.5250356

[32] Yu, S., Gao, B., Fang, Z., Yu, H., Kang, J., Wong, H.S.P.: "A neuromorphic visual system using RRAM synaptic devices with sub-pJ energy and tolerance to variability: experimental characterization and large-scale modeling". In: *IEEE International Electron Devices Meeting (IEDM)*, pp. 4–10 (2012)

[33] Yu, S., Wu, Y., Jeyasingh, R., Kuzum, D., Wong, H.S.: "An electronic synapse device based on metal oxide resistive switching memory for neuromorphic computation". *IEEE Transactions on Electron Devices* **58**(8), 2729–2737 (2011)

Chapter 12

Neuromorphic devices and circuits

Dhireesha Kudithipudi[1], *Cory Merkel*[1], *and Santosh Kurinec*[2]

Over the past few decades, significant strides were made in understanding the processes in the brain and mapping these processes onto different computing substrates for efficient information processing. Neuromorphic researchers strive to abstract the functional and structural behavior of the mammalian brain onto custom hardware platforms with various degrees of complexity. To this end, memristor device technology has enabled efficient neuromorphic realizations owing to its nonvolatility, multiple conductance states, and small footprint. In this chapter, a comprehensive overview of the memristor devices and their role in designing neuromorphic circuit primitives will be presented.

12.1 Introduction

Computing systems that are designed inspired by the processes in the brain are often referred to as neuromorphic computing or brain-inspired computing systems. In contrast to conventional Von Neumann computer architectures, the human nervous system is inherently mixed-signal, massively parallel, approximate, and plastic, giving rise to its incredible processing ability, low power processing, and capacity for adaptation. This paradigm offers great potential for designing the next generation of energy efficient real-time information processing systems. Historically, the challenge with this approach was the lack of understanding of the different processes and their functional role in the brain. There are several recent milestones in the neuroscience research that provide a good foundation for the neuromorphic computing.

Looking historically, there were significant findings before Ren Descartes, but his proclamation Cogito, ergo sum (I think, therefore I am) in 1637 was the start of

[1]Nano Computing Research Laboratory, Department of Computer Engineering, Rochester Institute of Technology, Rochester, NY
[2]Department of Electrical and Microelectronic Engineering, Rochester Institute of Technology, Rochester, NY

a new era of understanding the role of brain. Since then, specific brain functions are discovered by several pioneering scientists, including language by Paul Broca (1862) and Carl Wernicke (1874), general charting of the brain areas by Korbinian Brodmann (1909), and a key finding about the role of individual synapses/interconnections between the neurons by Santiago Ramn y Cajal (1911). Donald Hebb's findings on the organizational behavior of the neurons (1949), "Neurons That Fire Together, Wire Together", was foundational to the neuromorphic computing paradigm. In 1957 Rosenblatt first proposed the perceptron model of a neural system, which is a template for supervised classification based on applying a thresholding function to the weighted sum of the inputs to the system. The simplicity of the perceptron has made it a very common and useful model for implementing neural networks in hardware. For example, a variant of the perceptron (ADALINE) was introduced by Widrow in 1960, which uses a memistor as an artificial synapse in hardware implementations of neural networks. However, these early works did not reach their full potential in hardware platforms, due to the fundamental device limitations. It is vital to understand the role of the devices and primitive circuits in these systems that enable efficient neuromorphic computing systems.The rest of this chapter provides a detailed discussion of the device landscape, the synapse, and neuron primitive circuits that we have developed, and applications of these circuits in a larger neuromorphic system.

12.2 Emerging memory technologies

Generally, "memory" technologies can be split into two categories, volatile and nonvolatile. Volatile memory will not retain data when power is turned off, conversely, nonvolatile memory will retain data once power is turned off. The dominating memory technologies in the industry today are SRAM, DRAM (volatile), and NAND flash (nonvolatile). The general technology requirements of memories are compatibility and integration with complementary metal oxide semiconductor (CMOS) platform, high functional bit density, high speed, low power dissipation, and low cost. The major technology barriers are stability, reliability, data retention, disturbance, on–off ratio, and endurance. There is a significant interplay between requirements and barriers, and optimized trade-offs between them are expected.

In the past decade, significant focus has been put on the emerging memories field to find possible contenders to displace either or both NAND flash and DRAM. Some of these newer emerging technologies include: MRAM (magnetic RAM), STT-RAM (Spin-Transfer Torque RAM), FeRAM (Ferroelectric RAM), PCRAM (Phase change RAM (PCRAM), RRAM (Resistive RAM), and Memristor-based RRAM (Figure 12.1).

In MRAM, the most common basic cell is composed of one NMOS transistor as the access device and one magnetic tunnel junction (MTJ) as the storage element (1T1J structure). The MTJ constitutes a pinned magnetic layer (e.g., CoFe or NiFe/CoFe) and a free magnetic layer (e.g., CoFe or NiFe/CoFe) separated by an insulating barrier (e.g., MgO). Information is stored in the magnetization direction of the

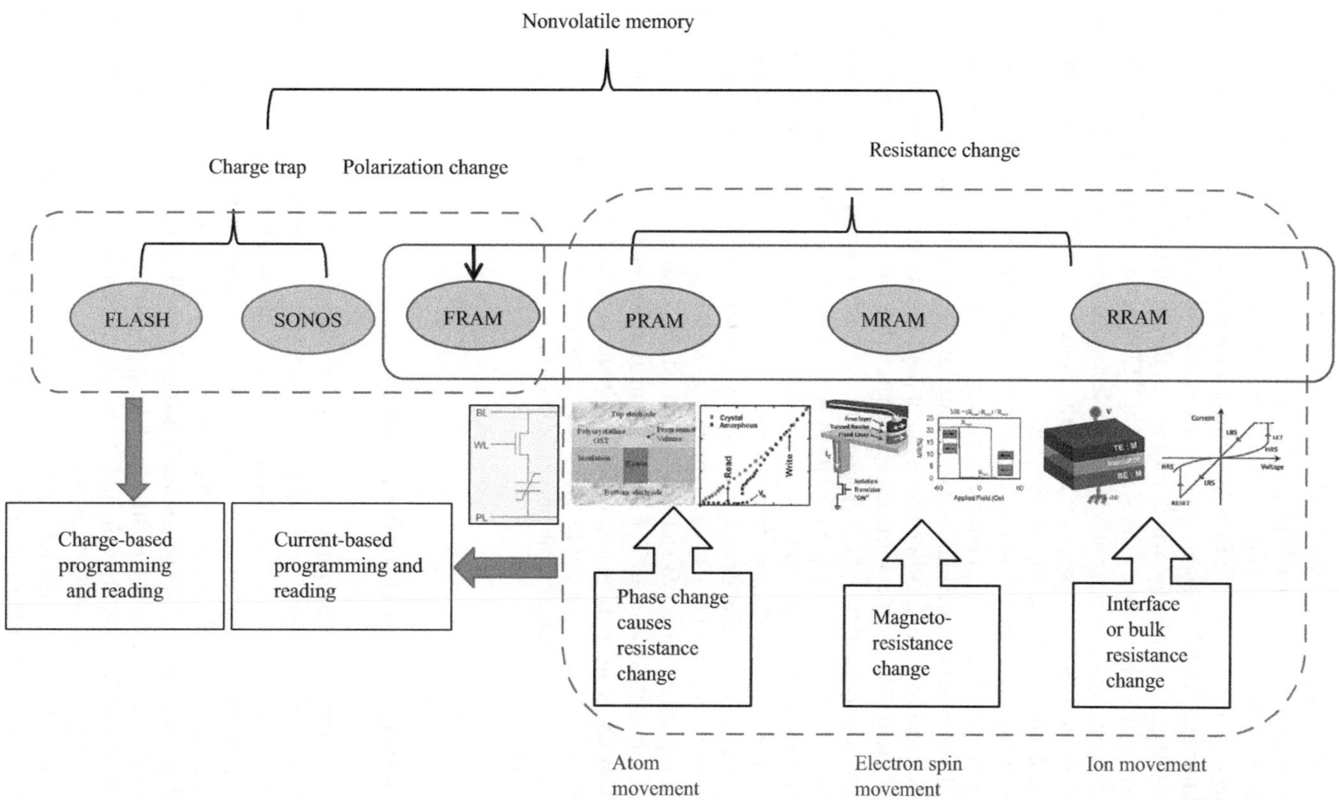

Figure 12.1 Various nonvolatile memory technologies classified in terms of the switching mechanisms involved in their operation
Source: Yole Development; Leti, 2012

free layer. By employing a magnetic field, the orientation of the free magnetic layer can be flipped in order to make the two layers parallel or antiparallel with each other. These two conditions correspond to high or low barrier conductance, respectively, and thus define the state of the memory bit. The latest MRAM technology is STT-RAM (spin torque transfer RAM). In STT-RAM, the direction of magnetization of the free layer is changed by directly passing spin-polarized currents through MTJs. STT-RAM has the advantage of scalability, which means that the threshold current to make the state reversal will scale down as the size of the MTJ becomes smaller [2]. FeRAM utilizes the permanent polarization of a ferroelectric material such as PZT (Lead-Zirconate-Titanate), SBT (Strontium–Bismuth–Tantalate) or BLT (La substituted-Bismuth–Tantalate) as the storing mechanism.

Phase change memory (PCM) is one of the leading candidates among alternatives to flash and DRAM [18]. This memory works based on the thermally induced reversible phase transition in a phase change materials that exhibit two stable material phases: a low-resistance crystalline phase and a high-resistance short range ordered amorphous phase. The most commonly used material is a chalcogenide, $Ge_2Sb_2Te_2$ (GST) that is widely used in optical storage devices such as compact discs and digital video discs, wherein heating by a laser beam enables the GST layer to switch between the two states. These two states have a distinct difference in optical reflectivity. A basic PCRAM cell consists of the phase change material layer sandwiched between two electrodes. The device is driven by a bipolar or field-effect transistor in a 1 transistor/1 resistor (1T1R) configuration or by a diode in a 1 diode/1 resistor (1D1R) configuration. The two states of the PCM are known as SET (crystalline) and RESET (amorphous) states. The RESET state is achieved by applying a pulse to heat the PCM above its melting point and rapidly quenching it to its high-resistance short range order state. To return to SET state, a longer pulse is applied to heat the PCM above its crystallization temperature but below its melting point allowing it to crystallize to its LRS.

12.3 Memristor and resistive memory

The resistance switching behavior of several materials has attracted considerable interest for its applications in nonvolatile memory commonly known as resistive random access memory (RRAM) [10, 23]. RRAM shares some similarities with PCM as both are considered to be types of memristor technologies—a passive two-terminal electronic device that is designed to express only the property of an electronic component that lets it recall the last resistance it had before being shut off (memristance, explained in the next section). Resistive switching phenomena has been observed in many materials and devices. These include (i) binary transition metal oxides (TMOs), (ii) perovskite type complex oxide, (iii) wide band gap dielectrics, (iv) resonant tunneling diodes, and (v) magnetic and ferroelectric tunneling junctions. Resistive switching has been observed in more than 50 material systems [6]. The mechanism of switching may be different in each of these materials and devices [4, 9].

12.3.1 Memristor

In 1971, Leon Chua proposed a possible existence of a fourth circuit element that would provide a missing relationship between the magnetic flux, ϕ and the electric charge, q described as

$$d\phi = M dq \tag{12.1}$$

where, M is the memristance having the same units as that of resistance [4]. The memristor acts like a resistor, by relating the voltage over the element and the current through it as

$$v = M(w)i \tag{12.2}$$

The memristance M depends on a parameter w, which is either q or the flux ϕ. Thus, M depends on the complete history of current passing through the element, which makes a memristor act like a resistor with memory for current (hence the name, memory resistance or memristor).

Chua and Kang [5] later described memristor as a broader class of systems as

$$\begin{aligned} v &= M(w,i)i \\ \frac{dw}{dt} &= f(w,i) \end{aligned} \tag{12.3}$$

where, w can be a controllable property and f is some function called the memristor equivalent learning rule, analogous to the learning rule in the synapses.

The basic construction of a memristive device is creating an element with two regions of different resistances R_{on} and R_{off}. The boundary between these two regions will shift due to applied voltages or currents, resulting in net change of resistance as described in Figure 12.2 showing a schematic of a TiO_2 -based memristor

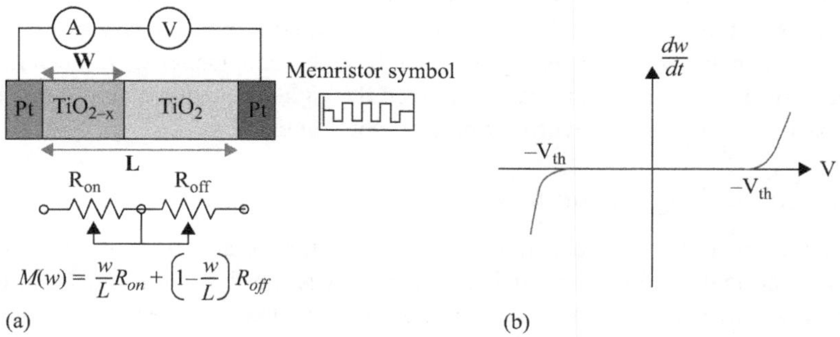

Figure 12.2 (a) Schematic of TiO_{x0}-based memristive device. Due to migration of oxygen vacancies, wall between low resistivity TiO_x and high-resistivity TiO_2 moves changing the resistance; (b) threshold characteristic rate of change of w for a general memristor

device depicting a low resistivity oxygen deficient TiO_{2-x} region separated by a high resistivity TiO_2 region. The memristance can be expressed as a function of w as

$$M(w) = \frac{w}{L}R_{on} + \left(1 - \frac{w}{L}\right)R_{off} \tag{12.4}$$

Initial applied voltage results in a low current but also a shift in w. The shift in w changes the memristance, and during the next voltage period the current is larger. The most characteristic attribute of a memristor device is its current–voltage curve that exhibits a pinched hysteresis loop that always passes through the origin. The most remarkable feature of a memristor is that it does not lose the value of the magnetic flux and the electric charge when both the voltage and current are zero at the instant when the power is switched off and retains the original value making it a nonvolatile memory element.

The learning rule of w is given to a linear approximation by

$$\frac{\mathrm{d}w}{\mathrm{d}t} = ai(t) \tag{12.5}$$

where, a is some constant dependent on device properties. A theoretical simplification of a general memristor system implementing the threshold and nonlinear region was proposed by Linares-Barranco and Serrano-Gotarredona [14]. In this model there is a dead zone where nothing changes, while w changes exponentially outside this region

$$\frac{\mathrm{d}w}{\mathrm{d}t} = I_0 \mathrm{sgn}(v)\left[\exp\left(\frac{v}{v_0}\right) - \exp\left(\frac{v_{th}}{v_0}\right)\right] \tag{12.6}$$

where, v_{th} describes the threshold, I_0 and v_0 are the parameters determining the slopes. This learning rule is illustrated in Figure 12.2(b).

The memristor offers many new advantages. It allows for analog-based data storage, rather than 0s and 1s. In binary digital circuits, memristors can be employed as switches, toggling between maximum and minimum resistance. If several intermediate resistance values could be distinguished reliably, then the information density can be increased to multiple bits per device. Some flash memory devices have already achieved this multilevel logic. This could evolve to a continuously varying resistance device that can operate as an analog device. It has inherent plasticity, with clear learning rules based on the current that has passed through the device making memristor as a possible contender for neuromorphic computing element.

12.3.2 Switching mechanisms

Typical memristor RRAM cells, have a metal-insulator-metal (MIM) structure, where two conducting electrodes sandwich a thin-film switching layer. Various MIM memristor stacks have been explored, and there are several ways to categorize them based on their material properties. In bistable resistive materials, like in some TMOs, the switching between high resistance state (HRS) and low resistance state (LRS) can be unipolar or bipolar as illustrated in Figure 12.3. In unipolar switching, the SET and RESET operations are independent of the voltage/current polarity. The SET voltage is always higher than the RESET voltage, and the RESET current is higher than the

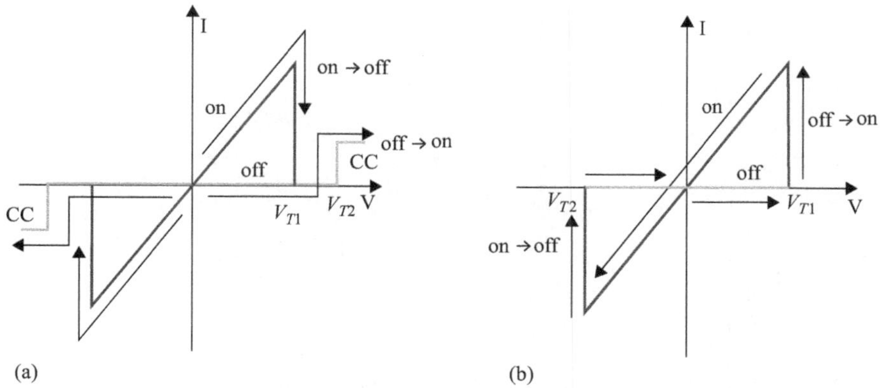

Figure 12.3 (a) Current–voltage characteristics for (a) unipolar switching, CC is current compliance when the device switches to LRS; and (b) bipolar switching

current compliance of the SET operation. In bipolar switching, the switching occurs depending on the polarity and magnitude of the applied signal. For example, the device can be changed into SET (LRS) state when a positive voltage larger than the threshold voltage (V_{T1}) is applied to the top electrode, while a negative voltage larger than another threshold voltage (V_{T2}) switches the device back to off-state (HRS). The device state is not affected if the applied bias is between two threshold voltages V_{T1} and V_{T2}, enabling a low-voltage read process. It can be observed that the bipolar switching behavior is analogous to the memristor pinched hysteresis curve.

There are numerous underlying switching mechanisms which can be chemical, thermal, stochastic, and localized in nature [23]. Resistive memristors traditionally operate by voltage-induced displacement of matter with different mechanisms. A robust and predictive understanding is essential to realize these devices in commercial applications. The switching mechanisms can be broadly classified into thermo-chemical electric, physical, or purely electronic. Figure 12.4 shows a range of thermo-chemical switching devices which can be 1, 2, or 3 dimensional switching with their polarity nature. Generally speaking, the switching tends to be bipolar if the electric field plays a significant role and unipolar if thermal effects are dominant.

There are also various physical phenomena where only physical changes are involved. These are, e.g., electronic, magnetic, ferroelectric, or resonant tunneling devices [15]. In some devices, resistance change can occur from charge trapping or detrapping at an electrode/insulator interface that results in increase or decrease of contact potential. These switches have been demonstrated in metal-doped polymers. Another type of electronic resistance switch relies on electronic phase change described by Mott transition. Purely electronic memristors based on physical phenomena have also emerged. These are based on resistance changes through electron-mediated phenomena such as those in magnetic and ferroelectric tunneling devices.

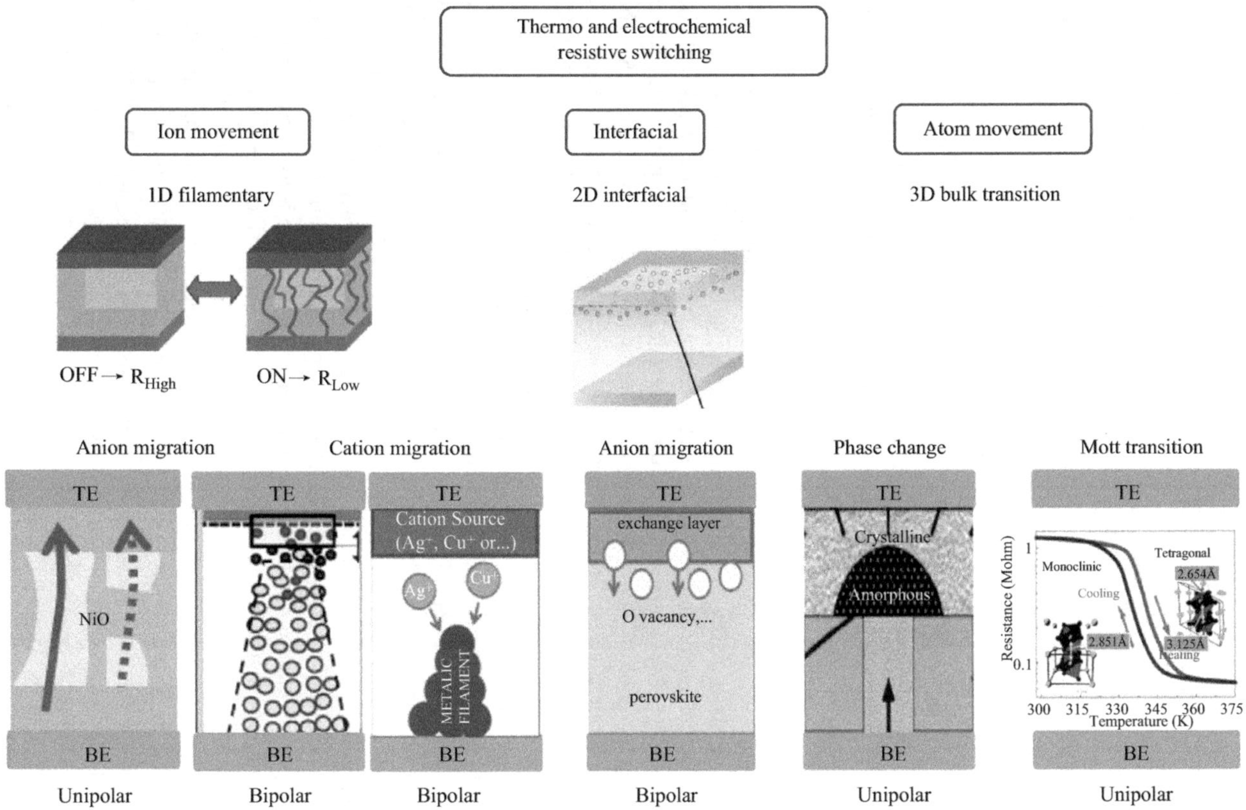

Figure 12.4 Fundamental characteristics of different two terminal thermo- and electrochemical switching devices

Table 12.1 Important features of key memristor technologies

	PCM	Red-Ox	STT	FTJ
Speed	50 ns	10 ns	25 ns	10 ns
Power consumption	6 pJ	<1 pJ	0.02–5 pJ	10 fJ
Feature size	6 F^2	(5/8) F^2	(20/40) F^2	(5/8) F^2
Switching	Unipolar	Both	Bipolar	Bipolar
Physical understanding	Yes	No	Yes	Yes
Prototypes	Commercial	Some	Yes	?

Since the 1990s, the development of MRAM has shown, in certain cases, memristive properties. The configuration known as a spin valve, the simplest structure for a MRAM bit, allows for state change. A spintronic memristor's resistance state uses magnetization to alter the spin direction of electrons in two different sections of a device [20, 21]. Two sections of different electron spin directions are kept separate based on a moving wall, controlled by magnetization, and the relation of the wall dividing the electron spins is what controls the devices overall resistance state. In STTRAM, the resistance in a memristive effective spin-torque transfer is controlled by a spin torque induced by current flowing through a magnetic junction, and is dependent on the difference in spin orientation between the two sides of the junction.

In tunnel junctions with a ferroelectric barrier, known as ferroelectric tunnel junctions (FTJ), switching the ferroelectric polarization induces the change in tunnel resistance with a resistant contrast of several orders of magnitude between the low-resistance ON state and the high-resistance OFF state generally defined as the giant tunnel electroresistance (analogous to giant magnetoresistance in magnetic tunnel junctions) [3]. In ferroelectric random access memories, these two binary states are utilized for data storage. As the ferroelectric barrier thickness is scaled down, the ferroelectric domain size scales down as the square root of the ferroelectric barrier thickness. For a tunneling thickness of the range of 5 nm, nanometer-sized domains are expected. The relative proportion of up- and down-oriented domains can result in a continuous change of resistance between the ON and the OFF states. Since the polarization reversal process in ferroelectric thin film depends on pulse amplitude and duration, these parameters can be used to achieve the desired resistance change a very promising feature for STDP-based learning implementation. It has been recently shown in $Co/BiFeO_3/Ca_{0.96}Ce_{0.04}MnO_3$ tunnel junction devices, a resistance change of 3 orders of magnitude with a 100 ns pulses of 2-V amplitude [1]. Tables 12.1 and 12.2 summarize some of the important features observed and materials systems being explored for developing new commercial memristor technologies.

12.3.3 Plasticity

Synaptic electronics aims at building artificial synaptic devices to emulate biological neural systems [12, 17, 19]. In biological neural networks, each nerve cell communicates with other cells through thousands of synapses which are functional

Table 12.2 Emerging memory device materials

Memory technology	Materials
Red-Ox RAM	TiO_x, HfO_2, TaO_x, etc.
STT-MRAM	CoFeB/MgO/CoFeB
Ferroelectric memory	$SrBi_2Ta_2O_9$, Doped HfO_2
Mott memory	VO_x, $Pr_{0.7}CaO_{0.3}MnO_3$, etc.
Macromolecular	Polymer with metal oxides
Molecular	Molecular monolayer

connections between neurons. A biological synapse typically consists of a small gap between the terminal end of the axon and the target cell. When the depolarizing signal caused by the release of positive sodium ions reaches the synapse, it triggers the release of signaling molecules called neurotransmitters, which are the signaling molecules used at the synapse to pass a signal from a neuron to its target cell. Synaptic plasticity is the ability of synapses to strengthen or weaken over time. Two neurons presenting a correlated activity reinforce their synaptic weight. This experience-dependent change in connectivity between neurons is believed to underlie learning and memory. A neuron activity can be defined in two ways: (1) rate coding: mean firing rate estimated on chosen time window; (2) temporal coding: assigning a single neuron activity to a single spike even at a given time with respect to other neurons in the network. Based on these two strategies, Hebbian learning has been proposed as spike rate dependent plasticity or the spike timing dependent plasticity (STDP). STDP has attracted more attraction because of its possible implementation in memristive devices based on overlapping pulses from the pre- and postneurons.

The memristor is basically a resistor, whose resistance depends on how much current has passed through the element in the past. As such, it has memory and is plastic. Using memristors in relatively simple circuits, which can be implemented in crossbar structures, may lead to associative memory. It was found that if one assumes two spiking neurons with specific kinds of spike shapes, connected by a memristor, various kinds of STDP automatically follow [14]. This has attracted tremendous interest as memristors could potentially revolutionize neuromorphic engineering and improve both the way we understand the role of plasticity in the brain and how we could apply this knowledge to actual circuits.

12.3.4 Memristor integration

With an ability to be scaled down to 8-nm technology nodes while maintaining its distinct features, memristor has been identified as one of the most promising candidates for future generation memory technology in the 2010 International Technology Roadmap for Semiconductor workgroup [7]. One of the many cited advantages of the memristor technology is that it is completely compatible with existing circuitry. Hybrid CMOS/memristor circuit design allows minimal cell size, but the interconnected passive network structure also leads to sneak leakage currents that

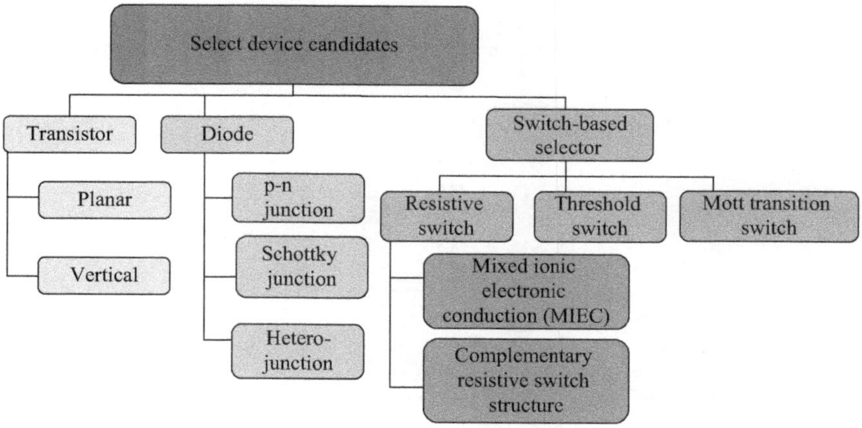

Figure 12.5 A flow chart showing possible select devices under research and development for memristor RAM. Each of these need to address scalability, compatibility, power, and reliability issues

can severely limit the output read margin [24]. To avoid this problem, each cell is serially connected to a nonlinear selector along with the storage element, forming the commonly known one-selector-one-resistor (1S1R) structure [10]. The choice of suitable selector device, among many under research (Figure 12.5), is based on multiple requirements of current drivability, low voltage drop, nonlinearity, scaling, and reliability [22].

Use of memristors has been also demonstrated in programmable analog circuits such as threshold comparators, programmable gain amplifiers, and digital potentiometers employing memristors.

The dynamical behavior of memristors has also prompted researchers to investigate the possibilities of designing electronic synapses and cellular neural networks using memristors [8, 16]. It can be used to make networks with plastic synapses, connected to conventional circuits.

12.4 Memristive synapse circuits: current-mode design

12.4.1 Overview

Synapse circuits provide the means for weighted communication in neuromorphic systems. Figure 12.6 shows a memristive synapse circuit that can achieve both positive and negative weight values, which generally improves neural network performance. The synapse's input current is the output current of the presynaptic neuron. Notice that both the diode-connected PMOSFET and the diode-connected NMOSFET from the presynaptic neuron are used to mirror the input in two places. The PMOS mirror has a 1:2 size ratio, so the output of the mirror is $2i_{in}$. Assuming this synapse connects a

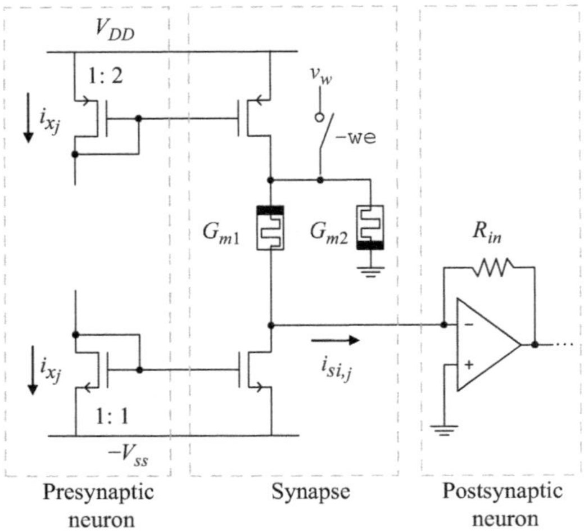

Figure 12.6 Current-mode memristive synapse circuit with a bipolar weight value

j^{th} presynaptic neuron to an ith postsynaptic neuron, then its weight value $w_{i,j}$ can be described as

$$\frac{i_{out}}{I_{max}} = \left(2\frac{G_{m1}}{G_{m1} + G_{m2}} - 1\right) x_j = w_{i,j}x_j \tag{12.7}$$

where, G_{m1} and G_{m2} are the conductances of the two memristors. If both memristors have a high G_{on}/G_{off} ratio, then $w_{i,j}$ will range approximately from -1 to 1.

The analysis above relies on the assumption that the synapse's output node is connected to a virtual ground. Figure 12.7(a) shows a simplified version of the synapse circuit, where the memristors have been replaced with resistors, the programming switch has been removed, and the inputs are ideal current sources. The small signal model of the input stage of the postsynaptic neuron is also shown. Generally, each neuron will have multiple synapses connected to its input, labeled as *common node* (this is also the negative input of the postsynaptic neuron's op amp). In this case, we are showing the synapse that connects the jth neuron in the network to the ith neuron. The total output current of all of the synapses, normalized to the maximum neuron output current is:

$$s_i = \sum_j w_{i,j}x_j \tag{12.8}$$

In the ideal case, when the op-amp gain is infinite, $w_{i,j}$ is defined in (12.7). However, in the general case, nodal analysis reveals

$$w_{i,j} = \left(2\frac{G_{m1j}}{G_{m1j} + G_{m2j}}\right)\left(\frac{1 + A_0 + \dfrac{G_{m2j}R_{in}i_{in}^*}{2i_{inj}}}{1 + A_0 + \sum_k \dfrac{G_{m1k}G_{m2k}R_{in}}{G_{m1k} + G_{m2k}}}\right) - 1, \tag{12.9}$$

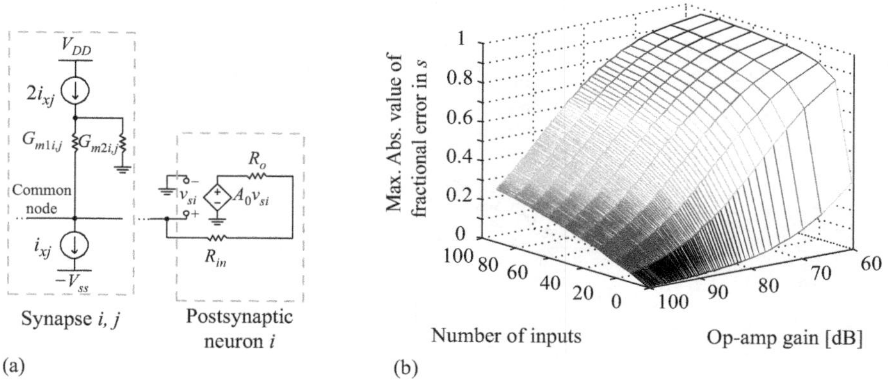

Figure 12.7 (a) Simplified model of the synapse connected to the input of a postsynaptic neuron. (b) Number of neuron inputs and op-amp gain versus the maximum absolute value of the fractional error between the total synaptic output current and the ideal total synaptic output current

where, the index k has the same range as j, and i_{in}^* is the sum of all of the individual synapse input currents. When the op-amp gain A_0 is large, $w_{i,j}$ reduces to our original definition. However, when the gain is smaller, the accuracy of the synapse circuit is degraded. To illustrate this, we have plotted the maximum absolute value of the fractional error in s_i versus the number of synapses connected to neuron i and the gain of the neuron's op amp. The fractional error is defined as $|(\bar{s}_i - s_i)/\bar{s}_i|$, where $\bar{s}_i$ is the expected ($A_0 = \infty$) sum of all synapse outputs divided by I_{max}. For each set of the independent parameters (number of inputs and op-amp gain), we picked 1000 sets of random parameters for each synapse's conductance and input current. The value of s_i was determined using (12.8) and (12.9), while $\bar{s}_i$ was determined using (12.8) and (12.7). The maximum value plotted in Figure 12.7(b) is the maximum over all 1000 sets of random parameters. For this work, we designed an op amp with gain $A_0 > 100$ dB, which allows us to have neurons with $\approx$50 synaptic inputs while keeping the fractional error below $\approx$20 %. Our op-amp design uses a high-gain folded cascade input stage and a common source output stage.

In Figure 12.8, we show the output of a single synapse versus the synapse's input at different values of the weight $w_{i,j}$. The synapse's output is connected to the input of our high-gain op amp. The weights were adjusted by changing the values of each synapses's memristor conductances G_{m1} and G_{m2}. Our synapse design can achieve weight values from -1.0 to 1.0 and has very good linearity.

12.4.2 Area and power consumption

In a neuromorphic system, synapses can outnumber neurons by a factor of N_x, where N_x is the total number of neurons. Consequently, it is imperative that synapse circuits have minimal area and power overhead. The circuit in Figure 12.6 uses only

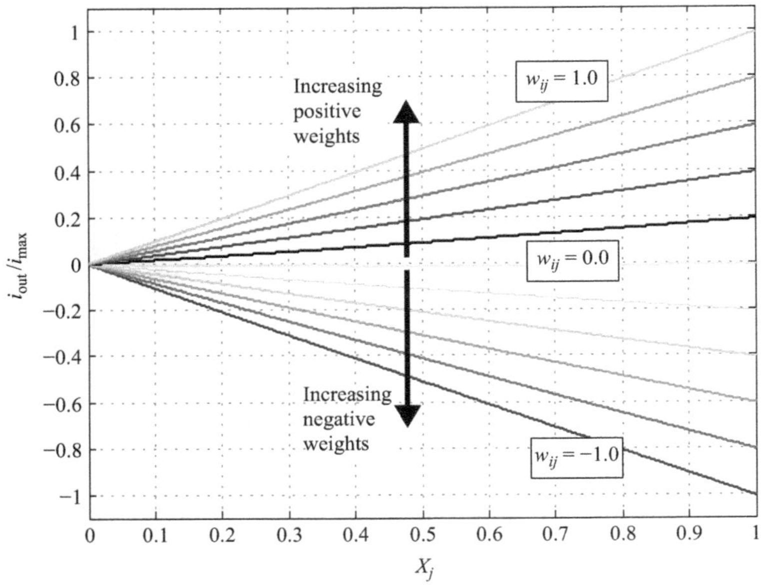

Figure 12.8 Synapse output versus input for different weight values. Results indicate excellent linearity and bipolar weights

two MOSFETs and two memristors to achieve a bipolar weight value. Note, however, that one of the MOSFETs will have a minimum area, while the other has twice the minimum area to achieve the 1:2 NMOS mirror ratio. Compare this to the voltage-mode bipolar memristor-based synapse proposed by Kim *et al.* [11], which requires four memristors and at least four transistors, once one considers the training circuit.

Now, let us consider the power consumed by the memristive synapse discussed above. Clearly, it will depend on the activity of the presynaptic neuron, which will be represented by η. Furthermore, the power consumption of the synapse will be dominated by static power. This is in stark contrast to voltage-mode circuits, where power consumption is typically dominated by switching. The synaptic power consumption is estimated as

$$P_{synapse} \approx \eta 3 V_{DD} I_{max}. \tag{12.10}$$

From (12.10), we see a number of ways to reduce the power consumption. The most interesting method, which is observed in biological brains, is the reduction of η. In other words, if information is represented by very sparse activity within a neuromorphic system, then it could drastically reduce power consumption without compromising on signal-to-noise ratio (i.e., reducing V_{DD} or I_{max}).

12.5 Application: image clustering

12.5.1 Algorithm overview

Clustering algorithms uncover structure in a set of m unlabeled input vectors $\{\mathbf{u}^{(p)}\}$ by identifying M groups or clusters of vectors that are similar in some way. In one common approach, each cluster is represented by its centroid, so the clustering algorithm is reduced to finding each of the M centroids. This can be achieved through a simple competitive learning algorithm: Initialize M vectors $\mathbf{w}_i$ by assigning them to randomly chosen input vectors. These will be referred to as weight vectors. Then, for each input vector, move the closest weight vector a little closer. After several iterations, the algorithm should converge with the weight vectors lying at (or close to) the centroids. Of course, there are several parameters which must be defined, including a distance metric for measuring closeness. The most obvious choice is the ℓ^2-norm. However, computing this is expensive in terms of hardware because it requires units for calculating squares and square roots. In addition, as we will discuss later, it is easy to use a high-density memristor circuit called a crossbar to compute dot products between input and weight vectors. Therefore, it is preferred to use a dot product as a distance metric. For example, if all of the vectors are normalized ($\|\mathbf{u}^{(p)}\| = \|\mathbf{w}_i\| = 1$), then $\mathbf{w}_{i*} \cdot \mathbf{u}^{(p)} > \mathbf{w}_i \cdot \mathbf{u}^{(p)} \forall \mathbf{w}_i \neq \mathbf{w}_{i*}$, where, $\mathbf{w}_{i*}$ is the closest weight vector to $\mathbf{u}^{(p)}$. However, the constraint that $\|\mathbf{u}^{(p)}\| = \|\mathbf{w}_i\| = 1$ creates a large overhead, because every input vector has to be normalized and every weight vector has to be renormalized each time it is updated.

We propose the following solution: map each input vector to the vertex of a hypercube centered about the origin: $\mathbf{u}^{(p)} \in \{-1, 1\}^N$, where N is the dimensionality of the input space. Now, $\mathbf{w}_i \cdot \mathbf{u}^{(p)}$ will yield a scalar value $d^*_{i,p}$ between $-N$ and $+N$. Moreover, this scalar value can be linearly transformed to a distance $d_{i,p}$ which is the ℓ^1-norm, or Manhattan distance, between the weight vector and the input:

$$d_{i,p} \equiv N - d^*_{i,p} = \sum_{j=1}^{N} |w_{i,j} - u_j^{(p)}|. \tag{12.11}$$

Using this distance metric, we do not ever need to renormalize the weight vectors. Furthermore, mapping input vectors to hypercube vertices can usually be accomplished by thresholding. For example, grayscale images can be mapped by assigning -1 to pixel values from 0 to 127 and $+1$ to pixel values from 128 to 255. Algorithm 12.1 summarizes the algorithm. The first two lines are initialization steps. Within the double `for` loop x_i is 1 when i corresponds to the index of the closest vector (called the winner) and 0 otherwise. Then, the weight components of the winner are moved closer to the current input vector using a Hebbian update rule. The prefactor α, which is called the learning rate, determines how far the weight vectors move each time they win. Notice that this algorithm is completely unsupervised, so there are no labeled input vectors.

Algorithm 12.1 Proposed clustering algorithm

Map inputs to hypercube vertices.
Initialize weight vectors to random input vectors.
for *epoch* = 1:N_{epochs}
  for p = 1:m
    $d^*_{i,p} = \mathbf{w}_i \cdot \mathbf{u}^{(p)} \quad \forall i = 1, 2, \ldots, M$
    $x_i = \begin{cases} 1, & d^*_{i,p} = \max(d^*_{i,p}) \\ 0, & \text{otherwise} \end{cases} \quad \forall i = 1, 2, \ldots, M$
    $\Delta w_{i,j} = \alpha x_i u_j^{(p)} \quad \forall i = 1, 2, \ldots, M \quad \forall j = 1, 2, \ldots, m$
  end for
end for

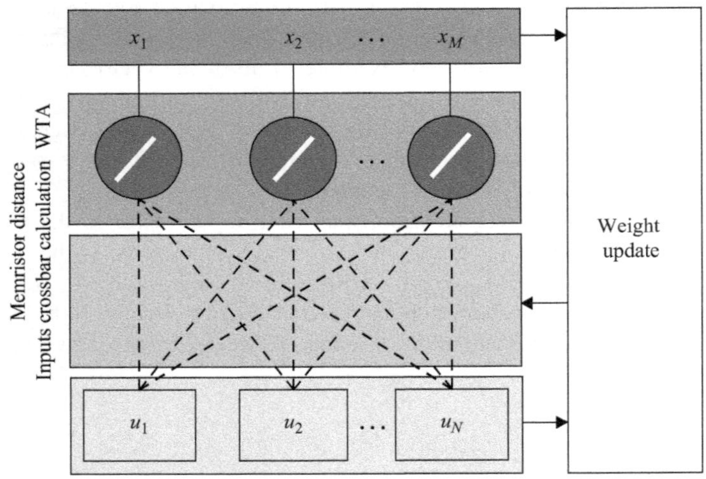

Figure 12.9 Block diagram of proposed neuromorphic system for unsupervised clustering

12.5.2 Hardware design

The unsupervised clustering algorithm discussed in the previous section can be implemented efficiently in a neuromorphic system by representing weight vectors as memristor conductances. A block diagram of the proposed design is shown in Figure 12.9. The inputs, which are represented as positive and negative currents, are fed through M crossbar circuits. Together with a noninverting summing amplifier, (represented as a circle), each crossbar computes the distance between the current input and the weight vector represented by its memristors' conductances.

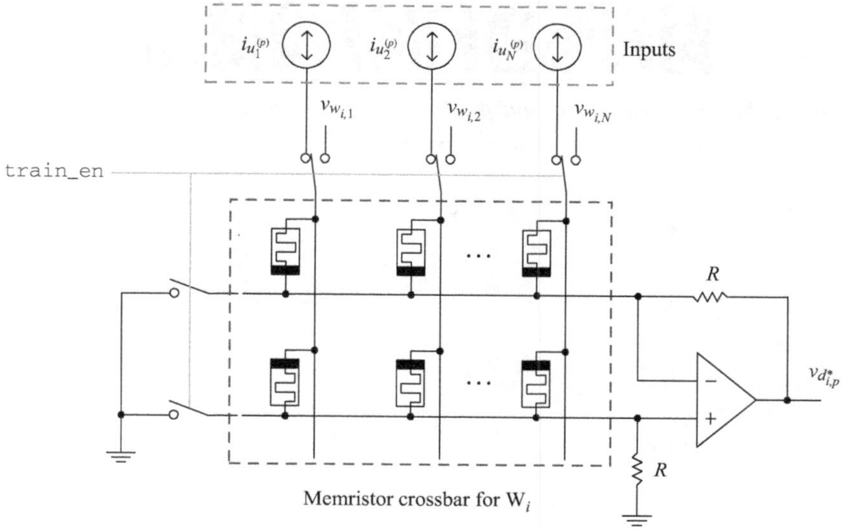

Figure 12.10 Crossbar and summing amplifier circuit for computing the distance between the input and a weight vector

The configuration of the crossbar and summing amplifier is shown in Figure 12.10. Memristors in the top row inhibit or contribute a negative component to the output, while memristors in the bottom row excite or contribute a positive component to the output. Therefore, each crossbar column represents one component of one weight vector $\mathbf{w}_i$, which can be positive or negative. If we assume that the op amp has a high open loop gain and the wire resistances are small, then

$$v_{d^*_{i,p}} = \sum_{j=1}^{N} i_{u_j^{(p)}} R \left(\frac{G_2 - G_1}{G_1 + G_2} \right)_{i,j}, \tag{12.12}$$

where, G_1 and G_2 are the top and bottom memristors in each column, respectively. The output of the circuit is a voltage representation of the distance between the current input and the weight vector represented by the crossbar. The weight vectors are modified by connecting them to write voltages $v_{w_{i,j}}$ using a training enable signal `train_en`. The write voltages are determined by the value of $\Delta w_{i,j}$ in line 7 of Algorithm 12.1. Specifically, if $\Delta w_{i,j}$ is negative, then $v_{w_{i,j}}$ will be a negative voltage below the memristor's write threshold, and if $\Delta w_{i,j}$ is positive, then $v_{w_{i,j}}$ will be a positive voltage above the memristor's write threshold. Otherwise, the write voltage is zero.

So far, we have only discussed the memristor crossbar and distance calculation parts of Figure 12.9 (line 5 in Algorithm 12.1). The winner-takes-all circuit (line 6 in Algorithm 12.1) can be implemented in a number of ways. In this work, we used the

Figure 12.11 10 cluster centroids found in a set of 1000 MNIST images using the proposed neuromorphic system

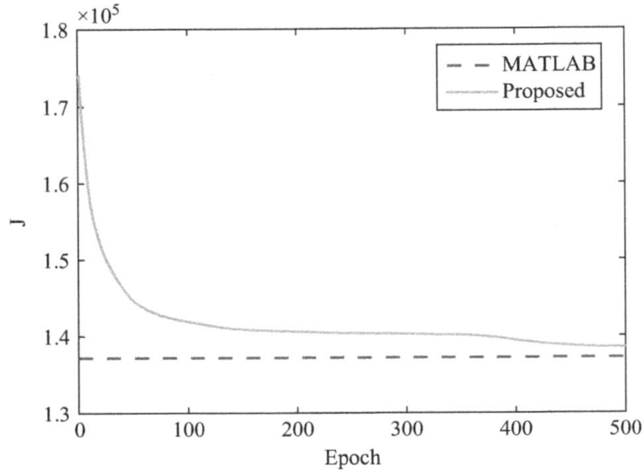

Figure 12.12 Cost function versus epoch while clustering MNIST images using the proposed neuromorphic system

current-mode design described in reference 13. Finally, the weight update (line 7 in Algorithm 12.1) can be computed using simple combinational logic circuits.

12.5.3 Clustering MNIST images

One exciting application of the proposed hardware is automatically identifying clusters in sets of images. We took 1000 images (m=1000) from the MNIST handwritten digit data set and clustered them using a behavioral model of the neuromorphic system described in the last section. Each image was originally 20×20 grayscale pixels (N=400). They were mapped to hypercube vertices using the thresholding approach discussed earlier. In addition, we used 10 clusters (M=10), 500 training epochs (N_{train}=500), and α=0.005. The results are shown in Figure 12.11. Here, we have plotted the weight vectors representing the centroid of each cluster. Figure 12.12 shows the cost versus the training epoch, where the cost is defined as

$$J = \sum_{p=1}^{m} \left(\min d_{i,p} \forall i \right). \tag{12.13}$$

We see that the cost function for the proposed neuromorphic system approaches that of MATLAB's built-in k-means clustering after 500 epochs.

12.6 Summary

Inspite of these promising developments in the device and circuit design space, there are still quite a few challenges to be addressed in neuromorphic devices and circuits. For example, (i) There are no standardized models (SPICE or high-level) or design methodologies for realistic comparisons of different designs, (ii) device isolation techniques are getting lot of attention now; however, none of the proposed techniques are scalable, (iii) auxiliary circuits to support the memristive circuits consume significant power and area, and (iv) multilevel conductance devices are still under development and have a long way to go. However, there is a lot of potential to design simplified massively parallel neuromorphic systems using these devices and circuits.

Acknowledgment

This research is partly supported by NSF EAGER #1445386.

References

[1] Boyn, S., Girod, S., Garcia, V., *et al.*, "High-performance ferroelectric memory based on fully patterned tunnel junctions". *Applied Physics Letters* **104**(5), 052,909 (2014)

[2] Burr, G.W., Breitwisch, M.J., Franceschini, M., *et al.*, "Phase change memory technology". *Journal of Vacuum Science & Technology B* **28**(2), 223–262 (2010)

[3] Chanthbouala, A., Garcia, V., Cherifi, R.O., *et al.*, "A ferroelectric memristor". *Nature Materials* **11**(10), 860–864 (2012)

[4] Chua, L.O., "Memristor-the missing circuit element". *IEEE Transactions on Circuit Theory* **18**(5), 507–519 (1971)

[5] Chua, L.O., Kang, S.M., "Memristive devices and systems". *Proceedings of the IEEE* **64**(2), 209–223 (1976)

[6] Gale, E., "TiO_2-based memristors and ReRAM: materials, mechanisms and models (a review)". *Semiconductor Science and Technology* **29**(10), 104,004 (2014)

[7] Hutchby, J., Garner, M., "Assessment of the potential & maturity of selected emerging research memory technologies". In: *Workshop & ERD/ERM Working Group Meeting* (April 6–7, 2010). [Online]. Available at: http://www. itrs. net/Links/2010ITRS/2010Update/ToPost/ERD ERM 2010FINALReport-MemoryAssessment ITRS. pdf (2010)

[8] Itoh, M., Chua, L.O., "Autoassociative memory cellular neural networks". *International Journal of Bifurcation and Chaos* **20**(10), 3225–3266 (2010)

[9] Jeong, D.S., Thomas, R., Katiyar, R., *et al.*, "Emerging memories: resistive switching mechanisms and current status". *Reports on Progress in Physics* **75**(7), 076,502 (2012)

[10] Jo, S.H., "Recent progress in RRAM, materials and devices". *SEMICON, SEMI* (2015)

[11] Kim, H., Sah, M.P., Yang, C., Roska, T., Chua, L.O., "Neural synaptic weighting with a pulse-based memristor circuit". *IEEE Transactions on Circuit Theory* **59**(1), 148–158 (2012)

[12] Kuzum, D., Yu, S., Wong, H.P., "Synaptic electronics: materials, devices and applications". *Nanotechnology* **24**(38), 382,001 (2013)

[13] Lazzaro, J., Ryckebusch, S., Mahowald, M.A., Mead, C.A., "Winner-take-all networks of O(N) complexity". In: *Advances in Neural Information Processing Systems, IEEE*, pp. 703–711 (1988)

[14] Linares-Barranco, B., Serrano-Gotarredona, T., "Memristance can explain spike-time-dependent-plasticity in neural synapses". *Nature Proceedings* **1**, 2009, 1–4 (2009)

[15] Mazady, A., *Modeling, Fabrication, and Characterization of Memristors, Uconn Digital Commons* (2014)

[16] Pershin, Y.V., Di Ventra, M., "Experimental demonstration of associative memory with memristive neural networks". *Neural Networks* **23**(7), 881–886 (2010)

[17] Saïghi, S., Mayr, C.G., Serrano-Gotarredona, T., *et al.*, "Plasticity in memristive devices for spiking neural networks". *Frontiers in Neuroscience* **9**, 1–16 (2015)

[18] Strukov, D., Kohlstedt, H., "Resistive switching phenomena in thin films: materials, devices, and applications". *MRS Bulletin* **37**(02), 108–114 (2012)

[19] Thomas, A., "Memristor-based neural networks". *Journal of Physics D: Applied Physics* **46**(9), 093,001 (2013)

[20] Wang, L., Yang, C., Wen, J., Gai, S., Peng, Y., "Overview of emerging memristor families from resistive memristor to spintronic memristor". *Journal of Materials Science: Materials in Electronics* **26**, 1–11 (2015)

[21] Wang, X., Chen, Y., Xi, H., Li, H., Dimitrov, D., "Spintronic memristor through spin-torque-induced magnetization motion". *IEEE Electron Device Letters* **30**(3), 294–297 (2009)

[22] Wouters, D., "Scaling challenges for 2-terminal select devices". In: *ITRS ERD Selector Workshop* (2012)

[23] Yang, J.J., Strukov, D.B., Stewart, D.R., "Memristive devices for computing". *Nature Nanotechnology* **8**(1), 13–24 (2013)

[24] Zhou, J., Kim, K.H., Lu, W., "Crossbar RRAM arrays: selector device requirements during read operation". *IEEE Transactions on Electron Devices* **61**(5), 1369–1376 (2014)

Index

acoustic phonon 107
ADALINE 338
add buffer 296, 300–1, 303
add driver 297, 302, 304
add wire 294–6, 299, 301, 303
advanced ALD process 4–5
advanced lithography techniques 35
Al_2O_3 13–14, 65–6, 72, 79, 81–4, 87, 89–90
Al_2O_3/HfO_2 bilayer structures 15–16
Al_2O_3 MIM capacitors 66, 70, 72–3, 89
ALD HfO_2 14–15, 17
AlSb 42, 46, 51
  growth and characterization of, on silicon 44–7
  nucleation layer 42, 46, 47, 57
AlTiO 74–6, 78, 81–2, 91
ambipolar field effect 104
ammonium pentaborate (APB) 67
analog and mixed signal (AMS) IC technologies 62, 64, 84
analog stochastic switching behavior 325
  of a TiO_2 memristor 328
analog-to-digital (AD) converters 62
annealing 14–15, 20, 23, 63, 66
anodic alumina MIM capacitors 66
  capacitance and voltage linearity 67–70
  fabrication process flow and crystalline properties 67
  leakage characteristics and conduction mechanisms 70–3
anodic bilayer MIM capacitors 79
  capacitance, voltage linearity, and leakage characteristics 82–4
  fabrication process flow 80
  formation of bilayer and crystallization 81–2
anodic bilayer oxides, nucleation/crystallization of 82
anodic titania MIM capacitors 73
  capacitance, voltage linearity, and leakage characteristics 76–9
  fabrication process, oxide formation, and crystallization 74–6
anodization 61, 64–7, 70, 73–6
anodization time 67
anodization voltages 67, 74–6, 80–2, 89–91
antiphase boundaries (APBs) 36, 48
antiphase domains (APDs) 36, 47–50
arithmetic logic 249
armchair nanotubes 288
atomic layer deposition (ALD) 2, 4, 6–7, 14, 18–19, 66, 70, 73
auto-correlation function (ACF) 319
automatic measurement programs 8

band structure, of graphene 101–3
barrier type oxide 64–5
BDD Decomposition System based on Majority decomposition (BDS-MAJ) 251
Beaumont and Jacobs's electrode polarization model 76–8
BiCMOS technology 159–60, 166
Biconditional Binary Decision Diagrams (BBDDs) 250

bipolar charge-plasma transistor (BCPT) 160–6
bipolar junction transistor (BJT) 146, 159–66
BitLines (BLs) 255
BLT (La substituted-Bismuth–Tantalate) 340
Boltzmann statistics 86
Boltzmann transport equation (BTE) 115
brain-inspired computing systems 337
branch merge 296–7, 302
buffer insertion 287, 289–90, 298
bundled SWCNTs 289
  Elmore delay model for 294
  parameters of 304
  quantum capacitance for 293
  resistance for 292
  schematic of 291

capacitance voltage (*C–V*) measurements 8, 67–8, 76, 82, 89
capacitor 61
  anodic alumina MIM capacitors 66–73
  anodic bilayer MIM capacitors 79–84
  anodic titania MIM capacitors 73–9
  high-κ MIM capacitors, modeling of 84–90
  metal-insulator-metal (MIM) capacitors 63–4, 66–7, 74, 84–5, 89–91
  polysilicon–insulator–polysilicon (PIP) 62–3
  RF-AMS applications of 62
carbon nanotubes (CNTs) 265, 287–90
  for 2D power delivery network 269
    branch analysis with CNTs 270–4
  for 3D power delivery network 274
    TSV CNT analysis 278–80
    voltage drop analysis on a 3D PDN 280–4
  buffering 289
    algorithm 298
    example 299–302
    library 300
  capacitance for 292
  contact resistance 292
  experimental results 304–8
  experimental setup 302–4
  interconnects 290–1
  modeling of 266–9
  problem formulation 290
  resistance for 291–2
  timing buffering for CNT 294
    add buffer 296
    add driver 297
    add wire 294–6
    branch merge 296–7
    pruning 298
carrier density, of graphene 103–4, 106–7
Cascade Micro-chamber 8
CGs (connected to the gate outputs) 253
chalcogenide 340
charge neutrality point (CNP) 110, 121, 123–4
charge-plasma (CP) diode 146
chemical vapor deposition (CVD) 2, 65, 100, 176
chiral nanotubes 288
classical transport 114–15
Clausius–Mossotti equation 88
Clausius–Mossotti model 86
clustering algorithms 351–2
Colt scheme 232
complementary metal-oxide semiconductor (CMOS) 1, 35, 99, 159, 338
computing systems 337
conductivity 104–6
constant voltage stress
  effect of 22–4
  measurement 72
  reliability study by 9
    effect of Zr addition and SPA plasma on TDDB 12–13

impact of stress on flat-band voltage, leakage current, and interface state density 10–12
contact resistance 112–13, 266–7, 278, 292, 296, 304, 308
continuous device scaling 1
Control Gate (CG) 238–9
copper 169–71, 266, 274, 287, 289, 303
-based interconnects 274
buffering 289
Coulomb scattering 105–6
critical thickness 37–9
CUDA core 216–17
cumulative probability function (CDF) 325
current voltage (I–V) measurements 8
cyclic deposition and plasma treatment (DSDS) process 5, 7, 10–13, 27

DC gate voltage 8
Debye length 69, 148
Deep Reactive Ion Etching (DRIE) process step 240
device reliability 2
dielectric polarization 66, 69, 108
Differential Cascade Voltage Switch Logic (DCVSL) 252–3
digital-to-analog (DA) converters 62
dipolar/orientation polarization 86
domain annihilation 48, 50
doping-free junctionless field-effect transistor (DL-JLFET) 146
dopingless FET 146–8
double-gate FinFET (DG-FinFET) 171, 178
-based NAND gate 201
-based technology 169, 190
GIDL of FinFET 190
propagation delay 190–1
sub-threshold leakage power 190
with independent gates (IGs) 178
modeling 180–3
structure 178–80
transistors, used in logic gates 199–204
with unified gates 178
Drude model 104
dual-threshold voltage operations 251–2
dynamic power 126
HKMG-based technology 187
dynamic random-access memory (DRAM) 62, 64
dynamic switching process 325–7

effective oxide thickness (EOT) 157
electrode polarization model 69–70, 76, 86
electron beam lithography (EBL) 178, 314
electronic polarization 85–6
electron movement in high-κ/poly-Si gate and HKMG 175
electron-phonon scattering 105, 107
electrostatic coupling 268
embedded power-gating 252–4
epitaxial layers 36, 43, 51
epitaxial liftoff (ELO) technique 53–4
equivalent oxide thickness (EOT) 1–2, 177
comparison with flat-band voltage and J_g 20–2
downscaling 8, 14
etch stop layer 51–3
Exclusive-OR (XOR) operator 238

FeRAM (ferroelectric RAM) 338, 340
Fermi-Level pinning 113, 175
Fermi velocity 102–3, 105, 268
Fermi's Golden rule 115
ferroelectric tunnel junctions (FTJ) 345
filament conduction phenomenon 316
Fin field-effect-transistor (FinFET) 213
background 215
NBTI degradation mechanism 215–16
target GPU architecture 216–17

double-gate FinFET 171–2, 178, 187, 199
experimental setup 222–4
hybrid-device sequential-access L2 cache 221–2
hybrid-device warp scheduler 217
opportunity for improvement 217–19
two-stage scheduling 219–21
I–V characteristics of 181
related work 232
characterization of FinFET reliability 233
hybrid-device design 233
NBTI mitigation 232–3
result analysis 224
L2 cache 228–32
warp scheduler 224
improvement on reliability 224–6
performance overhead 227–8
transistor structure 216
Fin Pitch 179
flat-band voltage 8–9, 16–17, 23, 26
comparison with equivalent oxide thickness 20–2
impact of stress on 10–12
4156B Semiconductor Parameter Analyzer 8
4284A LCR meter 8
four-transistor pseudo-SRAM cells 255–7
fractional error 349
Fully-Depleted Silicon-On-Insulator (FDSOI) 246

GaAs 52–3
growth, on silicon 36–7
III-Sb on 39–42
InAs and InGaSb channels on 54–7
InGaSb channels on 54–7
GaSb membranes 51
ELO technique 53–4
substrate removal technique 51–3
Gate-All-Around (GAA) structures 237
gated resistors 139, 141
gate-induced-drain leakage (GIDL) 170, 178, 204
of FinFET 190
HKMG-based technology 188
gate stack reliability 2
$Ge_2Sb_2Te_2$ (GST) 340
giant tunnel electroresistance 345
GPGPU simulator 217, 222
GPU architecture 216–17, 234
GPUWattch 222
graphene 99
fabrication of 100–1
GNR FET
device performance metrics 123–8
device structure 123
graphene nanoribbon 119–23
modeling and simulation
classical transport 114–15
quantum transport 117–19
semiclassical transport 115
properties of 101
ambipolar field effect 104
band structure 101–3
carrier density 103–4
conductivity 104–6
contact resistance 112–13
high-field transport 108–9
Joule heating 112
low-field mobility 109–10
quantum capacitance 113–14
scattering mechanism 106
substrate and gate dielectrics 110–11
Raman spectroscopy of 100–1
transistors 99
graphene nanoribbon (GNR) 99, 119–23
graphene nanoribbon field effect transistors (GNR FETs) 99–100
device performance metrics 123–8
device structure 123
graphene nanoribbon 119–23
graphene pseudospin 102

grazing incident in plane X-ray diffraction (GIIXRD) 3
Green's functions 117, 176

H_2O as oxidant 5
Hardware Description Language (HDL) 174
Haswell central processing unit 214
hexagonal boron nitride (h-BN) 111–12
$HfAlO_x$ alloy structures 14–15
$HfAlO_x$ dielectrics, physical properties of 19–20
Hf-based high-κ dielectrics 1, 27
HfO_2 2–4, 7, 15–17, 20, 23, 27
  Al incorporation into 13–14
    Al_2O_3/HfO_2 bilayer structures 15–16
    extremely low Al incorporation in HfO_2 17–26
    $HfAlO_x$ alloy structures 14–15
    problems with excess Al incorporation 16–17
  -based dielectric materials 1
high-field transport 108–9
high-κ MIM capacitors, modeling of 84
  macroscopic model 86–8
  microscopic model 88–9
  model verification 89–90
  voltage linearity 85–6
high mobility n and p channels 35
  GaSb membranes 51
    ELO technique 53–4
    substrate removal technique 51–3
  III-Sb on GaAs substrates 39–42
  III-Sb on silicon substrates 42
    antiphase domains (APDs) 47–50
    lattice mismatch solution 43–7
    thermal expansion coefficient 50–1
  IMF versus pseudomorphic growth 38–9
  InAs and InGaSb channels on GaAs 54–7
high performance computation (HPC) 221
high-κ dielectrics and device reliability 1
  advanced ALD process 4–5
  Al incorporation into HfO_2 13–14
    Al_2O_3/HfO_2 bilayer structures 15–16
    extremely low Al incorporation in HfO_2 17–26
    $HfAlO_x$ alloy structures 14–15
    problems with excess Al incorporation 16–17
  alloying HfO_2 and ZrO_2 2–4
  cyclic deposition and SPA plasma treatment to ALD $Hf_{1-x}Zr_xO_2$ 5–8
    impact of Zr addition and SPA plasma on electrical properties 8–9
    reliability study by constant voltage stress 9–13
  reliability 2
  SPA plasma 2, 5
high-κ/metal-gate (HKMG)-based technology 187
  dynamic power 187
  GIDL power 188
  propagation delay 188–9
  sub-threshold leakage power 187
high-κ/metal-gate (HKMG) bulk MOSFET 174
  modeling 176–8
  structure 175–6
high-κ/metal gate (HK/MG) interface 17
high-κ/metal-gate (HKMG) logic cells at room temperature 191–9
high-κ/metal gate (HK/MG) stack 1, 17
high-resolution XRD measurements 39
hold SNM (HSNM) 158
HotSpot 5.0 222
HP4155C semiconductor parameter analyzer 67, 76
hybrid-device 2-stage scheduler 224, 226–7
  architecture of 220

normalized IPC on the GPU with 227
hybrid-device design 233
hybrid-device sequential-access L2 cache 221–2
workflow of 222
hybrid-device warp scheduler 217
opportunity for improvement 217–19
two-stage scheduling 219–21

image clustering 351
algorithm overview 351
clustering MNIST images 354
hardware design 352–4
InAs and InGaSb channels on GaAs 54–7
InAs n-channels 54
independent gates (IGs), DG-FinFET with 178
InGaSb channels 54–7
integrated circuits (ICs) 237
Intel 214
inter-die process 184
interface state density 8, 24–5
impact of stress on 10–12
interfacial misfit (IMF)
-grown bulk GaSb 39, 41
versus pseudomorphic growth 38–9
interfacial plasmon–phonons (IPP) 111
interfacial polarization 86
International Technology Roadmap for Semiconductor (ITRS) 1, 61, 64, 76, 84–5, 90–1, 100, 124, 126, 346
intra-die process 184
inversion mode (IM) 140
inversion mode FET (IMFET) 141, 143
inverter, statistical state-dependent data for 193
ionic polarization 61, 65–6, 78, 86, 90–1
Ivy Bridge central processing unit 214

Jonscher response 69
Joule heating 112
junction and doping-free transistors 139
dopingless FET 146–8
junction and doping-free FET 148
dopingless BJT 159–62
junction and doping-free DG FET 149–59
junctionless field-effect transistors (JLFETs) 139
limitations 143
junctionless field-effect transistors (JLFETs) 139
challenges 142
versus inversion mode FET (IMFET) 141–2
limitations 143–5

Kapitza resistance 112

L2 cache 221–2, 228–32
Langevin function 87
lattice-matched epitaxy 38
lattice mismatch in crystal growth 36, 39
leakage current, impact of stress on 10–12
leakage current density 16, 72–3, 82
leakage power 251–2
sub-threshold 187, 190
line edge roughness (LER) 314, 321
logic library creation, proposed methodology for 183
sources of variation and nature of variability 183–4
statistical logic library characterization flow 184–7
Lomer dislocations: *see* misfit dislocations
longest-warp chain 227–8
low-field mobility 109–10
Luttinger liquid theory 268–9

magnetic tunnel junction (MTJ) 338, 340

Majority-Inverter-Graph (MIG) 238, 251
  MIG-SiNWFETs 258
McLaurin series expansion 189
memory technologies 338–40
memristance 314, 318–19, 330, 341–2
memristive synapse circuits 347
  area and power consumption 349–50
  overview 347–9
memristor 313
  metal oxide–based memristor 314–15
  modeling 313
  neuromorphic system, robustness of 329–32
  properties of 314
  and resistive memory 337, 340–2
    memristor integration 346–7
    plasticity 345–6
    switching mechanisms 342–5
  static modeling 315–16
    memristor static (bulk) model 316
    TiO_2 thin-film memristor 315–16
  statistical modeling 316–24
    3D device sample generation flow 319–22
    impact of process variations 322–4
    theoretical analysis 316–19
  stochastic modeling 324–9
    dynamic switching process 325–7
    ON and OFF static states 324
    stochastic model verification 327–9
  TiO_2-based memristor 315
memristor-based RRAM 338
memristor equivalent learning rule 341
memristor static (bulk) model 316
metal-insulator-metal (MIM) capacitors 61, 63–4
  anodic alumina MIM capacitors 66
    capacitance and voltage linearity 67–70
    fabrication process flow and crystalline properties 67
    leakage characteristics and conduction mechanisms 70–3
  anodic bilayer MIM capacitors 79
    capacitance, voltage linearity, and leakage characteristics 82–4
    fabrication process flow 80
    formation of bilayer and crystallization 81–2
  anodic titania MIM capacitors 73
    capacitance, voltage linearity, and leakage characteristics 76–9
    fabrication process, oxide formation, and crystallization 74–6
  anodization for nanoelectronics 64–6
  high-κ MIM capacitors, modeling of 84
    macroscopic model 86–8
    microscopic model 88–9
    model verification 89–90
    voltage linearity 85–6
metal-insulator-metal (MIM) structure 342
metallic carbon nanotubes (m-CNTs) 265, 277
metal-organic chemical vapor deposition (MOCVD) 36, 176
metal oxide–based memristor 314–15
metal-oxide semiconductor (MOS) capacitor 27, 61–3, 65
metal-oxide semiconductor (MOS) device 2
metal oxide semiconductor capacitors (MOSCAPs) 7
metal-oxide semiconductor field-effect transistors (MOSFETs) 1, 139, 169, 237, 350
  conventional scaling 155
  effective oxide thickness (EOT) 157
  fabrication, incorporating anodic alumina in 65
microelectronics 1, 65
misfit dislocations 38–9, 46

MNIST images, clustering 354
mobile electronics 169
modeling and simulation 114
  classical transport 114–15
  quantum transport 117–19
  semiclassical transport 115
molecular beam epitaxy (MBE) 36–7, 39
Monte Carlo simulations 185, 187, 321–2
Moore's Law 35, 99, 237
Motorola 36–7
Mott transition 343
MRAM (magnetic RAM) 338, 340, 345
MTJ (magnetic tunneling junction) 314, 338, 340
multi-gate transistors 170
multilayer metals, anodization of 80
multiple-independent-gate field-effect transistor (MIGFET) 237, 246, 252, 255, 257–9
  implementation of a Polar code decoder with 257
    improvement evaluation and discussions 258–9
    methodology 257–8
multiple-independent-gate nanowire transistors 237
  circuit design opportunities 247
    advanced low-power techniques 251–4
    compact data path design 249–51
    dual-threshold voltage operations 251–2
    embedded power-gating 252–4
    four-transistor pseudo-SRAM cells 255–7
    generalities 247–9
    implementation of a Polar code decoder with MIGFETs (case study) 257–9
    improvement evaluation and discussions 258–9
    memory opportunities 254–7
    methodology 257–8
    TSPC flip-flops 254–5
  multiple-independent-gate field-effect transistors 238
    device fabrication and electrical characterization 240–3
    performance predictions 245–7
    physical understanding 243–5
    TIG device overview and operation 238–40
multi-wall CNTs (MWCNTs) bundle 101, 265–8, 270, 272, 288
  branch capacitance of 272
  branch inductance of 273
  branch resistivity of 271

NAND, statistical state-dependent data for 193
nano–complementary metal-oxide semiconductor 145
nanoelectronics, anodization for 64–6
nanoscale CMOS 170–1, 184
nanotubes, capacitance of 268
negative bias temperature instability (NBTI) 213–14
  degradation mechanism 213, 215–16, 225
  mitigation 232–3
  parameter values for computing 224
neural networks 329–30, 338
neuromorphic devices and circuits 337
  emerging memory technologies 338–40
  image clustering 351
    algorithm overview 351
    clustering MNIST images 354
    hardware design 352–4
  memristive synapse circuits 347
    area and power consumption 349–50
    overview 347–9
  memristor and resistive memory 340–2
    memristor integration 346–7
    plasticity 345–6

switching mechanisms 342–5
neuromorphic system, robustness of 329–32
neurotransmitters 346–7, 349, 352, 354
NMOSFET 347
non-equilibrium Green's function (NEGF) 117–18, 176
nonpolar dielectrics 86, 91
nonpolar oxides 86
nonvolatile memory 338–40
NOR circuit 182
n-type FinFET 180–1
n-type junctionless transistor 141
nucleation/crystallization of anodic bilayer oxides 82

off-state current suppression in HVT configuration 243
ON and OFF static states 324
one-selector-one-resistor (1S1R) structure 347
on-state currents of both HVT and LVT modes 242–3
operating temperature 171
optical phonon 107–8, 109
overhead indicators 228
over tune, resistance shifting due to 326–7
oxygen-deficient titanium dioxide 315
oxygen vacancies 69, 326

PCRAM (Phase change RAM (PCRAM) 338, 340
PFT mechanism 70–2
Phase change memory (PCM) 340
phonon scattering 106–7
plasticity 345–6
PMOSFET 347
PMOS transistor 215
p-n diode 146
Polar code decoder 238, 257–9
Polarity Gate at Drain 239
Polarity Gate at Source (PG_S) 238
polysilicon–insulator–polysilicon (PIP) capacitors 62–3
Poole–Frenkel (PF) emission 70–1
porous type anodic oxides 65
post-deposition annealing (PDA) 4, 7–8
power delivery networks 269
power distribution networks 269
power-gating 252
power spectral density (PSD) 319
Predictive Technology Model (PTM) 177
probabilistic-CMOS 174
probability density function (PDF) function 174, 185, 324
process, voltage, and temperature (PVT)-aware logic level characterization 204–5
process, voltage, and temperature (PVT)-aware statistical logic library 186
process-induced interface traps 2
process variation 171, 204
  in nanoscale circuits 183–4
progressive breakdown (PBD) 12
propagation delay
  DG-FinFET-based technology 190–1
  HKMG-based technology 188–9
pruning 298–9
pseudo-SRAM cells, four-transistor 255–7
p-type FinFET 181
pure edge dislocations: *see* misfit dislocations
PVT variability 172
PZT (Lead-Zirconate-Titanate) 340

quadratic coefficient of capacitance 89
quantum capacitance 113–14, 268
quantum transport 117–19
quasi-elastic scattering 107

radical flow nitridation (RFN) process 6
random discrete doping (RDD) 314
random dopant fluctuations (RDFs) 139
reactive sputtering 176
read static noise margin (RSNM) 158
"recovery" stage 215

reflection high-energy electron diffraction (RHEED) pattern 39–40, 44, 46
relaxation 38–9, 78
RESET operation 342
RESET state 340
resistive memristors 343
resistive random access memory (RRAM) 340
resistive switching phenomena 340
RF-AMS applications 61–2
RRAM (Resistive RAM) 338, 340

SBT (Strontium–Bismuth–Tantalate) 340
scanning tunneling microscopy (STM) 100
scattering mechanism 106
  acoustic phonon 107
  of graphene 106
    acoustic phonon 107
    long- and short-range scattering 106–7
    optical phonon 107–8
    surface polar phonons 108
  long- and short-range scattering 106–7
  optical phonon 107–8
  surface polar phonons 108
scheduler activity 219, 226
Schottky-Barrier (SB) NWFETs 237
Schottky emission (SE) 70, 72
Schottky junctions 239, 241, 243
secondary ion mass spectrometry (SIMS) 74, 81
selective area epitaxy (SAE) 36
self-aligned-gate process 169
semiclassical transport 115
  Boltzmann transport equation 115
  top-of-the-barrier approach 115–17
Sentaurus Device 243
SET operation 342–3
SET state 340
shell-to-shell coupling capacitance 268
short-channel effects (SCEs) 170–1
silicon, growth of GaAs on 36
silicon dioxide (SiO_2) 178, 240
  dielectric transistors 174
silicon–hydrogen (Si-H) bonds 215–16
Silicon NanoWires (SiNW) 237
silicon nitride (SiN_3) 178
silicon nitride spacers 241
silicon-on-insulator (SOI) 140, 170, 178
  substrate 240
  thickness 180
silicon substrates, III-Sb on 42
  antiphase domains (APDs) 47–50
  lattice mismatch solution 43
  thermal expansion coefficient 50–1
single-wall CNTs (SWCNTs) CNTs 265–8, 288
  bundled
    Elmore delay model for 294
    resistance for 292
  inductive impact 293
  isolated
    capacitance for 292–3
    resistance for 291–2
Si wafers (100) 67
slot-plane-antenna (SPA) plasma 2, 4–5
  treatment to ALD $Hf_1-xZr_xO_2$ 5–8
    impact of Zr addition and SPA plasma on electrical properties 8–9
    reliability study by constant voltage stress 9–13
soft breakdown (SBD) 12, 27
SOI-CMOS technology 160
space charge polarization 86
special function units (SFUs) 216
spectroscopic ellipsometry (SE) 7
SPICE model 177, 204
spike timing dependent plasticity (STDP) 346
static modeling, of memristor 315
  memristor static (bulk) model 316
  TiO_2 thin-film memristor 315–16
static noise margin (SNM) 158

static random access memory (SRAM) 158, 238, 255–7
6T-SRAM cell 159
static states 327
statistical logic library characterization flow 184–7
statistical modeling, of memristor 316–24
3D device sample generation flow 319–22
impact of process variations 322–4
theoretical analysis 316–19
step-graded metamorphic buffers 37
stochastic modeling, of memristor 324–9
dynamic switching process 325–7
ON and OFF static states 324
stochastic model verification 327–9
streaming multiprocessors (SMs) 216, 218
stress-induced flat-band voltage shifts 9, 11, 22–3, 27
stress-induced interface state generation 9, 25, 27
stress-induced leakage currents (SILCs) 9, 22–3, 25–6
"stress" phase 215
STT-RAM (spin torque transfer RAM) 338, 340, 345
substrate and gate dielectrics 110–11
substrate removal technique 51–3
sub-threshold leakage power
DG-FinFET-based technology 190
HKMG-based technology 187
supply voltage 158, 171, 187, 189
surface accumulation layer transistor (SALTran) effect 163
surface polar phonons (SPP) 108
switching
probability, time and voltage dependency of 325–6, 328
symmetric bipolar charge-plasma transistor 160
synaptic electronics 345
synaptic plasticity 346
System-on-Chip (SoC) designs 171

tag array 222, 230
TAT mechanisms 70, 84
TELTrias™cleanroom tool 5
temperature coefficient of capacitance (TCC) 68–9
tensile strain 38, 50
tertragonal distortion 38
tetrakis (ethylmethylamido) hafnium (TEMAH) 5, 18
tetrakis (ethylmethylamido) zirconium 5
thermal evaporation techniques 66
thermal expansion coefficient 50–1
Thermo Fisher Theta Probe™XPS system 7, 19
threading dislocation density (TDD) 36, 42, 55–6
threading dislocations 36, 38
III-V compound semiconductors 43, 50
integration, with silicon 36
III-Sb–based compound semiconductors 37
on GaAs substrates 39–42
on silicon substrates 42
antiphase domains (APDs) 47–50
lattice mismatch solution 43–7
thermal expansion coefficient 50–1
3D device sample generation flow 319–22
3D power delivery network (3D PDN), CNTs for 274
TSV CNT analysis 278–80
voltage drop analysis 280–4
three-independent-gate (TIG) SiNWFETs 238, 240, 243, 248, 258
conceptual sketch of 239
SEM image of 242
Three-Independent-Gate Field-Effect Transistor (TIGFET) 237–8, 251

TIGFET TSPC flip-flop transient simulation 255
threshold voltage 187, 204
through-silicon-vias (TSVs) 265, 275, 278–80
time-dependent dielectric breakdown (TDDB) 12
characteristics 25–6
time-to-breakdown (TBD) 72
TiO_2-based memristor 315
TiO_2 thin-film memristor 315–16
equivalent circuit 315
simulation results of 323
static modeling 315
structure 315
TiO_{x0}-based memristive device 341
titania: *see* titanium oxide
titanium, anodization of 74
titanium oxide 73–4
crystallization of 75–6
top-of-the-barrier approach 115–17
trap-assisted tunneling (TAT) 11, 70
tri-gate transistor 214
trimethylaluminum 18
True-Single Phase Clock (TSPC) flip-flops 238, 254–5
using TIGFETs compactness 254
2D power delivery network, CNTs for 269
branch analysis with CNTs 270–4
2-stage warp scheduler 224, 226–7, 229
architecture of 220
normalized IPC on the GPU with 227

unified gates, DG-FinFET with 178
unified programming language 216

volatile memory 338–9
voltage coefficient of capacitance (VCC) 62, 68–9, 79
voltage drop, computing 272
voltage drop analysis on a 3D PDN 280–4
voltage linearity 76
capacitance and 67–70
modeling 85–6

warp scheduler 218–19, 224
architecture of 218
filter rate on the first stage of 226
improvement on reliability 224–6
performance overhead 227–8
power consumed by 226
steady temperature on 225
Weibull slope 13, 26–7
weight vectors 351, 353
Wentzel–Kramers–Brillouin approximation 117, 243

X-ray diffraction (XRD) measurements 7, 14, 75, 82
X-ray photoelectron spectroscopy (XPS) 7, 19
X-ray reflectivity (XRR) 7

zigzag nanotubes 288